Daniel W. Stroock

A Concise Introduction to the Theory of Integration

Third Edition

Birkhäuser
Boston • Basel • Berlin

Daniel W. Stroock
Department of Mathematics
Massachusetts Institute of Technology
Cambridge, MA 02139

Library of Congress Cataloging-in-Publication Data

Stroock, Daniel W.
A concise introduction to the theory of integration / Daniel W. Stroock. — 3rd ed.
p. cm.
Includes bibliographical references and index.
ISBN 0-8176-4073-8 (acid-free paper). — ISBN 3-7643-4073-8 (acid-free paper)
1. Integrals, Generalized. 2. Measure Theory. I. Title.
QA312.S78 1999
512'.42—dc21 98-42436
CIP

AMS Subject Classifications: 26-01, 26A42, 26B15, 28-01, 42A38, 42B10, 44A12

Printed on acid-free paper.

Birkhäuser 

ISBN 0-8176-4073-8
ISBN 3-7643-4073-8

Typeset by the author in LaTeX.
Printed and bound by Hamilton Printing Co., Rensselaer, NY.
Printed in the United States of America.

9 8 7 6 5 4 3 2 1

The book is dedicated to my Colorado colleagues

L. Baggett, R. Holley, and W. Fulks

Contents

Preface to the Third Edition

Aside from the ceaseless displacement of old errors to make room for new ones, the book has changed from the second edition mostly by the addition of a chapter about Fourier analysis. In the spirit of Norbert Wiener (and quantum field theorists), I have chosen to base the development on Hermite functions. In particular, I first use the Hermite functions to diagonalize the Fourier transform, and then derive Parseval's identity and the Fourier inversion as a consequence. The other substantial change results from my decision to provide solutions to *all* the problems and to publish them under the same cover.

Unfortunately, these changes have resulted in a book which appears to be less "concise" than the original. Thus, for example, it is probably no longer realistic to think that everything can be covered in a single semester. On the other hand, there is no need to cover everything. For example, many students will have seen the material in Chapter V elsewhere, and others will be exposed to Fourier analysis later. Thus, I expect that most instructors will make a selection from the topics covered in Chapters V and VII.

Finally, and significantly, in several places the presentation has benefited from feedback which I received from J.J. Duistermaat, who subjected himself and his students to the book for two consecutive years, thereby replacing Burckel as the world's leading authority on its shortcomings. In addition, it was Duistermaat who convinced me to admit that this is not the only book on the topic and to suggest an alternate source. As it turns out, there are several books which approach measure theory from a similar point of view. Among these are *Measure and Integral* by R.L. Wheeden and A. Zygmund, published by Marcel Dekker in 1977, and *Lebesgue Integration on Euclidean Space* by Frank Jones, published in 1993 by Jones & Bartlett.

Daniel W. Stroock, Cambridge, MA, July, 1998

Preface to the Second Edition

It is four years since the first edition of this book appeared, and, in that time, there has been little, if any, change in either the basic material covered or my attitude toward that material. On the other hand, experience has taught me that my presentation of several points could be considerably refined and that the inclusion of some additional topics would be desirable. Thus, although they may not be immediately apparent, changes have been made throughout. Among those which are obvious are the addition of two new sections: Section 4.2, in which I prove the isodiametric inequality and discuss Lebesgue measure from the Hausdorff measure point of view, and Section 3.4, in which I have given a proof (based on the Hardy–Littlewood maximal inequality) of Lebesgue's Differentiation Theorem for $\mathbb{R}$. These additions made it desirable to reorganize the table of contents, with the result that now product measures appear in Chapter IV and succeeding chapters have been renumbered accordingly. Besides these new sections, the exposition, particularly in what are now Chapters V and VII, has been, I hope, improved. In addition, even where substantive alterations are slight, I have made a great effort to remove some of the more egregious errors with which the first edition was riddled. (In particular, I believe that I have, at last, mastered the spelling of Lebesgue's name.) Finally, at the behest of my students, I have attempted to solve some of the exercises, and the fruits of my labor appear at the end of the book.

If I have been successfully eliminated many of the errors in the first edition, most of the credit should go to R.B. Burckel, who was kind enough to send me a (five page) list of those which he found. In addition, I am indebted to Ann Kostant at Birkhäuser for her efforts, without which this second edition would probably not have appeared.

Daniel W. Stroock, Cambridge, MA, December, 1993

Preface to the First Edition

This little book is the outgrowth of a one semester course which I have taught for each of the past four years at M.I.T. Although this class used to be one of the standard courses taken by essentially every first year graduate student of mathematics, in recent years (at least in those when I was the instructor), the clientele has shifted from first year graduate students of mathematics to more advanced graduate students in other disciplines. In fact, the majority of my students have been from departments of engineering (especially electrical engineering) and most of the rest have been economists. Whether this state of affairs is a reflection on my teaching, the increased importance of mathematical analysis in other disciplines, the superior undergraduate preparation of students coming to M.I.T in mathematics, or simply the lack of enthusiasm that these students have for analysis, I have preferred not to examine too closely. On the other hand, the situation did force me to do a certain amount of thinking about what constitutes an appropriate course for a group of non-mathematicians who are courageous (foolish?) enough to sign up for an introduction to integration theory offered by the department of mathematics. In particular, I had to figure out what to do about that vast body of material which, in standard mathematics offerings, is "assumed to have been covered in your advanced calculus course". Aspiring young mathematicians seldom challenge even the most ridiculous declarations of this sort: the good ones look it up, and the others trust that "it will not appear on the exam". On the other hand, students who are not heading into mathematics are less easily shamed into accepting such claims; in fact, as I soon discovered, many of them were attending my course for the express purpose of learning what mathematicians call "advanced calculus".

In view of the preceding comments about the origins of this text, it should come as no surprise that the contents of this book are somewhat different from that of many modern introductions to measure theory. Indeed, I believe that nothing has been done here "in complete generality"! On the other hand, greater space than usual has been given to the properties of Lebesgue's measure on $\mathbb{R}^N$. In particular, the whole of Chapter IV [now, Chapter V] is devoted to applications of Lebesgue's measure to topics which are customarily "assumed to have been covered in your advanced calculus course". As a consequence, what has emerged is a kind of hybrid in which both modern integration theory and advanced calculus are represented. Because none of the many existing books on integration theory contained precisely the mix for which I was looking, I decided to add my own version of the subject to the long list of books for the next guy to reject.

Cambridge, MA, January, 1990

Chapter I

The Classical Theory

1.1 Riemann Integration

We begin by recalling a few basic facts about the integration theory which is usually introduced in advanced calculus. We do so not only for purposes of later comparison with the modern theory but also because it is the theory with which most computations are actually performed.

Let $N \in \mathbb{Z}^+$ (throughout $\mathbb{Z}^+$ will denote the positive integers). A **rectangle** in $\mathbb{R}^N$ is a subset I of $\mathbb{R}^N$ which can be written as the Cartesian product $\prod_1^N [a_k, b_k]$ of closed intervals $[a_k, b_k]$, where it is assumed that $a_k \leq b_k$ for each $1 \leq k \leq N$. If I is such a rectangle, we call

$$\operatorname{diam}(I) \equiv \left(\sum_{k=1}^{N}(b_k - a_k)^2\right)^{\frac{1}{2}} \quad \text{and} \quad \operatorname{vol}(I) \equiv \prod_{k=1}^{N}(b_k - a_k)$$

the **diameter** and the **volume** of I, respectively. For purposes of this exposition, it will be convenient to also take the empty set to be a rectangle with diameter and volume 0.

Given a collection $\mathcal{C}$, we will say that $\mathcal{C}$ is **non-overlapping** if distinct elements of $\mathcal{C}$ have disjoint interiors. The following *obvious* fact is surprisingly difficult to prove.

1.1.1 Lemma. *If $\mathcal{C}$ is a non-overlapping, finite collection of rectangles each of which is contained in the rectangle J, then $\operatorname{vol}(J) \geq \sum_{I \in \mathcal{C}} \operatorname{vol}(I)$. On the other hand, if $\mathcal{C}$ is any finite collection of rectangles and J is a rectangle which is covered by $\mathcal{C}$ (i.e., $J \subseteq \bigcup \mathcal{C}$), then $\operatorname{vol}(J) \leq \sum_{I \in \mathcal{C}} \operatorname{vol}(I)$.*

PROOF: Without loss of generality, we will assume throughout that each of the rectangles $I \in \mathcal{C}$ is contained in the rectangle J. Indeed, $I \subseteq J$ has already been assumed in the first part of the lemma, and, in the second part, if $I \not\subseteq J$, then one can replace I by $I \cap J$.

We begin with the observation that, if $J = \prod_1^N [a_k, b_k]$, $a_k \leq c \leq b_k$ for some $1 \leq k \leq N$, and J^+ and J^- are the rectangles obtained by replacing the kth side of J by $[a_k, c]$ and $[c, b_k]$, respectively, then $\operatorname{vol}(J) = \operatorname{vol}(J^+) + \operatorname{vol}(J^-)$. More generally, if, for each $1 \leq k \leq N$, $a_k = c_{k,0} \leq \cdots \leq c_{k,n_k} = b_k$ and $\mathcal{S}$ is the collection of $\prod n_k$ rectangles of the form

$$R(m_1, \dots, m_N) \equiv \prod_{k=1}^{N} \left[c_{k,m_k-1}, c_{k,m_k}\right]$$

with $1 \le m_k \le n_k$, then an obvious induction argument on the n_k's shows that $\mathrm{vol}(J) = \sum_{R \in \mathcal{S}} \mathrm{vol}(R)$.

Next suppose that $\mathcal{C}$ is a finite collection of rectangles I contained in $J = \prod [a_k, b_k]$. For each $1 \le k \le N$, choose $a_k = c_{k,0} \le \cdots \le c_{k,n_k} = b_k$ so that each $I \in \mathcal{C}$ has the form

$$\prod_{k=1}^{N} \left[c_{k,m_k}, c_{k,m'_k}\right]$$

for some choice of $0 \le m_k < m'_k \le n_k$, and determine the collection $\mathcal{S}$ accordingly, as in the preceding paragraph. Then, by the observation just made, $\mathrm{vol}(J) = \sum_{R \in \mathcal{S}} \mathrm{vol}(R)$. For the same reason, if, for a given $I \in \mathcal{C}$, $\mathcal{S}(I)$ is the collection of those $R \in \mathcal{S}$ with $R \subseteq I$, then $\mathrm{vol}(I) = \sum_{R \in \mathcal{S}(I)} \mathrm{vol}(R)$.

Now add the assumption that $\mathcal{C}$ is a non-overlapping collection, and let I and I' be different elements of $\mathcal{C}$. Then, in order for $R = R(m_1, \dots, m_N)$ to be an element of both $\mathcal{S}(I)$ and of $\mathcal{S}(I')$, it is necessary that $c_{k,m_k-1} = c_{k,m_k}$ for at least one $1 \le k \le N$. But this means that $\mathrm{vol}(R) = 0$, and so we now know that

$$\mathrm{vol}(J) = \sum_{R \in \mathcal{S}} \mathrm{vol}(R) \ge \sum_{I \in \mathcal{C}} \sum_{R \in \mathcal{S}(I)} \mathrm{vol}(R) = \sum_{I \in \mathcal{C}} \mathrm{vol}(I).$$

Finally, if we drop the non-overlapping assumption but instead assume that $\mathcal{C}$ covers J, then every $R \in \mathcal{S}$ is in $\mathcal{S}(I)$ for at least one $I \in \mathcal{C}$. Hence, in this case,

$$\mathrm{vol}(J) = \sum_{R \in \mathcal{S}} \mathrm{vol}(R) \le \sum_{I \in \mathcal{C}} \sum_{R \in \mathcal{S}(I)} \mathrm{vol}(R) = \sum_{I \in \mathcal{C}} \mathrm{vol}(I). \quad \square$$

Given a collection $\mathcal{C}$ of rectangles I, we say that $\xi : \mathcal{C} \longrightarrow \bigcup \mathcal{C}$ is a *choice map* for $\mathcal{C}$ if $\xi(I) \in I$ for each $I \in \mathcal{C}$, and we use $\Xi(\mathcal{C})$ to denote the set of all such maps. Given a finite collection $\mathcal{C}$, a choice map $\xi \in \Xi(\mathcal{C})$, and a bounded function $f : \bigcup \mathcal{C} \longrightarrow \mathbb{R}$, we define the **Riemann sum of f over $\mathcal{C}$ relative to ξ** to be

$$\mathcal{R}(f; \mathcal{C}, \xi) \equiv \sum_{I \in \mathcal{C}} f(\xi(I)) \mathrm{vol}\,(I). \tag{1.1.2}$$

Finally, if J is a rectangle and $f : J \longrightarrow \mathbb{R}$ is a function, we say that f is **Riemann integrable on J** if there is a number $A \in \mathbb{R}$ with the property that, for all $\epsilon > 0$, there is a $\delta > 0$ such that

$$|\mathcal{R}(f; \mathcal{C}, \xi) - A| < \epsilon \tag{1.1.3}$$

whenever $\xi \in \Xi(\mathcal{C})$ and $\mathcal{C}$ is a non-overlapping, finite, **exact cover** of J (i.e., $J = \bigcup \mathcal{C}$) whose **mesh size**

$$\|\mathcal{C}\| \equiv \max\{\text{diam}(I) : I \in \mathcal{C}\} < \delta.$$

When f is Riemann integrable on J, we call the associated number A in (1.1.3) the **Riemann integral of f on** J and we will use

$$(\text{R}) \int_J f(x)\,dx$$

to denote A.

It is a relatively simple matter to see that any $f \in C(J)$ (the space of continuous real-valued functions on J) is Riemann integrable on J. However, in order to determine when more general functions are Riemann integrable, it is useful to introduce the **Riemann upper sum**

$$\mathcal{U}(f;\mathcal{C}) \equiv \sum_{I \in \mathcal{C}} \sup_{x \in I} f(x) \text{vol}\,(I)$$

and the **Riemann lower sum**

$$\mathcal{L}(f;\mathcal{C}) \equiv \sum_{I \in \mathcal{C}} \inf_{x \in I} f(x) \text{vol}\,(I).$$

Clearly, one always has

$$\mathcal{L}(f;\mathcal{C}) \leq \mathcal{R}(f;\mathcal{C},\xi) \leq \mathcal{U}(f;\mathcal{C})$$

for any $\mathcal{C}$ and $\xi \in \Xi(\mathcal{C})$. Also, it is clear that a bounded f is Riemann integrable if and only if

$$\varliminf_{\|\mathcal{C}\| \to 0} \mathcal{L}(f;\mathcal{C}) \geq \varlimsup_{\|\mathcal{C}\| \to 0} \mathcal{U}(f;\mathcal{C})$$

where the limits are taken over non-overlapping, finite, exact covers of J. What we want to show now is that the preceding can be replaced by the condition[1]

$$\sup_{\mathcal{C}} \mathcal{L}(f;\mathcal{C}) \geq \inf_{\mathcal{C}} \mathcal{U}(f;\mathcal{C}) \tag{1.1.4}$$

where $\mathcal{C}$'s run over all non-overlapping, finite, exact covers of J.

To this end, we partially order the covers $\mathcal{C}$ by *refinement*. That is, we say that $\mathcal{C}_2$ is **more refined** than $\mathcal{C}_1$ and write $\mathcal{C}_1 \leq \mathcal{C}_2$, if, for every $I_2 \in \mathcal{C}_2$, there is an $I_1 \in \mathcal{C}_1$ such that $I_2 \subseteq I_1$. Note that, for every pair $\mathcal{C}_1$ and $\mathcal{C}_2$, the *least* common refinement $\mathcal{C}_1 \vee \mathcal{C}_2$ is given by

$$\mathcal{C}_1 \vee \mathcal{C}_2 = \{I_1 \cap I_2 : I_1 \in \mathcal{C}_1 \text{ and } I_2 \in \mathcal{C}_2 \text{ with } I_1 \cap I_2 \neq \emptyset\}.$$

1.1.5 Lemma. *For any pair of non-overlapping, finite, exact covers $\mathcal{C}_1$ and $\mathcal{C}_2$ of J and any function $f : J \longrightarrow \mathbb{R}$, $\mathcal{L}(f;\mathcal{C}_1) \leq \mathcal{U}(f;\mathcal{C}_2)$. Moreover, if $\mathcal{C}_1 \leq \mathcal{C}_2$, then $\mathcal{L}(f;\mathcal{C}_1) \leq \mathcal{L}(f;\mathcal{C}_2)$ and $\mathcal{U}(f;\mathcal{C}_1) \geq \mathcal{U}(f;\mathcal{C}_2)$.*

---

[1] In many texts, this condition is adopted as the definition of Riemann integrability. Obviously, since, as we are about to show, it is equivalent to the definition which we have given, there is no harm in doing so. However, when working in the more general setting studied in § 1.2, the distinction between these two definitions does make a difference. See, Exercise 1.2.26 for a little more information.

PROOF: We begin by proving the second statement. Noting that

$$\mathcal{L}(f;\mathcal{C}) = -\mathcal{U}(-f;\mathcal{C}), \tag{1.1.6}$$

we see that it suffices to check that $\mathcal{U}(f;\mathcal{C}_1) \geq \mathcal{U}(f;\mathcal{C}_2)$ if $\mathcal{C}_1 \leq \mathcal{C}_2$. But, for each $I_1 \in \mathcal{C}_1$,

$$\sup_{x\in I_1} f(x)\mathrm{vol}\,(I_1) \geq \sum_{\{I_2\in\mathcal{C}_2:I_2\subseteq I_1\}} \sup_{x\in I_2} f(x)\mathrm{vol}\,(I_2),$$

where we have used Lemma 1.1.1 to see that

$$\mathrm{vol}\,(I_1) = \sum_{\{I_2\in\mathcal{C}_2:\,I_2\subseteq I_1\}} \mathrm{vol}\,(I_2).$$

After summing the above over $I_1 \in \mathcal{C}_1$, one arrives at the required result.

Given the preceding, the rest is immediate. Namely, for any $\mathcal{C}_1$ and $\mathcal{C}_2$,

$$\mathcal{L}(f;\mathcal{C}_1) \leq \mathcal{L}(f;\mathcal{C}_1 \vee \mathcal{C}_2) \leq \mathcal{U}(f;\mathcal{C}_1 \vee \mathcal{C}_2) \leq \mathcal{U}(f;\mathcal{C}_2). \quad \square$$

Lemma 1.1.5 really depends only on properties of our order relation and not on the properties of vol(I). In contrast, the next lemma depends on the continuity of volume with respect to side-lengths of rectangles.

1.1.7 Lemma. *Let $\mathcal{C}$ be a non-overlapping, finite, exact cover of the rectangle J and $f : J \longrightarrow \mathbb{R}$ a bounded function. Then, for each $\epsilon > 0$, there is a $\delta > 0$ such that*

$$\mathcal{U}(f;\mathcal{C}') \leq \mathcal{U}(f;\mathcal{C}) + \epsilon \quad \text{and} \quad \mathcal{L}(f;\mathcal{C}') \geq \mathcal{L}(f;\mathcal{C}) - \epsilon$$

whenever $\mathcal{C}'$ is a non-overlapping, finite, exact cover of J with the property that $\|\mathcal{C}'\| < \delta$.

PROOF: In view of (1.1.6), we need only consider the Riemann upper sums.

Let $J = \prod_1^N [c_k, d_k]$. Given a $\delta > 0$ and a rectangle $I = \prod_1^N [a_k, b_k]$, define $I_k^-(\delta)$ and $I_k^+(\delta)$ to be the rectangles

$$[c_1, d_1] \times \cdots \times [a_k - \delta, a_k + \delta] \times \cdots \times [c_N, d_N]$$

and

$$[c_1, d_1] \times \cdots \times [b_k - \delta, b_k + \delta] \times \cdots \times [c_N, d_N],$$

respectively. Then, for any rectangle $I' \subseteq J$ with $\mathrm{diam}(I') < \delta$, either $I' \subseteq I$ for some $I \in \mathcal{C}$ or, for some $I \in \mathcal{C}$ and $1 \leq k \leq N$, $I' \subseteq I_k^+(\delta)$ or $I' \subseteq I_k^-(\delta)$.

Now let $\mathcal{C}'$ with $\|\mathcal{C}'\| < \delta$ be given. Then, by an application of Lemma 1.1.1, we can write

$$\mathcal{U}(f;\mathcal{C}') = \sum_{I'\in\mathcal{C}'} \sup_{I'} f \operatorname{vol}(I') = \sum_{I\in\mathcal{C}} \sum_{I'\in\mathcal{C}'} \sup_{I'} f \operatorname{vol}(I\cap I')$$
$$= \sum_{I\in\mathcal{C}} \sum_{I'\in\mathcal{C}'} \sup_{I\cap I'} f \operatorname{vol}(I\cap I') + \sum_{I\in\mathcal{C}} \sum_{I'\in\mathcal{C}'} \left(\sup_{I'} f - \sup_{I\cap I'}\right) \operatorname{vol}(I\cap I').$$

But clearly

$$\sum_{I\in\mathcal{C}} \sum_{I'\in\mathcal{C}'} \sup_{I\cap I'} f \operatorname{vol}(I\cap I') \le \sum_{I\in\mathcal{C}} \sum_{I'\in\mathcal{C}'} \sup_{I} f \operatorname{vol}(I\cap I') = \mathcal{U}(f;\mathcal{C}),$$

where, in the final step, we have again used Lemma 1.1.1. Thus, it remains to estimate

$$\sum_{I\in\mathcal{C}} \sum_{I'\in\mathcal{C}'} \left(\sup_{I'} f - \sup_{I\cap I'} f\right) \operatorname{vol}(I\cap I').$$

However, by the discussion in the preceding paragraph, for each $I' \in \mathcal{C}'$, either $I' \subseteq I$ for some $I \in \mathcal{C}$, in which case

$$\sum_{I\in\mathcal{C}} \left(\sup_{I'} f - \sup_{I\cap I'} f\right) \operatorname{vol}(I\cap I') = 0,$$

or, for some $I \in \mathcal{C}$ and $1 \le k \le N$, $I' \subseteq I_k^+(\delta)$ or $I' \subseteq I_k^-(\delta)$. Thus, if we set

$$\mathcal{B}(k,I)^{\pm} = \{I' \in \mathcal{C} : I' \subseteq I_k^{\pm}(\delta)\},$$

then

$$\sum_{I\in\mathcal{C}} \sum_{I'\in\mathcal{C}'} \left(\sup_{I'} f - \sup_{I\cap I'} f\right) \operatorname{vol}(I\cap I')$$
$$\le 2\|f\|_{\mathrm{u}} \sum_{k=1}^{N} \sum_{I\in\mathcal{I}} \left(\sum_{I'\in\mathcal{B}(k,I)^+} \operatorname{vol}(I\cap I') + \sum_{I'\in\mathcal{B}(k,I)^-} \operatorname{vol}(I\cap I') \right).$$

(In the preceding, we have introduced the notation, to be used throughout, that $\|f\|_{\mathrm{u}}$ denotes the **uniform norm of** f: the supremum of $|f|$ over the set on which f is defined.) Finally, by Lemma 1.1.1, for each $1 \le k \le N$ and $I \in \mathcal{C}$,

$$\sum_{I'\in\mathcal{B}(k,I)^{\pm}} \operatorname{vol}(I\cap I') \le \operatorname{vol}\big(I_k^{\pm}(\delta)\big) \le 2\delta \frac{\operatorname{vol}(J)}{d_k - c_k},$$

and so we now proved that, whenever $\|\mathcal{C}'\| \le \delta$,

$$\mathcal{U}(f;\mathcal{C}') - \mathcal{U}(f;\mathcal{C}) \le \sum_{I\in\mathcal{C}} \sum_{I'\in\mathcal{C}'} \left(\sup_{I'} f - \sup_{I\cap I'}\right) \operatorname{vol}(I\cap I') \le K\|f\|_{\mathrm{u}}\delta,$$

where

$$K \equiv 4N\text{card}(\mathcal{C}) \max_{1\le k\le N} \frac{\text{vol}(J)}{d_k - c_k}. \quad \square$$

As an essentially immediate consequence of Lemma 1.1.7, we have the following theorem.

1.1.8 Theorem. *Let $f : J \longrightarrow \mathbb{R}$ be a bounded function on the rectangle J. Then*

$$\lim_{\|\mathcal{C}\|\to 0} \mathcal{L}(f;\mathcal{C}) = \sup_{\mathcal{C}} \mathcal{L}(f;\mathcal{C}) \quad \text{and} \quad \lim_{\|\mathcal{C}\|\to 0} \mathcal{U}(f;\mathcal{C}) = \inf_{\mathcal{C}} \mathcal{U}(f;\mathcal{C}),$$

where $\mathcal{C}$ runs over non-overlapping, finite, exact covers of J. In particular, (1.1.4) *is a necessary and sufficient condition that a bounded f on J be Riemann integrable. Moreover, if* (1.1.4), *then*

$$(\text{R})\int_J f(x)\,dx = \sup_{\mathcal{C}} \mathcal{L}(f;\mathcal{C}) = \inf_{\mathcal{C}} \mathcal{U}(f;\mathcal{C}).$$

Exercises

1.1.9 Exercise: Prove Theorem 1.1.8. Next, suppose that f and g are Riemann integrable functions on J. Show that $f \vee g \equiv \max\{f,g\}$, $f \wedge g \equiv \min\{f,g\}$, and, for any α, $\beta \in \mathbb{R}$, $\alpha f + \beta g$ are all Riemann integrable on J. In addition, check that

$$(\text{R})\int_J (f\vee g)(x)\,dx \ge \left((\text{R})\int_J f(x)\,dx\right) \vee \left((\text{R})\int_J g(x)\,dx\right),$$

$$(\text{R})\int_J (f\wedge g)(x)\,dx \le \left((\text{R})\int_J f(x)\,dx\right) \wedge \left((\text{R})\int_J g(x)\,dx\right),$$

and

$$(\text{R})\int_J (\alpha f + \beta g)(x)\,dx = \alpha\left((\text{R})\int_J f(x)\,dx\right) + \beta\left((\text{R})\int_J g(x)\,dx\right).$$

Conclude, in particular, that if f and g are Riemann integrable on J and $f \le g$, then $(\text{R})\int_J f(x)\,dx \le (\text{R})\int_J g(x)\,dx$.

1.1.10 Exercise: Show that if f is a bounded real-valued function on the rectangle J, then f is Riemann integrable if and only if, for each $\epsilon > 0$, there is a $\delta > 0$ such that

$$\sum_{\{I\in\mathcal{C}:\sup_I f - \inf_I f > \epsilon\}} \text{vol}\,(I) < \epsilon \tag{1.1.11}$$

whenever $\|\mathcal{C}\| < \delta$. (We use $\sup_I f$ and $\inf_I f$ to denote $\sup_{x\in I} f(x)$ and $\inf_{x\in I} f(x)$, respectively.) As a consequence, show that a bounded f on J is Riemann integrable if it is continuous on J at all but a finite number of points. (See Section 4.1 for more information on this subject.)

1.1.12 Exercise: Show that the condition in Exercise 1.1.10 can be replaced by the condition that, for each $\epsilon > 0$, there exists some $\mathcal{C}$ for which (1.1.11) holds.

1.2 Riemann–Stieltjes Integration

In Section 1.1, we developed the classical integration theory with respect to the standard notion of Euclidean volume. In the present section, we will extend the classical theory, at least for integrals in one dimension, to cover more general notions of volume.

Let $J = [a,b]$ be an interval in $\mathbb{R}$ and φ and ψ a pair of real-valued functions on J. Given a non-overlapping, finite, exact cover $\mathcal{C}$ of J by closed intervals I and a choice map $\xi \in \Xi(\mathcal{C})$, define the **Riemann sum of φ over $\mathcal{C}$ with respect to ψ relative to ξ** to be

$$\mathcal{R}(\varphi|\psi;\mathcal{C},\xi) = \sum_{I\in\mathcal{C}} \varphi(\xi(I))\Delta_I\psi,$$

where $\Delta_I\psi \equiv \psi(I^+) - \psi(I^-)$ and I^+ and I^- denote, respectively, the right and left hand end-points of the interval I. Obviously, when $\psi(x) = x, x \in J$, $\mathcal{R}(\varphi|\psi;\mathcal{C},\xi) = \mathcal{R}(\varphi;\mathcal{C},\xi)$. Thus, it is consistent for us to say that φ is **Riemann integrable on J with respect to ψ**, or, more simply, **ψ-Riemann integrable on J**, if there is a number A with the property that, for each $\epsilon > 0$, there is a $\delta > 0$ such that

$$\sup_{\xi\in\Xi(\mathcal{C})} |\mathcal{R}(\varphi|\psi;\mathcal{C},\xi) - A| < \epsilon \tag{1.2.1}$$

whenever $\mathcal{C}$ is a non-overlapping, finite, exact cover of J satisfying $\|\mathcal{C}\| < \delta$. Assuming that φ is ψ-Riemann integrable on J, we will call the number A in (1.2.1) the **Riemann–Stieltjes integral of φ on J with respect to ψ**, and we will use

$$(\mathrm{R})\int_J \varphi(x)\,d\psi(x)$$

to denote A.

1.2.2 Examples: The following examples may help to explain what is going on here.

(i) If $\varphi \in C(J)$ and $\psi \in C^1(J)$ (i.e., ψ is continuously differentiable on J), then one can use the Mean Value Theorem to check that φ is ψ-Riemann integrable on J and that

$$(\mathrm{R})\int_J \varphi(x)\,d\psi(x) = (\mathrm{R})\int_J \varphi(x)\psi'(x)\,dx. \tag{1.2.3}$$

(ii) If there exist $a = a_0 < a_1 < \cdots < a_n = b$ such that ψ is constant on each of the intervals (a_{m-1}, a_m), then every $\varphi \in C([a,b])$ is ψ-Riemann integrable on $[a,b]$, and

$$\text{(R)} \int_{[a,b]} \varphi(x)\, d\psi(x) = \sum_{m=0}^{n} \varphi(a_m) d_m, \tag{1.2.4}$$

where $d_0 = \psi(a+) - \psi(a)$, $d_m = \psi(a_m+) - \psi(a_m-)$ for $1 \le m \le n-1$, and $d_n = \psi(b) - \psi(b-)$. (We have used $f(x+)$ and $f(x-)$ to denote the right and left limits of f at x.)

(iii) If both (R) $\int_J \varphi_1(x)\, d\psi(x)$ and (R) $\int_J \varphi_2(x)\, d\psi(x)$ exist (i.e., φ_1 and φ_2 are both ψ-Riemann integrable on J), then, for all real numbers α and β, $(\alpha\varphi_1 + \beta\varphi_2)$ is ψ-Riemann integrable on J and

$$\begin{aligned} &\text{(R)} \int_J (\alpha\varphi_1 + \beta\varphi_2)(x)\, d\psi(x) \\ &\quad = \alpha\left(\text{(R)} \int_J \varphi_1(x)\, d\psi(x)\right) + \beta\left(\text{(R)} \int_J \varphi_2(x)\, d\psi(x)\right). \end{aligned} \tag{1.2.5}$$

(iv) If $J = J_1 \cup J_2$ where $\mathring{J}_1 \cap \mathring{J}_2 = \emptyset$ and if φ is ψ-Riemann integrable on J, then φ is ψ-Riemann integrable on both J_1 and J_2, and

$$\text{(R)} \int_J \varphi(x)\, d\psi(x) = \text{(R)} \int_{J_1} \varphi(x)\, d\psi(x) + \text{(R)} \int_{J_2} \varphi(x)\, d\psi(x). \tag{1.2.6}$$

All the assertions made in Examples 1.2.2 are reasonably straightforward consequences of the definition of Riemann integrability. Not so obvious, but terribly important, is the following theorem which shows that the notion of Riemann integrability is *symmetric* in φ and ψ.

1.2.7 Theorem (Integration by Parts). *If φ is ψ-Riemann integrable on $J = [a,b]$, then ψ is φ-Riemann integrable on J and*

$$\text{(R)} \int_J \psi(x)\, d\varphi(x) = \psi(b)\varphi(b) - \psi(a)\varphi(a) - \text{(R)} \int_J \varphi(x)\, d\psi(x). \tag{1.2.8}$$

PROOF: Let $\mathcal{C} = \{[\alpha_{m-1}, \alpha_m] : 1 \le m \le n\}$, where $a = \alpha_0 \le \cdots \le \alpha_n = b$; and let $\xi \in \Xi(\mathcal{C})$ with $\xi([\alpha_{m-1}, \alpha_m]) = \beta_m \in [\alpha_{m-1}, \alpha_m]$. Set $\beta_0 = a$ and

$\beta_{n+1} = b$. Then

$$
\begin{aligned}
\mathcal{R}(\psi|\varphi;\mathcal{C},\xi) &= \sum_{m=1}^{n} \psi(\beta_m)\big(\varphi(\alpha_m) - \varphi(\alpha_{m-1})\big) \\
&= \sum_{m=1}^{n} \psi(\beta_m)\varphi(\alpha_m) - \sum_{m=0}^{n-1} \psi(\beta_{m+1})\varphi(\alpha_m) \\
&= \psi(\beta_n)\varphi(\alpha_n) - \sum_{m=1}^{n-1} \varphi(\alpha_m)\big(\psi(\beta_{m+1}) - \psi(\beta_m)\big) - \psi(\beta_1)\varphi(\alpha_0) \\
&= \psi(b)\varphi(b) - \psi(a)\varphi(a) - \sum_{m=0}^{n} \varphi(\alpha_m)\big(\psi(\beta_{m+1}) - \psi(\beta_m)\big) \\
&= \psi(b)\varphi(b) - \psi(a)\varphi(a) - \mathcal{R}(\varphi|\psi;\mathcal{C}',\xi'),
\end{aligned}
$$

where $\mathcal{C}' = \big\{[\beta_{m-1},\beta_m] : 1 \le m \le n+1\big\}$ and $\xi' \in \Xi(\mathcal{C}')$ is defined by $\xi'([\beta_m,\beta_{m+1}]) = \alpha_m$ for $0 \le m \le n$. Noting that $\|\mathcal{C}'\| \le 2\|\mathcal{C}\|$, one now sees that if φ is ψ-Riemann integrable, then ψ is φ-Riemann integrable and that (1.2.8) holds. □

It is hardly necessary to point out, but notice that when $\psi \equiv 1$ and φ is continuously differentiable, (1.2.8) becomes *the Fundamental Theorem of Calculus*.

Although the preceding theorem indicates that it is natural to consider φ and ψ as playing symmetric roles in the theory of Riemann–Stieltjes integration, it turns out that, in practice, one wants to impose a condition on ψ which will guarantee that every $\varphi \in C(J)$ is Riemann integrable with respect to ψ and that, in addition, (recall that $\|\psi\|_{\mathrm{u}}$ is the uniform norm of ψ)

$$
\left|(\mathrm{R})\int_J \varphi(x)\,d\psi(x)\right| \le K_\psi \|\varphi\|_{\mathrm{u}} \tag{1.2.9}
$$

for some $K_\psi < \infty$ and all φ which are ψ-Riemann integrable on J. Example (**i**) in Examples 1.2.2 tells us that one condition on ψ which guarantees the ψ-Riemann integrability of every continuous φ is that $\psi \in C^1(J)$. Moreover, from (1.2.3), it is an easy matter to check that in this case (1.2.9) holds with $K_\psi = \|\psi'\|_{\mathrm{u}}(b-a)$. On the other hand, example (**ii**) makes it clear that ψ need not be even continuous, much less differentiable, in order that Riemann integration with respect to ψ has the above properties. The following result emphasizes this same point.

1.2.10 Theorem. *Let ψ be non-decreasing on J. Then every $\varphi \in C(J)$ is ψ-Riemann integrable on J. In addition, if φ is non-negative and ψ-Riemann integrable on J, then* (R) $\int_J \varphi(x)\,d\psi(x) \ge 0$. *In particular,* (1.2.9) *holds with $K_\psi = \Delta_J\psi$.*

PROOF: The fact that (R) $\int_J \varphi(x)\,d\psi(x) \geq 0$ if φ is a non-negative function which is ψ-Riemann integrable on J follows immediately from the fact the $\mathcal{R}(\varphi|\psi;\mathcal{C},\xi) \geq 0$ for any $\mathcal{C}$ and $\xi \in \Xi(\mathcal{C})$. Applying this to the function $\|\varphi\|_{\mathrm{u}} - \varphi$ and using the linearity property in **(iii)** of Examples 1.2.2, we conclude that (1.2.9) holds with $K_\psi = \Delta_J\psi$. Thus, all that we have to do is check that every $\varphi \in C(J)$ is ψ-Riemann integrable on J.

Let $\varphi \in C(J)$ be given and define

$$\mathcal{U}(\varphi|\psi;\mathcal{C}) = \sum_{I\in\mathcal{C}} (\sup_I \varphi)\Delta_I\psi \quad \text{and} \quad \mathcal{L}(\varphi|\psi;\mathcal{C}) = \sum_{I\in\mathcal{C}} (\inf_I \varphi)\Delta_I\psi$$

for $\mathcal{C}$ and $\xi \in \Xi(\mathcal{C})$. Then, just as in Section 1.1,

$$\mathcal{L}(\varphi|\psi;\mathcal{C}) \leq \mathcal{R}(\varphi|\psi;\mathcal{C},\xi) \leq \mathcal{U}(\varphi|\psi;\mathcal{C})$$

for any $\xi \in \Xi(\mathcal{C})$. In addition (cf. Lemma 1.1.5), for any pair $\mathcal{C}_1$ and $\mathcal{C}_2$, one has that $\mathcal{L}(\varphi|\psi;\mathcal{C}_1) \leq \mathcal{U}(\varphi|\psi;\mathcal{C}_2)$. Finally, for any $\mathcal{C}$,

$$\mathcal{U}(\varphi|\psi;\mathcal{C}) - \mathcal{L}(\varphi|\psi;\mathcal{C}) \leq \omega(\|\mathcal{C}\|)\Delta_J\psi,$$

where

$$\omega(\delta) \equiv \sup\left\{|\varphi(y) - \varphi(x)| : x,y \in J \text{ and } |y-x| \leq \delta\right\}$$

is the **modulus of continuity** of φ. Hence[2]

$$\lim_{\|\mathcal{C}\|\to 0}\big(\mathcal{U}(\varphi|\psi;\mathcal{C}) - \mathcal{L}(\varphi|\psi;\mathcal{C})\big) = 0.$$

But this means that, for every $\epsilon > 0$, there is a $\delta > 0$ such that

$$\mathcal{U}(\varphi|\psi;\mathcal{C}) - \mathcal{U}(\varphi|\psi;\mathcal{C}') \leq \mathcal{U}(\varphi|\psi;\mathcal{C}) - \mathcal{L}(\varphi|\psi;\mathcal{C}) < \epsilon$$

no matter what $\mathcal{C}'$ is chosen as long as $\|\mathcal{C}\| < \delta$. From the above it is clear that both

$$\lim_{\|\mathcal{C}\|\to 0} \mathcal{U}(\varphi|\psi;\mathcal{C}) \quad \text{and} \quad \lim_{\|\mathcal{C}\|\to 0} \mathcal{L}(\varphi|\psi;\mathcal{C})$$

exist and are equal. □

One obvious way to extend the preceding result is to note that if φ is Riemann integrable on J with respect to both ψ_1 and ψ_2, then it is Riemann integrable on J with respect to $\psi \equiv \psi_2 - \psi_1$ and

$$(\mathrm{R})\int_J \varphi(x)\,d\psi(x) = (\mathrm{R})\int_J \varphi(x)\,d\psi_2(x) - (\mathrm{R})\int_J \varphi(x)\,d\psi_1(x).$$

(This can be seen directly or as a consequence of Theorem 1.2.7 combined with **(iii)** in Examples 1.2.2.) In particular, we have the following corollary to Theorem 1.2.10.

1.2.11 Corollary. *If $\psi = \psi_2 - \psi_1$ where ψ_1 and ψ_2 are non-decreasing functions on J, then every $\varphi \in C(J)$ is Riemann integrable with respect to ψ and* (1.2.9) *holds with $K_\psi = \Delta_J\psi_1 + \Delta_J\psi_2$.*

---

[2] Recall that a continuous function on a compact set is uniformly continuous there. Hence, $\omega(\delta) \searrow 0$ as $\delta \searrow 0$.

We are now going to embark on a program which will show that, at least among ψ's that are right continuous on $\mathring{J}$ and have left limits at each point in $J \setminus \{J^-\}$, the ψ's in Corollary 1.2.11 are the only ones with the properties that every $\varphi \in C(J)$ is ψ-Riemann integrable on J and (1.2.9) holds for some $K_\psi < \infty$. The first step is to provide an alternative description of those ψ's which can be expressed as the difference between two non-decreasing functions. To this end, let ψ be a real-valued function on J and define

$$\mathcal{S}(\psi; \mathcal{C}) = \sum_{I \in \mathcal{C}} |\Delta_I \psi|$$

for any non-overlapping, finite, exact cover $\mathcal{C}$ of J. Clearly

$$\mathcal{S}(\alpha\psi; \mathcal{C}) = |\alpha| \mathcal{S}(\psi; \mathcal{C}) \quad \text{for all } \alpha \in \mathbb{R},$$

$$\mathcal{S}(\psi_1 + \psi_2; \mathcal{C}) \leq \mathcal{S}(\psi_1; \mathcal{C}) + \mathcal{S}(\psi_2; \mathcal{C}) \quad \text{for all } \psi_1 \text{ and } \psi_2,$$

and

$$\mathcal{S}(\psi; \mathcal{C}) = |\Delta_J \psi|$$

if ψ is monotone on J. Moreover, if $\mathcal{C}$ is given and $\mathcal{C}'$ is obtained from $\mathcal{C}$ by replacing one of the I's in $\mathcal{C}$ by a pair $\{I_1, I_2\}$, where $I = I_1 \cup I_2$ and $\mathring{I}_1 \cap \mathring{I}_2 = \emptyset$, then, by the triangle inequality,

$$\begin{aligned}&\mathcal{S}(\psi; \mathcal{C}') - \mathcal{S}(\psi; \mathcal{C}) \\ &\qquad = |\psi(I_1^+) - \psi(I_1^-)| + |\psi(I_2^+) - \psi(I_2^-)| - |\psi(I^+) - \psi(I^-)| \geq 0.\end{aligned}$$

Hence we see that

$$\mathcal{S}(\psi; \mathcal{C}) \leq \mathcal{S}(\psi; \mathcal{C}') \quad \text{for} \quad \mathcal{C} \leq \mathcal{C}'.$$

We now define the **variation of ψ on J** to be the number (possibly infinite)

$$\operatorname{Var}(\psi; J) \equiv \sup_{\mathcal{C}} \mathcal{S}(\psi; \mathcal{C}), \tag{1.2.12}$$

where the $\mathcal{C}$'s run over all non-overlapping, finite, exact covers of J. Also, we say that ψ has **bounded variation on J** if $\operatorname{Var}(\psi; J) < \infty$. It should be clear that if $\psi = \psi_2 - \psi_1$ for non-decreasing ψ_1 and ψ_2 on J, then ψ has bounded variation on J and $\operatorname{Var}(\psi; J) \leq \Delta_J \psi_1 + \Delta_J \psi_2$. What is less obvious is that every ψ having bounded variation on J can be expressed as the difference between two non-decreasing functions. In order to prove this, we introduce

$$\mathcal{S}_+(\psi; \mathcal{C}) = \sum_{I \in \mathcal{C}} (\Delta_I \psi)^+$$

and

$$\mathcal{S}_-(\psi; \mathcal{C}) = \sum_{I \in \mathcal{C}} (\Delta_I \psi)^-,$$

where $a^+ \equiv \alpha \vee 0$ and $\alpha^- \equiv -(\alpha \wedge 0)$ for $\alpha \in \mathbb{R}$. Also, we call

$$\operatorname{Var}_+(\psi; J) \equiv \sup_{\mathcal{C}} \mathcal{S}_+(\psi; \mathcal{C}) \quad \text{and} \quad \operatorname{Var}_-(\psi; J) \equiv \sup_{\mathcal{C}} \mathcal{S}_-(\psi; \mathcal{C})$$

the **positive variation** and the **negative variation** of ψ on J. Noting that

$$\begin{aligned} 2\mathcal{S}_\pm(\psi; \mathcal{C}) &= \mathcal{S}(\psi; \mathcal{C}) \pm \Delta_J\psi \\ \mathcal{S}_+(\psi; \mathcal{C}) - \mathcal{S}_-(\psi; \mathcal{C}) &= \Delta_J\psi \\ \mathcal{S}_+(\psi; \mathcal{C}) + \mathcal{S}_-(\psi; \mathcal{C}) &= \mathcal{S}(\psi; \mathcal{C}) \end{aligned} \tag{1.2.13}$$

for any $\mathcal{C}$, we see that

$$\mathcal{S}_\pm(\psi; \mathcal{C}) \le \mathcal{S}_\pm(\psi; \mathcal{C}'), \qquad \mathcal{C} \le \mathcal{C}',$$

and that

$$\operatorname{Var}_+(\psi; J) < \infty \iff \operatorname{Var}(\psi; J) < \infty \iff \operatorname{Var}_-(\psi; J) < \infty.$$

1.2.14 Lemma. *If* $\operatorname{Var}(\psi; J) < \infty$, *then*

$$\operatorname{Var}_+(\psi; J) + \operatorname{Var}_-(\psi; J) = \operatorname{Var}(\psi; J) \tag{1.2.15}$$

and

$$\operatorname{Var}_+(\psi; J) - \operatorname{Var}_-(\psi; J) = \Delta_J\psi \tag{1.2.16}$$

PROOF: By the middle relation in (1.2.13), we see that

$$\mathcal{S}_\pm(\psi; \mathcal{C}) \le \operatorname{Var}_\mp(\psi; J) \pm \Delta_J\psi.$$

Hence

$$\operatorname{Var}_\pm(\psi; J) \le \operatorname{Var}_\mp(\psi; J) \pm \Delta_J\psi;$$

and so (1.2.16) has been proved. Moreover, (1.2.16) combined with the middle relation in (1.2.13) leads to

$$\operatorname{Var}_+(\psi; J) - \mathcal{S}_+(\psi; \mathcal{C}) = \operatorname{Var}_-(\psi; J) - \mathcal{S}_-(\psi; \mathcal{C})$$

for any $\mathcal{C}$. In particular, there is a sequence $\{\mathcal{C}_n\}_1^\infty$ such that $\mathcal{S}_+(\psi; \mathcal{C}_n) \longrightarrow \operatorname{Var}_+(\psi; J)$ as $n \to \infty$ and, at the same time, $\mathcal{S}_-(\psi; \mathcal{C}_n) \longrightarrow \operatorname{Var}_-(\psi; J)$. Hence, by the last relation in (1.2.13), we see that

$$\operatorname{Var}_+(\psi; J) + \operatorname{Var}_-(\psi; J) \le \varlimsup_{n\to\infty} \mathcal{S}(\psi; \mathcal{C}_n) \le \operatorname{Var}(\psi; J).$$

At the same time, by that same relation in (1.2.13),

$$\mathcal{S}(\psi; \mathcal{C}) = \mathcal{S}_+(\psi; \mathcal{C}) + \mathcal{S}_-(\psi; \mathcal{C}) \le \operatorname{Var}_+(\psi; J) + \operatorname{Var}_-(\psi; J)$$

for every $\mathcal{C}$. When combined with the preceding, this completes the proof of (1.2.15). □

1.2.17 Lemma. *If* ψ *has bounded variation on* $[a, b]$ *and* $a < c < b$, *then*

$$\operatorname{Var}_\pm(\psi; [a, b]) = \operatorname{Var}_\pm(\psi; [a, c]) + \operatorname{Var}_\pm(\psi; [c, b]),$$

and therefore also $\operatorname{Var}(\psi; [a, b]) = \operatorname{Var}(\psi; [a, c]) + \operatorname{Var}(\psi; [c, b])$.

PROOF: Because of (1.2.15) and (1.2.16), we see that it suffices to check the equality only for "Var" itself. But if $\mathcal{C}_1$ and $\mathcal{C}_2$ are non-overlapping, finite, exact covers of $[a,c]$ and $[c,b]$, then $\mathcal{C} = \mathcal{C}_1 \cup \mathcal{C}_2$ is a non-overlapping, finite, exact cover of $[a,b]$; and so

$$\mathcal{S}(\psi;\mathcal{C}_1) + \mathcal{S}(\psi;\mathcal{C}_2) = \mathcal{S}(\psi;\mathcal{C}) \le \mathrm{Var}(\psi;[a,b]).$$

Hence $\mathrm{Var}(\psi;[a,c]) + \mathrm{Var}(\psi;[c,b]) \le \mathrm{Var}(\psi;[a,b])$. On the other hand, if $\mathcal{C}$ is a non-overlapping, finite, exact cover of $[a,b]$, then it is easy to construct non-overlapping, finite, exact covers $\mathcal{C}_1$ and $\mathcal{C}_2$ of $[a,c]$ and $[c,b]$ such that $\mathcal{C} \le \mathcal{C}_1 \cup \mathcal{C}_2$. Hence,

$$\mathcal{S}(\psi;\mathcal{C}) \le \mathcal{S}(\psi;\mathcal{C}_1 \cup \mathcal{C}_2) = \mathcal{S}(\psi;\mathcal{C}_1) + \mathcal{S}(\psi;\mathcal{C}_2) \le \mathrm{Var}(\psi;[a,c]) + \mathrm{Var}(\psi;[c,b]).$$

Since this is true for every $\mathcal{C}$, the asserted equality is now proved. □

We have now proved the following decomposition theorem for functions having bounded variation.

1.2.18 Theorem. *Let $\psi : J \longrightarrow \mathbb{R}$ be given. Then ψ has bounded variation on J if and only if there exist non-decreasing functions ψ_1 and ψ_2 on J such that $\psi = \psi_2 - \psi_1$. In fact, if ψ has bounded variation on $J = [a,b]$ and we define $\psi_\pm(x) = \mathrm{Var}_\pm(\psi;[a,x])$ for $x \in J$, then ψ_+ and ψ_- are non-decreasing and $\psi(x) = \psi(a) + \psi_+(x) - \psi_-(x)$, $x \in J$. Finally, if ψ has bounded variation on J, then every $\varphi \in C(J)$ is Riemann integrable on J with respect to ψ and*

$$\left|(\mathrm{R})\int_J \varphi(x)\,d\psi(x)\right| \le \mathrm{Var}(\psi;J)\|\varphi\|_{\mathrm{u}}. \tag{1.2.19}$$

In order to complete our program, we need one more elementary fact.

1.2.20 Lemma. *If $\psi : J \longrightarrow \mathbb{R}$ has a right limit in $\mathbb{R}$ at every $x \in J \setminus \{J^+\}$ and a left limit in $\mathbb{R}$ at every $x \in J \setminus \{J^-\}$, then ψ is bounded and*

$$\mathrm{card}\Big(\Big\{x \in \mathring{J} : |\psi(x) - \psi(x+)| \vee |\psi(x) - \psi(x-)| \ge \epsilon\Big\}\Big) < \infty \quad \text{for each } \epsilon > 0.$$

In particular, ψ has at most countably many discontinuities. Also, if $\tilde{\psi}(x) \equiv \psi(x+)$ for $x \in \mathring{J}$ and $\tilde{\psi}(x) = \psi(x)$ for $x \in \{J^-, J^+\}$, then $\tilde{\psi}$ is right-continuous on $\mathring{J}$, has a left limit in $\mathbb{R}$ at every $x \in J \setminus \{J^-\}$, and coincides with ψ at all points where ψ is continuous. Thus, if φ is Riemann integrable on J with respect to both ψ and $\tilde{\psi}$, then $(\mathrm{R})\int_J \varphi(x)\,d\tilde{\psi}(x) = (\mathrm{R})\int_J \varphi(x)\,d\psi(x)$. Finally, if $\varphi \in C(J)$ is Riemann integrable on J with respect to ψ, then it is also Riemann integrable on J with respect to $\tilde{\psi}$.

PROOF: Suppose that ψ were unbounded. Then we could find $\{x_n\}_1^\infty \subseteq J$ so that $|\psi(x_n)| \longrightarrow \infty$ as $n \to \infty$; and clearly there is no loss of generality if we assume that $x_{n+1} < x_n$ for all $n \ge 1$. But this would mean that $|\psi(x+)| = \infty$, where $x = \lim_{n\to\infty} x_n$, and so no such sequence can exist. Thus ψ must be bounded. The proof that $\operatorname{card}\bigl(\{x \in \mathring{J} : |\psi(x) - \psi(x+)| \vee |\psi(x) - \psi(x-)| \geq \epsilon\}\bigr) < \infty$ is similar. Namely, if not, we could assume that there exists a strictly decreasing sequence $\{x_n\} \subseteq \mathring{J}$ with limit $x \in J$ such that $|\psi(x_n) - \psi(x_n+)| \vee |\psi(x_n) - \psi(x_n-)| \geq \epsilon$ for each $n \geq 1$. But then, for each $n \geq 1$, we could find $x'_n \in (x, x_n)$ and $x''_n \in (x_n, x_n + \frac{1}{n}) \cap \mathring{J}$ so that $|\psi(x_n) - \psi(x'_n)| \vee |\psi(x_n) - \psi(x''_n)| \geq \frac{\epsilon}{2}$; clearly this contradicts the existence in $\mathbb{R}$ of $\psi(x+)$.

The preceding makes it obvious that ψ can be discontinuous at only countably many points. In addition it is clear that $\tilde{\psi}(x\pm) = \psi(x\pm)$ for all $x \in \mathring{J}$. To prove the equality of Riemann integrals with respect to ψ and $\tilde{\psi}$ of φ's which are Riemann integrable with respect to both, note that, because ψ coincides with $\tilde{\psi}$ on $\{J^-, J^+\}$ as well as on a dense subset of $\mathring{J}$, we can always evaluate these integrals using Riemann sums which are the same whether they are computed with respect to ψ or to $\tilde{\psi}$.

Finally, we must show that if $\varphi \in C(J)$ is Riemann integrable with respect to ψ, then it also is with respect to $\tilde{\psi}$. To do this, it clearly suffices to show that for any $\mathcal{C}$, choice map $\xi \in \Xi(\mathcal{C})$, and $\epsilon > 0$, there is a $\mathcal{C}'$ and a $\xi' \in \Xi(\mathcal{C}')$ such that $\|\mathcal{C}'\| \leq 2\|\mathcal{C}\|$ and $|\mathcal{R}(\varphi|\tilde{\psi}; \mathcal{C}, \xi) - \mathcal{R}(\varphi|\psi; \mathcal{C}', \xi')| < \epsilon$. To this end, let $J^- = c_0 < \cdots < c_{n+1} = J^+$ be chosen so that $\mathcal{C} = \{[c_k, c_{k+1}] : 0 \leq k \leq n\}$. For $0 < \alpha < \min_{1\leq k\leq n+1} c_{k+1} - c_k$, set

$$c_{k,\alpha} = \begin{cases} c_k & \text{if} \quad k \in \{0, n+1\} \\ c_k + \alpha & \text{if} \quad 1 \leq k \leq n, \end{cases}$$

and let $\mathcal{C}_\alpha = \bigl\{[c_{k,\alpha}, c_{k+1,\alpha}] : 0 \leq k \leq n\bigr\}$. Clearly $\|\mathcal{C}_\alpha\| \leq 2\|\mathcal{C}\|$. Moreover, $c_{k,\alpha} \leq c_{k+1} \leq c_{k+1,\alpha}$ for each $0 \leq k \leq n$, and so we can take

$$\xi_\alpha\bigl([c_{k,\alpha}, c_{k+1,\alpha}]\bigr) = c_{k,\alpha} \vee \xi\bigl([c_k, c_{k+1}]\bigr).$$

Then, because φ is continuous and $\tilde{\psi}$ is right-continuous on $\mathring{J}$,

$$\mathcal{R}(\varphi|\tilde{\psi}; \mathcal{C}, \xi) = \lim_{\alpha\searrow 0} \mathcal{R}(\varphi|\tilde{\psi}; \mathcal{C}_\alpha, \xi_\alpha).$$

At the same time,

$$\mathcal{R}(\varphi|\tilde{\psi}; \mathcal{C}_\alpha, \xi_\alpha) = \mathcal{R}(\varphi|\psi; \mathcal{C}_\alpha, \xi_\alpha)$$

for all but a countable number of α's. Thus, for any $\epsilon > 0$, there is an $\alpha > 0$ for which $|\mathcal{R}(\varphi|\tilde{\psi}; \mathcal{C}, \xi) - \mathcal{R}(\varphi|\tilde{\psi}; \mathcal{C}_\alpha, \xi_\alpha)| < \epsilon$ and $\mathcal{R}(\varphi|\tilde{\psi}, \mathcal{C}_\alpha, \xi_\alpha) = \mathcal{R}(\varphi|\psi, \mathcal{C}_\alpha, \xi_\alpha)$. □

1.2.21 Remark: As a consequence of Theorem 1.2.18, any ψ having bounded variation on J certainly satisfies the hypotheses of Lemma 1.2.20. Moreover, if $\tilde{\psi}$ is defined accordingly, then one can check that $\operatorname{Var}(\tilde{\psi}; J) \leq \operatorname{Var}(\psi; J)$.

1.2.22 Theorem. *Let ψ be a function on J which satisfies the hypotheses of* Lemma *1.2.20, and define $\tilde{\psi}$ accordingly. If every $\varphi \in C(J)$ is Riemann integrable on J with respect to ψ, and if there is a $K < \infty$ such that*

$$\left|(\mathrm{R})\int_J \varphi(x)\, d\psi(x)\right| \leq K\|\varphi\|_{\mathrm{u}}, \qquad \varphi \in C(J), \tag{1.2.23}$$

then $\tilde{\psi}$ has bounded variation on J and

$$
\begin{aligned}
\operatorname{Var}(\tilde{\psi};J) &= \sup\left\{(\mathrm{R})\int_J \varphi(x)\,d\psi(x) : \varphi\in C(J) \text{ and } \|\varphi\|_{\mathrm{u}}=1\right\}\\
&= \sup\left\{(\mathrm{R})\int_J \varphi(x)\,d\tilde{\psi}(x) : \varphi\in C(J) \text{ and } \|\varphi\|_{\mathrm{u}}=1\right\}.
\end{aligned}
\tag{1.2.24}
$$

In particular, if ψ itself is right-continuous on $\mathring{J}$, then ψ has bounded variation on J if and only if every $\varphi\in C(J)$ is Riemann integrable on J with respect to ψ and (1.2.23) holds for some $K<\infty$, in which case $\operatorname{Var}(\psi;J)$ is the optimal choice of K.

PROOF: In view of what we already know, all that we have to do is check that for each $\mathcal{C}$ and $\epsilon>0$ there is a $\varphi\in C(J)$ such that $\|\varphi\|_{\mathrm{u}}=1$ and $\mathcal{S}(\tilde{\psi};\mathcal{C})\le(\mathrm{R})\int_J\varphi(x)\,d\psi(x)+\epsilon$. Moreover, because $\tilde{\psi}$ is right continuous, we may and will assume that $\mathcal{C}=\{[c_k,c_{k+1}]:0\le k\le n\}$ where $J^-=c_0<\cdots<c_{n+1}=J^+$ and c_k is a point of continuity of ψ for each $1\le k\le n$.

Given $0<\alpha<\min_{1\le k\le n}\frac{c_{k+1}-c_k}{2}$, define $\varphi_\alpha\in C(J)$ so that

$$
\varphi_\alpha(x)=\begin{cases}
\operatorname{sgn}\bigl(\Delta_{[c_0,c_1]}\psi\bigr) & \text{for } x\in[c_0,c_1-\alpha],\\
\operatorname{sgn}\bigl(\Delta_{[c_k,c_{k+1}]}\psi\bigr) & \text{for } x\in[c_k+\alpha,c_{k+1}-\alpha] \text{ and } 1\le k<n,\\
\operatorname{sgn}\bigl(\Delta_{[c_n,c_{n+1}]}\psi\bigr) & \text{for } x\in[c_n+\alpha,c_{n+1}],
\end{cases}
$$

and φ_α is linear on each of the intervals $[c_k-\alpha,c_k+\alpha]$, $1\le k\le n$. (The **signum function** $t\in\mathbb{R}\longmapsto\operatorname{sgn}(t)$ is defined so that $\operatorname{sgn}(t)$ is -1 or 1 according to whether $t<0$ or $t\ge0$.) Then, by (**iv**) in Examples 1.2.2,

$$
\begin{aligned}
&(\mathrm{R})\int_J\varphi_\alpha(x)\,d\psi(x)-\mathcal{S}(\tilde{\psi};\mathcal{C})\\
&\quad=\sum_{k=0}^n(\mathrm{R})\int_{[c_k,c_{k+1}]}\bigl(\varphi_\alpha(x)-\operatorname{sgn}(\Delta_{[c_k,c_{k+1}]}\psi)\bigr)\,d\psi(x)\\
&\quad=\sum_{k=1}^n\Biggl[(\mathrm{R})\int_{[c_k-\alpha,c_k]}\bigl(\varphi_\alpha(x)-\varphi_\alpha(c_k-\alpha)\bigr)\,d\psi(x)\\
&\qquad\qquad+(\mathrm{R})\int_{[c_k,c_k+\alpha]}\bigl(\varphi_\alpha(x)-\varphi_\alpha(c_k+\alpha)\bigr)\,d\psi(x)\Biggr].
\end{aligned}
$$

For each $1\le k\le n$, either $\varphi_\alpha\equiv\varphi_\alpha(c_k-\alpha)$ on $[c_k-\alpha,c_k+\alpha]$, in which case the corresponding term does not contribute to the preceding sum, or $\varphi_\alpha(c_k)=0$ and $\varphi_\alpha'\equiv\bigl(\varphi_\alpha(c_k+\alpha)-\varphi_\alpha(c_k-\alpha)\bigr)/2\alpha$ on $[c_k-\alpha,c_k+\alpha]$. In the latter case, we apply Theorem 1.2.7 and equation (1.2.3) to show that

$$\begin{aligned}(\mathrm{R})\int_{[c_k-\alpha,c_k]}&\big(\varphi_\alpha(x)-\varphi_\alpha(c_k-\alpha)\big)\,d\psi(x)\\ &+(\mathrm{R})\int_{[c_k,c_k+\alpha]}\big(\varphi_\alpha(x)-\varphi_\alpha(c_k+\alpha)\big)\,d\psi(x)\\ =\big[\varphi_\alpha(c_k+\alpha)&-\varphi_\alpha(c_k-\alpha)\big]\psi(c_k)\\ &-\frac{\varphi_\alpha(c_k+\alpha)-\varphi_\alpha(c_k-\alpha)}{2\alpha}(\mathrm{R})\int_{[c_k-\alpha,c_k+\alpha]}\psi(x)\,dx,\end{aligned}$$

which, since ψ is continuous at c_k, clearly tends to 0 as $\alpha\searrow 0$. In other words, we now see that

$$\mathcal{S}(\tilde{\psi};\mathcal{C})=\lim_{\alpha\searrow 0}(\mathrm{R})\int_J\varphi_\alpha(x)\,d\psi(x),$$

which is all that we had to prove. □

Exercises

1.2.25 Exercise: Check all of the assertions in Examples 1.2.2. The only one which presents a challenge is the assertion in (**iv**) that φ is Riemann integrable on both J_1 and J_2 with respect to ψ.

1.2.26 Exercise: If ψ is non-decreasing on J, show that a bounded function φ is Riemann integrable on J with respect to ψ if and only if, for every $\epsilon>0$, there is a $\delta>0$ such that

$$\sum_{\{I\in\mathcal{C}:\sup_I\varphi-\inf_I\varphi\geq\epsilon\}}\Delta_I\psi<\epsilon \tag{1.2.27}$$

whenever $\mathcal{C}$ is a non-overlapping, finite, exact cover of J satisfying $\|\mathcal{C}\|<\delta$. Also, show that when, in addition, $\psi\in C(J)$, the preceding can be replaced by the condition that, for each $\epsilon>0$, (1.2.27) holds for some $\mathcal{C}$. (**Hint**: for the last part, compare the situation here to the one handled in Lemma 1.1.7.)

1.2.28 Exercise: If $\psi\in C(J)$, show that

$$\mathrm{Var}_\pm(\psi;J)=\lim_{\|\mathcal{C}\|\to 0}\mathcal{S}_\pm(\psi;\mathcal{C})\ \big(\in[0,\infty]\big)$$

and conclude that $\mathrm{Var}(\psi,J)=\lim_{\|\mathcal{C}\|\to 0}\mathcal{S}(\psi;\mathcal{C})$. Also, show that if $\psi\in C^1(J)$, then

$$\mathrm{Var}_\pm(\psi;J)=(\mathrm{R})\int_J\psi'(x)^\pm\,dx,$$

and therefore $\mathrm{Var}(\psi;J)=(\mathrm{R})\int_J|\psi'(x)|\,dx$.

1.2.29 Exercise: Let ψ be a function of bounded variation on the interval $J = [c, d]$, and define the non-decreasing functions ψ_+ and ψ_- accordingly, as in Theorem 1.2.18. Given any other pair of non-decreasing functions ψ_1 and ψ_2 on J satisfying $\psi = \psi_2 - \psi_1$, show that $\psi_2 - \psi_+$ and $\psi_1 - \psi_-$ are both non-decreasing functions. In particular, this means that $\psi_+ \leq \psi_2 - \psi_2(c)$ and $\psi_- \leq \psi_1 - \psi_1(c)$ whenever ψ_2 and ψ_1 are non-decreasing functions for which $\psi = \psi_2 - \psi_1$. Using Lemma 1.2.17 and the preceding, show that

$$\psi_\pm(x+) - \psi_\pm(x) = \Big(\psi(x+) - \psi(x)\Big)^\pm, \quad x \in [c, d),$$

and

$$\psi_\pm(x) - \psi_\pm(x-) = \Big(\psi(x) - \psi(x-)\Big)^\pm, \quad x \in (c, d].$$

Conclude, in particular, that the jumps in $x \longmapsto \mathrm{Var}(\psi; [a, x])$, from both the right and left, coincide with the absolute value of the corresponding jumps in ψ. Hence, ψ is continuous if $x \in J \longmapsto \mathrm{Var}\big(\psi; [c, x]\big)$ is; and if ψ is continuous, then so are ψ_+ and ψ_- and therefore also $\mathrm{Var}\big(\psi; [c, \cdot\,]\big)$.

Hint: In order to handle the last part, show that it is enough to check that $\psi_+(c+) = \big(\psi(c+) - \psi(c)\big)^+$. Next, show that this comes down to checking that $\beta \equiv \psi_+(c+) \wedge \psi_-(c+) = 0$. Finally, define ψ_1 and ψ_2 on $[c, d]$ so that $\psi_1(c) = 0$, $\psi_2(c) = \psi(c)$, and, for $x \in (c, d]$, $\psi_1(x) = \psi_-(x) - \beta$ while $\psi_2(x) = \psi(c) + \psi_+(x) - \beta$; and apply the first part of this exercise to see that $\psi_- \leq \psi_1$.

1.2.30 Exercise: Construct an example of a $\psi \in C([0, 1])$ which is not of bounded variation. Also, give an example of a ψ having bounded variation on $[0, 1]$ for which

$$\sup\left\{ (\mathrm{R}) \int_{[0,1]} \varphi(x)\, d\psi(x) : \varphi \in C(J) \text{ and } \|\varphi\|_{\mathrm{u}} = 1 \right\} < \mathrm{Var}(\psi; J).$$

Chapter II
Lebesgue Measure

2.0 The Idea

In this chapter we construct Lebesgue's measure on $\mathbb{R}^N$, and in the following chapter we will develop his method of integration. To avoid getting lost in the details, it will be important to keep in mind what it is we are attempting to do. For this reason, we begin with a brief summary of our goals.

The essence of any theory of integration is a *divide and conquer* strategy. That is, given a space E and a family $\mathcal{B}$ of subsets $\Gamma \subseteq E$ for which one has a *reasonable notion* of *measure* assignment $\Gamma \in \mathcal{B} \longmapsto \mu(\Gamma) \in [0, \infty]$, the *integral* of a function $f : E \longrightarrow \mathbb{R}$ should be computed by a prescription containing the following ingredients. In the first place, one has to choose a partition $\mathcal{P}$ of the space E into subsets $\Gamma \in \mathcal{B}$. Secondly, given $\mathcal{P}$, one has to select for each $\Gamma \in \mathcal{P}$ a *typical* value a_Γ of f on Γ. Thirdly, given both the partition $\mathcal{P}$ and the selection

$$\Gamma \in \mathcal{P} \longmapsto a_\Gamma \in \mathrm{Range}(f \restriction \Gamma),$$

one forms the sum

$$\sum_{\Gamma \in \mathcal{P}} a_\Gamma \, \mu(\Gamma). \tag{2.0.1}$$

Finally, using a limit procedure if necessary, one removes the ambiguity (inherent in the notion of *typical*) by choosing the partitions $\mathcal{P}$ in such a way that the restriction of f to each Γ is increasingly close to a constant.

Obviously, even if we ignore all questions of convergence, the only way in which we can make sense out of (2.0.1) is if we restrict ourselves to either finite or, at worst, countable partitions $\mathcal{P}$. Hence, in general, the final limit procedure will be essential. Be that as it may, when E is itself countable and $\{x\} \in \mathcal{B}$ for every $x \in E$, there is an *obvious* way to avoid the limit step, namely one chooses $\mathcal{P} = \big\{\{x\} : x \in E\big\}$ and takes

$$\sum_{x \in E} f(x)\mu(\{x\}) \tag{2.0.2}$$

to be the *integral.* (We will ignore, for the present, all problems arising from questions of convergence.) Clearly, this is the idea on which Riemann based his theory of integration. On the other hand, Riemann's is not the only *obvious* way to proceed, even in the case of countable spaces E. For example, again assuming that E is countable, let $f : E \longrightarrow \mathbb{R}$ be given. Then Range(f) is countable and, assuming that $\Gamma(a) \equiv \{x \in E : f(x) = a\} \in \mathcal{B}$ for every $a \in \mathbb{R}$, Lebesgue would say that

$$\sum_{a \in \text{Range}(f)} a\, \mu(\Gamma(a)) \tag{2.0.3}$$

is an equally *obvious* candidate for the *integral* of f.

In order to reconcile these two *obvious* definitions, one has to examine the assignment $\Gamma \in \mathcal{B} \longmapsto \mu(\Gamma) \in [0, \infty]$ of *measure*. Indeed, even if E is countable and $\mathcal{B}$ contains every subset of E, (2.0.2) and (2.0.3) give the same answer only if one knows that, for any countable collection $\{\Gamma_n\} \subseteq \mathcal{B}$,

$$\mu\left(\bigcup_n \Gamma_n\right) = \sum_n \mu(\Gamma_n) \quad \text{when } \Gamma_m \cap \Gamma_n = \emptyset \text{ for } m \neq n. \tag{2.0.4}$$

When E is countable, (2.0.4) is equivalent to taking

$$\mu(\Gamma) = \sum_{x \in \Gamma} \mu(\{x\}), \quad \Gamma \subseteq E.$$

However, when E is uncountable, the property in (2.0.4) becomes highly non-trivial. In fact, it is unquestionably Lebesgue's most significant achievement to have shown that there are non-trivial *assignments of measure* which enjoy this property.

Having compared Lebesgue's ideas to Riemann's in the countable setting, we close this introduction to Lebesgue's theory with a few words about the same comparison for uncountable spaces. For this purpose, we will suppose that $E = [0, 1]$ and, without worrying about exactly which subsets of E are included in $\mathcal{B}$, we will suppose that $\mathcal{B}$ contains not only all open and closed subsets of E but also all the sets which can be obtained, starting from open and closed sets, by countable, set-theoretic operations. Further, we assume that $\Gamma \in \mathcal{B} \longmapsto \mu(\Gamma) \in [0, 1]$ is a mapping which satisfies (2.0.4).

Now let $f : [0, 1] \longrightarrow \mathbb{R}$ be given. In order to integrate f, Riemann says that we should divide up $[0, 1]$ into small intervals, choose a representative value of f from each interval, form the associated Riemann sum, and then take the limit as the mesh size of the division tends to 0. As we know, his procedure works beautifully as long as the function f respects the topology of the real line; that is, as long as f is sufficiently continuous. However, Riemann's procedure is doomed to failure when f does not respect the topology of $\mathbb{R}$. The problem is, of course, that Riemann's partitioning procedure is tied to the topology of the reals and is therefore too rigid to accommodate functions which pay little or no attention to that topology. To get around this problem, Lebesgue tailors his partitioning procedure to the particular function f under consideration. Thus, for a given function f, Lebesgue might consider the sequence of partitions $\mathcal{P}_n$, $n \in \mathbb{N}$, consisting of the sets

$$\Gamma_{n,k} = \left\{x \in E : f(x) \in \left[\frac{k}{2^n}, \frac{k+1}{2^n}\right)\right\}, \quad k \in \mathbb{Z}.$$

Obviously, no two values that f takes on any one of the $\Gamma_{n,k}$'s can differ by more than $\frac{1}{2^n}$. Hence, assuming that $\Gamma_{n,k} \in \mathcal{B}$ for every $n \in \mathbb{N}$ and $k \in \mathbb{Z}$ and ignoring convergence problems,

$$\lim_{n\to\infty} \sum_{k\in\mathbb{Z}} \frac{k}{2^n}\,\mu(\Gamma_{n,k})$$

simply must be the *integral* of f!

When one hears Lebesgue's ideas for the first time, one may well wonder what there is left to be done. On the other hand, after a little reflection, some doubts begin to emerge. For example, what is so sacrosanct about Lebesgue's partitioning suggested in the preceding paragraph and, for instance, why should one have not done the same thing relative to 3^n instead of 2^n? The answer is, of course, that there is nothing to recommend 2^n over 3^n and that it should make no difference which of them is used. Thus, one has to check that it really does not matter, and, once again, the verification entails repeated application of the property in (2.0.4). In fact, it will become increasingly evident that Lebesgue's entire program rests on (2.0.4).

With the preceding comments in mind, it should be clear why we initiate Lebesgue's program with his proof that an interesting μ satisfying (2.0.4) actually exists. To be more precise, the rest of this chapter is devoted to showing that we can define such a μ on a rich class of subsets of $\mathbb{R}^N$ so that $\mu(I) = \operatorname{vol}(I)$ whenever I is a rectangle.

2.1 Existence

Given a countable (possibly overlapping) cover $\mathcal{C}$ of a subset $\Gamma \subseteq \mathbb{R}^N$ by rectangles I, define $\Sigma(\mathcal{C}) = \sum_{I\in\mathcal{C}} \operatorname{vol}(I) \in [0,\infty]$. We call

$$|\Gamma|_e = \inf\left\{\Sigma(\mathcal{C}) : \Gamma \subseteq \bigcup \mathcal{C}\right\}$$

the **outer** or **exterior Lebesgue measure** of Γ. What we are going to do is describe a family $\overline{\mathcal{B}}_{\mathbb{R}^N}$ for which the map

$$\Gamma \in \overline{\mathcal{B}}_{\mathbb{R}^N} \longmapsto |\Gamma|_e$$

satisfies (2.0.4). (The notation here, in particular the *bar*, will be explained in (**ii**) of Examples 3.1.5 and Exercise 3.1.9.) However, before starting on this project, we first check that $|I|_e = \operatorname{vol}(I)$ for rectangles I.

2.1.1 Lemma. *If $\Gamma = \bigcup_1^n J_m$ where the J_m's are non-overlapping rectangles, then $|\Gamma|_e = \sum_1^n \operatorname{vol}(J_m)$.*

PROOF: Obviously $|\Gamma|_e \leq \sum_1^n \text{vol}(J_m)$. To prove the opposite inequality, let $\mathcal{C} = \{I_\ell\}_1^\infty$ be a cover of J. Given an $\epsilon > 0$, choose I'_ℓ for each $\ell \in \mathbb{Z}^+$ so that $I_\ell \subseteq \mathring{I}'_\ell$ and $\text{vol}(I'_\ell) \leq \text{vol}(I_\ell) + 2^{-\ell}\epsilon$. Because Γ is compact, there exists an $L \in \mathbb{Z}^+$ such that $\{I'_1, \dots, I'_L\}$ covers Γ. In particular, by Lemma 1.1.1,

$$\begin{aligned}\sum_{m=1}^n \text{vol}(J_m) &\leq \sum_{m=1}^n \sum_{\ell=1}^L \text{vol}(J_m \cap I'_\ell) \\ &\leq \sum_{\ell=1}^L \text{vol}(I'_\ell) \leq \sum_{\ell=1}^\infty \text{vol}(I'_\ell) \leq \Sigma(\mathcal{C}) + \epsilon.\end{aligned}$$

(In the preceding, we have used the fact that, for any pair of rectangles I and J, $I \cap J$ is again a rectangle.) After first letting $\epsilon \searrow 0$ and then taking the infimum over $\mathcal{C}$'s, we get the required result. □

In view of Lemma 2.1.1, we are justified in replacing $\text{vol}(I)$ by $|I|_e$ for rectangles I.

Our next result shows that half the equality in (2.0.4) is automatic, even before we restrict to Γ's from $\overline{\mathcal{B}}_{\mathbb{R}^N}$.

2.1.2 Lemma. *If $\Gamma_1 \subseteq \Gamma_2$, then $|\Gamma_1|_e \leq |\Gamma_2|_e$. In addition, if $\Gamma \subseteq \bigcup_1^\infty \Gamma_n$, then $|\Gamma|_e \leq \sum_1^\infty |\Gamma_n|_e$. In particular, if $\Gamma \subseteq \bigcup_1^\infty \Gamma_n$ and $|\Gamma_n|_e = 0$ for each $n \geq 1$, then $|\Gamma|_e = 0$; and so $|\partial I|_e = 0$ for any rectangle I. (Here, and throughout, ∂S will denote the boundary $\overline{S} \setminus \mathring{S}$ of the set S.) Finally, if Γ_1 and Γ_2 are subsets of $\mathbb{R}^N$ for which*

$$\text{dist}(\Gamma_1, \Gamma_2) \equiv \inf\{|y - x| : x \in \Gamma_1 \text{ and } y \in \Gamma_2\} > 0,$$

then $|\Gamma_1 \cup \Gamma_2|_e = |\Gamma_1|_e + |\Gamma_2|_e$.

PROOF: The first assertion follows immediately from the fact that every cover of Γ_2 is also a cover of Γ_1.

In order to prove the second assertion, let $\epsilon > 0$ be given, and choose for each $n \geq 1$ a cover $\mathcal{C}_n$ so that $\Sigma(\mathcal{C}_n) \leq |\Gamma_n|_e + 2^{-n}\epsilon$. It is obvious that $\mathcal{C} \equiv \bigcup_1^\infty \mathcal{C}_n$ is a countable cover of Γ. Hence

$$|\Gamma|_e \leq \Sigma(\mathcal{C}) \leq \sum_{n=1}^\infty \sum_{I \in \mathcal{C}_n} \text{vol}(I) \leq \sum_{n=1}^\infty \left(|\Gamma_n|_e + 2^{-n}\epsilon\right) \leq \sum_{n=1}^\infty |\Gamma_n|_e + \epsilon.$$

Given the preceding, one proves that $|\partial I|_e = 0$ by observing that ∂I is the union of $2N$ degenerate rectangles of the form $[a_1, b_1] \times \cdots \times \{c_k\} \times \cdots \times [a_N, b_N]$, each of which has exterior measure 0.

Turning to the final assertion, set $\delta = \text{dist}(\Gamma_1, \Gamma_2)$, and let $\mathcal{C}$ be a countable cover of $\Gamma_1 \cup \Gamma_2$ by rectangles I. Without loss of generality, we will assume that $\text{diam}(I) < \delta$ for all $I \in \mathcal{C}$. (If this is not already the case, it can be brought about by replacing $\mathcal{C}$ by a new covering $\mathcal{C}'$ in which each $I \in \mathcal{C}$ with $\text{diam}(I) > \delta$

has been replaced with rectangles obtained by repeatedly subdividing I. As an application of Lemma 1.1.1, $\Sigma(\mathcal{C}) = \Sigma(\mathcal{C}')$.) Next, set

$$\mathcal{C}_i = \{I \in \mathcal{C} : I \cap \Gamma_i \neq \emptyset\},$$

and observe that, on the one hand, $\mathcal{C}_i$ covers Γ_i, while, on the other hand, $\mathcal{C}_1 \cap \mathcal{C}_2 = \emptyset$. Hence,

$$|\Gamma_1|_e + |\Gamma_2|_e \leq \Sigma(\mathcal{C}_1) + \Sigma(\mathcal{C}_2) \leq \Sigma(\mathcal{C}),$$

and so, after taking the infimum over $\mathcal{C}$'s, we see that $|\Gamma_1|_e + |\Gamma_2|_e \leq |\Gamma_1 \cup \Gamma_2|_e$. Since the opposite inequality is trivial, this completes the proof. □

2.1.3 Remark: We will use $\mathfrak{G}$ to denote the class of all open subsets in a topological space. Thus, in the present context, $\mathfrak{G}$ stands for the class of open subsets of $\mathbb{R}^N$. In this connection, we also introduce the class $\mathfrak{G}_\delta$ which consists of all subsets which can be written as the intersection of a countable number of open subsets. Note that $\mathfrak{G} \cup \mathfrak{F} \subseteq \mathfrak{G}_\delta$, where we use $\mathfrak{F}$ to stand for the class of all closed subsets. (To see this, let $F \in \mathfrak{F} \setminus \{\emptyset\}$ be given and write $F = \bigcap_1^\infty G_n$ where G_n is the set of $x \in \mathbb{R}^N$ for which there exists a $y \in F$ with $|x - y| < \frac{1}{n}$.) Finally, note that $\Gamma \in \mathfrak{G}_\delta$ if and only if its complement $\Gamma\complement$ is an element of $\mathfrak{F}_\sigma$, the class of subsets which can be written as the union of a countable number of closed sets.

2.1.4 Lemma. *For any* $\Gamma \subseteq \mathbb{R}^N$

$$|\Gamma|_e = \inf\{|G|_e : \Gamma \subseteq G \in \mathfrak{G}\}. \tag{2.1.5}$$

In particular, for each $\Gamma \subseteq \mathbb{R}^N$ *there is a* $B \in \mathfrak{G}_\delta$ *such that* $\Gamma \subseteq B$ *and* $|\Gamma|_e = |B|_e$.

PROOF: Obviously the left hand side of (2.1.5) is dominated by the right hand side. To prove the opposite inequality, assume that $|\Gamma|_e < \infty$, let $\epsilon > 0$ be given, and choose $\mathcal{C} = \{I_n\}_1^\infty$ to be a cover of Γ for which $|\Gamma|_e \geq \Sigma(\mathcal{C}) - \frac{\epsilon}{2}$. Next, for each $n \geq 1$, let I'_n be a rectangle satisfying $I_n \subseteq \mathring{I}'_n$ and $|I'_n|_e \leq |I_n|_e + 2^{-n-1}\epsilon$. Then $G \equiv \bigcup_1^\infty \mathring{I}'_n$ is certainly open, it contains Γ, and

$$|G|_e \leq \sum_1^\infty |I'_n|_e \leq |\Gamma|_e + \epsilon.$$

Having proved the first assertion, the second one follows by choosing a sequence $\{G_n\}_1^\infty \subseteq \mathfrak{G}$ so that $\Gamma \subseteq G_n$ and $|G_n|_e \leq |\Gamma|_e + \frac{1}{n}$ for each $n \geq 1$. Clearly the set $B \equiv \bigcap_1^\infty G_n$ will then serve. □

We are now ready to describe the class $\overline{\mathcal{B}}_{\mathbb{R}^N}$ (alluded to at the beginning of this section), although it will not be immediately clear why it has the properties which we want. Be that as it may, we will say that $\Gamma \subseteq \mathbb{R}^N$ is **Lebesgue measurable** (or, when it is clear that we are discussing Lebesgue measure, simply **measurable**), and we will write $\Gamma \in \overline{\mathcal{B}}_{\mathbb{R}^N}$ if, for each $\epsilon > 0$, there is an open $G \supseteq \Gamma$ such that $|G \setminus \Gamma|_e < \epsilon$. In order to distinguish $|\cdot|_e$ from its restriction to $\overline{\mathcal{B}}_{\mathbb{R}^N}$, we will use $|\Gamma|$ instead of $|\Gamma|_e$ when Γ is measurable, and we will call $|\Gamma|$ the **Lebesgue measure** (or simply, the **measure**) of Γ.

2.1.6 Remark: At first, one might be tempted to say that, in view of Lemma 2.1.4, every subset Γ is measurable. This is because one is inclined to think that $|G|_e = |G\setminus\Gamma|_e + |\Gamma|_e$ when, in fact, $|G|_e \le |G\setminus\Gamma|_e + |\Gamma|_e$ is all that we know. Therein lies the subtlety of the definition! Nonetheless, it is clear that every open G is measurable. Furthermore, if $|\Gamma|_e = 0$, then Γ is measurable since we can choose, for any $\epsilon > 0$, an open $G \supseteq \Gamma$ such that $|G\setminus\Gamma|_e \le |G| < \epsilon$. Finally, if Γ is measurable, then there is a $B \in \mathfrak{G}_\delta$ such that $\Gamma \subseteq B$ and $|B \setminus \Gamma|_e = 0$. Indeed, simply choose $\{G_n\}_1^\infty \subseteq \mathfrak{G}$ so that $\Gamma \subseteq G_n$ and $|G_n \setminus \Gamma|_e < \frac{1}{n}$, and take $B = \bigcap_1^\infty G_n$.

Our next result shows that many more sets are measurable.

2.1.7 Lemma. *If $\{\Gamma_n\}_1^\infty$ is a sequence of measurable sets, then $\Gamma = \bigcup_1^\infty \Gamma_n$ is also measurable and, of course (cf. Lemma 2.1.2),*

$$|\Gamma| \le \sum_1^\infty |\Gamma_n|. \tag{2.1.8}$$

In particular, every rectangle I is measurable.

Proof: For each $n \ge 1$, choose $\Gamma_n \subseteq G_n \in \mathfrak{G}$ so that $|G_n \setminus \Gamma_n|_e < 2^{-n}\epsilon$. Then $G \equiv \bigcup_1^\infty G_n$ is open, contains Γ, and (by Lemma 2.1.2) satisfies

$$|G\setminus\Gamma|_e \le \left|\bigcup_1^\infty (G_n \setminus \Gamma_n)\right|_e \le \sum_1^\infty |G_n \setminus \Gamma_n|_e < \epsilon.$$

Finally, by writing a rectangle $I = \mathring{I} \cup \partial I$, we see that every rectangle is measurable. □

Knowing that $\overline{\mathcal{B}}_{\mathbb{R}^N}$ is closed under countable unions, our next goal is to prove that it is also closed under complementation. In doing so, we will be simultaneously coming closer to showing that (2.0.4) holds for $|\cdot|$ on $\overline{\mathcal{B}}_{\mathbb{R}^N}$. Our proof will turn on an elementary fact about the topology of $\mathbb{R}^N$. Recall that a **cube** Q in $\mathbb{R}^N$ is a rectangle all of whose sides have the same length.

2.1.9 Lemma. *If G is an open set in $\mathbb{R}$, then G is the union of a countable number of mutually disjoint open intervals. More generally, if G is an open set in $\mathbb{R}^N$, then, for each $\delta > 0$, G admits a countable, non-overlapping, exact cover $\mathcal{C}$ by cubes Q with $\operatorname{diam}(Q) < \delta$.*

Proof: If $G \subseteq \mathbb{R}$ is open and $x \in G$, let $\mathring{I}_x$ be the open connected component of G containing x. Then $\mathring{I}_x$ is an open interval and, for any $x, y \in G$, either $\mathring{I}_x \cap \mathring{I}_y = \emptyset$ or $\mathring{I}_x = \mathring{I}_y$. Hence, $\mathcal{C} \equiv \{\mathring{I}_x : x \in G \cap \mathbb{Q}\}$ ($\mathbb{Q}$ denotes the set of rational numbers) is the required cover.

To handle the second assertion, set $Q_n = \left[0, 2^{-n}\right]^N$ and $\mathcal{K}_n = \{\frac{\mathbf{k}}{2^n} + Q_n : \mathbf{k} \in \mathbb{Z}^N\}$. Note that if $m \le n$, $Q \in \mathcal{K}_m$, and $Q' \in \mathcal{K}_n$, then either $Q' \subseteq Q$ or $\mathring{Q} \cap \mathring{Q}' = \emptyset$. Now let $G \subseteq \mathbb{R}^N$ and $\delta > 0$ be given. Let n_0 be the smallest

$n \in \mathbb{Z}$ such that $2^{-n}\sqrt{N} < \delta$, and set $\mathcal{C}_{n_0} = \{Q \in \mathcal{K}_{n_0} : Q \subseteq G\}$. Next, define $\mathcal{C}_n$ inductively for $n > n_0$ so that

$$\mathcal{C}_{n+1} = \left\{ Q' \in \mathcal{K}_{n+1} : Q' \subseteq G \text{ and } \mathring{Q}' \cap \mathring{Q} = \emptyset \text{ for each } Q \in \bigcup_{m=n_0}^{n} \mathcal{C}_m \right\}.$$

Note that if $m \le n$, $Q \in \mathcal{C}_m$, and $Q' \in \mathcal{C}_n$, then either $Q = Q'$ or $\mathring{Q} \cap \mathring{Q}' = \emptyset$. Hence $\mathcal{C} \equiv \bigcup_{n=n_0}^{\infty} \mathcal{C}_n$ is non-overlapping, and certainly $\bigcup \mathcal{C} \subseteq G$. Finally, if $x \in G$, choose $n \ge n_0$ and $Q' \in \mathcal{K}_n$ so that $x \in Q' \subseteq G$. If $Q' \notin \mathcal{C}_n$, then there is an $n_0 \le m < n$ and a $Q \in \mathcal{C}_m$ such that $\mathring{Q} \cap \mathring{Q}' \neq \emptyset$. But this means that $Q' \subseteq Q$ and therefore that $x \in Q \subseteq \bigcup \mathcal{C}$. Thus $\mathcal{C}$ covers G. □

2.1.10 Lemma. *If Γ is measurable, then so is its complement $\Gamma\complement$.*

PROOF: We first check that every compact set K is measurable. To this end, let $\epsilon > 0$ be given, and choose an open set $G \supseteq K$ so that $|G| - |K|_e < \epsilon$. Set $H = G \setminus K$ and choose a non-overlapping sequence $\{Q_n\}_1^\infty$ of cubes for which $H = \bigcup_1^\infty Q_n$. By Lemma 2.1.1, $\sum_1^n |Q_m| = |\bigcup_1^n Q_m|$. Moreover, since K and $\bigcup_1^n Q_m$ are disjoint compact sets (and are therefore a positive distance apart), the last part of Lemma 2.1.2 says that $\left|\left(\bigcup_1^n Q_m\right) \cup K\right|_e = |\bigcup_1^n Q_m| + |K|_e$. Hence

$$|G| \ge \left| \left(\bigcup_1^n Q_m \right) \cup K \right|_e = \left| \bigcup_1^n Q_m \right| + |K|_e = \sum_1^n |Q_m| + |K|_e,$$

and so $\sum_1^n |Q_m| \le |G| - |K|_e < \epsilon$ for all $n \ge 1$. As a consequence, we now see that $|H|_e \le \sum_1^\infty |Q_m| \le \epsilon$; and so K is measurable.

We next show that every closed set F is measurable. To this end, simply write $F = \bigcup_1^\infty (F \cap \overline{B(0,n)})$, where $B(x,r)$ denotes the open Euclidean ball* $\{y \in \mathbb{R}^N : |y - x| < r\}$ of radius r around the point x. Since each $F \cap \overline{B(0,n)}$ is compact, it follows from the preceding and Lemma 2.1.7 that F is measurable.

To complete the proof, first observe that, after another application of Lemma 2.1.7, we know that (cf. Remark 2.1.3) $\mathcal{F}_\sigma \subseteq \overline{\mathcal{B}}_{\mathbb{R}^N}$. Next (cf. Remark 2.1.6) choose $B \in \mathfrak{G}_\delta$ so that $\Gamma \subseteq B$ and $|B \setminus \Gamma|_e = 0$. Then, since $B\complement \in \mathfrak{F}_\sigma$ and $|B \setminus \Gamma|_e = 0$, $\Gamma\complement = B\complement \cup (B \setminus \Gamma)$ is measurable. □

We are now very close to our goal. However, we still need the following simple fact about double sums of non-negative numbers.

2.1.11 Lemma. *If $\{a_{m,n} : m, n \in \mathbb{Z}^+\} \subseteq [0,\infty)$, then*

$$\sum_{m=1}^{\infty} \sum_{n=1}^{\infty} a_{m,n} = \sum_{n=1}^{\infty} \sum_{m=1}^{\infty} a_{m,n}.$$

---

* In general, if (E,ρ) is a metric space and $a \in E$ and $r > 0$, we will use $B(a,r)$ to denote the open ball $\{x \in E : \rho(a,x) < r\}$.

PROOF: For each $M,\ N \in \mathbb{Z}^+$,

$$\sum_{m=1}^{\infty}\sum_{n=1}^{\infty} a_{m,n} \geq \sum_{m=1}^{M}\sum_{n=1}^{N} a_{m,n} = \sum_{n=1}^{N}\sum_{m=1}^{M} a_{m,n}.$$

Hence, by letting first $M \nearrow \infty$ and then $N \nearrow \infty$, we conclude that

$$\sum_{m=1}^{\infty}\sum_{n=1}^{\infty} a_{m,n} \geq \sum_{n=1}^{\infty}\sum_{m=1}^{\infty} a_{m,n}.$$

The opposite inequality is checked by reversing the roles of m and n in the preceding. □

2.1.12 Theorem. *The class $\overline{\mathcal{B}}_{\mathbb{R}^N}$ contains $\mathfrak{G}$, is closed under countable unions, complementation, and therefore also under differences and countable intersections. Hence, $\mathfrak{G}_\delta \cup \mathfrak{F}_\sigma \subseteq \overline{\mathcal{B}}_{\mathbb{R}^N}$; in fact, $\Gamma \in \overline{\mathcal{B}}_{\mathbb{R}^N}$ if and only if there exist $A \in \mathfrak{F}_\sigma$ and $B \in \mathfrak{G}_\delta$ such that $A \subseteq \Gamma \subseteq B$ and $|B \setminus A| = 0$, in which case $|\Gamma| = |A| = |B|$. Finally, for any $\{\Gamma_n\}_1^\infty \subseteq \overline{\mathcal{B}}_{\mathbb{R}^N}$,*

$$\left|\bigcup_1^\infty \Gamma_n\right| = \sum_1^\infty |\Gamma_n| \quad \textit{if } \Gamma_m \cap \Gamma_n = \emptyset \textit{ for } m \neq n. \tag{2.1.13}$$

PROOF: The first assertion follows immediately from what we already know together with the trivial manipulation of set-theoretic operations, and clearly the fact that $\mathfrak{G}_\delta \cup \mathcal{F}_\sigma \subseteq \overline{\mathcal{B}}_{\mathbb{R}^N}$ is a consequence of the first assertion. Next suppose that Γ is measurable. By the final part of Remark 2.1.6 applied to Γ and $\Gamma\complement$, we can find $A \in \mathfrak{F}_\sigma$ and $B \in \mathfrak{G}_\delta$ such that $\Gamma\complement \subseteq A\complement$, $\Gamma \subseteq B$, $|B\setminus\Gamma| = 0$, and $|\Gamma \setminus A| = |A\complement \setminus \Gamma\complement| = 0$, from which $|B \setminus A| = 0$ is immediate. On the other hand, if there exist $A \in \mathcal{F}_\sigma$ and $B \in \mathfrak{G}_\delta$ such that $A \subseteq \Gamma \subseteq B$ and $|B \setminus A| = 0$, then $\Gamma = A \cup (\Gamma \setminus A)$ is measurable because $|\Gamma \setminus A|_e \leq |B \setminus A| = 0$. Hence, it remains only to check (2.1.13).

We first prove (2.1.13) under the additional assumption that each of the Γ_n's is bounded. Given $\epsilon > 0$, choose open sets G_n so that $\Gamma_n\complement \subseteq G_n$ and $|G_n \setminus \Gamma_n\complement| < 2^{-n}\epsilon$. Then $K_n \equiv G_n\complement \subseteq \Gamma_n$ is compact and $|\Gamma_n \setminus K_n| < 2^{-n}\epsilon$. Since $K_m \cap K_n = \emptyset$ and therefore $\operatorname{dist}(K_m, K_n) > 0$ for $m \neq n$, we have, by the last part of Lemma 2.1.2, that $|\bigcup_1^n K_m| = \sum_1^n |K_m|$ for every $n \geq 1$. Hence

$$\sum_{m=1}^{\infty} |\Gamma_m| < \sum_{m=1}^{\infty} |K_m| + \epsilon = \lim_{n\to\infty}\left|\bigcup_{m=1}^{n} K_m\right| + \epsilon \leq \left|\bigcup_{m=1}^{\infty} \Gamma_m\right| + \epsilon.$$

That is, $\sum_1^\infty |\Gamma_m| \leq |\bigcup_1^\infty \Gamma_m|$. Since the opposite inequality always holds, (2.1.13) is now proved for bounded Γ_n's.

Finally, to handle the general case, set

$$A_1 = B(0,1) \quad \text{and} \quad A_{n+1} = B(0, n+1) \setminus B(0,n).$$

Then

$$(\Gamma_m \cap A_n) \cap (\Gamma_{m'} \cap A_{n'}) = \emptyset \quad \text{if} \quad (m,n) \neq (m',n').$$

Hence, by the preceding and Lemma 2.1.11,

$$\begin{aligned}\sum_{m=1}^{\infty} |\Gamma_m| &= \sum_{m=1}^{\infty}\sum_{n=1}^{\infty} |\Gamma_m \cap A_n| = \sum_{n=1}^{\infty}\sum_{m=1}^{\infty} |\Gamma_m \cap A_n| \\ &= \sum_{n=1}^{\infty} \left|\bigcup_{m=1}^{\infty} (\Gamma_m \cap A_n)\right| = \sum_{n=1}^{\infty} \left|\left(\bigcup_{m=1}^{\infty} \Gamma_m\right) \cap A_n\right| \\ &= \left|\bigcup_{n=1}^{\infty}\left[\left(\bigcup_{m=1}^{\infty} \Gamma_m\right) \cap A_n\right]\right| = \left|\bigcup_{m=1}^{\infty} \Gamma_m\right|. \quad \square\end{aligned}$$

2.1.14 Remark: Although it seems hardly necessary to point out, exterior Lebesgue measure has an obvious but extremely important property: it is **invariant under translation**. That is, $|x + \Gamma|_e = |\Gamma|_e$ for all $x \in \mathbb{R}^N$ and all $\Gamma \subseteq \mathbb{R}^N$. As a consequence, we also see that $x + \Gamma$ is measurable whenever $x \in \mathbb{R}^N$ and Γ itself is measurable.

Before concluding this preliminary discussion of Lebesgue measure, it may be appropriate to examine whether there are any non-measurable sets. It turns out that the existence of non-measurable sets brings up some extremely delicate points about the foundations of mathematics. Indeed, if one is willing to abandon the full axiom of choice, then R. Solovay has shown that there is a model of mathematics in which *every* subset of $\mathbb{R}^N$ is Lebesgue measurable. However, if one accepts the full axiom of choice, then the following argument, due to Vitali, shows that there are sets which are not Lebesgue measurable. The use of the axiom of choice comes in Lemma 2.1.16 below; it is not used in the proof of the next lemma, a result which is interesting in its own right.

2.1.15 Lemma. *If Γ is a measurable subset of $\mathbb{R}$ and $|\Gamma| > 0$, then the set $\Gamma - \Gamma \equiv \{y - x : x, y \in \Gamma\}$ contains the open interval $(-\delta, \delta)$ for some $\delta > 0$.*

PROOF: Without loss of generality, we assume that $|\Gamma| < \infty$.

Choose an open set $G \supseteq \Gamma$ so that $|G \setminus \Gamma| < \frac{1}{3}|\Gamma|$, and let $\mathcal{C}$ be a countable collection of mutually disjoint, non-empty, open intervals $\mathring{I}$ whose union is G (cf. the first part of Lemma 2.1.9). Then

$$\sum_{\mathring{I} \in \mathcal{C}} |\mathring{I} \cap \Gamma| = |\Gamma| \geq \frac{3}{4}|G| = \frac{3}{4}\sum_{\mathring{I} \in \mathcal{C}} |\mathring{I}|.$$

Hence, there must be an $\mathring{I} \in \mathcal{C}$ such that $|\mathring{I} \cap \Gamma| \geq \frac{3}{4}|\mathring{I}|$. Set $A = \mathring{I} \cap \Gamma$. If $d \in \mathbb{R}$ and $(d + A) \cap A = \emptyset$, then

$$2|A| = |d + A| + |A| = |(d + A) \cup A| \leq |(d + \mathring{I}) \cup \mathring{I}|.$$

At the same time, $(d+\mathring{I})\cup\mathring{I} \subseteq (I^-, d+I^+)$ if $d \geq 0$ and $(d+\mathring{I})\cup\mathring{I} \subseteq (d+I^-, I^+)$ if $d < 0$; and so, in either case, $|(d+\mathring{I})\cup\mathring{I}| \leq |d| + |\mathring{I}|$. Hence, if $(d+A)\cap A = \emptyset$, then $2|A| \leq |d| + |\mathring{I}|$, from which we deduce that $|d| \geq \frac{1}{2}|\mathring{I}|$. In other words, if $|d| < \frac{1}{2}|\mathring{I}|$, then $(d+A)\cap A \neq \emptyset$. But this means that for every $d \in \left(-\frac{1}{2}|\mathring{I}|, \frac{1}{2}|\mathring{I}|\right)$ there exist $x, y \in A \subseteq \Gamma$ such that $d = y - x$. □

2.1.16 Lemma. *Assuming the axiom of choice, there is a subset A of $\mathbb{R}$ such that $(A - A) \cap \mathbb{Q} = \{0\}$ and yet $\mathbb{R} = \bigcup_{q\in\mathbb{Q}}(q + A)$.*

PROOF: Write $x \sim y$ if $y - x \in \mathbb{Q}$. Then "$\sim$" is an equivalence relation on $\mathbb{R}$, and, for each $x \in \mathbb{R}$, the equivalence class $[x]^\sim$ of x is $x + \mathbb{Q}$. Now, using the axiom of choice, let A be a set which contains precisely one element from each of the equivalence classes $[x]^\sim$, $x \in \mathbb{R}$. It is then clear that A has the required properties. □

2.1.17 Theorem. *Assuming the axiom of choice, every $\Gamma \subseteq \mathbb{R}$ with $|\Gamma|_e > 0$ contains a non-measurable subset.*

PROOF: Let A be the set constructed in Lemma 2.1.16. Then $0 < |\Gamma|_e \leq \sum_{q\in\mathbb{Q}} |\Gamma\cap(q+A)|_e$, and so there must exist a $q \in \mathbb{Q}$ such that $|\Gamma\cap(q+A)|_e > 0$. Hence, if $\Gamma \cap (q + A)$ were measurable, then, by Lemma 2.1.15, we would have that $(-\delta, \delta) \subseteq \{y - x : x, y \in (q + A)\} \subseteq \{0\} \cup \mathbb{Q}\complement$ for some $\delta > 0$. □

Exercises

2.1.18 Exercise: Let Γ_1 and Γ_2 be measurable subsets in $\mathbb{R}^N$. If $\Gamma_1 \subseteq \Gamma_2$ and $|\Gamma_1| < \infty$, show that $|\Gamma_2 \setminus \Gamma_1| = |\Gamma_2| - |\Gamma_1|$. More generally, show that if $|\Gamma_1 \cap \Gamma_2| < \infty$, then $|\Gamma_1 \cup \Gamma_2| = |\Gamma_1| + |\Gamma_2| - |\Gamma_1 \cap \Gamma_2|$.

2.1.19 Exercise: Let $\{\Gamma_n\}_1^\infty$ be a sequence of measurable sets in $\mathbb{R}^N$. Assuming that $|\Gamma_m \cap \Gamma_n| = 0$ for $m \neq n$, show that $|\bigcup_1^\infty \Gamma_n| = \sum_1^\infty |\Gamma_n|$.

2.1.20 Exercise: It is clear that any countable set has Lebesgue measure zero. However, it is not so immediately clear that there are uncountable subsets of $\mathbb{R}$ whose Lebesgue measure is zero. We will show here how to construct such a set. Namely, start with the set $C_0 = [0, 1]$ and let C_1 be the set obtained by removing the open middle third of C_0 (i.e., $C_1 = C_0 \setminus \left(\frac{1}{3}, \frac{2}{3}\right) = \left[0, \frac{1}{3}\right] \cup \left[\frac{2}{3}, 1\right]$). Next, let C_2 be the set obtained from C_1 after removing the open middle third of each of the (two) intervals of which C_1 is the disjoint union. More generally, given C_k (which is the union of 2^k disjoint, closed intervals), let C_{k+1} be the set which one gets from C_k by removing the open middle third of each of the intervals of which C_k is the disjoint union. Finally, set $C = \bigcap_{k=0}^\infty C_k$. The set C is called the **Cantor set**, and it turns out to be an extremely useful source of examples. Here we will show that it is an example of an uncountable set of Lebesgue measure zero.

(i) Note that C is closed and that $|C| \le |C_k| = \left(\frac{2}{3}\right)^k$, $k \ge 0$. Conclude that $C \in \overline{\mathcal{B}}_{\mathbb{R}}$ and that $|C| = 0$.

(ii) Let $\mathcal{A}$ denote the set of $\boldsymbol{\alpha} \in \{0,1,2\}^{\mathbb{N}}$ with the properties that:

(a) $\alpha_0 \in \{0,1\}$ and $\alpha_0 = 1$ only if $\alpha_k = 0$ for all $k \in \mathbb{Z}^+$;

(b) $\alpha_k \in \{0,1\}$ for infinitely many $k \in \mathbb{Z}^+$.

Check that the map

$$\boldsymbol{\alpha} \in \mathcal{A} \longmapsto \sum_{k \in \mathbb{N}} \frac{\alpha_k}{3^k} \in [0,1]$$

is an one-to-one and onto, and let $x \in [0,1] \longmapsto \boldsymbol{\alpha}(x) \in \mathcal{A}$ denote the inverse mapping. Next, define $\mathcal{A}_0$ to be the set of $\boldsymbol{\alpha} \in \mathcal{A}$ such that $\alpha_k = 0$ for all but a finite number of $k \in \mathbb{N}$; and, for $\boldsymbol{\alpha} \in \mathcal{A}_0$, define $\ell(\boldsymbol{\alpha}) = \max\{k \in \mathbb{N} : \alpha_k \ne 0\}$. Show that

$$\partial C_\ell = \Big\{x : \boldsymbol{\alpha}(x) \in \mathcal{A}_0,\ \ell(\boldsymbol{\alpha}(x)) \le \ell \ \&\ \alpha_k(x) \in \{0,2\} \text{ for } 0 \le k < \ell(\alpha(x))\Big\},$$

and conclude that

$$\mathring{C}_\ell = \Big\{x : \alpha_k(x) \in \{0,2\} \text{ for every } 0 \le k \le \ell \\ \text{and } \alpha_k(x) \ne 0 \text{ for some } k > \ell\Big\}.$$

Finally, define

$$\mathring{\mathcal{A}} = \Big\{\boldsymbol{\alpha} \in \mathcal{A} \setminus \mathcal{A}_0 : \alpha_k \in \{0,2\} \text{ for all } k \in \mathbb{N}\Big\},$$

and show that

$$\bigcap_{\ell=0}^{\infty} \mathring{C}_\ell = \Big\{x : \boldsymbol{\alpha}(x) \in \mathring{\mathcal{A}}\Big\},$$

while

$$C \setminus \bigcap_{\ell=0}^{\infty} \mathring{C}_\ell = \Big\{x : \boldsymbol{\alpha}(x) \in \mathcal{A}_0 \text{ and } \alpha_k(x) \in \{0,2\} \text{ for every } 0 \le k < \ell(\boldsymbol{\alpha}(x))\Big\}.$$

(iii) To see that C is not countable, suppose that it were. Using **(ii)** and the countability of $\mathcal{A}_0$, show that one would then have a way of counting $\{0,2\}^{\mathbb{Z}^+}$. Finally, recall Cantor's famous *anti-diagonalization procedure* for showing that $\{0,2\}^{\mathbb{Z}^+}$ cannot be counted.

2.2 Euclidean Invariance

Although the property of translation invariance was built into our construction of Lebesgue measure, it is not immediately obvious how Lebesgue measure reacts to rotations of $\mathbb{R}^N$. One suspects that, as *the natural* measure on $\mathbb{R}^N$, Lebesgue measure should be invariant under the full group of Euclidean transformations (i.e., rotations as well as translations). However, because our definition of the measure was based on rectangles and the rectangles were inextricably tied to a fixed set of coordinate axes, rotation invariance is not as clear as translation invariance. In the present section we will see how Lebesgue measure transforms under an arbitrary linear transformation of $\mathbb{R}^N$, and rotation invariance will follow as an immediate corollary.

We begin with a results about the behavior of measurable sets under general transformations.

2.2.1 Lemma. *Let $F \subseteq \mathbb{R}^M$ be closed and $\Phi : F \longrightarrow \mathbb{R}^N$ continuous. Then $\Phi(\Gamma \cap F) \in \mathfrak{F}_\sigma$ whenever $\Gamma \in \mathfrak{F}_\sigma$. Furthermore, if in addition, $|\Phi(\Gamma \cap F)|_e = 0$ whenever $|\Gamma|_e = 0$, then $\Phi(\Gamma \cap F)$ is measurable whenever Γ is. In particular, if $M \leq N$ and Φ is Lipschitz continuous with Lipschitz constant L (i.e., $|\Phi(y) - \Phi(x)| \leq L|y - x|$ for all $x, y \in F$), then $|\Phi(\Gamma \cap F)|_e \leq (2\sqrt{N}L)^N |\Gamma|_e$ and therefore Φ takes measurable subsets of F into measurable sets in $\mathbb{R}^N$.*

PROOF: Remember that functions preserve unions. Hence, the class of sets Γ for which $\Phi(\Gamma \cap F) \in \mathfrak{F}_\sigma$ is closed under countable unions. Next note that if K is compact, then, by continuity, so is $\Phi(K \cap F)$. But every closed set in $\mathbb{R}^N$ is the countable union of compact sets, and therefore we see that $\Phi(\Gamma \cap F) \in \mathfrak{F}_\sigma$ for every closed Γ. Finally, since every $\Gamma \in \mathfrak{F}_\sigma$ is a countable union of closed sets, the first assertion is proved.

Next assume, in addition, that $|\Phi(\Gamma \cap F)|_e = 0$ whenever $|\Gamma|_e = 0$. Given a measurable Γ, choose $A \in \mathcal{F}_\sigma$ so that $A \subseteq \Gamma$ and $|\Gamma \setminus A| = 0$. Then $\Phi(\Gamma \cap F) = \Phi(A \cap F) \cup \Phi((\Gamma \setminus A) \cap F)$ is measurable because $\Phi(A \cap F) \in \mathfrak{F}_\sigma$ and $\left|\Phi\big((\Gamma \setminus A) \cap F\big)\right|_e = 0$.

We now show that if Φ is Lipschitz continuous with Lipschitz constant L, then $|\Phi(\Gamma \cap F)|_e \leq (2\sqrt{N}L)^N |\Gamma|_e$. But clearly it suffices to do this when Γ is a cube Q with diameter less than 1. Indeed, if we knew it in this case and were given Γ an arbitrary subset of $\mathbb{R}^M$ with $|\Gamma|_e < \infty$, then, by (2.1.5) and Lemma 2.1.9, we could find, for any $\epsilon > 0$, a countable collection $\mathcal{C}$ of non-overlapping cubes Q with diameter less than 1 such that $\Gamma \subseteq \bigcup \mathcal{C}$ and $\sum_{Q \in \mathcal{C}} |Q| \leq |\Gamma|_e + \epsilon$. Hence, we could conclude that

$$\begin{aligned} \left|\Phi(\Gamma \cap F)\right|_e &\leq \left|\bigcup_{Q \in \mathcal{C}} \Phi(Q \cap F)\right|_e \leq \sum_{Q \in \mathcal{C}} \left|\Phi(Q \cap F)\right|_e \\ &\leq (2\sqrt{N}L)^N \sum_{Q \in \mathcal{C}} |Q| \leq (2\sqrt{N}L)^N \big(|\Gamma|_e + \epsilon\big). \end{aligned}$$

Thus, let Q be a cube in $\mathbb{R}^M$ with diameter $D < 1$. If $Q \cap F = \emptyset$, there is nothing to do. If $x \in Q \cap F$, note that $\Phi(Q \cap F)$ must be a subset of the ball in $\mathbb{R}^N$ of radius LD around $\Phi(p)$. Hence, $\Phi(Q \cap F)$ is contained in the cube

$$\prod_1^N [\Phi(x)_k - LD, \Phi(x)_k + LD],$$

and therefore

$$|\Phi(Q \cap F)|_{\mathrm{e}} \leq (2LD)^N = (2\sqrt{N}L)^N |Q|^{\frac{N}{M}} \leq (2\sqrt{N}L)^N |Q|. \quad \square$$

Given an $N \times N$ matrix A of real numbers a_{ij}, we will use T_A to denote the linear transformation of $\mathbb{R}^N$ which A determines relative to the standard basis $\{\mathbf{e}_1, \ldots, \mathbf{e}_N\}$. That is,

$$T_A \mathbf{x} = \begin{bmatrix} \sum_{j=1}^N a_{1j} x^j \\ \vdots \\ \sum_{j=1}^N a_{Nj} x^j \end{bmatrix} \quad \text{for } \mathbf{x} = \begin{bmatrix} x^1 \\ \vdots \\ x^N \end{bmatrix} \in \mathbb{R}^N.$$

Since T_A is obviously Lipschitz continuous, T_A takes measurable sets into measurable sets and sets of measure 0 into sets of measure 0. The main result of this section is the following important fact about Lebesgue measure.

2.2.2 Theorem. *Given a real $N \times N$ matrix A, T_A takes measurable sets into measurable sets and $|T_A(\Gamma)|_{\mathrm{e}} = |\det(A)||\Gamma|_{\mathrm{e}}$ for all $\Gamma \subseteq \mathbb{R}^N$. (We use $\det(A)$ to denote the determinant of A.)*

PROOF: There are several steps.
Step 1: *For any $\mathbf{c} \in \mathbb{R}^N$, $\lambda \in \mathbb{R}$, and $\Gamma \subseteq \mathbb{R}^N$, $|(\mathbf{c} + \lambda\Gamma)|_{\mathrm{e}} = |\lambda|^N |\Gamma|_{\mathrm{e}}$, where $\lambda\Gamma \equiv \{\lambda x : x \in \Gamma\}$* By translation invariance, we may and will assume that $c = 0$. Moreover, there is nothing to prove when $\lambda = 0$. Finally, it is clear that $|\lambda I| = |\lambda|^N |I|$ for any $\lambda \neq 0$ and rectangle I. Hence, since $\mathcal{C}$ is a countable cover of Γ by rectangles I if and only if $\{\lambda I : I \in \mathcal{C}\}$ is a countable cover of $\lambda\Gamma$ by rectangles, we are done.

Step 2: *For any linear transformation T and all cubes Q, $|T(Q)| = \alpha(T)|Q|$ where $\alpha(T) \equiv |T(Q_0)|$ and $Q_0 = [0,1]^N$.* Since every cube $Q = \mathbf{c} + \lambda Q_0$ for some $\mathbf{c} \in \mathbb{R}^N$ and λ satisfying $|\lambda|^N = |Q|$, Step 1 plus the linearity of T yields $|T(Q)| = |(T(\mathbf{c}) + \lambda T(Q_0)| = |\lambda|^N |T(Q_0)| = \alpha(T)|Q|$.

Step 3: *For any linear transformation T and open G,*

$$|T(G)| \leq \alpha(T)|G|.$$

Moreover, equality holds if T is non-singular. Let (cf. Lemma 2.1.9) $\mathcal{C}$ be a countable, exact cover of G by non-overlapping cubes Q. Then

$$|T(G)| \leq \sum_{Q \in \mathcal{C}} |T(Q)| = \alpha(T) \sum_{Q \in \mathcal{C}} |Q| = \alpha(T)|G|.$$

Now suppose that T is non-singular. Then

$$T(Q) \setminus T(\mathring{Q}) = T(Q \setminus \mathring{Q})$$

has measure 0, and

$$T(\mathring{Q}) \cap T(\mathring{Q}') = \emptyset \quad \text{for distinct } Q,\, Q' \in \mathcal{C}.$$

Hence $|T(Q)| = |T(\mathring{Q})|$ and (cf. Exercise 2.1.19)

$$|T(G)| \geq \sum_{Q \in \mathcal{C}} |T(\mathring{Q})| = \sum_{Q \in \mathcal{C}} |T(Q)| = \alpha(T) \sum_{Q \in \mathcal{C}} |Q| = \alpha(T)|G|.$$

Step 4: *For any non-singular linear transformation T and all $\Gamma \subseteq \mathbb{R}^N$, $|T(\Gamma)|_e = \alpha(T)|\Gamma|_e$.* Since $\Gamma \subseteq G \in \mathfrak{G}$ if and only if $T(\Gamma) \subseteq T(G) \in \mathfrak{G}$, this step is an immediate consequence of (2.1.5) and Step 3.

Step 5: *If S and T linear transformations and S is non-singular, then $\alpha(S \circ T) = \alpha(S)\alpha(T)$.* Simply note that, by Step 4,

$$\alpha(S \circ T) = |S \circ T(Q_0)| = |S(T(Q_0))| = \alpha(S)|T(Q_0)| = \alpha(S)\alpha(T).$$

Step 6: *If A is an orthogonal matrix, then $\alpha(T_A) = 1$.* Because A is orthogonal, $B(0,1) = T_A(B(0,1))$ and therefore $|B(0,1)| = \alpha(T_A)|B(0,1)|$.

Step 7: *If A is non-singular and symmetric, then $\alpha(T_A) = |\det(A)|$.* If A is already diagonal, then it is clear that $\alpha(T_A) = |T_A(Q_0)| = |\lambda_1 \cdots \lambda_N|$, where λ_k is the kth diagonal entry. Hence, the assertion is obvious in this case. On the other hand, in the general case, we can find an orthogonal matrix $\mathcal{O}$ such that $A = \mathcal{O}\Lambda\mathcal{O}^{\mathrm{T}}$, where Λ is a diagonal matrix whose diagonal entries are the eigenvalues of A and $\mathcal{O}^{\mathrm{T}}$ is the transpose of $\mathcal{O}$. Hence, by Steps 5 and 6, $\alpha(A) = \alpha(\mathcal{O})\alpha(\Lambda)\alpha(\mathcal{O}^{\mathrm{T}}) = \alpha(\Lambda) = |\det(A)|$.

Step 8: *For every non-singular matrix A, $\alpha(T_A) = |\det(A)|$.* Set $B = (AA^{\mathrm{T}})^{\frac{1}{2}}$. Then B is symmetric and $\det(B) = |\det(A)|$. Next set $\mathcal{O} = B^{-1}A$ and note that $\mathcal{O}^{\mathrm{T}} = A^{\mathrm{T}}B^{-1}$ and so $\mathcal{O}\mathcal{O}^{\mathrm{T}} = B^{-1}AA^{\mathrm{T}}B^{-1} = B^{-1}B^2B^{-1} = \mathbf{I}_{\mathbb{R}^N}$, where $\mathbf{I}_{\mathbb{R}^N}$ denotes the identity matrix. In other words, $\mathcal{O}$ is orthogonal. Since $A = B\mathcal{O}$, we now have that $\alpha(A) = \alpha(B) = |\det(A)|$.

Step 9: *If A is singular, then $\alpha(T_A) = 0$.* Choose $\mathbf{y} \in \mathbb{R}^N$ with unit length so that $\mathbf{y} \perp \mathrm{Range}(T_A)$. Next, choose an orthogonal $\mathcal{O}$ so that $\mathbf{e}_1 = T_{\mathcal{O}}(\mathbf{y})$. Then $\mathbf{e}_1 \perp \mathrm{Range}(T_{\mathcal{O}} \circ T_A)$ and so there a rectangle $\hat{I}$ in $\mathbb{R}^{N-1}$ such that $T_{\mathcal{O}} \circ T_A(Q_0) \subseteq \{0\} \times \hat{I}$. But $\{0\} \times \hat{I}$ has measure 0, and therefore $\alpha(T_A) = \alpha(T_{\mathcal{O}} \circ T_A) = 0$. $\square$

Exercises

2.2.3 Exercise: Here are two rather easy applications of Theorem 2.2.2.

(**i**) If H is a **hyperplane** in $\mathbb{R}^N$ (i.e., $H = \{y \in \mathbb{R}^N : y - c \perp \ell\}$ for some $c \in \mathbb{R}^N$ and $\ell \in \mathbb{R}^N \setminus \{0\}$), show that $|H| = 0$.

(**ii**) If $B_{\mathbb{R}^N}(c,r)$ is the open ball in $\mathbb{R}^N$ of radius r and center c, show that

$$\big|B_|RN(c,r)\big| = \big|\overline{B_{\mathbb{R}^N}(c,r)}\big| = \Omega_N r^N \quad \text{where } \Omega_N \equiv \big|B_{\mathbb{R}^N}(0,1)\big|.$$

2.2.4 Exercise: If $\mathbf{v}_1, \cdots,$ and $\mathbf{v}_N$ are vectors in $\mathbb{R}^N$, the **parallelepiped spanned by** $\{\mathbf{v}_1, \cdots, \mathbf{v}_N\}$ is the set

$$P(\mathbf{v}_1, \dots, \mathbf{v}_N) \equiv \left\{\sum_1^N x^i \mathbf{v}_i : x^i \in [0,1] \text{ for all } 1 \le i \le N\right\}.$$

When $N \ge 2$, the classical prescription for computing the *volume* of a parallelepiped is to take the product of *the area of any one side* times the length of the corresponding *altitude.* In analytic terms, this means that the volume is 0 if the vectors $\mathbf{v}_1, \dots, \mathbf{v}_N$ are linearly dependent and that otherwise the volume of $P(\mathbf{v}_1, \dots, \mathbf{v}_N)$ can be computed by taking the product of the volume of $P(\mathbf{v}_1, \dots, \mathbf{v}_{N-1})$, thought of as a subset of the hyperplane $H(\mathbf{v}_1, \dots, \mathbf{v}_{N-1})$ spanned by $\mathbf{v}_1, \dots, \mathbf{v}_{N-1}$, times the distance between the vector $\mathbf{v}_N$ and the hyperplane $H(\mathbf{v}_1, \dots, \mathbf{v}_{N-1})$. Using Theorem 2.2.2, show that this prescription is correct when the *volume* of a set is interpreted as the Lebesgue measure of that set.

Chapter III

Lebesgue Integration

3.1 Measure Spaces

In Chapter II we constructed Lebesgue measure on $\mathbb{R}^N$. The result of our efforts was a proof that there is a class $\overline{\mathcal{B}}_{\mathbb{R}^N}$ of subsets of $\mathbb{R}^N$ and a map $\Gamma \in \overline{\mathcal{B}}_{\mathbb{R}^N} \longmapsto |\Gamma| \in [0,\infty]$ such that: $\overline{\mathcal{B}}_{\mathbb{R}^N}$ contains all open sets; $\overline{\mathcal{B}}_{\mathbb{R}^N}$ is closed under both complementation and countable unions; $|I| = \text{vol}(I)$ for all rectangles I; and $|\bigcup_1^\infty \Gamma_n| = \sum_1^\infty |\Gamma_n|$ whenever $\{\Gamma_n\}_1^\infty$ is a sequence of mutually disjoint elements of $\overline{\mathcal{B}}_{\mathbb{R}^N}$. What we are going to do in this section is discuss a few of the general properties which are possessed by such structures.

Given a set E, we will use $\mathcal{P}(E)$ to denote the **power set** of E, that is $\mathcal{P}(E) \equiv \{\Gamma : \Gamma \subseteq E\}$. An **algebra over** E is an $\mathcal{A} \subseteq \mathcal{P}(E)$ with the properties that

(a) $\emptyset \in \mathcal{A}$,
(b) $\Gamma \in \mathcal{A} \implies \Gamma\complement \in \mathcal{A}$,
(c) $\Gamma_1, \Gamma_2 \in \mathcal{A} \implies \Gamma_1 \cup \Gamma_2 \in \mathcal{A}$.

By elementary set-theoretic manipulations, one sees that an algebra is also closed under differences as well as finite unions and intersections. A **σ-algebra over** E is an algebra $\mathcal{B}$ which is closed under countable unions. Of course, σ-algebras are also closed under differences as well as countable intersections.

3.1.1 Examples: Here are two, somewhat trivial, examples.

(i) For any E, $\{\emptyset, E\}$ is the *smallest* algebra over E in the sense that every algebra contains this one.

(ii) For any E, $\mathcal{P}(E)$ is the *largest* algebra over E in the sense that it contains every other one.

In fact, both $\{\emptyset, E\}$ and $\mathcal{P}(E)$ are σ-algebras over E. Of course, most of the interesting algebras and σ-algebras lie somewhere in between these two extreme examples, to wit, the σ-algebra $\overline{\mathcal{B}}_{\mathbb{R}^N}$ over $\mathbb{R}^N$.

3.1.2 Lemma. *The intersection of any collection of algebras or σ-algebras is again an algebra or a σ-algebra. In particular, given any non-empty $\mathcal{C} \subseteq \mathcal{P}(E)$, there is a unique minimal algebra $\mathcal{A}(E;\mathcal{C})$ and a unique minimal σ-algebra $\sigma(E;\mathcal{C})$ over E containing $\mathcal{C}$.*

PROOF: The first assertion is easily checked. Given the first assertion, the second one is handled by considering the collection of all algebras or all σ-algebras over E containing the given $\mathcal{C}$. Noting that neither of these collections can be empty ($\mathcal{P}(E)$ being an element of both), one sees that $\mathcal{A}(E;\mathcal{C})$ and $\sigma(E;\mathcal{C})$ can be constructed by taking intersections. □

The σ-algebra $\sigma(E;\mathcal{C})$ is called the σ-algebra **generated by** $\mathcal{C}$. Perhaps the most important examples of σ-algebras which are described in terms of a generating set are those that arise in connection with topological spaces. Namely, if E is a topological space and $\mathfrak{G}$ denotes the class of all open sets in E, then $\mathcal{B}_E \equiv \sigma(E;\mathfrak{G})$ is called the **Borel σ-algebra** or **Borel field** over E, and the elements of $\mathcal{B}_E$ are called the **Borel measurable** subsets of E. (For those who are struck by the similarity between $\mathcal{B}_{\mathbb{R}^N}$ and $\overline{\mathcal{B}}_{\mathbb{R}^N}$, a complete explanation will be forthcoming shortly, in **(ii)** of Examples 3.1.5 below. In the meantime, suffice it to say that, by Theorem 2.1.12, $\Gamma \in \overline{\mathcal{B}}_{\mathbb{R}^N}$ if and only if there exist $A, B \in \mathcal{B}_{\mathbb{R}^N}$ such that $A \subseteq \Gamma \subseteq B$ and $|B \setminus A| = 0$.)

Usually the class which generates a σ-algebra is not itself even an algebra. Nonetheless, it often has the property that it is closed under finite intersections. For example, this is the case when the generators are the open sets of some topological space. It was also true of the collection of all rectangles in $\mathbb{R}^N$. In the future, we will call a collection $\mathcal{C} \subseteq \mathcal{P}(E)$ a **π-system** if it is closed under finite intersections. As we will see below, it is useful to know what additional properties a π-system must possess in order to be a σ-algebra. For this reason we introduce a notion which complements that of a π-system. Namely, we will say that $\mathcal{H} \subseteq \mathcal{P}(E)$ is a **λ-system over E** if

(a) $E \in \mathcal{H}$,
(b) $\Gamma_1, \Gamma_2 \in \mathcal{H}$ and $\Gamma_1 \cap \Gamma_2 = \emptyset \implies \Gamma_1 \cup \Gamma_2 \in \mathcal{H}$,
(c) $\Gamma_1, \Gamma_2 \in \mathcal{H}$ and $\Gamma_1 \subseteq \Gamma_2 \implies \Gamma_2 \setminus \Gamma_1 \in \mathcal{H}$,
(d) $\{\Gamma_n\}_1^\infty \subseteq \mathcal{H}$ and $\Gamma_n \nearrow \Gamma \implies \Gamma \in \mathcal{H}$.

The sense in which λ-systems and π-systems constitute complementary notions is explained in the following useful lemma.*

3.1.3 Lemma. *The intersection of an arbitrary collection of π-systems or of λ-systems is again a π-system or a λ-system. Moreover, $\mathcal{B} \subseteq \mathcal{P}(E)$ is a σ-algebra over E if and only if it is both a π-system as well as being a λ-system over E. Finally, if $\mathcal{C} \subseteq \mathcal{P}(E)$ is a π-system, then $\sigma(E;\mathcal{C})$ is the smallest λ-system over E containing $\mathcal{C}$.*

PROOF: The first assertion requires no comment. To prove the second one, it suffices to prove that if $\mathcal{B}$ is both a π-system and a λ-system over E, then it

*The author learned these ideas from E. B. Dynkin's treatise on Markov processes. In fact, the λ- and π-system scheme is often attributed to Dynkin, who certainly deserves the credit for its exploitation by a whole generation of probabilists. On the other hand, the author has been told that their origins go back to Minkowski, although no corroborating reference was ever provided.

is a σ-algebra over E. To this end, first note that $A\complement = E \setminus A \in \mathcal{B}$ for every $A \in \mathcal{B}$ and therefore $\emptyset \in \mathcal{B}$ and $\mathcal{B}$ is closed under complementation. Second, if $\Gamma_1, \Gamma_2 \in \mathcal{B}$, then $\Gamma_1 \cup \Gamma_2 = \Gamma_1 \cup (\Gamma_2 \setminus \Gamma_3)$ where $\Gamma_3 = \Gamma_1 \cap \Gamma_2$. Hence $\mathcal{B}$ is an algebra over E. Finally, if $\{\Gamma_n\} \subseteq \mathcal{B}$, set $A_n = \bigcup_1^n \Gamma_m$ for $n \geq 1$. Then $\{A_n\}_1^\infty \subseteq \mathcal{B}$ and $A_n \nearrow \bigcup_1^\infty \Gamma_m$. Hence $\bigcup_1^\infty \Gamma_m \in \mathcal{B}$, and so $\mathcal{B}$ is a σ-algebra.

To prove the final assertion, let $\mathcal{C}$ be a π-system and $\mathcal{H}$ the smallest λ-system over E containing $\mathcal{C}$. Clearly $\sigma(E;\mathcal{C}) \supseteq \mathcal{H}$; and so all that we have to do is show that $\mathcal{H}$ is π-system over E. To this end, first set

$$\mathcal{H}_1 = \{\Gamma \subseteq E : \Gamma \cap \Delta \in \mathcal{H} \text{ for all } \Delta \in \mathcal{C}\}.$$

It is then easy to check that $\mathcal{H}_1$ is a λ -system over E. Moreover, since $\mathcal{C}$ is a π-system, $\mathcal{C} \subseteq \mathcal{H}_1$, and therefore $\mathcal{H} \subseteq \mathcal{H}_1$. In other words, $\Gamma \cap \Delta \in \mathcal{H}$ for all $\Gamma \in \mathcal{H}$ and $\Delta \in \mathcal{C}$. Next set

$$\mathcal{H}_2 = \{\Gamma \subseteq E : \Gamma \cap \Delta \in \mathcal{H} \text{ for all } \Delta \in \mathcal{H}\}.$$

Again it is clear that $\mathcal{H}_2$ is a λ-system. Also, by the preceding, $\mathcal{C} \subseteq \mathcal{H}_2$. Hence we have shown that $\mathcal{H}$ is a π-system. □

Given a set E and a σ-algebra $\mathcal{B}$ over E, we call the pair $(E, \mathcal{B})$ a **measurable space**. The reason for introducing measurable spaces is that they are the natural place on which to define measures. Namely, if $(E, \mathcal{B})$ is a measurable space, we say that the map $\mu : \mathcal{B} \longrightarrow [0, \infty]$ is a **measure** on $(E, \mathcal{B})$ if $\mu(\emptyset) = 0$ and μ is **countably additive** in the sense that for, $\{\Gamma_n\}_1^\infty \subseteq \mathcal{B}$,

$$\mu\left(\bigcup_1^\infty \Gamma_n\right) = \sum_1^\infty \mu(\Gamma_n) \quad \text{if } \Gamma_m \cap \Gamma_n = \emptyset \text{ for } m \neq n. \tag{3.1.4}$$

When $\mu(E) < \infty$, μ is said to be a **finite measure**, and when $\mu(E) = 1$ it is called a **probability measure**. Given a measurable space $(E, \mathcal{B})$ and a measure μ on $(E, \mathcal{B})$, the triple $(E, \mathcal{B}, \mu)$ is called a **measure space**. The measure space $(E, \mathcal{B}, \mu)$ is said to be a **finite measure space** or a **probability space** according to whether μ is a finite measure or a probability measure on $(E, \mathcal{B})$.

3.1.5 Examples: As we will show in Chapter VII, there is a general method for producing lots of measures. However, at the moment we will have to settle for the following examples.

(i) Our basic examples of measures are those constructed by Lebesgue. Namely, when $E = \mathbb{R}^N$, $\mathcal{B} = \overline{\mathcal{B}}_{\mathbb{R}^N}$, and $\mu = \lambda_{\mathbb{R}^N}$, where $\lambda_{\mathbb{R}^N}$ is the measure defined by $\lambda_{\mathbb{R}^N}(\Gamma) = |\Gamma|$ for $\Gamma \in \overline{\mathcal{B}}_{\mathbb{R}^N}$.

(ii) Given a measure space $(E, \mathcal{B}, \mu)$, one can always extend μ as a measure $\overline{\mu}$ on the σ-algebra $\overline{\mathcal{B}}^\mu$ of sets $\Gamma \subseteq E$ with the property that there exist A, $B \in \mathcal{B}$ such that $A \subseteq \Gamma \subseteq B$ and $\mu(B \setminus A) = 0$; indeed, one simply defines $\overline{\mu}(\Gamma) = \mu(A)$. The σ-algebra $\overline{\mathcal{B}}^\mu$ is called the **completion of $\mathcal{B}$ with respect to** μ, and the resulting measure space $(E, \overline{\mathcal{B}}^{\mu}, \overline{\mu})$ is said to be **complete**. In this connection, note that what we have been denoting by $\overline{\mathcal{B}}_{\mathbb{R}^N}$ is the completion of the Borel algebra $\mathcal{B}_{\mathbb{R}^N}$ over $\mathbb{R}^N$ with respect to the restriction of Lebesgue measure $\lambda_{\mathbb{R}^N}$ to $\mathcal{B}_{\mathbb{R}^N}$. Thus we really should have been using the hideous notation $\overline{\mathcal{B}}_{\mathbb{R}^N}^{\lambda_{\mathbb{R}^N}}$, but, for obvious reasons of aesthetics, we will continue to reserve $\overline{\mathcal{B}}_{\mathbb{R}^N}$ for the completion of $\mathcal{B}_{\mathbb{R}^N}$ with respect to Lebesgue measure.

(iii) An easy and useful source of examples of measure spaces are those in which E is a countable set, $\mathcal{B} = \mathcal{P}(E)$, and $\mu(\Gamma) = \sum_{x \in \Gamma} \mu_x$, where $\{\mu_x : x \in E\} \subseteq [0, \infty]$. A closely related example is the counting measure on an arbitrary set E. In this example, $\mathcal{B}$ is again $\mathcal{P}(E)$ and $\mu(\Gamma) = \text{card}(\Gamma)$ when Γ is finite and ∞ otherwise.

(iv) As a final example, we point out that measure spaces give rise to other measure spaces by means of restriction. Namely, if $(E, \mathcal{B}, \mu)$ is a measure space and $E' \in \mathcal{B}$ is given, define $\mathcal{B}[E'] = \{\Gamma \cap E' : \Gamma \in \mathcal{B}\}$. Then $\mathcal{B}[E']$ is a σ-algebra over E' and $(E', \mathcal{B}[E'], \mu \restriction \mathcal{B}[E'])$ is a measure space. (See Section 5.3 for a refinement of this procedure.)

The following theorem gives some of the basic consequences of (3.1.4).

3.1.6 Theorem. *Let $(E, \mathcal{B}, \mu)$ be a measure space. If $\Gamma_1, \Gamma_2 \in \mathcal{B}$ and $\Gamma_1 \subseteq \Gamma_2$ then $\mu(\Gamma_1) \le \mu(\Gamma_2)$ and, when $\mu(\Gamma_1) < \infty$, $\mu(\Gamma_2 \setminus \Gamma_1) = \mu(\Gamma_2) - \mu(\Gamma_1)$. Moreover, for $\{\Gamma_n\}_1^\infty \subseteq \mathcal{B}$:*

$$\text{(i)} \qquad \mu(\Gamma_n) \nearrow \mu(\Gamma) \quad \textit{if} \quad \Gamma_n \nearrow \Gamma,$$

$$\text{(ii)} \qquad \mu(\Gamma_n) \searrow \mu(\Gamma) \quad \textit{if } \Gamma_n \searrow \Gamma \textit{ and } \mu(\Gamma_1) < \infty,$$

$$\text{(iii)} \qquad \mu\left(\bigcup_1^\infty \Gamma_n\right) \le \sum_1^\infty \mu(\Gamma_n),$$

and

$$\text{(iv)} \qquad \mu\left(\bigcup_1^\infty \Gamma_n\right) = \sum_1^\infty \mu(\Gamma_n) \quad \textit{if } \mu(\Gamma_m \cap \Gamma_n) = 0 \textit{ for } m \ne n.$$

PROOF: If $\Gamma_1 \subseteq \Gamma_2$, then $\mu(\Gamma_2) = \mu(\Gamma_1) + \mu(\Gamma_2 \setminus \Gamma_1)$, since $\Gamma_2 = \Gamma_1 \cup (\Gamma_2 \setminus \Gamma_1)$. The initial assertions follow immediately from this.

To prove **(i)**, set $\Gamma_0 = \emptyset$ and define $A_{n+1} = \Gamma_{n+1} \setminus \Gamma_n$ for $n \ge 0$. Then $A_m \cap A_n = \emptyset$ for $m \ne n$, $\Gamma_n = \bigcup_1^n A_m$, and $\Gamma = \bigcup_1^\infty A_n$. Hence

$$\mu(\Gamma_n) = \sum_1^n \mu(A_m) \nearrow \sum_1^\infty \mu(A_m) = \mu(\Gamma).$$

The proof of **(ii)** is accomplished by taking $\Delta_n = \Gamma_1 \setminus \Gamma_n$ and applying the preceding to $\{\Delta_n\}$. (One needs $\mu(\Gamma_1) < \infty$ in order to subtract it from both sides.)

To prove (**iii**) and (**iv**), again set $\Gamma_0 = \emptyset$ and take $A_{n+1} = \Gamma_{n+1} \setminus \bigcup_1^n \Gamma_m$ for $n \geq 0$. Then $\Gamma_n = A_n \cup D_n$, where $D_n = \bigcup_{m=1}^{n-1} (\Gamma_n \cap \Gamma_m)$, and $\bigcup_1^\infty \Gamma_n = \bigcup_1^\infty A_n$. Hence, since $A_m \cap A_n = \emptyset$ for $m \neq n$,

$$\mu\left(\bigcup_1^\infty \Gamma_n\right) = \mu\left(\bigcup_1^\infty A_n\right) = \sum_1^\infty \mu(A_n) \leq \sum_1^\infty \mu(\Gamma_n) \leq \sum_1^\infty \big(\mu(A_n) + \mu(D_n)\big).$$

This proves that the inequality in (**iii**) always holds and that the equality in (**iv**) holds when $\mu(D_n) \leq \sum_{m=1}^{n-1} \mu(\Gamma_n \cap \Gamma_m) = 0$ for all $n \geq 2$. □

Exercises

3.1.7 Exercise: The decomposition of the properties of a σ-algebra in terms of π-systems and λ-systems is not the traditional one. Instead, most of the early books on measure theory used algebras instead of π-systems as the usual source of generating sets. In this case the complementary notion is that of a **monotone class**: $\mathcal{M}$ is said to be a monotone class if $\Gamma \in \mathcal{M}$ whenever there exists $\{\Gamma_n\}_1^\infty \subseteq \mathcal{M}$ such that either $\Gamma_n \nearrow \Gamma$ or $\Gamma_n \searrow \Gamma$. Show that $\mathcal{B}$ is a σ-algebra over E if and only if it is both an algebra over E and a monotone class. In addition, show that if $\mathcal{A}$ is an algebra over E, then $\sigma(E;\mathcal{A})$ is the smallest monotone class containing $\mathcal{A}$.

3.1.8 Exercise: Let $(E,\mathcal{B})$ be a measurable space with $\mathcal{B} = \sigma(E;\mathcal{C})$. Suppose that μ and ν are a pair of measures on $(E,\mathcal{B})$ such that $\mu(E) = \nu(E) < \infty$ and $\mu(\Gamma) = \nu(\Gamma)$ for all $\Gamma \in \mathcal{C}$. Assuming that $\mathcal{C}$ is a π-system over E, show that $\mu = \nu$ on $\mathcal{B}$. (**Hint**: Consider the collection $\mathcal{H}$ of $\Gamma \in \mathcal{B}$ for which $\mu(\Gamma) = \nu(\Gamma)$, and apply Lemma 3.1.3.)

3.1.9 Exercise: Let (E,ρ) be a metric space and suppose that μ is a finite measure on $(E,\mathcal{B}_E)$. Show that the completion $\overline{\mathcal{B}}_E^\mu$ of the Borel field $\mathcal{B}_E$ with respect to μ coincides with the class of all $\Gamma \subseteq E$ such that for each $\epsilon > 0$ there exists a closed set F and an open set G with the properties that $F \subseteq \Gamma \subseteq G$ and $\mu(G \setminus F) < \epsilon$.

3.1.10 Exercise: Suppose that $(E_1,\mathcal{B}_1)$ and $(E_2,\mathcal{B}_2)$ are two measurable spaces and that $\Phi : E_1 \longrightarrow E_2$ has the property that $\Phi^{-1}(\Gamma) \in \mathcal{B}_1$ for every element Γ in a collection $\mathcal{C}$ which generates $\mathcal{B}_2$. Show that $\Phi^{-1}(\Gamma) \in \mathcal{B}_1$ for every $\Gamma \in \mathcal{B}_2$. In particular, if E_1 and E_2 are topological spaces and $\mathcal{B}_1$ and $\mathcal{B}_2$ are the corresponding Borel algebras, show that $\Phi^{-1}(\Gamma) \in \mathcal{B}_1$ for every $\Gamma \in \mathcal{B}_2$ if Φ is continuous. Conclude from this that $x + \Gamma \in \mathcal{B}_{\mathbb{R}^N}$ for all $x \in \mathbb{R}^N$ and $\Gamma \in \mathcal{B}_{\mathbb{R}^N}$.

3.1.11 Exercise: Let μ be a measure on $(\mathbb{R}^N, \mathcal{B}_{\mathbb{R}^N})$ which is translation invariant (i.e., $\mu(x + \Gamma) = \mu(\Gamma)$). In addition, assume that $\mu([0,1]^N) = 1$. Show that $\mu = \lambda_{\mathbb{R}^N}$ on $\mathcal{B}_{\mathbb{R}^N}$.

Hint: First check that $\mu(\partial Q) = 0$ for any cube Q. Second, show that $\mu([0, m\lambda]^N) = m^N \mu([0, \lambda])$ for any $m \in \mathbb{Z}^+$ and $\lambda \in \mathbb{R}$. From these, conclude that $\mu(Q) = |Q|$ for all cubes Q. Finally, deduce the required result.

3.1.12 Exercise: Given sets Γ_n for $n \geq 1$, define

$$\overline{\lim_{n\to\infty}}\, \Gamma_n = \bigcap_{m=1}^{\infty} \bigcup_{n=m}^{\infty} \Gamma_n \quad \text{and} \quad \underline{\lim_{n\to\infty}}\, \Gamma_n = \bigcup_{m=1}^{\infty} \bigcap_{n=m}^{\infty} \Gamma_n.$$

Observe that

$$\overline{\lim_{n\to\infty}}\, \Gamma_n = \left\{ x : x \in \Gamma_n \text{ for infinitely many } n \in \mathbb{Z}^+ \right\}$$

and that

$$\underline{\lim_{n\to\infty}}\, \Gamma_n \subseteq \overline{\lim_{n\to\infty}}\, \Gamma_n \tag{3.1.13}$$

with equality holding when $\{\Gamma_n\}_1^\infty$ is monotone. One says that the limit $\lim_{n\to\infty} \Gamma_n$ exists if equality holds in (3.1.13), in which case $\lim_{n\to\infty} \Gamma_n \equiv \underline{\lim}_{n\to\infty} \Gamma_n$.

Let $(E, \mathcal{B}, \mu)$ be a measure space and $\{\Gamma_n\}_1^\infty \subseteq \mathcal{B}$. Prove each of the following.

(i)
$$\mu\left(\underline{\lim_{n\to\infty}}\, \Gamma_n\right) \leq \underline{\lim_{n\to\infty}}\, \mu(\Gamma_n).$$

and

(ii)
$$\overline{\lim_{n\to\infty}}\, \mu(\Gamma_n) \leq \mu\left(\overline{\lim_{n\to\infty}}\, \Gamma_n\right) \quad \text{if} \quad \mu\left(\bigcup_1^\infty \Gamma_n\right) < \infty.$$

In particular, under the condition in **(ii)**, conclude that

(iii)
$$\lim_{n\to\infty} \mu(\Gamma_n) = \mu\left(\lim_{n\to\infty} \Gamma_n\right) \quad \text{if} \quad \lim_{n\to\infty} \Gamma_n \text{ exists.}$$

Finally, show that

(iv)
$$\mu\left(\overline{\lim_{n\to\infty}}\, \Gamma_n\right) = 0 \quad \text{if} \quad \sum_1^\infty \mu(\Gamma_n) < \infty.$$

The result in **(iv)** is often called the **Borel–Cantelli Lemma**, and it has many applications in probability theory.

3.1.14 Exercise: Let $(E, \mathcal{B}, \mu)$ be a measure space.

(i) Assuming that $\Gamma_1, \Gamma_2 \in \mathcal{B}$ and that $\mu(\Gamma_1 \cap \Gamma_2) < \infty$, show that

$$\mu(\Gamma_1 \cup \Gamma_2) = \mu(\Gamma_1) + \mu(\Gamma_2) - \mu(\Gamma_1 \cap \Gamma_2).$$

(ii) Let $\{\Gamma_m\}_1^n \subseteq \mathcal{B}$ and assume that $\max_{1 \le m \le n} \mu(\Gamma_m) < \infty$. Show that

$$\mu(\Gamma_1 \cup \cdots \cup \Gamma_n) = -\sum_F (-1)^{\operatorname{card}(F)} \mu(\Gamma_F),$$

where the summation is over non-empty subsets F of $\{1, \ldots, n\}$ and $\Gamma_F \equiv \bigcap_{i \in F} \Gamma_i$.

(iii) Although the formula in **(ii)** above is seldom used except in the case when $n = 2$, the following is an interesting application of the general result. Let E be the group of permutations on $\{1, \ldots, n\}$, $\mathcal{B} = \mathcal{P}(E)$, and $\mu(\{\pi\}) = \frac{1}{n!}$ for each $\pi \in E$. Denote by A the set of $\pi \in E$ such that $\pi(i) \neq i$ for any $1 \le i \le n$. Then one can interpret $\mu(A)$ as the probability that, when the numbers $1, \ldots, n$ are randomly ordered, none of them is placed in the correct position. On the basis of this interpretation, one might suspect that $\mu(A)$ should tend to 0 as $n \to \infty$. However, by direct computation, one can see that this is not the case. Indeed, let Γ_i be the set of $\pi \in E$ such that $\pi(i) = i$. Then $A = (\Gamma_1 \cup \cdots \cup \Gamma_n)\complement$. Hence,

$$\mu(A) = 1 - \mu(\Gamma_1 \cup \cdots \cup \Gamma_n) = 1 + \sum_F (-1)^{\operatorname{card}(F)} \mu(\Gamma_F).$$

Show that $\mu(\Gamma_F) = \frac{(n-m)!}{n!}$ if $\operatorname{card}(F) = m$, and conclude from this that $\mu(A) = \sum_0^n \frac{(-1)^m}{m!} \longrightarrow \frac{1}{e}$ as $n \to \infty$.

3.2 Construction of Integrals

We are now very close to the point at which we can return to the problem of integrating functions on a measure space $(E, \mathcal{B}, \mu)$. Recall, from the introduction to Chapter II, that Lebesgue's procedure entails the use of sums like

$$\sum_k \frac{k}{2^n} \mu\left(\left\{x \in E : f(x) \in \left[\frac{k}{2^n}, \frac{k+1}{2^n}\right)\right\}\right)$$

to approximate the integral of f on E with respect to μ. In particular, we have to be dealing with functions f for which sets of the form

$$\{f \in \Delta\} \equiv \{x \in E : f(x) \in \Delta\}$$

are in $\mathcal{B}$ when Δ is an interval; and it is only reasonable that we should call such a function measurable. More generally, given measurable spaces $(E_1, \mathcal{B}_1)$ and $(E_2, \mathcal{B}_2)$, we will say that $\Phi : E_1 \longrightarrow E_2$ is a **measurable map** on $(E_1, \mathcal{B}_1)$ into $(E_2, \mathcal{B}_2)$ if

$$\{\Phi \in \Gamma\} \equiv \{x \in E_1 : \Phi(x) \in \Gamma\} = \Phi^{-1}(\Gamma) \in \mathcal{B}_1 \quad \text{for each} \quad \Gamma \in \mathcal{B}_2.$$

Notice the analogy between the definitions of measurability and continuity. In particular, it is clear that if Φ is a measurable map on $(E_1, \mathcal{B}_1)$ into $(E, \mathcal{B}_2)$ and Ψ is a measurable map on $(E_2, \mathcal{B}_2)$ into $(E_3, \mathcal{B}_3)$, then $\Psi \circ \Phi$ is a measurable map on $(E_1, \mathcal{B}_1)$ into $(E_3, \mathcal{B}_3)$. The following lemma is simply a restatement of Exercise 3.1.10 in the language of measurable functions.

3.2.1 Lemma. *Let $(E_1, \mathcal{B}_1)$ and $(E_2, \mathcal{B}_2)$ be measurable spaces and suppose that $\mathcal{B}_2 = \sigma(E_2; \mathcal{C})$ for some $\mathcal{C} \subseteq \mathcal{P}(E_2)$. If $\Phi : E_1 \longrightarrow E_2$ has the property that $\Phi^{-1}(\Gamma) \in \mathcal{B}_1$ for every $\Gamma \in \mathcal{C}$, then Φ is a measurable map on $(E_1, \mathcal{B}_1)$ into $(E_2, \mathcal{B}_2)$. In particular, if E_1 and E_2 are topological spaces and $\mathcal{B}_i = \mathcal{B}_{E_i}$ for $i \in \{1, 2\}$, then every continuous map on E_1 into E_2 is a measurable map on $(E_1, \mathcal{B}_1)$ into $(E_2, \mathcal{B}_2)$.*

In order to handle certain measurability questions, we introduce at this point a construction to which we will return in Section 4.1. Namely, given measurable spaces $(E_1, \mathcal{B}_1)$ and $(E_2, \mathcal{B}_2)$, we define the **product of** $(E_1, \mathcal{B}_1)$ **times** $(E_2, \mathcal{B}_2)$ to be the measurable space $(E_1 \times E_2, \mathcal{B}_1 \times \mathcal{B}_2)$ where[1]

$$\mathcal{B}_1 \times \mathcal{B}_2 \equiv \sigma\Big(E_1 \times E_2;\ \{\Gamma_1 \times \Gamma_2 : \Gamma_1 \in \mathcal{B}_1 \text{ and } \Gamma_2 \in \mathcal{B}_2\}\Big).$$

Also, if Φ_i is a measurable map on $(E_0, \mathcal{B}_0)$ into $(E_i, \mathcal{B}_i)$ for $i \in \{1, 2\}$, then we define the **tensor product of** Φ_1 **times** Φ_2 to be the map $\Phi_1 \otimes \Phi_2 : E_0 \longrightarrow E_1 \times E_2$ given by $\Phi_1 \otimes \Phi_2(x) = \big(\Phi_1(x), \Phi_2(x)\big)$, $x \in E_0$.

3.2.2 Lemma. *Referring to the preceding, suppose that Φ_i is a measurable map on $(E_0, \mathcal{B}_0)$ into $(E_i, \mathcal{B}_i)$ for $i \in \{1, 2\}$. Then $\Phi_1 \otimes \Phi_2$ is a measurable map on $\big(E_0, \mathcal{B}_0\big)$ into $\big(E_1 \times E_2, \mathcal{B}_1 \times \mathcal{B}_2\big)$. Moreover, if E_1 and E_2 are second countable topological spaces[2], then $\mathcal{B}_{E_1} \times \mathcal{B}_{E_2} = \mathcal{B}_{E_1 \times E_2}$.*

PROOF: To prove the first assertion, we need only note that if $\Gamma_i \in \mathcal{B}_i$, $i \in \{1, 2\}$, then ${\Phi_1 \otimes \Phi_2}^{-1}(\Gamma_1 \times \Gamma_2) = \Phi_1^{-1}(\Gamma_1) \cap \Phi_2^{-1}(\Gamma_2) \in \mathcal{B}_0$. As for the second assertion, first note that $G_1 \times G_2$ is open in $E_1 \times E_2$ for every pair of open sets

[1] Strictly speaking, the notation $\mathcal{B}_1 \times \mathcal{B}_2$ is incorrect and should be replace by something even more hideous like $\sigma(\mathcal{B}_\times \mathcal{B}_2)$. Nonetheless, for the sake of aesthetics if nothing else, we have chosen to stick with $\mathcal{B}_1 \times \mathcal{B}_2$.

[2] A topological space is said to be second countable if it possesses a countable collection $\mathfrak{B}$ of open subsets which form a basis in the sense that, for every open subset G and $x \in G$, there is a $B \in \mathfrak{B}$ with $x \in B \subseteq G$. For example, every separable metric space is second countable. Indeed, let $\{p_m\}_1^\infty$ be a countable dense subset, and take $\mathfrak{B}$ to be the set of open balls centered at some p_m and having a radius n^{-1} for some $n \geq 1$.

G_1 in E_1 and G_2 in E_2. Hence, even without second countability, $\mathcal{B}_{E_1} \times \mathcal{B}_{E_2} \subseteq \mathcal{B}_{E_1 \times E_2}$. At the same time, with second countability, one can write every open G in $E_1 \times E_2$ as the countable union of sets of the form $G_1 \times G_2$ where G_i is open in E_i. Hence, in this case, the opposite inclusion also holds. □

In measure theory one is most interested in real-valued functions. However, for reasons of convenience, it is often handy to allow functions to take values in the **extended real line** $\overline{\mathbb{R}} \equiv [-\infty, \infty]$. Unfortunately, the introduction of $\overline{\mathbb{R}}$ involves some annoying problems. We have already encountered such a problem when we wrote the equation $\mu(\Gamma_2 \setminus \Gamma_1) = \mu(\Gamma_2) - \mu(\Gamma_1)$ in Theorem 3.1.6. Our problem there, and the basic one which we want to discuss here, stems from the difficulty of extending the arithmetic operations to include ∞ and $-\infty$. Thus, in an attempt to lay all such technical difficulties to rest once and for all, we will spend a little time discussing them here.

To begin with, we point out that $\overline{\mathbb{R}}$ the *natural* notion of convergence in $\overline{\mathbb{R}}$ which says that $\{x_n\}_1^\infty \subseteq \overline{\mathbb{R}}$ converges to $a \in \mathbb{R}$ if $\{x_{m+n}\}_{n=1}^\infty \subseteq \mathbb{R}$ for some $m \geq 0$ and $x_{m+n} \longrightarrow a$ in $\mathbb{R}$ as $n \to \infty$ and that $\{x_n\}_1^\infty$ converges to $\pm\infty$ if, for every $R > 0$, there is an $m_R \geq 0$ such that $\pm x_n \geq R$ for all $n \geq m_R$. Second, we introduce a metric on $\overline{\mathbb{R}}$ for which this *natural* notion is *the* notion of convergence and under which $\overline{\mathbb{R}}$ is compact. Namely, define

$$\overline{\rho}(\alpha, \beta) = \frac{2}{\pi}|\arctan(y) - \arctan(x)|,$$

where $\arctan(\pm\infty) \equiv \pm\frac{\pi}{2}$. Clearly $(\overline{\mathbb{R}}, \overline{\rho})$ is a compact metric space: it is homeomorphic to $[-1,1]$ under the map $t \in [-1,1] \longmapsto \tan\left(\frac{\pi t}{2}\right)$. Moreover, $\mathbb{R}$, with its usual topology, is imbedded in $\overline{\mathbb{R}}$ as a dense open set. In particular, $\mathcal{B}_{\mathbb{R}} = \mathcal{B}_{\overline{\mathbb{R}}}[\mathbb{R}]$ (cf. **(iv)** in Examples 3.1.5). Having put a topology and measurable structure on $\overline{\mathbb{R}}$, we will next adopt the following extension to $\overline{\mathbb{R}}$ of multiplication: $(\pm\infty) \cdot 0 = 0 \cdot (\pm\infty) = 0$ and $(\pm\infty) \cdot \alpha = \alpha \cdot (\pm\infty) = \operatorname{sgn}(\alpha)\infty$ if $\alpha \in \overline{\mathbb{R}} \setminus \{0\}$. Although $(\alpha, \beta) \in \overline{\mathbb{R}}^2 \longmapsto \alpha \cdot \beta \in \overline{\mathbb{R}}$ is not continuous, one can easily check that it is a measurable map on $\left(\overline{\mathbb{R}}^2, \mathcal{B}_{\overline{\mathbb{R}}^2}\right)$ into $(\overline{\mathbb{R}}, \mathcal{B}_{\overline{\mathbb{R}}})$. Unfortunately, the extension of addition presents a knottier problem. Indeed, because we do not know how to interpret $\pm\infty \mp \infty$ in general, we will simply avoid doing so at all by restricting the domain of addition to the set $\widehat{\mathbb{R}^2}$ consisting of $\mathbb{R}^2$ with the two points (∞, ∞) and $(-\infty, -\infty)$ appended. Clearly $\widehat{\mathbb{R}^2}$ is an open subset of $\overline{\mathbb{R}}^2$, and so $\mathcal{B}_{\widehat{\mathbb{R}^2}} = \mathcal{B}_{\overline{R}^2}\left[\widehat{\mathbb{R}^2}\right]$. We define addition $(\alpha, \beta) \in \widehat{\mathbb{R}^2} \longmapsto \alpha + \beta \in \overline{\mathbb{R}}$ so that $(\pm\infty) + \alpha = \alpha + (\pm\infty) = \pm\infty$ if $\alpha \neq \mp\infty$. It is then easy to see that $(\alpha, \beta) \longmapsto \alpha + \beta$ is continuous on $\widehat{\mathbb{R}^2}$ into $\overline{\mathbb{R}}$; and therefore it is certainly a measurable map on $\left(\widehat{\mathbb{R}^2}, \mathcal{B}_{\widehat{\mathbb{R}^2}}\right)$ into $\left(\overline{\mathbb{R}}, \mathcal{B}_{\overline{\mathbb{R}}}\right)$. Finally, we complete our discussion of $\overline{\mathbb{R}}$ by pointing out that the lattice operations "$\vee$" and "$\wedge$" both admit unique continuous extensions as maps from $\overline{\mathbb{R}}^2$ into $\overline{\mathbb{R}}$ and are therefore not a source of concern.

Having adopted these conventions, we see that, for any pair of measurable functions f_1 and f_2 on a measurable space $(E, \mathcal{B})$ into $\left(\overline{\mathbb{R}}, \mathcal{B}_{\overline{\mathbb{R}}}\right)$, Lemma 3.2.2

guarantees that the measurability of the $\overline{\mathbb{R}}$-valued maps

$$x \in E \longmapsto f_1 \cdot f_2(x) \equiv f_1(x)\, f_2(x),$$
$$x \in E \longmapsto (f_1 + f_2)(x) \equiv f_1(x) + f_2(x) \in \overline{\mathbb{R}} \quad \text{if } \mathrm{Range}(f_1 \otimes f_2) \subseteq \widehat{\mathbb{R}^2},$$
$$x \in E \longmapsto f_1 \vee f_2(x) \equiv f_1(x) \vee f_2(x),$$

and

$$x \in E \longmapsto f_1 \wedge f_2(x) \equiv f_1(x) \wedge f_2(x).$$

Thus, of course, if f is measurable on $(E, \mathcal{B})$ into $(\overline{\mathbb{R}}, \mathcal{B}_{\overline{\mathbb{R}}})$, then so are $f^+ \equiv f \vee 0$, $f^- \equiv -(f \wedge 0)$, and $|f| = f^+ + f^-$. Finally, from now on we will call a measurable map on $(E, \mathcal{B})$ into $(\overline{\mathbb{R}}, \mathcal{B}_{\overline{\mathbb{R}}})$ a **measurable function** on $(E, \mathcal{B})$.

From the measure-theoretic standpoint, the most elementary functions are those which take on only a finite number of distinct values; thus, we will say that such a function is **simple**. Note that the class of simple functions is closed under the lattice operations "$\vee$" and "$\wedge$", multiplication, and, when the sum is defined, under addition. Aside from constant functions, the simplest of the simple functions are those which take their values in $\{0, 1\}$. Clearly there is a one-to-one correspondence between $\{0, 1\}$-valued functions and subsets of E. Namely, with $\Gamma \subseteq E$ we associate the function $\mathbf{1}_\Gamma$ defined by

$$\mathbf{1}_\Gamma(x) = \begin{cases} 1 & \text{if } \; x \in \Gamma \\ 0 & \text{if } \; x \notin \Gamma. \end{cases}$$

The function $\mathbf{1}_\Gamma$ is called the **indicator** (or **characteristic**) **function** of the set Γ.

The reason why simple functions play such a central role in measure theory is that their integrals are the easiest to describe. To be precise, let $(E, \mathcal{B}, \mu)$ be a measure space and f a non-negative (i.e., a $[0, \infty]$-valued) measurable function on $(E, \mathcal{B})$ which is simple. We then define the **Lebesgue integral of f on E** to be

$$\sum_{\alpha \in \mathrm{Range}(f)} \alpha \mu(f = \alpha),$$

where $\mu(f = \alpha)$ is shorthand for $\mu(\{f = \alpha\})$ which, in turn, is shorthand for $\mu(\{x \in E : f(x) = \alpha\})$. There are various ways in which we will denote the Lebesgue integral of f, depending on how many details the particular situation demands. The various expressions which we will use are, in decreasing order of information conveyed:

$$\int_E f(x)\, \mu(dx), \quad \int_E f\, d\mu, \quad \text{and} \quad \int f\, d\mu.$$

Further, for $\Gamma \in \mathcal{B}$ we will use

$$\int_\Gamma f(x)\, \mu(dx) \quad \text{or} \quad \int_\Gamma f\, d\mu$$

to denote the Lebesgue integral of $\mathbf{1}_\Gamma f$ on E. Observe that this notation is completely consistent, since we would get precisely the same number by restricting f to Γ and computing the Lebesgue integral of $f \restriction \Gamma$ relative to $(\Gamma, \mathcal{B}[\Gamma], \mu \restriction \mathcal{B}[\Gamma])$ (cf. (**iv**) of Examples 3.1.5).

It will turn out that all the basic properties of the Lebesgue integral rest on consistency results about the definition of the integral. The following lemma is the first such consistency result.

3.2.3 Lemma. *Let $(E, \mathcal{B}, \mu)$ be a measure space and f a non-negative simple measurable function on $(E, \mathcal{B})$. If $f = \sum_{\ell=1}^n \beta_\ell \mathbf{1}_{\Delta_\ell}$ where $\{\beta_1, \dots, \beta_n\} \subseteq [0, \infty]$ and $\{\Delta_1, \dots, \Delta_n\} \subseteq \mathcal{B}$, then $\int f\, d\mu = \sum_{\ell=1}^n \beta_\ell \mu(\Delta_\ell)$.*

PROOF: Let $\{\alpha_1, \dots, \alpha_m\}$ denote the distinct values of f and, for $1 \le k \le m$, set $\Gamma_k = \{f = \alpha_k\}$. Since $\Gamma_k \cap \Gamma_{k'} = \emptyset$ for $k \ne k'$,

$$\sum_{\ell=1}^n \beta_\ell \mu(\Delta_\ell) = \sum_{\ell=1}^n \beta_\ell \sum_{k=1}^m \mu(\Delta_\ell \cap \Gamma_k) = \sum_{k=1}^m \sum_{\ell=1}^n \beta_\ell \mu(\Delta_\ell \cap \Gamma_k);$$

and so all we need to show is that $\sum_{\ell=1}^n \beta_\ell \mu(\Delta_\ell \cap \Gamma_k) = \alpha_k \mu(\Gamma_k)$ for each $1 \le k \le m$. Since $\sum_{\ell=1}^n \beta_\ell \mathbf{1}_{\Delta_\ell \cap \Gamma_k} = \alpha_k \mathbf{1}_{\Gamma_k}$ for each $1 \le k \le m$, we now see that it suffices to treat the case when $f = \alpha \mathbf{1}_\Gamma$ for some $\alpha \in [0, \infty]$, $\Gamma \in \mathcal{B}$, and $\{\Delta_1, \dots, \Delta_n\} \subseteq \mathcal{B}[\Gamma]$. Further, it is clear that we may assume that $\alpha \ne 0$, since the only way in which one could have $\sum_{\ell=1}^n \beta_\ell \mathbf{1}_{\Delta_\ell} \equiv 0$ is if $\beta_\ell = 0$ whenever $\Delta_\ell \ne \emptyset$. In other words, what we still have to show is that, for any $\alpha \in (0, \infty]$, $\Gamma \in \mathcal{B}$, and $\{\Delta_1, \dots, \Delta_n\} \subseteq \mathcal{B}[\Gamma]$,

$$\sum_{\ell=1}^n \beta_\ell \mu(\Delta_\ell) = \alpha \mu(\Gamma) \quad \text{if} \quad \sum_{\ell=1}^n \beta_\ell \mathbf{1}_{\Delta_\ell} = \alpha \mathbf{1}_\Gamma.$$

Set $\mathcal{I} = (\{0,1\})^n$, and, for $\boldsymbol{\eta} = (\eta_1, \dots, \eta_n) \in \mathcal{I}$, define $\beta_{\boldsymbol{\eta}} = \sum_1^n \eta_\ell \beta_\ell$ and $\Delta_{\boldsymbol{\eta}} = \bigcap_{\ell=1}^n \Delta_\ell^{(\eta_\ell)}$, where $\Delta^{(1)} \equiv \Delta$ and $\Delta^{(0)} \equiv \Delta\complement$. Then $\Delta_{\boldsymbol{\eta}} \cap \Delta_{\boldsymbol{\eta}'} = \emptyset$ if $\boldsymbol{\eta} \ne \boldsymbol{\eta}'$, and, because

$$\Delta_\ell = \bigcup_{\{\boldsymbol{\eta} \in \mathcal{I} : \eta_\ell = 1\}} \Delta_{\boldsymbol{\eta}}$$

for each $1 \le \ell \le n$,

$$\sum_{\boldsymbol{\eta} \in \mathcal{I}} \beta_{\boldsymbol{\eta}} \mathbf{1}_{\Delta_{\boldsymbol{\eta}}} = \sum_{\boldsymbol{\eta} \in \mathcal{I}} \sum_{\ell=1}^n \eta_\ell \beta_\ell \mathbf{1}_{\Delta_{\boldsymbol{\eta}}} = \sum_{\ell=1}^n \beta_\ell \mathbf{1}_{\Delta_\ell} = \alpha \mathbf{1}_\Gamma.$$

In particular, $\Gamma = \bigcup_{\boldsymbol{\eta} \in \mathcal{I}} \Delta_{\boldsymbol{\eta}}$ and $\beta_{\boldsymbol{\eta}} = \alpha$ if $\Delta_{\boldsymbol{\eta}} \ne \emptyset$, from which it is clear that

$$\Gamma = \bigcup_{\boldsymbol{\eta} \in \mathcal{I}'} \Delta_{\boldsymbol{\eta}} \quad \text{and} \quad \boldsymbol{\eta} \in \mathcal{I}' \implies \beta_{\boldsymbol{\eta}} = \alpha,$$

when $\mathcal{I}' \equiv \{\boldsymbol{\eta} \in \mathcal{I} : \Delta_{\boldsymbol{\eta}} \neq \emptyset\}$. Hence,

$$\sum_{\ell=1}^{n} \beta_\ell \mu(\Delta_\ell) = \sum_{\ell=1}^{n} \beta_\ell \sum_{\{\boldsymbol{\eta}:\eta_\ell=1\}} \mu(\Delta_{\boldsymbol{\eta}}) = \sum_{\boldsymbol{\eta}\in\mathcal{I}'} \beta_{\boldsymbol{\eta}} \mu(\Delta_{\boldsymbol{\eta}}) = \alpha\mu(\Gamma). \quad \square$$

The importance of Lemma 3.2.3 is already apparent in the next lemma.

3.2.4 Lemma. *Let f and g be non-negative simple measurable functions on $(E, \mathcal{B}, \mu)$. Then, for any $\alpha, \beta \in [0, \infty]$, $\alpha f + \beta g$ is a non-negative simple function and*

$$\int (\alpha f + \beta g)\, d\mu = \alpha \int f\, d\mu + \beta \int g\, d\mu.$$

In particular, if $f \leq g$, then $\int f\, d\mu \leq \int g\, d\mu$. In fact, if $f \leq g$ and $\int f\, d\mu < \infty$, then

$$\int (g - f)\, d\mu = \int g\, d\mu - \int f\, d\mu.$$

PROOF: Clearly it suffices to prove the first assertion. But if $\{\alpha_1, \ldots, \alpha_m\}$ and $\{\beta_1, \ldots, \beta_n\}$ are the distinct values of f and g, respectively, then $\alpha f + \beta g = \sum_{k=1}^{m+n} \gamma_k \mathbf{1}_{\Delta_k}$ where $\gamma_k = \alpha\alpha_k$ and $\Delta_k = \{f = \alpha_k\}$ for $1 \leq k \leq m$ and $\gamma_k = \beta\beta_{k-m}$ and $\Delta_k = \{g = \beta_{k-m}\}$ for $m + 1 \leq k \leq m + n$. Hence the required result follows immediately from Lemma 3.2.3. $\square$

In order to extend our definition of the Lebesgue integral to arbitrary non-negative measurable functions, we want to use a limit procedure. The idea is to approximate such a function by ones which are simple. For example, if f is a non-negative measurable function on $(E, \mathcal{B}, \mu)$, then we might take

$$\varphi_n = \sum_{k=0}^{4^n-1} 2^{-n} k \mathbf{1}_{\{2^n f \in [k,k+1))\}} + 2^n \mathbf{1}_{\{f \geq 2^n\}}$$

for $n \geq 1$. Then each φ_n is a non-negative, measurable, simple function, and $\varphi_n \nearrow f$ as $n \to \infty$. In fact, $\varphi_n \longrightarrow f$ uniformly in $(\overline{\mathbb{R}}, \overline{\rho})$. Thus it would seem reasonable to define the integral of f as

$$\text{(3.2.5)} \qquad \lim_{n\to\infty} \int \varphi_n\, d\mu = \lim_{n\to\infty} \left[\sum_{k=1}^{4^n-1} \tfrac{k}{2^n} \mu\big(k \leq 2^n f < k + 1\big) + 2^n \mu\big(f \geq 2^n\big) \right].$$

Indeed, since $\varphi_n \leq \varphi_{n+1}$, Lemma 3.2.4 guarantees that this limit exists. However, before adopting this definition, we must first check that the definition is not too dependent on the choice of the approximating sequence. In fact, at the moment, it is not even clear that this definition would coincide with the one we have already given for simple f's. This brings us to our second consistency result, where, as distinguished from Lemma 3.2.3, we must use *countable*, as opposed to *finite*, additivity.

3.2.6 Lemma. *Let $(E, \mathcal{B}, \mu)$ be a measure space, and suppose that $\{\varphi_n\}_1^\infty$ and ψ are non-negative measurable simple functions on $(E, \mathcal{B})$. If $\varphi_n \le \varphi_{n+1}$ for all $n \ge 1$ and if $\psi \le \lim_{n\to\infty} \varphi_n$, then $\int \psi\, d\mu \le \lim_{n\to\infty} \int \varphi_n\, d\mu$. In particular, for any non-negative, measurable function f and any non-decreasing sequence $\{\psi_n\}_1^\infty$ of non-negative, measurable, simple functions ψ_n which tend to f as $n \longrightarrow \infty$, $\lim_{n\to\infty} \int \psi_n\, d\mu$ is the same as the limit in (3.2.5).*

PROOF: Note that the final statment is indeed an easy consequence of what precedes it. Namely, if the simple functions φ_n are those given above and $\{\psi_n\}_1^\infty$ is any other sequence of simple functions satisfying $\psi_n \nearrow f$, then, for every $m \ge 1$,

$$\varphi_m \le \lim_{n\to\infty} \psi_n \quad \text{and} \quad \psi_m \le \lim_{n\to\infty} \varphi_n.$$

Thus, by the first part,

$$\lim_{m\to\infty} \int \varphi_m\, d\mu \le \lim_{n\to\infty} \int \psi_n\, d\mu \quad \text{and} \quad \lim_{m\to\infty} \int \psi_m\, d\mu \le \lim_{n\to\infty} \int \varphi_n\, d\mu.$$

In order to prove the first part, we treat three cases.

Case 1: $\mu(\psi = \infty) > 0$. Then, for each $M < \infty$,

$$\mu(\varphi_n > M) \nearrow \mu\left(\bigcup_{n=1}^\infty \{\varphi_n > M\}\right) \ge \mu(\psi > M) \ge \mu(\psi = \infty) = \epsilon$$

for some $\epsilon > 0$, and so

$$\lim_{n\to\infty} \int \varphi_n\, d\mu \ge \lim_{n\to\infty} M\mu(\varphi_n > M) \ge M\epsilon$$

for all $M < \infty$. Hence, in this case, $\lim_{n\to\infty} \int \varphi_n\, d\mu = \infty = \int \psi\, d\mu$.

Case 2: $\mu(\psi > 0) = \infty$. Here, because ψ is simple, there is an $\epsilon > 0$ such that $\psi > \epsilon$ whenever $\psi > 0$. Hence,

$$\mu(\varphi_n > \epsilon) \nearrow \mu\left(\bigcup_{n=1}^\infty \{\varphi_n > \epsilon\}\right) \ge \mu(\psi > 0) = \infty,$$

which, because $\varphi_n \ge \epsilon \mathbf{1}_{\{\varphi_n \ge \epsilon\}}$, means that

$$\lim_{n\to\infty} \int \varphi_n\, d\mu \ge \lim_{n\to\infty} \epsilon\mu(\varphi_n > \epsilon) = \infty = \int \psi\, d\mu.$$

Case 3: $\mu(\psi = \infty) = 0$ and $\mu(\psi > 0) < \infty$. Set $\hat{E} = \{0 < \psi < \infty\}$. Under the present conditions, $\mu(\hat{E}) < \infty$, $\int \psi\, d\mu = \int_{\hat{E}} \psi\, d\mu$, and $\int \varphi_n\, d\mu \ge \int_{\hat{E}} \varphi_n\, d\mu$ for all $n \ge 1$. Hence, without loss of generality, we will assume that $E = \hat{E}$. But then $\mu(E) < \infty$, and, because ψ is simple, there exist $\epsilon > 0$ and $M < \infty$ such that $\epsilon \le \psi \le M$. Now let $0 < \delta < \epsilon$ be given, and define $E_n = \{\varphi_n \ge \psi - \delta\}$. Then $E_n \nearrow E$ and so

$$\begin{aligned}\lim_{n\to\infty} \int \varphi_n\, d\mu \ge \lim_{n\to\infty} \int_{E_n} \varphi_n\, d\mu &\ge \lim_{n\to\infty} \left[\int_{E_n} \psi\, d\mu - \delta\mu(E_n)\right] \\ &= \lim_{n\to\infty} \left[\int \psi\, d\mu - \int_{E_n\complement} \psi\, d\mu - \delta\mu(E_n)\right] \\ &\ge \int \psi\, d\mu - M \lim_{n\to\infty} \mu(E_n\complement) - \delta\mu(E) = \int \psi\, d\mu - \delta\mu(E),\end{aligned}$$

since $\mu(E) < \infty$ and therefore (cf. (**ii**) in Theorem 3.1.6) $\mu(E_n\complement) \searrow 0$. Because this holds for arbitrarily small $\delta > 0$, we get our result upon letting $\delta \searrow 0$. □

The Lemma 3.2.6 allows us to complete the definition of the Lebesgue integral for non-negative, measurable functions. Namely, if f on $(E, \mathcal{B}, \mu)$ is a non-negative, measurable function, then we define the **Lebesgue integral of f on E with respect to μ** to be the number in (3.2.5); and we will continue to use the same notation to denote integrals. Not only does Lemma 3.2.6 guarantee that this definition is consistent with our earlier one for simple f's, but it also makes clear that the value of $\int f\, d\mu$ does not depend on the particular way in which one chooses to approximate f by a non-decreasing sequence of non-negative, measurable, simple functions. Thus, for example, the following extension of Lemma 3.2.4 is trivial.

3.2.7 Lemma. *If f and g are non-negative, measurable functions on the measure space $(E, \mathcal{B}, \mu)$, then for every $\alpha, \beta \in [0, \infty]$,*

$$\int (\alpha f + \beta g)\, d\mu = \alpha \int f\, d\mu + \beta \int g\, d\mu.$$

In particular, if $f \le g$, then $\int f\, d\mu \le \int g\, d\mu$ and $\int (g - f)\, d\mu = \int g\, d\mu - \int f\, d\mu$ as long as $\int f\, d\mu < \infty$.

Obviously $\int f\, d\mu$ reflects the size of a non-negative measurable f. The result which follows makes this statement somewhat more quantitative.

3.2.8 Theorem (Markov's Inequality). *If f is a non-negative measurable function on $(E, \mathcal{B}, \mu)$, then*

$$\mu(f \ge \lambda) \le \frac{1}{\lambda} \int_{\{f \ge \lambda\}} f\, d\mu \le \frac{1}{\lambda} \int f\, d\mu, \quad \lambda > 0. \tag{3.2.9}$$

In particular, $\int f\, d\mu = 0$ if and only if $\mu(f > 0) = 0$; and $\mu(f = \infty) = 0$ if $\int f\, d\mu < \infty$.

PROOF: To prove (3.2.9), simply note that $\lambda \mathbf{1}_{\{f \ge \lambda\}} \le \mathbf{1}_{\{f \ge \lambda\}} f \le f$. Clearly (3.2.9) implies that $\mu(f > 0) = \lim_{\epsilon \searrow 0} \mu(f \ge \epsilon) = 0$ if $\int f\, d\mu = 0$. Similarly, if $M = \int f\, d\mu < \infty$, then $\mu(f \ge \lambda) \le \frac{M}{\lambda}$ for all $\lambda > 0$; and therefore,

$$\mu(f = \infty) \le \lim_{\lambda\to\infty} \mu(f \ge \lambda) = 0.$$

Finally, if $\int f\,d\mu = 0$, then, by the (3.2.9), $\mu(f \geq n^{-1}) = 0$ for even $n \geq 1$ and therefore $\mu(f > 0) = \lim_{n\to\infty} \mu(f \geq n^{-1}) = 0$. On the other hand, because $0 \leq f \leq 0\mathbf{1}_{\{f=0\}} + \infty\mathbf{1}_{\{f>0\}}$,

$$\mu(f > 0) = 0 \implies 0 \leq \int f\,d\mu \leq 0\mu(f = 0) + \infty\mu(f > 0) = 0. \quad \square$$

The final step in the definition of the Lebesgue integral is to extend the definition so that it covers measurable functions which can take both signs. To this end, let f be a measurable function on the measure space $(E, \mathcal{B}, \mu)$. Then both $\int f^+\,d\mu$ and $\int f^-\,d\mu$ (recall that $f^\pm \equiv (\pm f) \vee 0$) are defined; and, if we want our integral to be linear, we can do nothing but take $\int f\,d\mu$ to be the difference between these two. However, before doing so, we must make sure that this difference is well-defined. With this consideration in mind, we now say that $\int f\,d\mu$ **exists** if $\int f^+\,d\mu \wedge \int f^-\,d\mu < \infty$, in which case we define

$$\int_E f(x)\,\mu(dx) = \int f\,d\mu = \int f^+\,d\mu - \int f^-\,d\mu$$

to be the **Lebesgue integral of f on E**. Observe that if $\int f\,d\mu$ exists, then so does $\int_\Gamma f\,d\mu$ for every $\Gamma \in \mathcal{B}$, and in fact

$$\int_{\Gamma_1 \cup \Gamma_2} f\,d\mu = \int_{\Gamma_1} f\,d\mu + \int_{\Gamma_2} f\,d\mu$$

if Γ_1 and Γ_2 are disjoint elements of $\mathcal{B}$. Also, it is clear that, when $\int f\,d\mu$ exists,

$$\left|\int f\,d\mu\right| \leq \int |f|\,d\mu. \tag{3.2.10}$$

In particular, $\int_\Gamma f\,d\mu = 0$ if $\mu(\Gamma) = 0$. Finally, when $\int f^+\,d\mu \wedge \int f^-\,d\mu = \infty$, we do not even attempt to define $\int f\,d\mu$.

Once again, we need a consistency result before we know for sure that our definition accomplishes what we wanted it to do; in this case, the preservation of linearity.

3.2.11 Lemma. *Let f and g be measurable functions on $(E, \mathcal{B}, \mu)$ for which $\int f\,d\mu$ and $\int g\,d\mu$ exist and $\left(\int f\,d\mu, \int g\,d\mu\right) \in \widehat{\mathbb{R}^2}$. Then $\mu(f \otimes g \notin \widehat{\mathbb{R}^2}) = 0$, $\int_{\{f\otimes g \in \widehat{\mathbb{R}^2}\}} (f + g)\,d\mu$ exists, and*

$$\int_{\{f\otimes g \in \widehat{\mathbb{R}^2}\}} (f + g)\,d\mu = \int f\,d\mu + \int g\,d\mu.$$

PROOF: Set $\hat{E} = \{x \in E : (f(x), g(x)) \in \widehat{\mathbb{R}^2}\}$.

Note that, under the stated conditions, either

$$\int f^+ \, d\mu \vee \int g^+ \, d\mu < \infty \quad \text{or} \quad \int f^- \, d\mu \vee \int g^- \, d\mu < \infty.$$

For definiteness, we will assume that $\int f^- \, d\mu \vee \int g^- \, d\mu < \infty$. As a consequence:

$$\mu(\hat{E}\complement) \leq \mu(f^- \vee g^- = \infty) \leq \mu(f^- = \infty) + \mu(g^- = \infty) = 0$$

and, because $(a+b)^- \leq a^- + b^-$ for $(a,b) \in \widehat{\mathbb{R}^2}$, $\int_{\hat{E}} (f+g)^- \, d\mu < \infty$. Hence, all that remains is to prove the asserted equality, and, while doing so, we may and will assume that $E = \hat{E}$.

Set $E^+ = \{f + g \geq 0\}$ and $E^- = \{f + g < 0\}$. Then

$$\int (f+g) \, d\mu \equiv \int (f+g)^+ \, d\mu - \int (f+g)^- \, d\mu = \int_{E^+} (f+g) \, d\mu - \int_{E^-} (f+g) \, d\mu.$$

By applying the last part of Lemma 3.2.7 to $(f^+ + g^+)\mathbf{1}_{E^+} - (f^- + g^-)\mathbf{1}_{E^+} = (f+g)\mathbf{1}_{E^+} \geq 0$, we see that

$$\int_{E^+} (f+g) \, d\mu = \int_{E^+} (f^+ + g^+) \, d\mu - \int_{E^+} (f^- + g^-) \, d\mu.$$

Thus, by the first part of Lemma 3.2.7,

$$\int_{E^+} (f+g) \, d\mu = \int_{E^+} f^+ \, d\mu + \int_{E^+} g^+ \, d\mu - \int_{E^+} f^- \, d\mu - \int_{E^+} g^- \, d\mu.$$

Similarly,

$$\int_{E^-} (f+g) \, d\mu = \int_{E^-} f^+ \, d\mu + \int_{E^-} g^+ \, d\mu - \int_{E^-} f^- \, d\mu - \int_{E^-} g^- \, d\mu.$$

Finally, after adding these two and again applying the first part of Lemma 3.2.7, we arrive at

$$\int (f+g) \, d\mu = \int f^+ \, d\mu - \int f^- \, d\mu + \int g^+ \, d\mu - \int g^- \, d\mu = \int f \, d\mu + \int g \, d\mu. \quad \square$$

Given a measurable function f on $(E, \mathcal{B}, \mu)$, define

$$\|f\|_{L^1(\mu)} = \int |f| \, d\mu.$$

We say that $f : E \longrightarrow \overline{\mathbb{R}}$ is **μ-integrable** if f is a measurable function on $(E, \mathcal{B})$ and $\|f\|_{L^1(\mu)} < \infty$; and we use $L^1(\mu) = L^1(E, \mathcal{B}, \mu)$ to denote the set of all $\mathbb{R}$-valued μ-integrable functions. Note that, from the integration theoretic standpoint, there is no loss of generality to assume that $f \in L^1(\mu)$ is $\mathbb{R}$-valued. Indeed, if f is a μ-integrable function, then $\mathbf{1}_{\{|f|<\infty\}} f \in L^1(\mu)$, $\|f - \mathbf{1}_{\{|f|<\infty\}} f\|_{L^1(\mu)} = 0$, and so integrals involving f and $\mathbf{1}_{\{|f|<\infty\}} f$ are indistinguishable. The main reason for insisting that f's in $L^1(\mu)$ be $\mathbb{R}$-valued is so that we have no problems taking linear combinations of them over $\mathbb{R}$. This simplifies the statement of results like the following.

3.2.12 Lemma. *For any measure space $(E, \mathcal{B}, \mu)$, $L^1(\mu)$ is a linear space and*

$$\|\alpha f + \beta g\|_{L^1(\mu)} \leq |\alpha| \, \|f\|_{L^1(\mu)} + |\beta| \, \|g\|_{L^1(\mu)} \tag{3.2.13}$$

whenever $\alpha, \beta \in \mathbb{R}$ and $f, g \in L^1(\mu)$.

PROOF: Simply note that $|\alpha f + \beta g| \le |\alpha|\,|f| + |\beta|\,|g|$. □

3.2.14 Remark: As an application of the preceding inequality, we have that if $f, g, h \in L^1(\mu)$, then

$$\|f - h\|_{L^1(\mu)} \le \|f - g\|_{L^1(\mu)} + \|g - h\|_{L^1(\mu)}. \tag{3.2.15}$$

To see this, take $\alpha = \beta = 1$ and replace f and g by $f - g$ and $g - h$ in (3.2.13). Thus $\|f - g\|_{L^1(\mu)}$ looks like a good candidate to be chosen as a metric on $L^1(\mu)$. On the other hand, although, from the standpoint of integration theory, a measurable f for which $\|f\|_{L^1(\mu)} = 0$ might as well be identically 0, there is, in general, no reason why f need be identically 0 as a function. This fact prevents $\|\cdot\|_{L^1(\mu)}$ from being a completely satisfactory measure of size. To overcome this problem, we *quotient out by the offending subspace*. Namely, denote by $\mathcal{N}(\mu)$ the set of $f \in L^1(\mu)$ such that $\mu(f \ne 0) = 0$ (cf. Exercise 3.2.16 below), and, for $f, g \in L^1(\mu)$, write $f \overset{\mu}{\sim} g$ if $g - f \in \mathcal{N}(\mu)$. Since it is clear that $\mathcal{N}(\mu)$ is a linear subspace of $L^1(\mu)$, one sees that $\overset{\mu}{\sim}$ is an equivalence relation and that the quotient space $L^1(\mu)/\overset{\mu}{\sim}$ is again a vector space over $\mathbb{R}$. To be precise, for $f \in L^1(\mu)$, we use $[f]^{\overset{\mu}{\sim}}$ to denote the $\overset{\mu}{\sim}$-equivalence class of f; and, for any $f, g \in L^1(\mu)$ and $\alpha, \beta \in \mathbb{R}$, we take

$$\alpha[f]^{\overset{\mu}{\sim}} + \beta[g]^{\overset{\mu}{\sim}} = [\alpha f + \beta g]^{\overset{\mu}{\sim}}.$$

Finally, since

$$f \overset{\mu}{\sim} g \implies |f| \overset{\mu}{\sim} |g| \implies \|f\|_{L^1(\mu)} = \|g\|_{L^1(\mu)},$$

we can define $\|[f]^{\overset{\mu}{\sim}}\|_{L^1(\mu)} = \|f\|_{L^1(\mu)}$ and thereby turn $L^1(\mu)/\overset{\mu}{\sim}$ into a bona fide metric space in which the distance between $[f]^{\overset{\mu}{\sim}}$ and $[g]^{\overset{\mu}{\sim}}$ is given by $\|f - g\|_{L^1(\mu)}$.

Having made this obeisance to rigor, we will now lapse into the usual, more casual practice of ignoring the niceties just raised. Thus, unless there is particular danger in doing so, we will not stress the distinction between f as a function and the equivalence class $[f]^{\overset{\mu}{\sim}}$, which f determines. For this reason, we will continue to write $L^1(\mu)$, even when we mean $L^1(\mu)/\overset{\mu}{\sim}$, and we will use f instead of $[f]^{\overset{\mu}{\sim}}$. In particular, $L^1(\mu)$ becomes is this way a vector space over $\mathbb{R}$ on which $\|f - g\|_{L^1(\mu)}$ is a metric. As we will see in the next section (cf. Corollary 3.3.13), this metric space is complete.

Exercises

3.2.16 Exercise: Let f be an $\overline{\mathbb{R}}$-valued function on the measurable space $(E, \mathcal{B})$. Show that f is measurable if and only if $\{f > a\} \in \mathcal{B}$ for every $a \in \mathbb{R}$ if and only if $\{f \ge a\} \in \mathcal{B}$ for every $a \in \mathbb{R}$. At the same time, check that "$>$"

and "$\leq$" can be replaced by "$<$" and "$\geq$", respectively. In fact, show that one can restrict one's attention to a's from a dense subset of $\mathbb{R}$. Finally, if g is a second $\overline{\mathbb{R}}$-valued measurable function on $(E, \mathcal{B})$, show that each of the sets $\{f < g\}$, $\{f \leq g\}$, $\{f = g\}$, and $\{f \neq g\}$ is an element of $\mathcal{B}$.

3.2.17 Exercise: Show that an integrable function is determined, up to a set of measure 0, by its integrals. To be more precise, let $(E, \mathcal{B}, \mu)$ be a measure space and f and g a pair of functions from $L^1(\mu)$. Show that $f \overset{\mu}{\sim} g$ if and only if $\int_\Gamma f\, d\mu = \int_\Gamma g\, d\mu$ for each $\Gamma \in \mathcal{B}$.

Hint: Reduce the problem to showing that if $\varphi \in L^1(\mu)$, then $\mu(\{\varphi < 0\}) = 0$ if and only if $\int_\Gamma \varphi\, d\mu \geq 0$ for every $\Gamma \in \mathcal{B}$.

3.2.18 Exercise: Referring to the example (**iii**) in Examples 3.1.5, show that if E is any set and μ is the counting measure on E, then $f : E \longrightarrow \mathbb{R}$ is μ-integrable if and only if $S(f) \equiv \{x \in E : f(x) \neq 0\}$ is countable and $\sum_{x \in S(f)} |f(x)| < \infty$.

3.3 Convergence of Integrals

One of the distinct advantages that Lebesgue's theory of integration has over Riemann's approach is that Lebesgue's integral is wonderfully continuous with respect to convergence of integrands. In the present section we will explore some of these continuity properties. We begin by showing that the class of measurable functions is closed under pointwise convergence.

3.3.1 Lemma. *Let $(E, \mathcal{B})$ be a measurable space and $\{f_n\}_1^\infty$ a sequence of measurable functions on $(E, \mathcal{B})$. Then $\sup_{n \geq 1} f_n$, $\inf_{n \geq 1} f_n$, $\overline{\lim}_{n\to\infty} f_n$, and $\underline{\lim}_{n\to\infty} f_n$ are all measurable functions. In particular,*

$$\Delta \equiv \left\{x \in E : \lim_{n\to\infty} f_n(x) \text{ exists}\right\} \in \mathcal{B},$$

and the function f given by

$$f(x) = \begin{cases} 0 & \text{if } x \notin \Delta \\ \lim_{n\to\infty} f_n(x) & \text{if } x \in \Delta \end{cases}$$

is measurable on $(E, \mathcal{B})$.

Proof: We first suppose that $\{f_n\}_1^\infty$ is non-decreasing. It is then clear that

$$\left\{\lim_{n\to\infty} f_n > a\right\} = \bigcup_{n=1}^\infty \{f_n > a\} \in \mathcal{B}, \qquad a \in \mathbb{R},$$

and therefore (cf. Exercise 3.2.16) that $\lim_{n\to\infty} f_n$ is measurable. By replacing f_n with $-f_n$, we see that the same conclusion holds in the case when $\{f_n\}_1^\infty$ is non-increasing.

Next, for an arbitrary sequence $\{f_n\}_1^\infty$ of measurable functions,

$$\{f_1 \vee \cdots \vee f_n : n \in \mathbb{Z}^+\}$$

is a non-decreasing sequence of measurable functions. Hence, by the preceding,

$$\sup_{n\geq 1} f_n = \lim_{n\to\infty} (f_1 \vee \cdots \vee f_n)$$

is measurable; and a similar argument shows that $\inf_{n\geq 1} f_n$ is measurable. Noting that $\inf_{n\geq m} f_n$ does not decrease as m increases, we also see that

$$\varliminf_{n\to\infty} f_n = \lim_{m\to\infty} \inf_{n\geq m} f_n$$

is measurable; and, of course, the same sort of reasoning leads to the measurability of $\varlimsup_{n\to\infty} f_n$.

Finally, since (cf. Exercise 3.2.16)

$$\Delta \equiv \left\{x \in E : \lim_{n\to\infty} f_n(x) \text{ exists}\right\} = \left\{\varlimsup_{n\to\infty} f_n = \varliminf_{n\to\infty} f_n\right\},$$

it is an element of $\mathcal{B}$; and from this it is clear that the function f described in the last part of the statement is measurable. □

We are now ready to prove the first of three basic continuity theorems about the Lebesgue integral. In some ways the first one is the least surprising in that it really only echoes the result obtained in Lemma 3.2.6 and is nothing more than the function version of (**i**) in Theorem 3.1.6.

3.3.2 Theorem (The Monotone Convergence Theorem). *If $\{f_n\}_1^\infty$ is a sequence of non-negative, measurable functions on the measure space $(E, \mathcal{B}, \mu)$ and if $f_n \nearrow f$ point-wise as $n \to \infty$, then $\int f\, d\mu = \lim_{n\to\infty} \int f_n\, d\mu$.*

PROOF: Obviously $\int f\, d\mu \geq \lim_{n\to\infty} \int f_n\, d\mu$. To prove the opposite inequality, for each $m \geq 1$ choose a non-decreasing sequence $\{\varphi_{m,n}\}_{n=1}^\infty$ of non-negative, measurable, simple functions so that $\varphi_{m,n} \nearrow f_m$ as $n \to \infty$. Next, define the non-negative, simple, measurable functions $\psi_n = \varphi_{1,n} \vee \cdots \vee \varphi_{n,n}$ for $n \geq 1$. One then has that

$$\psi_n \leq \psi_{n+1} \text{ and } \varphi_{m,n} \leq \psi_n \leq f_n \quad \text{for all } 1 \leq m \leq n;$$

and therefore

$$f_m \leq \lim_{n\to\infty} \psi_n \leq f \quad \text{for each } m \in \mathbb{Z}^+.$$

In particular, $\psi_n \nearrow f$, and therefore

$$\int f\,d\mu = \lim_{n\to\infty} \int \psi_n\,d\mu.$$

At the same time, because $\psi_n \le f_n$ for all $n \in \mathbb{Z}^+$, we know that

$$\int f\,d\mu \le \varliminf_{n\to\infty} \int f_n\,d\mu. \quad \square$$

Being an inequality instead of an equality, the second continuity result is often more useful than the other two. It is the function version of (**i**) and (**ii**) of Exercise 3.1.12.

3.3.3 Theorem (Fatou's Lemma). *Let $\{f_n\}_1^\infty$ be a sequence of functions on the measure space $(E, \mathcal{B}, \mu)$. If $f_n \ge 0$ for all $n \ge 1$, then*

$$\int \varliminf_{n\to\infty} f_n\,d\mu \le \varliminf_{n\to\infty} \int f_n\,d\mu.$$

In particular, if there exists a μ-integrable function g such that $f_n \le g$ for all $n \ge 1$, then

$$\int \varlimsup_{n\to\infty} f_n\,d\mu \ge \varlimsup_{n\to\infty} \int f_n\,d\mu.$$

PROOF: Assume that the f_n's are non-negative. To check the first assertion, set $h_m = \inf_{n\ge m} f_n$. Then $f_m \ge h_m \nearrow \varliminf_{n\to\infty} f_n$ and so, by the Monotone Convergence Theorem,

$$\int \varliminf_{n\to\infty} f_n\,d\mu = \lim_{m\to\infty} \int h_m\,d\mu \le \varliminf_{m\to\infty} \int f_m\,d\mu.$$

Next, drop the non-negativity assumption, but impose $f_n \le g$ for some μ-integrable g. Clearly, $f_n' \equiv g - f_n$ is non-negative,

$$\varliminf_{n\to\infty} f_n' = g - \varlimsup_{n\to\infty} f_n \quad\text{and}\quad \varliminf_{n\to\infty} \int f_n'\,d\mu = \int g\,d\mu - \varlimsup_{n\to\infty} \int f_n\,d\mu.$$

Hence, the required result follows from the first part applied to $\{f_n'\}_1^\infty$. $\square$

Before stating the third continuity result, we need to introduce a notion which is better suited to measure theory than ordinary pointwise equality. Namely, we will say that an x-dependent statement about quantities on the measure space $(E, \mathcal{B}, \mu)$ holds μ-**almost everywhere** if the set Δ of x for which the statement fails is an element of $\overline{\mathcal{B}}^\mu$ which has $\overline{\mu}$-**measure** 0 (i.e., $\overline{\mu}(\Delta) = 0$). Thus, if $\{f_n\}_1^\infty$ is a sequence of measurable functions on the measure space $(E, \mathcal{B}, \mu)$, we will say that $\{f_n\}_1^\infty$ **converges** μ-**almost everywhere** and will write $\lim_{n\to\infty} f_n$ exists (a.e., μ), if

$$\mu(\{x \in E : \lim_{n\to\infty} f_n(x) \text{ does not exist}\}) = 0;$$

also, we will say that $\{f_n\}_1^\infty$ **converges μ-almost everywhere** to f, and will write $f_n \longrightarrow f$ (a.e., μ), if $\mu(\{x \in E : f(x) \neq \lim_{n\to\infty} f_n(x)\}) = 0$. By Lemma 3.3.1, we see that if $\{f_n\}_1^\infty$ converges μ-almost everywhere, then there is a measurable f to which $\{f_n\}_1^\infty$ converges μ-almost surely. Similarly, if f and g are measurable functions, we write $f = g$ (a.e., μ), $f \leq g$ (a.e., μ), or $f \geq g$ (a.e., μ) if $\mu(f \neq g) = 0$, $\mu(f > g) = 0$, or $\mu(f < g) = 0$, respectively. Note that $f = g$ (a.e., μ) is the same statement as $f \overset{\mu}{\sim} g$, discussed in Remark 3.2.14.

The following can be thought of as the function version of (**iii**) of Exercise 3.1.12.

3.3.4 Theorem (**Lebesgue's Dominated Convergence Theorem**). *Suppose that $\{f_n\}_1^\infty$ is a sequence of measurable functions on $(E, \mathcal{B}, \mu)$, and let f be a measurable function to which $\{f_n\}_1^\infty$ converges μ-almost everywhere. If there is a μ-integrable function g for which $|f_n| \leq g$ (a.e., μ), $n \geq 1$, then f is integrable and $\lim_{n\to\infty} \int |f_n - f|\, d\mu = 0$. In particular, $\int f\, d\mu = \lim_{n\to\infty} \int f_n\, d\mu$.*

PROOF: Let $\widehat{E}$ be the set of $x \in E$ for which $f(x) = \lim_{n\to\infty} f_n(x)$ and $\sup_{n\geq 1} |f_n(x)| \leq g(x)$. Then $\widehat{E}$ is measurable and $\mu(\widehat{E}\complement) = 0$, and so integrals over $\widehat{E}$ are the same as those over E. Thus, without loss of generality, we will assume that all our assumptions hold for every $x \in E$. But then, $f = \lim_{n\to\infty} f_n$, $|f| \leq g$ and $|f - f_n| \leq 2g$. Hence, by the second part of Fatou's Lemma,

$$\varlimsup_{n\to\infty} \left| \int f\, d\mu - \int f_n\, d\mu \right| \leq \varlimsup_{n\to\infty} \int |f - f_n|\, d\mu \leq \int \varlimsup_{n\to\infty} |f - f_n|\, d\mu = 0. \quad \square$$

It is important to understand the role played by the *Lebesgue dominant*. Namely, it acts as an *umbrella* to keep everything under control. To see that such control is important, consider Lebesgue measure $\lambda_{[0,1]}$ on $\big([0,1], \mathcal{B}_{[0,1]}\big)$ and the functions $f_n = n\mathbf{1}_{(0,n^{-1})}$. Obviously, $f_n \longrightarrow 0$ everywhere on $[0,1]$, but $\|f_n\|_1 = 1$ for all $n \in \mathbb{Z}^+$. Unfortunately, in many circumstances, it is difficult to find an appropriate Lebesgue dominant, and, for this reason, results like the following variation on Fatou's Lemma are interesting and often useful. (See Exercise 3.3.20 for other variations.)

3.3.5 Theorem (**Lieb's Version of Fatou's Lemma**). *Let $(E, \mathcal{B}, \mu)$ be a measure space, $\{f_n\}_1^\infty \cup \{f\} \subseteq L^1(\mu)$, and assume that $f_n \longrightarrow f$ (a.e., μ). Then*

$$\begin{aligned} &\lim_{n\to\infty} \Big| \|f_n\|_{L^1(\mu)} - \|f\|_{L^1(\mu)} - \|f_n - f\|_{L^1(\mu)} \Big| \\ &\qquad = \lim_{n\to\infty} \int \Big| |f_n| - |f| - |f_n - f| \Big|\, d\mu = 0. \end{aligned} \tag{3.3.6}$$

In particular, if $\|f_n\|_{L^1(\mu)} \longrightarrow \|f\|_{L^1(\mu)} < \infty$, then $\|f - f_n\|_{L^1(\mu)} \longrightarrow 0$.

PROOF: Since

$$\Big|\|f_n\|_{L^1(\mu)} - \|f\|_{L^1(\mu)} - \|f_n - f\|_{L^1(\mu)}\Big| \le \int \big||f_n| - |f| - |f_n - f|\big|\, d\mu, \quad n \ge 1,$$

we need only check the second equality in (3.3.6). But, because

$$\big||f_n| - |f| - |f_n - f|\big| \longrightarrow 0 \quad \text{(a.e., } \mu)$$

and

$$\big||f_n| - |f| - |f_n - f|\big| \le \big||f_n| - |f_n - f|\big| + |f| \le 2|f|,$$

(3.3.6) follows from Lebesgue's Dominated Convergence Theorem. □

We now have a great deal of evidence that almost everywhere convergence of integrands often leads to convergence of the corresponding integrals. We next want to see what can be said about the converse implication. To begin with, we point out that $\|f_n\|_{L^1(\mu)} \longrightarrow 0$ *does not* imply that $f_n \longrightarrow 0$ (a.e., μ). Indeed, define the functions $\{f_n\}_1^\infty$ on $[0, 1]$ so that, for $m \ge 0$ and $0 \le \ell < 2^m$,

$$f_{2^m+\ell} = \mathbf{1}_{[2^{-m}\ell, 2^{-m}(\ell+1)]}.$$

It is then clear that these f_n 's are non-negative and measurable on $([0, 1], \mathcal{B}_{[0,1]})$ and that $\overline{\lim}_{n\to\infty} f_n(x) = 1$ for every $x \in [0, 1]$. On the other hand,

$$\int_{[0,1]} f_n(x)\, dx = 2^{-m} \quad \text{if } 2^m \le n < 2^{m+1},$$

and therefore $\int_{[0,1]} f_n(x)\, dx \longrightarrow 0$ as $n \to \infty$.

The preceding discussion makes it clear that it may be useful to consider other notions of convergence. Keeping in mind that we are looking for a type of convergence which can be tested using integrals, we take a hint from Markov's inequality (cf. Theorem 3.2.8) and say that the sequence $\{f_n\}_1^\infty$ of measurable functions on the measure space $(E, \mathcal{B}, \mu)$ **converges in μ-measure** to the measurable function f if $\mu(|f_n - f| \ge \epsilon) \longrightarrow 0$ as $n \to \infty$ for every $\epsilon > 0$, in which case we write $f_n \longrightarrow f$ **in μ-measure**. Note that, by Markov's inequality (3.2.9), if $\|f_n - f\|_{L^1(\mu)} \longrightarrow 0$ then $f_n \longrightarrow f$ in μ-measure. Hence, this sort of convergence can be easily tested with integrals (cf. Exercise 3.3.22 below); and, as a consequence, we see that convergence in μ-measure certainly does not imply convergence μ-almost everywhere. In fact, it takes a moment to see in what sense the limit is even uniquely determined by convergence in μ-measure. For this reason, suppose that $\{f_n\}_1^\infty$ converges to both f and to g in μ-measure. Then, for $\epsilon > 0$,

$$\mu(|f - g| \ge \epsilon) \le \mu\left(|f - f_n| \ge \tfrac{\epsilon}{2}\right) + \mu\left(|f_n - g| \ge \tfrac{\epsilon}{2}\right) \longrightarrow 0.$$

Hence,

$$\mu(f \ne g) = \lim_{\epsilon \searrow 0} \mu(|f - g| \ge \epsilon) = \lim_{n\to\infty} \mu(|f - g| \ge n^{-1}) = 0,$$

and so $f = g$ (a.e., μ). That is, convergence in μ-measure determines the limit function to precisely the same extent as does either μ-almost everywhere or $\|\cdot\|_{L^1(\mu)}$-convergence. In particular, from the standpoint of μ-integration theory, convergence in μ-measure has unique limits.

The following theorems should help to elucidate the notions of μ-almost everywhere convergence, convergence in μ-measure, and the relations between them.

3.3.7 Theorem. *Let $\{f_n\}_1^\infty$ be a sequence of $\mathbb{R}$-valued measurable functions on the measure space $(E, \mathcal{B}, \mu)$. Then there is an $\mathbb{R}$-valued, measurable function f for which*

$$\lim_{m\to\infty} \mu\left(\sup_{n\geq m} |f - f_n| \geq \epsilon\right) = 0, \quad \epsilon > 0, \tag{3.3.8}$$

if and only if

$$\lim_{m\to\infty} \mu\left(\sup_{n\geq m} |f_n - f_m| \geq \epsilon\right) = 0, \quad \epsilon > 0. \tag{3.3.9}$$

Moreover, (3.3.8) implies that $f_n \longrightarrow f$ both (a.e., μ) and in μ-measure. Finally, when $\mu(E) < \infty$, $f_n \longrightarrow f$ (a.e., μ) if and only if (3.3.8) holds. That is, μ-almost every convergence implies convergence in μ-measure on a finite measure space.

PROOF: Set

$$\Delta = \left\{x \in E : \lim_{n\to\infty} f_n(x) \text{ does not exist in } \mathbb{R}\right\}.$$

For $m \geq 1$ and $\epsilon > 0$, define $\Delta_m(\epsilon) = \left\{\sup_{n\geq m} |f_n - f_m| \geq \epsilon\right\}$. It is then easy to check (from Cauchy's convergence criterion for $\mathbb{R}$) that

$$\Delta = \bigcup_{\ell=1}^{\infty} \bigcap_{m=1}^{\infty} \Delta_m\left(\tfrac{1}{\ell}\right).$$

Since (3.3.9) implies that $\mu\left(\bigcap_{m=1}^\infty \Delta_m(\epsilon)\right) = 0$ for every $\epsilon > 0$, and, by the preceding,

$$\mu(\Delta) \leq \sum_{\ell=1}^{\infty} \mu\left(\bigcap_{m=1}^{\infty} \Delta_m\left(\tfrac{1}{\ell}\right)\right),$$

we see that (3.3.9) does indeed imply that $\{f_n\}_1^\infty$ converges μ-almost everywhere. In addition, if (cf. the last part of Lemma 3.3.1) f is an $\mathbb{R}$-valued, measurable function to which $\{f_n\}_1^\infty$ converges μ-almost everywhere, then

$$\sup_{n\geq m} |f_n - f| \leq \sup_{n\geq m} |f_n - f_m| + |f_m - f| \leq 2 \sup_{n\geq m} |f_n - f_m| \quad \text{(a.e., } \mu\text{)};$$

and so (3.3.9) leads to the existence of an f for which (3.3.8) holds.

Next, suppose that (3.3.8) holds for some f. Then it is obvious that $f_n \longrightarrow f$ both (a.e., μ) and in μ-measure. In addition, (3.3.9) follows immediately from

$$\mu\left(\sup_{n\geq m}|f_n - f_m| \geq \epsilon\right) \leq \mu\left(\sup_{n\geq m}|f_n - f| \geq \tfrac{\epsilon}{2}\right) + \mu\left(\sup_{n\geq m}|f - f_m| \geq \tfrac{\epsilon}{2}\right).$$

Finally, suppose that $\mu(E) < \infty$ and that $f_n \longrightarrow f$ (a.e, μ). Then, by **(ii)** of Theorem 3.1.6,

$$\lim_{m\to\infty}\mu\left(\sup_{n\geq m}|f_n - f| \geq \epsilon\right) = \mu\left(\bigcap_{m=1}^{\infty}\left\{\sup_{n\geq m}|f_n - f| \geq \epsilon\right\}\right) = 0$$

for every $\epsilon > 0$, and therefore (3.3.8) holds. In particular, this means that $f_n \longrightarrow f$ in μ-measure. □

Clearly, the first part of Theorem 3.3.7 provides a Cauchy criterion for μ-almost everywhere convergence. The following theorem gives a Cauchy criterion for convergence in μ-measure. In the process, it shows that, after passing to a subsequence, convergence in μ-measure leads to μ-almost everywhere convergence.

3.3.10 Theorem. *Again let $\{f_n\}_1^\infty$ be a sequence of $\mathbb{R}$-valued, measurable functions on the measure space $(E, \mathcal{B}, \mu)$. Then there is an $\mathbb{R}$-valued, measurable function f to which $\{f_n\}_1^\infty$ converges in μ-measure if and only if*

$$\lim_{m\to\infty}\sup_{n\geq m}\mu\big(|f_n - f_m| \geq \epsilon\big) = 0, \qquad \epsilon > 0. \tag{3.3.11}$$

Furthermore, if $f_n \longrightarrow f$ in μ-measure, then there is a subsequence $\{f_{n_j}\}_{j=1}^\infty$ with the property that

$$\lim_{i\to\infty}\mu\left(\sup_{j\geq i}|f - f_{n_j}| \geq \epsilon\right) = 0, \qquad \epsilon > 0;$$

and therefore $f_{n_i} \longrightarrow f$ (a.e., μ) as well as in μ-measure.

PROOF: To see that $f_n \longrightarrow f$ in μ-measure implies (3.3.11), simply note that

$$\mu\big(|f_n - f_m| \geq \epsilon\big) \leq \mu\left(|f - f_n| \geq \tfrac{\epsilon}{2}\right) + \mu\left(|f - f_m| \geq \tfrac{\epsilon}{2}\right).$$

Conversely, assume that (3.3.11) holds, and choose $1 \leq n_1 < \cdots < n_i < \cdots$ so that

$$\sup_{n\geq n_i}\mu\Big(|f_n - f_{n_i}| \geq 2^{-i-1}\Big) \leq 2^{-i-1}, \qquad i \geq 1.$$

Then

$$\begin{aligned}\mu\left(\sup_{j\geq i}|f_{n_j} - f_{n_i}| > 2^{-i}\right) &\leq \mu\left(\bigcup_{j\geq i}\left\{|f_{n_{j+1}} - f_{n_j}| \geq 2^{-j-1}\right\}\right)\\ &\leq \sum_{j=i}^{\infty}\mu\big(|f_{n_{j+1}} - f_{n_j}| \geq 2^{-j-1}\big) \leq 2^{-i}.\end{aligned}$$

From this it is clear that $\{f_{n_i}\}_{i=1}^\infty$ satisfies (3.3.9) and therefore that there is an f for which (3.3.8) holds with $\{f_n\}$ replaced by $\{f_{n_i}\}$. Hence, $f_{n_i} \longrightarrow f$ both μ-almost everywhere and in μ-measure. In particular, when combined with (3.3.11), this means that

$$\begin{aligned}\mu(|f_m - f| \geq \epsilon) &\leq \overline{\lim_{i\to\infty}}\, \mu\left(|f_m - f_{n_i}| \geq \tfrac{\epsilon}{2}\right) + \overline{\lim_{i\to\infty}}\, \mu\left(|f_{n_i} - f| \geq \tfrac{\epsilon}{2}\right) \\ &\leq \sup_{n\geq m} \mu\left(|f_n - f_m| \geq \tfrac{\epsilon}{2}\right) \longrightarrow 0\end{aligned}$$

as $m \to \infty$; and so $f_n \longrightarrow f$ in μ-measure.

Notice that the preceding argument proves the final statement as well. Namely, if $f_n \longrightarrow f$ in μ-measure, then (3.3.11) holds and therefore our argument shows that there exists a subsequence $\{f_{n_i}\}_{i=1}^\infty$ which satisfies (3.3.9). But this means that $\{f_{n_i}\}_{i=1}^\infty$ converges both (a.e., μ) and in μ-measure, and, as a subsequence of a sequence which is already converging in μ-measure to f, we conclude that f must be the function to which $\{f_{n_i}\}_1^\infty$ is converging (a.e., μ). □

Because it is quite important to remember the relationships between the various sorts of convergence discussed in Theorems 3.3.7 and 3.3.10, we will summarize them as follows:

$$\begin{aligned}&\|f_n - f\|_{L^1(\mu)} \longrightarrow 0 \Longrightarrow f_n \longrightarrow f \text{ in } \mu\text{-measure} \\ \Longrightarrow &\lim_{i\to\infty} \mu\left(\sup_{j\geq i} |f_{n_j} - f| \geq \epsilon\right) = 0,\ \epsilon > 0, \text{ for some subsequence } \{f_{n_i}\} \\ &\Longrightarrow f_{n_i} \longrightarrow f \text{ (a.e., } \mu)\end{aligned}$$

and

$$\mu(E) < \infty \text{ and } f_n \longrightarrow f \text{ (a.e., } \mu) \implies f_n \longrightarrow f \text{ in } \mu\text{-measure.}$$

Notice that, when $\mu(E) = \infty$, μ-almost everywhere convergence *does not* imply μ-convergence. For example, consider the functions $\mathbf{1}_{[n,\infty)}$ on $\mathbb{R}$ with Lebesgue measure.

We next show that, at least as far as Theorems 3.3.3 through 3.3.5 are concerned, convergence in μ-measure is just as good as μ-almost everywhere convergence.

3.3.12 Theorem. *Let f and $\{f_n\}_1^\infty$ all be measurable, $\mathbb{R}$-valued functions on the measure space $(E, \mathcal{B}, \mu)$, and assume that $f_n \longrightarrow f$ in μ-measure.*

Fatou's Lemma: *If $f_n \geq 0$ (a.e., μ) for each $n \geq 1$, then $f \geq 0$ (a.e., μ) and*

$$\int f\, d\mu \leq \underline{\lim_{n\to\infty}} \int f_n\, d\mu.$$

Lebesgue's Dominated Convergence Theorem: *If there is an integrable g on $(E, \mathcal{B}, \mu)$ such that $|f_n| \leq g$ (a.e., μ) for each $n \geq 1$, then f is integrable, $\lim_{n\to\infty} \|f_n - f\|_{L^1(\mu)} = 0$, and so $\int f_n\, d\mu \longrightarrow \int f\, d\mu$ as $n \to \infty$.*

Lieb's Version of Fatou's Lemma: *If $\sup_{n\geq 1}\|f_n\|_{L^1(\mu)} < \infty$, then f is integrable and*

$$\lim_{n\to\infty}\Big|\|f_n\|_{L^1(\mu)}-\|f\|_{L^1(\mu)}-\|f_n-f\|_{L^1(\mu)}\Big| = \lim_{n\to\infty}\big\||f_n|-|f|-|f_n-f|\big\|_{L^1(\mu)} = 0$$

In particular, $\|f_n - f\|_{L^1(\mu)} \longrightarrow 0$ if $\|f_n\|_{L^1(\mu)} \longrightarrow \|f\|_{L^1(\mu)} \in \mathbb{R}$.

PROOF: Each of these results is obtained via the same trick from the corresponding result for μ-almost everywhere convergent sequences. Thus we will prove the preceding statement of Fatou's Lemma and will leave the proofs of the other assertions to the reader.

Assume that the f_n 's are non-negative μ-almost everywhere and that $f_n \longrightarrow f$ in μ-measure. Choose a subsequence $\{f_{n_m}\}$ so that $\lim_{m\to\infty}\int f_{n_m}\,d\mu = \underline{\lim}_{n\to\infty}\int f_n\,d\mu$. Next, choose a subsequence $\{f_{n_{m_i}}\}$ of $\{f_{n_m}\}$ so that $f_{n_{m_i}} \longrightarrow f$ (a.e., μ). Because each of the $f_{n_{m_i}}$ is non-negative (a.e., μ), it is now clear that $f \geq 0$ (a.e., μ). In addition, by restricting all integrals to the set $\widehat{E}$ on which the $f_{n_{m_i}}$'s are non-negative and $f_{n_{m_i}} \longrightarrow f$, we can apply Theorem 3.3.3 to obtain

$$\int f\,d\mu = \int_{\hat{E}} f\,d\mu \leq \underline{\lim}_{i\to\infty}\int_{\hat{E}} f_{n_{m_i}}\,d\mu = \lim_{m\to\infty}\int f_{n_m}\,d\mu = \underline{\lim}_{n\to\infty}\int f_n\,d\mu. \quad \square$$

An important dividend of these considerations is the fact that $L^1(\mu)$ is a *complete metric space.* More precisely, we have the following corollary.

3.3.13 Corollary. *Let $\{f_n\}_1^\infty \subseteq L^1(\mu)$. If*

$$\lim_{m\to\infty}\sup_{n\geq m}\|f_n - f_m\|_{L^1(\mu)} = 0,$$

then there exists an $f \in L^1(\mu)$ such that $\|f_n - f\|_{L^1(\mu)} \longrightarrow 0$. In other words, $\left(L^1(\mu), \|\cdot\|_{L^1(\mu)}\right)$ is a complete metric space.

PROOF: By Markov's inequality, we see that (3.3.11) holds. Hence, we can find a measurable f such that $f_n \longrightarrow f$ in μ-measure; and so, by Fatou's Lemma,

$$\|f - f_m\|_{L^1(\mu)} \leq \underline{\lim}_{n\to\infty}\|f_n - f_m\|_{L^1(\mu)} \leq \sup_{n\geq m}\|f_n - f_m\|_{L^1(\mu)} \longrightarrow 0$$

as $m \to \infty$. Finally, since $\sup_{n\geq 1}\|f_n\|_{L^1(\mu)} < \infty$, we also see that f is μ-integrable and therefore may be assumed to be $\mathbb{R}$-valued. $\square$

Before closing this discussion, we want to prove a result which is not only useful but also helps to elucidate the structure of $L^1(\mu)$.

3.3.14 Theorem. *Let $(E, \mathcal{B}, \mu)$ be a measure space, and assume that $\mu(E) < \infty$. Given a π-system $\mathcal{C} \subseteq \mathcal{P}(E)$ which generates $\mathcal{B}$, denote by $\mathcal{S}$ the set of functions $\sum_{m=1}^n \alpha_m \mathbf{1}_{\Gamma_m}$, where $n \in \mathbb{Z}^+$, $\{\alpha_m\}_1^n \subseteq \mathbb{Q}$, and $\{\Gamma_m\}_1^n \subseteq \mathcal{C}\cup\{E\}$. Then $\mathcal{S}$ is dense in $L^1(\mu)$. In particular, if $\mathcal{C}$ is countable, then $L^1(\mu)$ is a separable metric space.*

PROOF: Denote by $\overline{\mathcal{S}}$ the closure in $L^1(\mu)$ of $\mathcal{S}$. It is then easy to see that $\overline{\mathcal{S}}$ is a vector space over $\mathbb{R}$. In particular, if $f \in L^1(\mu)$ and both f^+ and f^- are elements of $\overline{\mathcal{S}}$, then $f \in \overline{\mathcal{S}}$. Hence, we need only check that every non-negative $f \in L^1(\mu)$ is in $\overline{\mathcal{S}}$. Since every such f is the limit in $L^1(\mu)$ of simple elements of $L^1(\mu)$ and since $\overline{\mathcal{S}}$ is a vector space, we now see that it suffices to check that $\mathbf{1}_\Gamma \in \overline{\mathcal{S}}$ for every $\Gamma \in \mathcal{B}$. But it is easy to see that the class of $\Gamma \subseteq E$ for which $\mathbf{1}_\Gamma \in \overline{\mathcal{S}}$ is a λ-system over E, and, by hypothesis, it contains the π-system $\mathcal{C}$. Now apply Lemma 3.1.3. □

3.3.15 Corollary. *Let (E, ρ) be a metric space, and suppose that μ is a measure on $(E, \mathcal{B}_E)$ with the property that there exists a non-decreasing sequence of open sets E_n such that $\mu(E_n) < \infty$ for each $n \geq 1$ and $E = \bigcup_1^\infty E_n$. For each $n \in \mathbb{Z}^+$, denote by $\mathcal{K}_n$ the set of bounded, ρ-uniformly continuous functions φ which vanish identically off E_n, and set $\mathcal{K} = \bigcup_{n \in \mathbb{Z}^+} \mathcal{K}_n$. Then $\mathcal{K}$ is dense in $L^1(\mu)$.*

PROOF: We will show first that, for each $n \in \mathbb{Z}^+$,

$$\widetilde{\mathcal{K}}_n \equiv \{\varphi \upharpoonright E_n : \varphi \in \mathcal{K}_n\}$$

is dense in $L^1(E_n, \mathcal{B}_{E_n}, \mu_n)$, where μ_n denotes the restriction of μ to $\mathcal{B}_{E_n}$. In view of Theorem 3.3.13, this will follow as soon as we show that $\mathbf{1}_G$ is in the $\|\cdot\|_{L^1(\mu_n)}$-closure of $\widetilde{\mathcal{K}}_n$ for each open $G \subseteq E_n$. If $G \subseteq E_n$ is open, define

$$\varphi_m(x) = \left(\frac{\rho(x, G\complement)}{1 + \rho(x, G\complement)}\right)^{\frac{1}{m}}, \quad m \geq 1,$$

where $\rho(x, F) \equiv \inf\{\rho(x, y) : y \in F\} (\equiv \infty$ when $F = \emptyset)$. Since $\rho(\,\cdot\,, F)$ is ρ-uniformly continuous, we see that φ_m is ρ-uniformly continuous. In addition, it is easy to check that $0 \leq \varphi_m \nearrow \mathbf{1}_G$ as $m \to \infty$. Hence, by the Monotone Convergence Theorem, it follows that $\varphi_m \upharpoonright E_n \longrightarrow \mathbf{1}_G \upharpoonright E_n$ in $L^1(E_n, \mathcal{B}_{E_n}, \mu_n)$ as $m \to \infty$.

By the preceding, we now know that if $f \in L^1(\mu)$ vanishes identically off E_n for some $n \geq 1$, then there is a sequence $\{\varphi_m\}_1^\infty \subseteq \mathcal{K}_n$ such that $\|\varphi_m - f\|_{L^1(\mu)} \longrightarrow 0$ as $m \to \infty$. At the same time, it is clear, by Lebesgue's Dominated Convergence Theorem, that for any $f \in L^1(\mu)$, $\|f_n - f\|_{L^1(\mu)} \longrightarrow 0$ as $n \to \infty$, where $f_n \equiv \mathbf{1}_{E_n} f$. □

Notice that, when applied to Lebesgue's $\lambda_{\mathbb{R}^N}$ measure on $\mathbb{R}^N$, Corollary 3.3.15 says that for every $f \in L^1(\lambda_{\mathbb{R}^N})$ and $\epsilon > 0$ there is a continuous function φ such that φ vanishes off of a compact set and $\|f - \varphi\|_{L^1(\lambda_{\mathbb{R}^N})} < \epsilon$. This fact can be interpreted in either one of two ways: either measurable functions are not all that different from continuous ones or $\|\cdot\|_{L^1(\lambda_{\mathbb{R}^N})}$ provides a rather crude gauge of size. Experience indicates that the latter interpretation is the more accurate one.

Exercises

3.3.16 Exercise: Let f be a non-negative, integrable function on the measure space $(E, \mathcal{B}, \nu)$, and define $\mu(\Gamma) = \int_\Gamma f\, d\nu$ for $\Gamma \in \mathcal{B}$. Show that μ is a finite measure on $(E, \mathcal{B})$. In addition, show that μ is **absolutely continuous with respect to** ν in the sense that, for each $\epsilon > 0$, there is a $\delta > 0$ with the property that $\mu(\Gamma) < \epsilon$ whenever $\nu(\Gamma) < \delta$.[3] In particular, for any ν-integrable function f, and any sequence of $\lfloor$-measurable sets $\{\Gamma_n\}_1\infty$ satisfying $\nu(\Gamma_n) \longrightarrow 0$, show that $\int_{\Gamma_n} f\, d\nu \longrightarrow 0$.

3.3.17 Exercise: Let f be a non-negative, measurable function on the measure space $(E, \mathcal{B}, \mu)$. If f is integrable, show that

$$\lim_{\lambda \to \infty} \lambda \mu(f \geq \lambda) = 0. \tag{3.3.18}$$

Next, produce a non-negative measurable f on $\big([0,1], \mathcal{B}_{[0,1]}, \lambda_{[0,1]}\big)$ ($\lambda_{[0,1]}$ is used here to denote the restriction of $\lambda_{\mathbb{R}}$ to $\mathcal{B}_{[0,1]}$) such that (3.3.18) holds but f fails to be integrable. Finally, show that if f is a non-negative measurable function on the finite measure space $(E, \mathcal{B}, \mu)$, then f is integrable if and only if

$$\sum_{n=1}^{\infty} \mu(f > n) < \infty.$$

3.3.19 Exercise: Let J be a closed rectangle in $\mathbb{R}^N$ and $f : J \longrightarrow \mathbb{R}$ a continuous function. Show that the Riemann integral (R) $\int_J f(x)\, dx$ of f over J is equal to the Lebesgue integral $\int_J f(x)\, \lambda_{\mathbb{R}^N}(dx)$. Next, suppose that $f \in L^1(\lambda_{\mathbb{R}^N})$ is continuous, and use the preceding to show that

$$\int f(x)\, \lambda_{\mathbb{R}^N}(dx) = \lim_{J \nearrow \mathbb{R}^N} (\mathrm{R}) \int_J f(x)\, dx,$$

where the limit means that, for any $\epsilon > 0$, there exists a rectangle J_ϵ such that

$$\left| \int f(x)\, \lambda_{\mathbb{R}^N}(dx) - (\mathrm{R}) \int_J f(x)\, dx \right| < \epsilon$$

whenever J is a rectangle containing J_ϵ. For this reason, even when f is not continuous, it is conventional to use $\int f(x)\, dx$ instead of $\int f\, d\lambda_{\mathbb{R}^N}$ to denote the Lebesgue integral of f.

---

[3] This is not the usual definition of absolute continuity, the one introduced in § 8.2. At first sight, this one appears to demand much more. That in truth it is not is the content of Exercise 8.2.15.

3.3.20 Exercise: Let $(E, \mathcal{B}, \mu)$ be a measure space and let $\{f_n\}_1^\infty$ be a sequence of measurable functions on $(E, \mathcal{B})$. Next, suppose that $\{g_n\}_1^\infty \subseteq L^1(\mu)$ and that $g_n \longrightarrow g \in L^1(\mu)$ in $L^1(\mu)$. The following variants of Fatou's Lemma and Lebesgue's Dominated Convergence Theorem are often useful.

(i) If $f_n \leq g_n$ (a.e., μ) for each $n \geq 1$, show that

$$\varlimsup_{n\to\infty} \int f_n \, d\mu \leq \int \varlimsup_{n\to\infty} f_n \, d\mu.$$

(ii) If $f_n \longrightarrow f$ either in μ-measure or μ-almost everywhere and if $|f_n| \leq g_n$ (a.e., μ) for each $n \geq 1$, show that $\|f_n - f\|_{L^1(\mu)} \longrightarrow 0$ and therefore that $\lim_{n\to\infty} \int f_n \, d\mu = \int f \, d\mu$.

3.3.21 Exercise: Let $(E, \mathcal{B}, \mu)$ be a measure space. A family $\mathcal{K}$ of measurable functions f on $(E, \mathcal{B}, \mu)$ is said to be **uniformly μ-absolutely continuous** if, for each $\epsilon > 0$, there is a $\delta > 0$ such that $\int_\Gamma |f| \, d\mu \leq \epsilon$ for all $f \in \mathcal{K}$ whenever $\Gamma \in \mathcal{B}$ and $\mu(\Gamma) < \delta$; also, it is said to be **uniformly μ-integrable** if for each $\epsilon > 0$ there is an $R < \infty$ such that $\int_{|f|\geq R} |f| \, d\mu \leq \epsilon$ for all $f \in \mathcal{K}$.

(i) Show that $\mathcal{K}$ is uniformly μ-integrable if it is uniformly μ-absolutely continuous and

$$\sup_{f\in\mathcal{K}} \|f\|_{L^1(\mu)} < \infty.$$

Conversely, suppose that $\mathcal{K}$ is uniformly μ-integrable and show that it is then necessarily uniformly μ-absolutely continuous and, when $\mu(E) < \infty$, that $\sup_{f\in\mathcal{K}} \|f\|_{L^1(\mu)} < \infty$.

(ii) If $\sup_{f\in\mathcal{K}} \int |f|^{1+\delta} \, d\mu < \infty$ for some $\delta > 0$, show that $\mathcal{K}$ is uniformly μ-integrable.

(iii) Let $\{f_n\}_1^\infty \subseteq L^1(\mu)$ be given. If $f_n \longrightarrow f$ in $L^1(\mu)$, show that $\{f_n\}_1^\infty \cup \{f\}$ is uniformly μ-absolutely continuous and uniformly μ-integrable. Conversely, assuming that $\mu(E) < \infty$, show that $f_n \longrightarrow f$ in $L^1(\mu)$ if $f_n \longrightarrow f$ in μ-measure and $\{f_n\}_1^\infty$ is uniformly μ-integrable.

(iv) Assume that $\mu(E) = \infty$. We say that a family $\mathcal{K}$ of measurable functions f on $(E, \mathcal{B}, \mu)$ is **tight** if, for each $\epsilon > 0$, there is a $\Gamma \in \mathcal{B}$ such that $\mu(\Gamma) < \infty$ and $\sup_{f\in\mathcal{K}} \int_{\Gamma \complement} |f| \, d\mu \leq \epsilon$. Assuming that $\mathcal{K}$ is tight, show that $\mathcal{K}$ is uniformly μ-integrable if and only if it is uniformly μ-absolutely continuous and $\sup_{f\in\mathcal{K}} \|f\|_{L^1(\mu)} < \infty$. Finally, suppose that $\{f_n\}_1^\infty \subseteq L^1(\mu)$ is tight and that $f_n \longrightarrow f$ in μ-measure. Show that $\|f_n - f\|_{L^1(\mu)} \longrightarrow 0$ if and only if $\{f_n\}$ is uniformly μ-integrable.

3.3.22 Exercise: Let $(E, \mathcal{B}, \mu)$ be a finite measure space. Show that $f_n \longrightarrow f$ in μ-measure if and only if $\int |f_n - f| \wedge 1 \, d\mu \longrightarrow 0$.

3.3.23 Exercise: Let (E, ρ) be a metric space and $\{E_n\}_1^\infty$ a non-decreasing sequence of open subsets of E such that $E_n \nearrow E$. Let μ and ν be two measures on $(E, \mathcal{B}_E)$ with the properties that $\mu(E_n) \vee \nu(E_n) < \infty$ for every $n \geq 1$ and $\int \varphi \, d\mu = \int \varphi \, d\nu$ whenever φ is a bounded, ρ-uniformly continuous φ for which there is an $n \geq 1$ such that $\varphi \equiv 0$ off of E_n. Show that $\mu = \nu$ on $\mathcal{B}_E$.

3.3.24 Exercise: Although almost everywhere convergence does not follow from convergence in measure, it nearly does. Indeed, suppose that $\{f_n\}_1^\infty$ is a sequence of measurable, $\mathbb{R}$-valued functions on $(E, \mathcal{B}, \mu)$. Given an $\mathbb{R}$-valued, measurable function f, show that (3.3.8) holds, and therefore that $f_n \longrightarrow f$ both (a.e., μ) and in μ-measure, if

$$\sum_1^\infty \mu\big(|f_n - f| \geq \epsilon\big) < \infty \quad \text{for every } \epsilon > 0.$$

In particular, conclude that $f_n \longrightarrow f$ (a.e., μ) and in μ-measure if

$$\sum_1^\infty \|f_n - f\|_{L^1(\mu)} < \infty.$$

3.4 Lebesgue's Differentiation Theorem

Although it represents something of a departure from the spirit of this chapter, we return in this concluding section to Lebesgue measure on $\mathbb{R}$ and prove the following remarkable generalization of the Fundamental Theorem of Calculus. Namely, we will show that if f is any Lebesgue integrable function on $\mathbb{R}$, then (cf. the notation introduced in Exercise 3.3.19)

$$\text{(3.4.1)} \quad \lim_{\mathring{I}\searrow\{x\}} \frac{1}{|\mathring{I}|}\int_{\mathring{I}} \big|f(t) - f(x)\big|\,dt = 0 \quad \text{for (Lebesgue) almost every } x \in \mathbb{R},$$

where (3.4.1) is to be interpreted as the statement that, for almost every $x \in \mathbb{R}$, there exists, for each $\epsilon > 0$, a $\delta = \delta(x, \epsilon) > 0$ such that

$$\frac{1}{|\mathring{I}|}\int_{\mathring{I}} \big|f(t) - f(x)\big|\,dt < \epsilon \text{ whenever } \mathring{I} \ni x \text{ is an open interval with } |\mathring{I}| < \delta.$$

In other words, except on a set of Lebesgue measure 0, *an integrable function can be recovered by differentiating its indefinite integral.*

In order to understand the strategy behind our proof, first note that (3.4.1) is completely obvious when f is continuous. Hence, since (cf. Corollary 3.3.15) the continuous elements of $L^1(\mathbb{R}) \equiv L^1(\lambda_{\mathbb{R}})$ are dense in $L^1(\mathbb{R})$, it suffices for us to show that the set $\mathcal{G}$ of $f \in L^1(\mathbb{R})$ for which (3.4.1) holds is closed in $L^1(\mathbb{R})$. To this end, we introduce the **Hardy–Littlewood maximal function**

$$\text{(3.4.2)} \qquad Mf(x) \equiv \sup\left\{\frac{1}{|\mathring{I}|}\int_{I} \big|f(t)\big|\,dt : \mathring{I} \ni x\right\}, \qquad x \in \mathbb{R},$$

for $f \in L^1(\mathbb{R})$. Next, for each $f \in L^1(\mathbb{R})$ and $\epsilon > 0$ set

$$\Delta(f,\epsilon) = \left\{ x : \varlimsup_{\mathring{I} \searrow \{x\}} \left| \frac{1}{|\mathring{I}|} \int_{\mathring{I}} \big|f(t) - f(x)\big|\, dt \right| > \epsilon \right\}.$$

Clearly, (3.4.1) holds if and only if $\big|\Delta(f,\epsilon)\big|_e = 0$ for every $\epsilon > 0$. Moreover, for any $\epsilon > 0$ and any $f,\, g \in L^1(\mathbb{R})$:

$$\Delta(f,3\epsilon) \subseteq \big\{x : M(f-g)(x) > \epsilon\big\} \cup \Delta(g,\epsilon) \cup \big\{x : \big|g(x) - f(x)\big| > \epsilon\big\}.$$

In particular, this means that if $g \in \mathcal{G}$, then

$$\begin{aligned} \big|\Delta(f,3\epsilon)\big|_e &\le \big|\{M(f-g) > \epsilon\}\big|_e + \big|\Delta(g,\epsilon)\big|_e + \big|\{|f-g| > \epsilon\}\big| \\ &\le \big|\{M(f-g) > \epsilon\}\big|_e + \frac{1}{\epsilon}\|f-g\|_{L^1(\mathbb{R})}, \end{aligned}$$

where, in the passage to the last line we have used $\big|\Delta(g,\epsilon)\big|_e = 0$ and Markov's inequality. Finally, suppose that $\{g_n\}_1^\infty \subseteq \mathcal{G}$ and that $g_n \longrightarrow f$ in $L^1(\mathbb{R})$. Then, the preceding line of reasoning leads us to the conclusion that

$$\big|\Delta(f,3\epsilon)\big|_e \le \varliminf_{n\to\infty} \big|\{M(f-g_n) > \epsilon\}\big|_e,$$

and so we would be done if we knew that

$$\text{(3.4.3)} \qquad \varphi_n \longrightarrow 0 \text{ in } L^1(\mathbb{R}) \implies \big|\{M\varphi_n > \epsilon\}\big|_e \longrightarrow 0 \quad \text{for all } \epsilon > 0.$$

With the preceding in mind, we now turn our attention to the analysis of the Hardy–Littlewood maximal function. To begin with, we first note that Mf is measurable, and therefore that we can drop the subscript "e" in (3.4.3). To see this, observe that an alternative expression for Mf is

$$Mf(x) = \sup_{a,\, b \in (0,\infty)} \frac{1}{a+b} \int_{(-a,b)} \big|f(x+t)\big|\, dt$$

and that, for each $a,\, b \in (0,\infty)$ and $x < y$,

$$\left| \int_{(-a,b)} \big|f(x+t)\big|\, dt - \int_{(-a,b)} \big|f(y+t)\big|\, dt \right| \le \int_{(x-a,y-a] \cup [x+b,y+b)} \big|f(t)\big|\, dt.$$

Hence, by Exercise 3.3.16, we know that, for each $a,\, b \in (0,\infty)$,

$$x \in \mathbb{R} \longmapsto \frac{1}{a+b} \int_{(-a,b)} \big|f(x+t)\big|\, dt \in \mathbb{R}$$

is uniformly continuous and, therefore, for each $\alpha \in \mathbb{R}$, that

$$\{x : Mf(x) > \alpha\} = \bigcup_{a,\, b \in (0,\infty)} \left\{ x : \frac{1}{a+b} \int_{(-a,b)} \big|f(x+t)\big|\, dt > \alpha \right\}$$

is open. We next observe that control on the size of Mf will follow from control on the *one-sided* maximal functions

$$M_+f(x) \equiv \sup_{h>0} \frac{1}{h} \int_{[x,x+h)} |f(t)|\,dt \quad \text{and} \quad M_-f(x) \equiv \sup_{h>0} \frac{1}{h} \int_{(x-h,x]} |f(t)|\,dt.$$

Indeed, for any $\mathring{I} \ni x$, let h_+ and h_- be the lengths of $[x,\infty)\cap\mathring{I}$ and $(-\infty,x]\cap\mathring{I}$, respectively, and note that

$$\begin{aligned}\frac{1}{|\mathring{I}|}\int_{\mathring{I}} |f(t)|\,dt &= \frac{1}{|\mathring{I}|}\int_{(x-h_-,x]} |f(t)|\,dt + \frac{1}{|\mathring{I}|}\int_{[x,x+h_+)} |f(t)|\,dt \\ &\le \frac{h_-}{|\mathring{I}|} M_-f(x) + \frac{h_+}{|\mathring{I}|} M_+f(x) \le M_-f(x) \vee M_+f(x);\end{aligned}$$

and therefore, since it is essentially obvious that $M_+f \vee M_-f \le Mf$, we have

$$Mf = M_+f \vee M_-f. \tag{3.4.4}$$

Moreover, by precisely the same argument as we used to prove that $\{Mf > \alpha\}$ is open, we know that the same is true of both $\{M_+f > \alpha\}$ and $\{M_-f > \alpha\}$. Finally, note that $M_-f(x) = M_+\check{f}(-x)$, where $\check{f}(t) = f(-t)$, which means that we really need to learn how to control only M_+f.

For this purpose, let $f \in L^1(\mathbb{R})$ and $\epsilon > 0$ be given, and consider the function

$$F_\epsilon(x) = \int_{(-\infty,x]} |f(t)|\,dt - \epsilon x, \quad x \in \mathbb{R}.$$

By Exercise 3.3.16, F_ϵ is uniformly continuous, and, obviously, $\lim_{x\to\pm\infty} F_\epsilon(x) = \mp\infty$. Moreover, it is an elementary matter to see that

$$\{M_+f > \epsilon\} = \{x : \exists y > x\ F_\epsilon(y) > F_\epsilon(x)\}.$$

Hence, as we will see shortly, all that we need is the following wonderfully simple observation.

3.4.5 Lemma (Sunrise Lemma). * *Let $F : \mathbb{R} \longrightarrow \mathbb{R}$ be a continuous function with the property that $\lim_{x\to\pm\infty} F(x) = \mp\infty$. Set*

$$G = \{x : \exists y > x\ F(y) > F(x)\}.$$

Then G is an open, and each non-empty, open, connected component of G is a bounded interval (α,β) with $F(\alpha) \le F(\beta)$.

---

* The name derives from the following picture. The sun is rising infinitely far to the right in mountainous (one-dimensional) terrain, $F(x)$ is the elevation at x, and G is the region in shadow at the instant when the sun comes over the horizon.

PROOF: Clearly G is open. Next, suppose that (α,β) is a non-empty, connected, open component of G, and take $\gamma \in (\alpha,\beta)$. If either $\beta = \infty$ or $\beta < \infty$ and $F(\beta) < F(\gamma)$, then there exists a unique $x \in [\gamma,\beta)$ such that $F(x) = F(\gamma)$ and $F(y) < F(\gamma)$ for all $y > x$. But this would mean that, on the one hand, $x \in G$ and, on the other hand, $F(y) < F(x)$ for all $y > x$, which is impossible. Hence, we now know that $\beta < \infty$ and that $F(\gamma) \le F(\beta)$ for all $\gamma \in (\alpha,\beta)$. Finally, from this it is clear that $\alpha > -\infty$ and that $F(\alpha) \le F(\beta)$. □

Applying Lemma 3.4.5 to the function F_ϵ, we see that $G_\epsilon \equiv \{M_+f > \epsilon\}$ is either empty or (cf. the first part of Lemma 2.1.9) is the union of countably many disjoint, bounded, open intervals (α_n,β_n) satisfying

$$0 \le F_\epsilon(\beta_n) - F_\epsilon(\alpha_n) = \int_{(\alpha_n,\beta_n)} |f(t)|\,dt - \epsilon(\beta_n - \alpha_n)$$

for each n. Hence, either $|G_\epsilon| = 0$ or, after summing the preceding over n, we arrive at

$$\epsilon|G_\epsilon| = \epsilon\sum_n(\beta_n,\alpha_n) \le \int_{G_\epsilon} |f(t)|\,dt.$$

In other words, we have now proved first that

$$\big|\{M_+f > \epsilon\}\big| \le \epsilon^{-1}\int_{\{M_+f>\epsilon\}} |f(t)|\,dt, \quad \epsilon > 0 \text{ and } f \in L^1(\mathbb{R}),$$

and then, after taking left limits with respect to $\epsilon > 0$,

$$\text{(3.4.6)} \quad \big|\{M_+f \ge \epsilon\}\big| \le \epsilon^{-1}\int_{\{M_+f\ge\epsilon\}} |f(t)|\,dt \le \epsilon^{-1}\|f\|_{L^1(\mathbb{R})}, \quad f \in L^1(\mathbb{R}).$$

In fact, because $M_-f(x) = M_+\check{f}(-x)$, we also know that (3.4.6) continues to hold when M_+f is replaced by M_-f throughout. Finally, in conjunction with (3.4.4), these lead to

$$\epsilon\big|\{Mf \ge \epsilon\}\big| \le \int_{\{M_+f\ge\epsilon\}} |f(t)|\,dt + \int_{\{M_-f\ge\epsilon\}} |f(t)|\,dt \le 2\int_{\{Mf\ge\epsilon\}} |f(t)|\,dt,$$

which means that we have now proved the renowned **Hardy–Littlewood maximal inequality**,

$$\text{(3.4.7)} \qquad \big|\{Mf \ge \epsilon\}\big| \le \frac{2}{\epsilon}\int_{\{Mf\ge\epsilon\}} |f(t)|\,dt \le \frac{2}{\epsilon}\|f\|_{L^1(\mathbb{R})}, \quad f \in L^1(\mathbb{R}).$$

Since (3.4.7) certainly implies (3.4.3), and (3.4.3) was the only missing ingredient in the program with which this section began, the derivation of the following statement is complete.

3.4.8 Theorem (Lebesgue Differentiation Theorem). *For any Lebesgue integrable function f on $\mathbb{R}$, (3.4.1) holds. In particular,*

$$\lim_{\mathring{I}\searrow\{x\}} \frac{1}{|\mathring{I}|}\int_{\mathring{I}} f(t) = f(x) \quad \textit{for almost every } x \in \mathbb{R}. \tag{3.4.9}$$

Before closing this section, there are several comments which should be made. First, one should notice that the conclusions drawn in Theorem 3.4.8 remain true for any Lebesgue measurable f which is integrable on each compact subset of $\mathbb{R}$. Indeed, all the assertions there are completely local and therefore follow by replacing f with $f\mathbf{1}_{(-R,R)}$, restricting ones attention to $x \in (-R,R)$, and then letting $R \nearrow \infty$.

Second, one should notice that (3.4.7) would be a trivial consequence of Markov's inequality if we had the estimate $\|Mf\|_{L^1(\mathbb{R})} \leq 2\|f\|_{L^1(\mathbb{R})}$. Thus, it is reasonable to ask whether such an estimate is true. That the answer is a resounding *no* can be most easily seen from the observation that, if $\|f\|_{L^1(\mathbb{R})} \neq 0$, then $\alpha \equiv \int_{(-r,r)} |f(t)|\,dt > 0$ for some $r > 0$ and therefore $Mf(x) \geq \frac{\alpha}{|x|+r}$ for all $x \in \mathbb{R}$. That is, *if $f \in L^1(\mathbb{R})$ does not vanish almost everywhere, then Mf is not integrable.* (To see that the situation is even worse and that, in general, Mf need not be integrable over bounded sets, see Exercise 3.4.12 below.) Thus, in a very real sense, (3.4.7) is about as well as one can do. (See Exercise 6.2.27 for an interesting continuation of these considerations.) Because this sort of situation arises quite often, inequalities of the form in (3.4.7) have been given a special name: they are called *weak-type inequalities* to distinguish them from inequalities of the form $\|Mf\|_{L^1(\mathbb{R})} \leq C\|f\|_{L^1(\mathbb{R})}$, which would be called a *strong-type inequality.*

Finally, it should be clear that, except for the derivation of (3.4.7), the arguments given here would work equally well in $\mathbb{R}^N$. Thus, we would know that, for each Lebesgue integrable f on $\mathbb{R}^N$,

$$\lim_{B\searrow\{x\}} \frac{1}{|B|}\int_B \big|f(y) - f(x)\big|\,dy = 0 \quad \text{for almost every } x \in \mathbb{R}^N \tag{3.4.10}$$

if we knew that

$$\big|\{Mf \geq \epsilon\}\big| \leq \frac{C}{\epsilon}\|f\|_{L^1(\lambda_{\mathbb{R}^N})}, \tag{3.4.11}$$

where Mf is the Hardy–Littlewood maximal function

$$Mf(x) \equiv \sup_{B\ni x} \frac{1}{|B|}\int_B \big|f(y)\big|\,dy$$

and B denotes a generic open ball in $\mathbb{R}^N$. It turns out that (3.4.11), and therefore (3.4.10), are both true. However, the proof of (3.4.11) for $N \geq 2$ is somewhat more involved than the one which we have given of (3.4.7)*.

* See, for example, E.M. Stein's *Singular Integrals and Differentiability Properties of Functions*, published by Princeton Univ. Press (1970)

Exercises

3.4.12 Exercise: Define $f:\mathbb{R}\longrightarrow[0,\infty)$ so that $f(x)=\left(x(\log x)^2\right)^{-1}$ if $x\in(0,e^{-1})$ and $f(x)=0$ if $x\notin(0,e^{-1})$. Using Exercise 3.3.19 and the Fundamental Theorem of Calculus, check that $f\in L^1(\mathbb{R})$ and that

$$\int_{(0,x)} f(t)\,dt=\frac{-1}{\log x},\quad x\in(0,e^{-1}).$$

In particular, conclude that $\int_{(0,r)} Mf(x)\,dx=\infty$ for every $r>0$.

3.4.13 Exercise: Given $f\in L^1(\mathbb{R})$, define the **Lebesgue set** of f to be the set $\mathrm{Leb}(f)$ of those $x\in\mathbb{R}$ for which the limit in (3.4.1) is 0. Clearly, (3.4.1) is the statement that $\mathrm{Leb}(f)\complement$ has Lebesgue measure 0, and clearly $\mathrm{Leb}(f)$ is the set on which f is *well behaved* in the sense that the averages $\frac{1}{|\mathring{I}|}\int_{\mathring{I}} f(t)\,dt$ converge to $f(x)$ as $\mathring{I}\searrow\{x\}$ for $x\in\mathrm{Leb}(f)$. The purpose of this exercise is to show that, in the same sense, other averaging procedures converge to f on $\mathrm{Leb}(f)$. To be precise, let ρ be a bounded continuous function on $\mathbb{R}$ having one bounded, continuous derivative ρ'. Further, assume that $\rho\in L^1(\mathbb{R})$, $\int\rho(t)\,dt=1$, $\int|t\rho'(t)|\,dt<\infty$, and $\lim_{|t|\to\infty}\rho(t)=\lim_{|t|\to\infty}t\rho'(t)=0$. Next, for each $\epsilon>0$, set $\rho_\epsilon(t)=\epsilon^{-1}\rho(\epsilon^{-1}t)$ and define

$$f_\epsilon(x)=\int\rho_\epsilon(x-t)f(t)\,dt,\quad x\in\mathbb{R}\text{ and }f\in L^1(\mathbb{R}).$$

The purpose of this exercise is to show that

$$f_\epsilon(x)\longrightarrow f(x)\quad\text{as }\epsilon\searrow 0\text{ for each }x\in\mathrm{Leb}(f).\tag{3.4.14}$$

(i) Show that, for any $f\in L^1(\mathbb{R})$ and $x\in\mathrm{Leb}(f)$,

$$\lim_{\delta\searrow0}\frac{1}{\delta}\int_{[x,x+\delta)}f(t)\,dt=f(x)=\lim_{\delta\searrow0}\frac{1}{\delta}\int_{(x-\delta,x]}f(t)\,dt.$$

(ii) Assuming that f is continuous and vanishes off of a compact set, first show that

$$f_\epsilon(x)=\int_{[0,\infty)}\rho(-t)f(x+\epsilon t)\,dt+\int_{[0,\infty)}\rho(t)f(x-\epsilon t)\,dt,$$

and then (using Exercise 3.3.19 and Theorem 1.2.7) verify the following equalities:

$$\int_{[0,\infty)}\rho(-t)f(x+\epsilon t)\,dt=\int_{(0,\infty)}t\rho'(-t)\left(\frac{1}{\epsilon t}\int_{[x,x+\epsilon t)}f(s)\,ds\right)dt\tag{3.4.15}$$

and

$$\int_{[0,\infty)}\rho(t)f(x-\epsilon t)\,dt=-\int_{(0,\infty)}t\rho'(t)\left(\frac{1}{\epsilon t}\int_{(x-\epsilon t,x]}f(s)\,ds\right)dt.\tag{3.4.15'}$$

Next (using Corollary 3.3.15) argue that (3.4.15) and (3.4.15') continue to hold for every $f\in L^1(\mathbb{R})$ and $x\in\mathrm{Leb}(f)$.

(iii) Combining part (**i**) with (3.4.15) and (3.4.15'), conclude that

$$\lim_{\epsilon \searrow 0} f_\epsilon(x) = -f(x) \int t\rho'(t)\, dt, \quad \text{for } x \in \text{Leb}(f),$$

and, after another application of Exercise 3.3.19 and Theorem 1.2.7, note that

$$-\int t\rho'(t)\, dt = \int \rho(t)\, dt = 1.$$

Chapter IV
Products of Measures

4.1 Fubini's Theorem

Just before Lemma 3.2.2, we introduced the product $(E_1 \times E_2, \mathcal{B}_1 \times \mathcal{B}_2)$ of two measurable spaces $(E_1, \mathcal{B}_1)$ and $(E_2, \mathcal{B}_2)$. We now want to show that if μ_i, $i \in \{1, 2\}$, is a measure on $(E_i, \mathcal{B}_i)$, then, under reasonable conditions, there is a unique measure ν on $(E_1 \times E_2, \mathcal{B}_1 \times \mathcal{B}_2)$ with the property that $\nu(\Gamma_1 \times \Gamma_2) = \mu_1(\Gamma_1)\, \mu(\Gamma_2)$ for all $\Gamma_i \in \mathcal{B}_i$.

The key to the construction of ν is found in the following function analog of π- and λ-systems (cf. Lemma 3.1.3). Namely, given a space E, we will say that a collection $\mathcal{L}$ of functions $f : E \longrightarrow (-\infty, \infty]$ is a **semi-lattice** if both f^+ and f^- are in $\mathcal{L}$ whenever $f \in \mathcal{L}$. A sub collection $\mathcal{K} \subseteq \mathcal{L}$ will be called an **$\mathcal{L}$-system** if:

(**a**) $\mathbf{1} \in \mathcal{K}$;
(**b**) if f, $g \in \mathcal{K}$ and $\{f = \infty\} \cap \{g = \infty\} = \emptyset$, then $g - f \in \mathcal{K}$ whenever either $f \leq g$ or $g - f \in \mathcal{L}$;
(**c**) if α, $\beta \in [0, \infty)$ and $f, g \in \mathcal{K}$, then $\alpha f + \beta g \in \mathcal{K}$;
(**d**) if $\{f_n\}_1^\infty \subseteq \mathcal{K}$ and $f_n \nearrow f$, then $f \in \mathcal{K}$ whenever f is bounded or $f \in \mathcal{L}$.

The analog of Lemma 3.1.3 in this context is the following.

4.1.1 Lemma. *Let $\mathcal{C}$ be a π-system which generates the σ-algebra $\mathcal{B}$ over E, and let $\mathcal{L}$ be a semi-lattice of functions $f : E \longrightarrow (-\infty, \infty]$. If $\mathcal{K}$ is an $\mathcal{L}$-system and $\mathbf{1}_\Gamma \in \mathcal{K}$ for every $\Gamma \in \mathcal{C}$, then $\mathcal{K}$ contains every $f \in \mathcal{L}$ which is measurable on $(E, \mathcal{B})$.*

PROOF: First note that $\{\Gamma \subseteq E : \mathbf{1}_\Gamma \in \mathcal{K}\}$ is a λ-system which contains $\mathcal{C}$. Hence, by Lemma 3.1.3, $\mathbf{1}_\Gamma \in \mathcal{K}$ for every $\Gamma \in \mathcal{B}$. Combined with (**c**) above, this means that $\mathcal{K}$ contains every non-negative, measurable, simple function on $(E, \mathcal{B})$.

Next, suppose that $f \in \mathcal{L}$ is measurable on $(E, \mathcal{B})$. Then both f^+ and f^- have the same properties, and, by (**b**) above, it is enough to show that $f^+, f^- \in \mathcal{K}$ in order to know that $f \in \mathcal{K}$. Thus, without loss of generality, we assume that $f \in \mathcal{L}$ is a non-negative measurable function on $(E, \mathcal{B})$. But in that case f is the non-decreasing limit of non-negative measurable simple functions; and so $f \in \mathcal{K}$ by (**d**). $\square$

The power of Lemma 4.1.1 to handle questions involving products is already apparent in the following.

4.1.2 Lemma. *Let $(E_1, \mathcal{B}_1)$ and $(E_2, \mathcal{B}_2)$ be measurable spaces, and suppose that f is an $\overline{\mathbb{R}}$-valued measurable function on $(E_1 \times E_2, \mathcal{B}_1 \times \mathcal{B}_2)$. Then for each $x_1 \in E_1$ and $x_2 \in E_2$, $f(x_1, \cdot\,)$ and $f(\,\cdot\,, x_2)$ are measurable functions on $(E_2, \mathcal{B}_2)$ and $(E_1, \mathcal{B}_1)$, respectively. Next, suppose that μ_i, $i \in \{1,2\}$, is a finite measure on $(E_i, \mathcal{B}_i)$. Then for every measurable function f on $(E_1 \times E_2, \mathcal{B}_1 \times \mathcal{B}_2)$ which is either bounded or non-negative, the functions*

$$\int_{E_2} f(\,\cdot\,, x_2)\,\mu_2(dx_2) \quad \text{and} \quad \int_{E_1} f(x_1, \cdot\,)\,\mu_1(dx_1)$$

are measurable on $(E_1, \mathcal{B}_1)$ and $(E_2, \mathcal{B}_2)$, respectively.

PROOF: Clearly it is enough to check all these assertions when f is non-negative.

Let $\mathcal{L}$ be the collection of all non-negative functions on $E_1 \times E_2$, and define $\mathcal{K}$ to be those elements of $\mathcal{L}$ which have all the asserted properties. It is clear that $\mathbf{1}_{\Gamma_1 \times \Gamma_2} \in \mathcal{K}$ for all $\Gamma_i \in \mathcal{B}_i$. Moreover, it is easy to check that $\mathcal{K}$ is an $\mathcal{L}$-system. Hence, by Lemma 4.1.1 with $\mathcal{C} = \{\Gamma_1 \times \Gamma_2 : \Gamma_i \in \mathcal{B}_i\}$, we are done. □

4.1.3 Lemma. *Given a pair $(E_1, \mathcal{B}_1, \mu_1)$ and $(E_2, \mathcal{B}_2, \mu_2)$ of finite measure spaces, there exists a unique measure ν on $(E_1 \times E_2, \mathcal{B}_1 \times \mathcal{B}_2)$ such that*

$$\nu(\Gamma_1 \times \Gamma_2) = \mu_1(\Gamma_1)\,\mu_2(\Gamma_2) \quad \textit{for all} \quad \Gamma_i \in \mathcal{B}_i.$$

Moreover, for every non-negative, measurable function f on $(E_1 \times E_2, \mathcal{B}_1 \times \mathcal{B}_2)$,

$$\begin{aligned} &\int_{E_1 \times E_2} f(x_1, x_2)\,\nu(dx_1 \times dx_2) \\ (4.1.4) \qquad &= \int_{E_2} \left(\int_{E_1} f(x_1, x_2)\,\mu_1(dx_1) \right) \mu_2(dx_2) \\ &= \int_{E_1} \left(\int_{E_2} f(x_1, x_2)\,\mu_2(dx_2) \right) \mu_1(dx_1). \end{aligned}$$

PROOF: The uniqueness of ν is guaranteed by Exercise 3.1.8. To prove the existence of ν, define

$$\nu_{1,2}(\Gamma) = \int_{E_2} \left(\int_{E_1} \mathbf{1}_\Gamma(x_1, x_2)\,\mu_1(dx_1) \right) \mu_2(dx_2)$$

and

$$\nu_{2,1}(\Gamma) = \int_{E_1} \left(\int_{E_2} \mathbf{1}_\Gamma(x_1, x_2)\,\mu_2(dx_2) \right) \mu_1(dx_1)$$

for $\Gamma \in \mathcal{B}_1 \times \mathcal{B}_2$. Using the Monotone Convergence Theorem, one sees that both $\nu_{1,2}$ and $\nu_{2,1}$ are finite measures on $(E_1 \times E_2, \mathcal{B}_1 \times \mathcal{B}_2)$. Moreover, by the same sort of argument as was used to prove Lemma 4.1.2, for every non-negative measurable function f on $(E_1 \times E_2, \mathcal{B}_1 \times \mathcal{B}_2)$,

$$\int f\, d\nu_{1,2} = \int_{E_1} \left(\int_{E_2} f(x_1, x_2)\, \mu_1(dx_1) \right) \mu_2(dx_2)$$

and

$$\int f\, d\nu_{2,1} = \int_{E_2} \left(\int_{E_1} f(x_1, x_2)\, \mu_2(dx_2) \right) \mu_1(dx_1).$$

Finally, since $\nu_{1,2}(\Gamma_1 \times \Gamma_2) = \mu(\Gamma_1)\, \mu(\Gamma_2) = \nu_{2,1}(\Gamma_1 \times \Gamma_2)$ for all $\Gamma_i \in \mathcal{B}_i$, we see that both $\nu_{1,2}$ and $\nu_{2,1}$ fulfill the requirements placed on ν. Hence, not only does ν exist, but it is also equal to both $\nu_{1,2}$ and $\nu_{2,1}$; and so the preceding equalities lead to (4.1.4). $\square$

In order to extend the preceding construction to measures which need not be finite, we must (cf. Exercise 4.1.12) introduce a qualified notion of finiteness. Namely, we will say that the measure μ on $(E, \mathcal{B})$ is **σ-finite** and will call $(E, \mathcal{B}, \mu)$ a **σ-finite measure space** if E can be written as the union of a countable number of sets $\Gamma \in \mathcal{B}$ for each of which $\mu(\Gamma) < \infty$. Thus, for example, $(\mathbb{R}^N, \overline{\mathcal{B}}_{\mathbb{R}^N}, \lambda_{\mathbb{R}^N})$ is a σ-finite measure space.

4.1.5 Theorem (Tonelli's Theorem). *Let $(E_1, \mathcal{B}_1, \mu_1)$ and $(E_2, \mathcal{B}_2, \mu_2)$ be σ-finite measure spaces. Then there is a unique measure ν on $(E_1 \times E_2, \mathcal{B}_1 \times \mathcal{B}_2)$ such that $\nu(\Gamma_1 \times \Gamma_2) = \mu_1(\Gamma_1)\, \mu_2(\Gamma_2)$ for all $\Gamma_i \in \mathcal{B}_i$. In addition, for every non-negative measurable function f on $(E_1 \times E_2, \mathcal{B}_1 \times \mathcal{B}_2)$, $\int f(\,\cdot\,, x_2)\, \mu_2(dx_2)$ and $\int f(x_1, \,\cdot\,)\, \mu_1(dx_1)$ are measurable on $(E_1, \mathcal{B}_1)$ and $(E_2, \mathcal{B}_2)$, respectively, and (4.1.4) continues to hold.*

PROOF: Choose sequences $\{E_{i,n}\}_{n=1}^{\infty} \subseteq \mathcal{B}_i$ for $i \in \{1, 2\}$ so that $\mu_i(E_{i,n}) < \infty$ for each $n \geq 1$ and $E_i = \bigcup_{n=1}^{\infty} E_{i,n}$. Without loss of generality, we assume that $E_{i,m} \cap E_{i,n} = \emptyset$ for $m \neq n$. For each $n \in \mathbb{Z}^+$, define $\mu_{i,n}(\Gamma_i) = \mu_i(\Gamma_i \cap E_{i,n})$, $\Gamma_i \in \mathcal{B}_i$; and, for $(m, n) \in {\mathbb{Z}^+}^2$, let $\nu_{(m,n)}$ on $(E_1 \times E_2, \mathcal{B}_1 \times \mathcal{B}_2)$ be the measure constructed from $\mu_{1,m}$ and $\mu_{2,n}$ as in Lemma 4.1.3.

Clearly, by Lemma 4.1.2, for any non-negative measurable function f on $(E_1 \times E_2, \mathcal{B}_1 \times \mathcal{B}_2)$,

$$\int_{E_2} f(\,\cdot\,, x_2)\, \mu_2(dx_2) = \sum_{n=1}^{\infty} \int_{E_{2,n}} f(\,\cdot\,, x_2)\, \mu_{2,n}(dx_2)$$

is measurable on $(E_1, \mathcal{B}_1)$; and, similarly, $\int_{E_1} f(x_1, \,\cdot\,)\, \mu_1(dx_1)$ is measurable on $(E_2, \mathcal{B}_2)$. Finally, the map $\Gamma \in \mathcal{B}_1 \times \mathcal{B}_2 \longmapsto \sum_{m,n=1}^{\infty} \nu_{(m,n)}(\Gamma)$ defines a measure ν_0 on $(E_1 \times E_2, \mathcal{B}_1 \times \mathcal{B}_2)$, and it is easy to check that ν_0 has all the required properties. At the same time, if ν is any other measure on $(E_1 \times E_2, \mathcal{B}_1 \times \mathcal{B}_2)$ for which $\nu(\Gamma_1 \times \Gamma_2) = \mu_1(\Gamma_1)\, \mu_2(\Gamma_2)$, $\Gamma_i \in \mathcal{B}_i$, then, by the uniqueness assertion in Lemma 4.1.3, ν coincides with $\nu_{(m,n)}$ on $\mathcal{B}_1 \times \mathcal{B}_2\big[E_{1,m} \times E_{2,n}\big]$ for each $(m,n) \in \mathbb{Z}^{+2}$ and is therefore equal to ν_0 on $\mathcal{B}_1 \times \mathcal{B}_2$. $\square$

The measure ν constructed in Theorem 4.1.5 is called the **product of** μ_1 **times** μ_2 and is denoted by $\mu_1 \times \mu_2$.

4.1.6 Theorem (Fubini's Theorem). *Let $(E_1, \mathcal{B}_1, \mu_1)$ and $(E_2, \mathcal{B}_2, \mu_2)$ be σ-finite measure spaces and f a measurable function on $(E_1 \times E_2, \mathcal{B}_1 \times \mathcal{B}_2)$. Then the f is $\mu_1 \times \mu_2$-integrable if and only if*

$$\int_{E_1} \left(\int_{E_2} |f(x_1,x_2)|\, \mu_2(dx_2) \right) \mu_1(dx_1) < \infty$$

if and only if

$$\int_{E_2} \left(\int_{E_1} |f(x_1,x_2)|\, \mu_1(dx_1) \right) \mu_2(dx_2) < \infty.$$

Next, set

$$\Lambda_1 = \left\{ x_1 \in E_1 : \int_{E_2} |f(x_1,x_2)|\, \mu_2(dx_2) < \infty \right\}$$

and

$$\Lambda_2 = \left\{ x_2 \in E_2 : \int_{E_1} |f(x_1,x_2)|\, \mu_1(dx_1) < \infty \right\};$$

and define f_i on E_i, $i \in \{1,2\}$, by

$$f_1(x_1) = \begin{cases} \int_{E_2} f(x_1,x_2)\, \mu_2(dx_2) & \text{if } \; x_1 \in \Lambda_1 \\ 0 & \text{otherwise} \end{cases}$$

and

$$f_2(x_2) = \begin{cases} \int_{E_1} f(x_1,x_2)\, \mu_1(dx_1) & \text{if } \; x_2 \in \Lambda_2 \\ 0 & \text{otherwise.} \end{cases}$$

Then f_i is an $\mathbb{R}$-valued, measurable function on $(E_i, \mathcal{B}_i)$. Finally, if f is $\mu_1 \times \mu_2$-integrable, then $\mu_i(\Lambda_i\complement) = 0$, $f_i \in L^1(\mu_i)$, and

$$\int_{E_i} f_i(x_i)\, \mu_i(dx_i) = \int_{E_1 \times E_2} f(x_1,x_2)\, (\mu_1 \times \mu_2)(dx_1 \times dx_2) \tag{4.1.7}$$

for $i \in \{1,2\}$.

PROOF: The first assertion is an immediate consequence of Theorem 4.1.5. Moreover, since $\Lambda_i \in \mathcal{B}_i$, it is easy to check (from Lemma 4.1.2) that f_i is an $\mathbb{R}$-valued, measurable function on $(E_i, \mathcal{B}_i)$. Finally, if f is $\mu_1 \times \mu_2$-integrable, then, by the first assertion, $\mu_i(\Lambda_i\complement) = 0$ and $f_i \in L^1(\mu_i)$. Hence, by Theorem 4.1.5 applied to f^+ and f^-, we see that

$$\begin{aligned}
&\int_{E_1 \times E_2} f(x_1, x_2)\,(\mu_1 \times \mu_2)(dx_1 \times dx_2) \\
&\quad = \int_{\Lambda_1 \times E_2} f^+(x_1, x_2)\,(\mu_1 \times \mu_2)(dx_1 \times dx_2) \\
&\qquad\qquad - \int_{\Lambda_1 \times E_2} f^-(x_1, x_2)\,(\mu_1 \times \mu_2)(dx_1 \times dx_2) \\
&\quad = \int_{\Lambda_1} \left(\int_{E_2} f^+(x_1, x_2)\,\mu_1(dx_2) \right) \mu_1(dx_1) \\
&\qquad\qquad - \int_{\Lambda_1} \left(\int_{E_2} f^-(x_1, x_2)\,\mu_2(dx_2) \right) \mu_1(dx_1) \\
&\quad = \int_{\Lambda_1} f_1(x_1)\,\mu_1(dx_1);
\end{aligned}$$

and the same line of reasoning applies to f_2. $\square$

Exercises

4.1.8 Exercise: Let $(E, \mathcal{B}, \mu)$ be a σ-finite measure space. Given a non-negative measurable function f on $(E, \mathcal{B})$, define

$$\Gamma(f) = \{(x,t) \in E \times [0,\infty) : t \le f(x)\}$$

and

$$\widehat{\Gamma}(f) = \{(x,t) \in E \times [0,\infty) : t < f(x)\}.$$

Show that both $\Gamma(f)$ and $\widehat{\Gamma}(f)$ are elements of $\mathcal{B} \times \mathcal{B}_{[0,\infty)}$ and, in addition, that

$$\mu \times \lambda_{\mathbb{R}}\big(\widehat{\Gamma}(f)\big) = \int_E f\,d\mu = \mu \times \lambda_{\mathbb{R}}\big(\Gamma(f)\big). \tag{4.1.9}$$

Hint: In proving measurability, consider the function $(x,t) \in E \times [0,\infty) \longmapsto f(x) - t \in (-\infty, \infty]$; and get (4.1.9) as an application of Tonelli's Theorem.

Clearly (4.1.9) can be interpreted as the statement that *the integral of a non-negative function is the area under its graph.*

4.1.10 Exercise: Let $(E_1, \mathcal{B}_1, \mu_1)$ and $(E_2, \mathcal{B}_2, \mu_2)$ be σ-finite measure spaces and assume that, for $i \in \{1,2\}$, $\mathcal{B}_i = \sigma(E_i; \mathcal{C}_i)$, where $\mathcal{C}_i$ is a π-system containing a sequence $\{E_{i,n}\}_{n=1}^\infty$ such that $E_i = \bigcup_{n=1}^\infty E_{i,n}$ and $\mu_i(E_{i,n}) < \infty$, $n \ge 1$. Show that if ν is a measure on $(E_1 \times E_2, \mathcal{B}_1 \times \mathcal{B}_2)$ with the property that $\nu(\Gamma_1 \times \Gamma_2) = \mu_1(\Gamma_1)\,\mu_2(\Gamma_2)$ for all $\Gamma_i \in \mathcal{C}_i$, then $\nu = \mu_1 \times \mu_2$. Use this fact to show that, for any $M, N \in \mathbb{Z}^+$,

$$\lambda_{\mathbb{R}^{M+N}} = \lambda_{\mathbb{R}^M} \times \lambda_{\mathbb{R}^N}$$

on $\mathcal{B}_{\mathbb{R}^{M+N}} = \mathcal{B}_{\mathbb{R}^M} \times \mathcal{B}_{\mathbb{R}^N}$. (Cf. Lemma 3.2.2.)

4.1.11 Exercise: Let $(E_1, \mathcal{B}_1, \mu_1)$ and $(E_2, \mathcal{B}_2, \mu_2)$ be σ-finite measure spaces. Given $\Gamma \in \mathcal{B}_1 \times \mathcal{B}_2$, define

$$\Gamma_{(1)}(x_2) \equiv \Big\{x_1 \in E_1 : (x_1, x_2) \in \Gamma\Big\} \quad \text{for} \quad x_2 \in E_2$$

and

$$\Gamma_{(2)}(x_1) \equiv \Big\{x_2 \in E_2 : (x_1, x_2) \in \Gamma\Big\} \quad \text{for} \quad x_1 \in E_1.$$

Check both that $\Gamma_{(i)}(x_j) \in \mathcal{B}_i$ for each $x_j \in E_j$ and that $x_j \in E_j \longmapsto \mu_i(\Gamma_{(i)}(x_j)) \in [0,\infty]$ is measurable on $(E_j, \mathcal{B}_j)$ ($\{i,j\} = \{1,2\}$). Finally, show that $\mu_1 \times \mu_2(\Gamma) = 0$ if and only if $\mu_i\Big(\Gamma_{(i)}(x_j)\Big) = 0$ for μ_j-almost every $x_j \in E_j$; and, conclude that $\mu_1\Big(\Gamma_{(1)}(x_2)\Big) = 0$ for μ_2-almost every $x_2 \in E_2$ if and only if $\mu_2\Big(\Gamma_{(2)}(x_1)\Big) = 0$ for μ_1-almost every $x_1 \in E_1$. In other words, $\Gamma \in \mathcal{B}_1 \times \mathcal{B}_2$ has $\mu_1 \times \mu_2$-measure 0 if and only if μ_1-almost every *vertical slice* (μ_2-almost every *horizontal slice*) has μ_2-measure (μ_1-measure) 0.

4.1.12 Exercise: The condition that the measure spaces of which one is taking a product be σ-finite is essential if one wants to carry out the program in this section. To see this, let $E_1 = E_2 = (0,1)$ and $\mathcal{B}_1 = \mathcal{B}_2 = \mathcal{B}_{(0,1)}$. Define μ_1 on $(E_1, \mathcal{B}_1)$ so that $\mu_1(\Gamma)$ is the number of elements in Γ ($\equiv \infty$ if Γ is not a finite set) and show that μ_1 is a measure on $(E_1, \mathcal{B}_1)$. Next, take μ_2 to be Lebesgue measure $\lambda_{(0,1)}$ on $(E_2, \mathcal{B}_2)$. Show that there is a set $\Gamma \in \mathcal{B}_1 \times \mathcal{B}_2$ such that

$$\int_{E_2} \mathbf{1}_\Gamma(x_1, x_2)\,\mu_2(dx_2) = 0 \quad \text{for every } x_1 \in E_1$$

but

$$\int_{E_1} \mathbf{1}_\Gamma(x_1, x_2)\,\mu_1(dx_1) = 1 \quad \text{for every } x_2 \in E_2.$$

(**Hint**: Try $\Gamma = \{(x,x) : 0 < x < 1\}$.) In particular, there is no way that the second equality in (4.1.4) can be made to hold. Notice that what fails here is really the *uniqueness* and not the *existence* in Lemma 4.1.3.

4.1.13 Exercise: Let $(E, \mathcal{B})$ be a measurable space. Given $-\infty \le a < b \le \infty$ and a function $f : (a,b) \times E \longrightarrow \mathbb{R}$ with the properties that $f(\,\cdot\,, x) \in C((a,b))$ for every $x \in E$ and $f(t, \,\cdot\,)$ is measurable on $(E, \mathcal{B})$ for every $t \in (a,b)$, show that f is measurable on $\big((a,b) \times E, \mathcal{B}_{(a,b)} \times \mathcal{B}\big)$. Next, suppose that $f(\,\cdot\,, x) \in C^1((a,b))$ for each $x \in E$, set $f'(t,x) = \frac{d}{dt} f(t,x)$, $x \in E$, and show that f' is measurable on $\big((a,b) \times E, \mathcal{B}_{(a,b)} \times \mathcal{B}\big)$. Finally, suppose that μ is a measure on $(E, \mathcal{B})$ and that there is a $g \in L^1(\mu)$ such that $|f(t,x)| \vee |f'(t,x)| \le g(x)$ for all $(t,x) \in (a,b) \times E$. Show not only that $\int_E f(\,\cdot\,, x)\,\mu(dx) \in C^1((a,b))$ but also that

$$\frac{d}{dt} \int_E f(t,x)\,\mu(dx) = \int_E f'(t,x)\,\mu(dx).$$

4.2 Steiner Symmetrization and the Isodiametric Inequality

In order to provide an example which displays the power of Fubini's Theorem, we will prove in this section an elementary but important inequality about Lebesgue measure. Namely, we will show that, for any bounded subset $\Gamma \subset \mathbb{R}^N$,

$$|\Gamma|_e \le \Omega_N \mathrm{rad}(\Gamma)^N, \tag{4.2.1}$$

where Ω_N denotes the volume of the unit ball $B(0,1)$ in $\mathbb{R}^N$ and

$$\mathrm{rad}(\Gamma) \equiv \sup\left\{ \tfrac{|y-x|}{2} : x,\, y \in \Gamma \right\}$$

is the *radius* (i.e., half the *diameter*) of Γ. Notice (cf. (**ii**) in Exercise 2.2.3) that what (4.2.1) says is that, among all the subsets of $\mathbb{R}^N$ with a given diameter, the ball of that diameter has the largest volume; it is for this reason that (4.2.1) is called the **isodiametric inequality**.

At first glance one might be inclined to think that there is nothing to (4.2.1). Indeed, one might carelessly suppose that every Γ is a subset of a closed ball of radius $\mathrm{rad}(\Gamma)$ and therefore that (4.2.1) is trivial. This is true when $N = 1$. However, after a moment's thought, one realizes that, for $N > 1$, although Γ is always contained in a closed ball whose radius is equal to the diameter of Γ, it is not necessarily contained in one with the same radius as Γ. (For example, consider an equilateral triangle in $\mathbb{R}^2$.) Thus, the inequality $|\Gamma|_e \le \Omega_N \big(2\mathrm{rad}(\Gamma)\big)^N$ is trivial, but the inequality in (4.2.1) is not! On the other hand, there are many Γ's for which (4.2.1) is easy. In particular, if Γ is *symmetric* in the sense that $\Gamma = -\Gamma \equiv \{-x : x \in \Gamma\}$, then it is clear that

$$x \in \Gamma \implies 2|x| = |x + x| \le 2\mathrm{rad}(\Gamma) \quad \text{and therefore} \quad \Gamma \subseteq \overline{B\big(0, \mathrm{rad}(\Gamma)\big)}.$$

Hence (4.2.1) is trivial when Γ is symmetric, and so all that we have to do is devise a procedure to reduce the general case to the symmetric one.

The method with which we will perform this reduction is based on a famous construction known as the **Steiner symmetrization procedure**. To describe Steiner's procedure, we must first introduce a little notation. Given $\mathbf{v}$ from the unit $(N-1)$-**sphere** $\mathbf{S}^{N-1} \equiv \{x \in \mathbb{R}^N : |x| = 1\}$, let $\mathbf{L}(\mathbf{v})$ denote the line $\{t\mathbf{v} : t \in \mathbb{R}\}$, $\mathbf{P}(\mathbf{v})$ the $(N-1)$-dimensional subspace $\{x \in \mathbb{R}^N : x \perp \mathbf{v}\}$, and define

$$\mathcal{S}(\Gamma; \mathbf{v}) \equiv \left\{x + t\mathbf{v} : x \in \mathbf{P}(\mathbf{v}) \text{ and } |t| < \tfrac{1}{2}\ell(\Gamma; \mathbf{v}, x)\right\},$$

where

$$\ell(\Gamma; \mathbf{v}, x) \equiv \left|\{t \in \mathbb{R} : x + t\mathbf{v} \in \Gamma\}\right|_{\mathsf{e}}$$

is the length of the intersection of the line $x + \mathbf{L}(\mathbf{v})$ with Γ. Notice that, in the creation $\mathcal{S}(\Gamma; \mathbf{v})$ from Γ, we have taken the intersection of Γ with $x + \mathbf{L}(\mathbf{v})$, squashed it to remove all gaps, and then slid the resulting interval along $x + \mathbf{L}(\mathbf{v})$ until its center point falls at x. In particular, $\mathcal{S}(\Gamma; \mathbf{v})$ is the *symmetrization* of Γ with respect to the subspace $\mathbf{P}(\mathbf{v})$ in the sense that, for each $x \in \mathbf{P}(\mathbf{v})$,

$$x + t\mathbf{v} \in \mathcal{S}(\Gamma; \mathbf{v}) \iff x - t\mathbf{v} \in \mathcal{S}(\Gamma; \mathbf{v}); \tag{4.2.2}$$

what is only slightly less obvious is that $\mathcal{S}(\Gamma; \mathbf{v})$ possesses the properties proved in the next lemma.

4.2.3 Lemma. *Let Γ be a bounded element of $\mathcal{B}_{\mathbb{R}^N}$. Then, for each $\mathbf{v} \in \mathbf{S}^{N-1}$, $\mathcal{S}(\Gamma; \mathbf{v})$ is also a bounded element of $\mathcal{B}_{\mathbb{R}^N}$,* $\mathrm{rad}(\mathcal{S}(\Gamma; \mathbf{v})) \leq \mathrm{rad}(\Gamma)$, *and $|\mathcal{S}(\Gamma; \mathbf{v})| = |\Gamma|$. Finally, if $\mathbf{R} : \mathbb{R}^N \longrightarrow \mathbb{R}^N$ is a rotation for which $\mathbf{L}(\mathbf{v})$ and Γ are invariant (i.e., $\mathbf{R}(\mathbf{L}(\mathbf{v})) = \mathbf{L}(\mathbf{v})$ and $\mathbf{R}(\Gamma) = \Gamma$), then $\mathbf{R}\mathcal{S}(\Gamma; \mathbf{v}) = \mathcal{S}(\Gamma; \mathbf{v})$.*

PROOF: We begin with the observations that there is nothing to do when $N = 1$ and that, because none of the quantities under consideration depends on the particular choice of coordinate axes, we may and will assume not only that $N > 1$ but also that $\mathbf{v} = \mathbf{e}_N \equiv (0, \ldots, 0, 1)$. In particular, this means, by Lemma 4.1.2, that

$$\xi \in \mathbb{R}^{N-1} \longmapsto f(\xi) \equiv \frac{1}{2} \int_{\mathbb{R}} \mathbf{1}_\Gamma\big((\xi, t)\big)\, dt \in [0, \infty)$$

is $\mathcal{B}_{\mathbb{R}^{N-1}}$-measurable; and therefore, by Exercise 4.1.8, because $\mathcal{S}(\Gamma; \mathbf{e}_N)$ is equal to

$$\big\{(\xi, t) \in \mathbb{R}^{N-1} \times [0, \infty) : t < f(\xi)\big\} \cup \big\{(\xi, t) \in \mathbb{R}^{N-1} \times (-\infty, 0] : -t < f(\xi)\big\},$$

we know both that $\mathcal{S}(\Gamma; \mathbf{e}_N)$ is an element of $\mathcal{B}_{\mathbb{R}^N}$ and that

$$\big|\mathcal{S}(\Gamma; \mathbf{e}_N)\big| = 2 \int_{\mathbb{R}^{N-1}} f(\xi)\, d\xi = \int_{\mathbb{R}^{N-1}} \left(\int_{\mathbb{R}} \mathbf{1}_\Gamma\big((\xi, t)\big)\, dt\right) d\xi = |\Gamma|,$$

where, in the final step, we have applied Tonelli's Theorem.

We next turn to the proof that $\mathrm{rad}\big(\mathcal{S}(\Gamma;\mathbf{e}_N)\big)$ cannot be larger than $\mathrm{rad}(\Gamma)$; and, in doing so, we will, without loss of generality, add the assumption that Γ is compact. Now suppose that $x,\,y\in\mathcal{S}(\Gamma;\mathbf{e}_N)$ are given, and choose $\xi,\,\eta\in\mathbb{R}^{N-1}$ and $s,\,t\in\mathbb{R}$ so that $x=(\xi,s)$ and $y=(\eta,t)$. Next, set

$$M^\pm(x)=\pm\sup\{\tau:\ (\xi,\pm\tau)\in\Gamma\}\quad\text{and}\quad M^\pm(y)=\pm\sup\{\tau:\ (\eta,\pm\tau)\in\Gamma\},$$

and note that, because Γ is compact, all four of the points $X^\pm\equiv\big(\xi,M^\pm(x)\big)$ and $Y^\pm=\big(\eta,M^\pm(y)\big)$ are elements of Γ. Moreover, $2|s|\le M^+(x)-M^-(x)$ and $2|t|\le M^+(y)-M^-(y)$; and therefore

$$\begin{aligned}\big(M^+(y)-M^-(x)\big)&\vee\big(M^+(x)-M^-(y)\big)\\&\ge\frac{M^+(y)-M^-(x)}{2}+\frac{M^+(x)-M^-(y)}{2}\\&=\frac{M^+(y)-M^-(y)}{2}+\frac{M^+(x)-M^-(x)}{2}\ge|s|+|t|.\end{aligned}$$

In particular, this means that

$$\begin{aligned}|y-x|^2=|\eta-\xi|^2+|t-s|^2&\le|\eta-\xi|^2+\big(|s|+|t|\big)^2\\&\le|\eta-\xi|^2+\Big(\big(M^+(y)-M^-(x)\big)\vee\big(M^+(x)-M^-(y)\big)\Big)^2\\&\le\big(|Y^+-X^-|\vee|X^+-Y^-|\big)^2\le4\,\mathrm{rad}(\Gamma)^2.\end{aligned}$$

Finally, let $\mathbf{R}$ be a rotation. It is then an easy matter to check that $\mathbf{P}(\mathbf{R}\mathbf{v})=\mathbf{R}\big(\mathbf{P}(\mathbf{v})\big)$ and that $\ell(\mathbf{R}\Gamma;\mathbf{R}\mathbf{v},\mathbf{R}x)=\ell(\Gamma;\mathbf{v},x)$ for all $x\in\mathbf{P}(\mathbf{v})$. Hence, $\mathcal{S}(\mathbf{R}\Gamma,\mathbf{R}\mathbf{v})=\mathbf{R}\mathcal{S}(\Gamma,\mathbf{v})$. In particular, if $\Gamma=\mathbf{R}\Gamma$ and $\mathbf{L}(\mathbf{v})=\mathbf{R}\big(\mathbf{L}(\mathbf{v})\big)$, then $\mathbf{R}\mathbf{v}=\pm\mathbf{v}$, and so the preceding (together with (4.2.2)) leads to $\mathbf{R}\mathcal{S}(\Gamma,\mathbf{v})=\mathcal{S}(\Gamma,\mathbf{v})$. □

4.2.4 Theorem. *The inequality in* (4.2.1) *holds for every bounded* $\Gamma\subseteq\mathbb{R}^N$.

PROOF: Clearly it suffices to treat the case when Γ is compact and therefore Borel measurable. Thus, let a compact Γ be given, choose an orthonormal basis $\{\mathbf{e}_1,\dots,\mathbf{e}_N\}$ for $\mathbb{R}^N$, set $\Gamma_0=\Gamma$, and define $\Gamma_n=\mathcal{S}(\Gamma_{n-1};\mathbf{e}_n)$ for $1\le n\le N$. By repeated application of (4.2.2) and Lemma 4.2.3, we know that $|\Gamma_n|=|\Gamma|$, $\mathrm{rad}(\Gamma_n)\le\mathrm{rad}(\Gamma)$, and that $\mathbf{R}_m\Gamma_n=\Gamma_n$, $1\le m\le n\le N$, where $\mathbf{R}_m$ is the rotation given by $\mathbf{R}_m x=x-2(x,\mathbf{e}_m)_{\mathbb{R}^N}\mathbf{e}_m$ for each $x\in\mathbb{R}^N$. In particular, this means that $\mathbf{R}_m\Gamma_N=\mathbf{R}_m\Gamma_N$ for all $1\le m\le N$, hence $-\Gamma_N=\Gamma_N$, and therefore (cf. the discussion preceding the introduction of Steiner's procedure)

$$|\Gamma|=|\Gamma_N|\le\Omega_N\mathrm{rad}(\Gamma_N)^N\le\Omega_N\mathrm{rad}(\Gamma)^N.\quad\square$$

We will now use (4.2.1) to give a description, due to F. Hausdorff, of Lebesgue's measure which, as distinguished from the one given at the beginning of

Section 2.1, is completely coordinate free. Namely, we are going to show that, for all $\Gamma \subseteq \mathbb{R}^N$,

$$|\Gamma|_e = \mathbf{H}^N(\Gamma) \equiv \inf\left\{\sum_{C\in\mathcal{C}} \Omega_N \operatorname{rad}(C)^N : \mathcal{C} \text{ a countable cover of } \Gamma\right\}. \tag{4.2.5}$$

We emphasize that we have placed *no restriction* on the sets C making up the cover $\mathcal{C}$. On the other hand, it should be clear that $\mathbf{H}^N(\Gamma)$ would be unchanged if we were to restrict ourselves to coverings by closed sets or, for that matter, to coverings by open sets.

Directly from its definition, one sees that $\mathbf{H}^N$ is monotone and subadditive in the sense that

$$\mathbf{H}^N(\Gamma_1) \leq \mathbf{H}^N(\Gamma_2) \quad \text{whenever } \Gamma_1 \subseteq \Gamma_2$$

and

$$\mathbf{H}^N\left(\bigcup_1^\infty \Gamma_n\right) \leq \sum_1^\infty \mathbf{H}^N(\Gamma_n) \quad \text{for all } \{\Gamma_n\}_1^\infty \subseteq \mathcal{P}(\mathbb{R}^N).$$

Indeed, the first of these is completely trivial, and the second follows by choosing, for a given $\epsilon > 0$, $\{\mathcal{C}_m\}_1^\infty$ so that

$$\Gamma_m \subseteq \bigcup \mathcal{C}_m \quad \text{and} \quad \sum_{C\in\mathcal{C}_m} \Omega_N\big(\operatorname{rad}(C)\big)^N \leq \mathbf{H}^N(\Gamma_m) + 2^{-m}\epsilon$$

and noting that

$$\mathbf{H}^N\left(\bigcup_1^\infty \Gamma_m\right) \leq \sum_{m=1}^\infty \sum_{C\in\mathcal{C}_m} \Omega_N\big(\operatorname{rad}(C)\big)^N \leq \sum_1^\infty \mathbf{H}^N(\Gamma_n) + \epsilon.$$

Moreover, because

$$|\Gamma|_e \leq \sum_{C\in\mathcal{C}} |C|_e \quad \text{for any countable cover } \mathcal{C},$$

the inequality $|\Gamma|_e \leq \mathbf{H}^N(\Gamma)$ is an essentially trivial consequence of (4.2.1). In order to prove the opposite inequality, we will use the following lemma.

4.2.6 Lemma. *For any open set G in $\mathbb{R}^N$ with $|G| < \infty$, there exists a sequence $\{B_n\}_1^\infty$ of mutually disjoint closed balls contained in G with the property that*

$$\left|G \setminus \bigcup_1^\infty B_n\right| = 0. \tag{4.2.7}$$

PROOF: If $G = \emptyset$, there is nothing to do. Thus, assume that $G \neq \emptyset$, and set $G_0 = G$. Using Lemma 2.1.9, choose a countable, exact cover $\mathcal{C}_0$ of G_0 by non-overlapping cubes Q. Next, given $Q \in \mathcal{C}_0$, choose $x \in \mathbb{R}^N$ and $\delta \in [0,\infty)$ so that

$$Q = \prod_1^N [x^i - \delta, x^i + \delta] \quad \text{and set} \quad B_Q = \overline{B\left(x, \tfrac{\delta}{2}\right)}.$$

Clearly, the B_Q's are mutually disjoint closed balls. At the same time, there is a dimensional constant $\alpha_N \in (0,1)$ for which $|B_Q| \geq \alpha_N |Q|$; and therefore we can choose a finite subset $\{B_{0,1} \dots, B_{0,n_0}\} \subseteq \{B_Q : Q \in \mathcal{C}_0\}$ in such a way that

$$\left| G_0 \setminus \bigcup_{m=1}^{n_0} B_{0,m} \right| \leq \beta_N |G_0| \quad \text{where } \beta_N \equiv 1 - \tfrac{\alpha_N}{2} \in (0,1).$$

Now set $G_1 = G_0 \setminus \bigcup_1^{n_0} B_{0,m}$. Noting that G_1 is again non-empty and open, we can repeat the preceding argument to find a finite collection of mutually disjoint closed balls $B_{1,m} \subset G_1$, $1 \leq m \leq n_1$, in such a way that

$$\left| G_1 \setminus \bigcup_{m=1}^{n_1} B_{1,m} \right| \leq \beta_N |G_1|.$$

More generally, we can use induction on $\ell \in \mathbb{Z}^+$ to construct open sets $G_\ell \subseteq G_{\ell-1}$ and finite collections $B_{\ell,1}, \dots, B_{\ell,n_\ell}$ of mutually disjoint closed balls $B \subset G_\ell$ so that

$$\left| G_{\ell+1} \right| \leq \beta_N \left| G_\ell \right| \quad \text{where } G_{\ell+1} = G_\ell \setminus \bigcup_{m=1}^{n_\ell} B_{\ell,m};$$

clearly the collection

$$\left\{ B_{\ell,m} : \ell \in \mathbb{N} \text{ and } 1 \leq m \leq n_\ell \right\}$$

has the required properties. □

4.2.8 Theorem. *The equality in* (4.2.5) *holds for any set* $\Gamma \subseteq \mathbb{R}^N$.

PROOF: As we have already pointed out, the inequality $|\Gamma|_e \leq \mathbf{H}^N(\Gamma)$ is an immediate consequence of (4.2.1). To get the opposite inequality, first observe that $\mathbf{H}^N(\Gamma) = 0$ if $|\Gamma|_e = 0$. Indeed, if $|\Gamma|_e = 0$, then, for each $\epsilon > 0$ we can first find an open $G \supseteq \Gamma$ with $|G| < \epsilon$ and then, by Lemma 2.1.9, a countable, exact cover $\mathcal{C}$ of G by non-overlapping cubes Q, which means that

$$\mathbf{H}^N(\Gamma) \leq \mathbf{H}^N(G) \leq \sum_{Q \in \mathcal{C}} \Omega_N \operatorname{rad}(Q)^N \leq \left(\frac{\sqrt{N}}{2} \right)^N \Omega_N |G| < \left(\frac{\sqrt{N}}{2} \right)^N \Omega_N \epsilon.$$

Next, because $\mathbf{H}^N$ is countably subadditive, it suffices to prove that $\mathbf{H}^N(\Gamma) \leq |\Gamma|_e$ for bounded sets Γ. Finally, suppose that Γ is a bounded set, and let G be any open superset of Γ with $|G| < \infty$. By Lemma 4.2.6, we can find a sequence $\{B_n\}_1^\infty$ of mutually disjoint closed balls $B \subset G$ for which

$$|G \setminus A| = 0 \quad \text{where } A = \bigcup_1^\infty B_n.$$

Hence, because $\mathbf{H}^N(\Gamma) \leq \mathbf{H}^N(G) \leq \mathbf{H}^N(A) + \mathbf{H}^N(G \setminus A) = \mathbf{H}^N(A)$, we see that

$$\mathbf{H}^N(\Gamma) \leq \sum_1^\infty \Omega_N \mathrm{rad}(B_n)^N = \sum_1^\infty |B_n| = |A| = |G|$$

for every open $G \supseteq \Gamma$; and, after taking the infimum over such G's, we arrive at the desired conclusion. □

Exercises

4.2.9 Exercise: Using the definition of $\mathbf{H}^N$ given by the second relation in (4.2.5), give a direct (i.e., one which does not make use of the first relation in (4.2.5)) proof that $\mathbf{H}^N(\Gamma) = 0$ for every bounded subset of a hyperplane (cf. (**i**) in Exercise 2.2.3) of $\mathbb{R}^N$.

Chapter V
Changes of Variable

5.0 Introduction

We have now developed the basic theory of Lebesgue integration. However, thus far we have nearly no tools with which to compute the integrals which we have shown to exist. The purpose of the present chapter is to introduce a technique which often makes evaluation, or at least estimation, possible. The technique is that of changing variables. In this introduction, we describe the technique in complete generality. In ensuing sections we will give some examples of its applications.

Let $(E_1, \mathcal{B}_1)$ and $(E_2, \mathcal{B}_2)$ be a pair of measurable spaces. Given a measure μ on $(E_1, \mathcal{B}_1)$ and a measurable map Φ on $(E_1, \mathcal{B}_1)$ into $(E_2, \mathcal{B}_2)$, we define the **pushforward** or **image** $\Phi_*\mu = \mu \circ \Phi^{-1}$ of μ under Φ by $\Phi_*\mu(\Gamma) = \mu(\Phi^{-1}(\Gamma))$ for $\Gamma \in \mathcal{B}_2$. Because set theoretic operations are preserved by inverse maps, it is an easy matter to check that $\Phi_*\mu$ is a measure on $(E_2, \mathcal{B}_2)$.

5.0.1 Lemma. *For every non-negative measurable function φ on $(E_2, \mathcal{B}_2)$,*

$$\int_{E_2} \varphi \, d(\Phi_*\mu) = \int_{E_1} \varphi \circ \Phi \, d\mu. \tag{5.0.2}$$

Moreover, $\varphi \in L^1(E_2, \mathcal{B}_2, \Phi_*\mu)$ if and only if $\varphi \circ \Phi \in L^1(E_1, \mathcal{B}_1, \mu)$, and (5.0.2) holds for all $\varphi \in L^1(E_2, \mathcal{B}_2, \Phi_*\mu)$.

PROOF: Clearly it suffices to prove the first assertion. To this end, note that (5.0.2) holds, by definition, when f is the indicator of a set $\Gamma \in \mathcal{B}_2$. Hence, it also holds when f is a non-negative measurable simple function on $(E_2, \mathcal{B}_2)$. Thus, by the Monotone Convergence Theorem, it must hold for all non-negative measurable functions on $(E_2, \mathcal{B}_2)$. □

The reader should note that Lemma 5.0.1 is so close to the definition that it is hardly an honest theorem. It is only when a judicious choice of Φ has been made that one gets anything useful from (5.0.2).

Exercises

5.0.3 Exercise: Referring to Theorem 2.2.2, let A be a non-singular $N \times N$ matrix and T_A the associated linear transformation on $\mathbb{R}^N$. Show that $(T_A)_*\lambda_{\mathbb{R}^N} = |\det(A)|^{-1}\lambda_{\mathbb{R}^N}$.

5.0.4 Exercise: Let $-\infty < a < b < \infty$ and a right-continuous, non-decreasing $\psi : [a,b] \longrightarrow \mathbb{R}$ with $\psi(a) = 0$ be given. Set $L = \psi(b)$, and define $\psi^{-1} : [0,L] \longrightarrow [a,b]$ so that $\psi^{-1}(t) = \inf\{s \in [a,b] : \psi(s) \geq t\}$. Show that $\{t \in [0,L] : \psi^{-1}(t) \leq s\} = [0, \psi(s)]$ for every $s \in [a,b]$. Conclude that ψ^{-1} is $\mathcal{B}_{[0,L]}$-measurable. Also, if $\lambda_{[0,L]}$ denotes the restriction of Lebesgue measure to $\mathcal{B}_{[0,L]}$, and $\mu_\psi \equiv (\psi^{-1})_*\lambda_{[0,L]}$, show that $\mu_\psi([a,s]) = \mu_\psi((a,s]) = \psi(s)$ for each $s \in [a,b]$.

5.1 Lebesgue vs. Riemann Integrals

Our first important example of a change of variables will relate integrals over an arbitrary measure space to integrals on the real line. Namely, given a measurable $\overline{\mathbb{R}}$-valued function f on a measure space $(E, \mathcal{B}, \mu)$, define the **distribution of f under μ** to be the measure $\mu_f \equiv f_*\mu$ on $(\overline{\mathbb{R}}, \mathcal{B}_{\overline{\mathbb{R}}})$.[1] We then have that, for any non-negative measurable φ on $(\overline{\mathbb{R}}, \mathcal{B}_{\overline{\mathbb{R}}})$

$$\int_E \varphi \circ f(x)\,\mu(dx) = \int_{\overline{\mathbb{R}}} \varphi(t)\,\mu_f(dt). \tag{5.1.1}$$

The reason why it is often useful to make this change of variables is that the integral on the right hand side of (5.1.1) can often be evaluated as the limit of Riemann integrals to which all the fundamental facts of the calculus are applicable.

In order to see how the right hand side of (5.1.1) leads us to Riemann integrals, we will prove a general fact about the relationship between Lebesgue and Riemann integrals on the line. Perhaps the most interesting feature of this result is that it shows that a complete description of the class of Riemann integrable functions in terms of continuity properties defies a totally Riemannian solution and requires the Lebesgue notion of *almost everywhere*.

5.1.2 Theorem. *Let ν be a finite measure on $((a,b], \mathcal{B}_{(a,b]})$, where $-\infty < a < b < \infty$, and set $\psi(t) = \nu((a,t])$ for $t \in [a,b]$ ($\psi(a) = \nu(\emptyset) = 0$). Then, ψ is right-continuous on $[a,b)$, non-decreasing on $[a,b]$, $\psi(a) = 0$, and, for each $t \in (a,b]$, $\psi(t) - \psi(t-) = \nu(\{t\})$, where $\psi(t-) \equiv \lim_{s \nearrow t} \psi(s)$ is the left-limit*

[1] In probability theory, distributions take on particular significance. In fact, from the point of view of a probabilistic purist, it is the distribution of a function, as opposed to the function itself, which is its distinguishing feature.

of ψ at t. Furthermore, if φ is a bounded function on $[a,b]$, then φ is Riemann integrable on $[a,b]$ with respect to ψ if and only if φ is continuous (a.e., ν) on $(a,b]$; in which case, φ is measurable on $\big((a,b],\overline{\mathcal{B}}^{\nu}_{(a,b]}\big)$ and

$$\int_{(a,b]} \varphi\, d\overline{\nu} = (\mathrm{R})\int_{[a,b]} \varphi(t)\, d\psi(t). \tag{5.1.3}$$

(See Exercises 5.0.4 and 8.1.31 to learn how to go from a right-continuous, non-decreasing ψ to a measure μ.)

PROOF: It will be convenient to think of ν as being defined on $\big([a,b],\mathcal{B}_{[a,b]}\big)$ by $\nu(\Gamma)\equiv\nu(\Gamma\cap(a,b])$ for $\Gamma\in\mathcal{B}_{[a,b]}$. Thus, we will do so.

Obviously ψ is non-decreasing on $[a,b]$; and therefore (cf. Lemma 1.2.20) ψ has at most countably many points of discontinuity. Moreover, $\psi(a)=\nu(\emptyset)=0$, for each $s\in[a,b)$ $\psi(s)=\mu\big((a,s]\big)=\lim_{t\searrow s}\mu\big((a,t]\big)=\lim_{t\searrow s}\psi(t)$, and for each $t\in(a,b]$ $\psi(t)-\psi(t-)=\lim_{s\nearrow t}\nu\big((s,t]\big)=\nu(\{t\})$.

Assume that φ is Riemann integrable on $[a,b]$ with respect to ψ. To see that φ is continuous (a.e., ν) on $(a,b]$, choose, for each $n\geq 1$, a finite, non-overlapping, exact cover $\mathcal{C}_n$ of $[a,b]$ by intervals I such that $\|\mathcal{C}_n\|<\frac{1}{n}$ and ψ is continuous at I^- for every $I\in\mathcal{C}_n$. If $\Delta=\bigcup_{n=1}^{\infty}\{I^-: I\in\mathcal{C}_n\}$, then $\mu(\Delta)=0$. Given $m\geq 1$, let $\mathcal{C}_{m,n}$ be the set of those $I\in\mathcal{C}_n$ such that $\sup_I\varphi-\inf_I\varphi\geq\frac{1}{m}$. It is then easy to check that

$$\{t\in(a,b]\setminus\Delta:\varphi \text{ is not continuous at } t\}\subseteq\bigcup_{m=1}^{\infty}\bigcap_{n=1}^{\infty}\bigcup\mathcal{C}_{m,n}.$$

But, by Exercise 1.2.26,

$$\nu\left(\bigcap_{n=1}^{\infty}\bigcup\mathcal{C}_{m,n}\right)\leq\lim_{n\to\infty}\sum_{I\in\mathcal{C}_{m,n}}\Delta_I\psi=0,$$

and therefore $\nu\big(\bigcup_{m=1}^{\infty}\bigcap_{n=1}^{\infty}\bigcup\mathcal{C}_{m,n}\big)=0$. Hence, we have now shown that φ is continuous (a.e., ν) on $(a,b]$.

Conversely, suppose that φ is continuous (a.e., ν) on $(a,b]$. Let $\{\mathcal{C}_n\}_1^{\infty}$ be a sequence of finite, non-overlapping, exact covers of $[a,b]$ by intervals I such that $\|\mathcal{C}_n\|\longrightarrow 0$. For each $n\geq 1$, define $\overline{\varphi}_n(t)=\sup_I\varphi$ and $\underline{\varphi}_n(t)=\inf_I\varphi$ for $t\in I\setminus\{I^-\}$ and $I\in\mathcal{C}_n$. Clearly, both $\overline{\varphi}_n$ and $\underline{\varphi}_n$ are measurable on $\big((a,b],\mathcal{B}_{(a,b]}\big)$. Moreover,

$$\inf_{(a,b]}\varphi\leq\underline{\varphi}_n\leq\varphi\leq\overline{\varphi}_n\leq\sup_{(a,b]}\varphi$$

for all $n\geq 1$. Finally, since φ is continuous (a.e., ν),

$$\varphi=\lim_{n\to\infty}\underline{\varphi}_n=\lim_{n\to\infty}\overline{\varphi}_n\quad\text{(a.e., }\nu);$$

and so, not only is $\varphi = \underline{\lim}_{n\to\infty} \underline{\varphi}_n$ (a.e., ν) and therefore measurable on $\left((a,b], \overline{\mathcal{B}}^{\nu}_{(a,b]}\right)$, but also

$$\lim_{n\to\infty} \int_{(a,b]} \underline{\varphi}_n \, d\nu = \int_{(a,b]} \varphi \, d\overline{\nu} = \lim_{n\to\infty} \int_{(a,b]} \overline{\varphi}_n \, d\nu.$$

In particular, we conclude that

$$\sum_{I\in\mathcal{C}_n} \left(\sup_I \varphi - \inf_I \varphi\right) \Delta_I \psi = \int_{(a,b]} \left(\overline{\varphi}_n - \underline{\varphi}_n\right) d\nu \longrightarrow 0$$

as $n \to \infty$. From this it is clear both that φ is Riemann integrable on $[a,b]$ with respect to ψ and that (5.1.3) holds. $\square$

We are now ready to prove the main result to which this section is devoted.

5.1.4 Theorem. *Let $(E, \mathcal{B}, \mu)$ be a measure space and f a non-negative, measurable function on $(E, \mathcal{B})$. Then $t \in (0,\infty) \longmapsto \mu(f > t) \in [0,\infty]$ is a right-continuous, non-increasing function. In particular, it is measurable on $\left((0,\infty), \mathcal{B}_{(0,\infty)}\right)$ and has at most a countable number of discontinuities. Next, assume that $\varphi \in C\left([0,\infty)\right) \cap C^1\left((0,\infty)\right)$ is a non-decreasing function satisfying $\varphi(0) = 0 < \varphi(t)$, $t > 0$, and set $\varphi(\infty) = \lim_{t\to\infty} \varphi(t)$. Then*

$$\int_E \varphi \circ f(x)\, \mu(dx) = \int_{(0,\infty)} \varphi'(t) \mu(f > t)\, \lambda_{\mathbb{R}}(dt). \tag{5.1.5}$$

Hence, either $\mu(f > \delta) = \infty$ for some $\delta > 0$, in which case both sides of (5.1.5) are infinite, or, for each $0 < \delta < r < \infty$, the map $t \in [\delta, r] \longmapsto \varphi'(t)\mu(f > t)$ is Riemann integrable and

$$\int_E \varphi \circ f(x)\, \mu(dx) = \lim_{\substack{\delta \searrow 0 \\ r \nearrow \infty}} (\mathrm{R}) \int_{[\delta,r]} \varphi'(t) \mu(f > t)\, dt.$$

PROOF: It is clear that $t \in (0,\infty) \longmapsto \mu(f > t)$ is right-continuous and non-increasing. Hence, if $\delta = \sup\{t \in (0,\infty) : \mu(f > t) = \infty\}$, then $\mu(f > t) = \infty$ for $t \in (0,\delta)$ and $t \in (\delta,\infty) \longmapsto \mu(f > t)$ has at most a countable number of discontinuities. Furthermore, if

$$h_n(t) = \mu\left(f > \tfrac{k+1}{n}\right) \quad \text{for} \quad t \in \left(\tfrac{k}{n}, \tfrac{k+1}{n}\right], k \geq 0, \text{ and } n \geq 1,$$

then each h_n is clearly measurable on $\left((0,\infty), \mathcal{B}_{(0,\infty)}\right)$ and $h_n(t) \longrightarrow \mu(f > t)$ for each $t \in (0,\infty)$. Hence, $t \in (0,\infty) \longmapsto \mu(f > t)$ is measurable on $\left((0,\infty), \mathcal{B}_{(0,\infty)}\right)$.

We turn next to the proof of (5.1.5). Since (cf. Exercise 3.3.19)

$$\lim_{\alpha\searrow 0} \int_{(\alpha,\delta]} \varphi'(t)\, dt = \lim_{\alpha\searrow 0} (\mathrm{R}) \int_\alpha^\delta \varphi'(t)\, dt = \varphi(\delta)$$

and therefore

$$\left(\int_E \varphi\circ f\,d\mu\right)\wedge\left(\int_{(0,\infty)}\varphi'(t)\mu(f>t)\,\lambda_{\mathbb{R}}(dt)\right)\geq\varphi(\delta)\mu(f>\delta),$$

it is clear that both sides of (5.1.5) are infinite when $\mu(f>\delta)=\infty$ for some $\delta>0$. Thus we will assume that $\mu(f>\delta)<\infty$ for every $\delta>0$. Then the restriction of μ_f to $\mathcal{B}_{[\delta,\infty)}$ is a finite measure for every $\delta>0$. Given $\delta>0$, set $\psi_\delta(t)=\mu_f((\delta,t])$ for $t\in[\delta,\infty)$ and apply Theorem 5.1.2 and Theorem 1.2.7 to see that

$$\begin{aligned}\int_{\{\delta<f\leq r\}}\varphi\circ f\,d\mu&=\int_{(\delta,r]}\varphi\,d\mu_f=(\mathrm{R})\int_{[\delta,r]}\varphi(t)\,d\psi_\delta(t)\\&=\varphi(r)\psi_\delta(r)-(\mathrm{R})\int_{[\delta,r]}\psi_\delta(t)\varphi'(t)\,dt\\&=\varphi(\delta)\psi_\delta(r)+(\mathrm{R})\int_{[\delta,r]}\big(\psi_\delta(r)-\psi_\delta(t)\big)\varphi'(t)\,dt\\&=\varphi(\delta)\mu(\delta<f\leq r)+(\mathrm{R})\int_{[\delta,r]}\mu(t<f\leq r)\varphi'(t)\,dt\\&=\varphi(\delta)\mu(\delta<f\leq r)+\int_{(\delta,r]}\mu(t<f\leq r)\varphi'(t)\,\lambda_{\mathbb{R}}(dt)\end{aligned}$$

for each $r\in(\delta,\infty)$. Hence, after simple arithmetic manipulation and an application of the Monotone Convergence Theorem, we get

$$\int_{\{\delta<f<\infty\}}\big(\varphi\circ f-\varphi(\delta)\big)\,d\mu=\int_{(\delta,\infty)}\varphi'(t)\mu(t<f<\infty)\,\lambda_{\mathbb{R}}(dt)$$

after $r\nearrow\infty$. At the same time, it is clear that

$$\begin{aligned}\int_{\{f=\infty\}}\big(\varphi\circ f-\varphi(\delta)\big)\,d\mu&=\big[\varphi(\infty)-\varphi(\delta)\big]\mu(f=\infty)\\&=\int_{(\delta,\infty)}\mu(f=\infty)\varphi'(t)\,\lambda_{\mathbb{R}}(dt);\end{aligned}$$

and, after combining these, we now arrive at

$$\int_{\{f>\delta\}}\big(\varphi\circ f(x)-\varphi(\delta)\big)\,\mu(dx)=\int_{(\delta,\infty)}\mu(f>t)\varphi'(t)\,\lambda_{\mathbb{R}}(dt).$$

Thus, (5.1.5) will be proved once we show that

$$\lim_{\delta\searrow 0}\int_{\{f>\delta\}}\big(\varphi\circ f(x)-\varphi(\delta)\big)\,\mu(dx)=\int_E\varphi\circ f(x)\,\mu(dx).$$

But if $0 < \delta_1 < \delta_2 < \infty$, then $0 \le (\varphi \circ f - \varphi(\delta_2))\mathbf{1}_{\{f>\delta_2\}} \le (\varphi \circ f - \varphi(\delta_1))\mathbf{1}_{\{f>\delta_1\}}$, and so the required convergence follows by the Monotone Convergence Theorem.

Finally, to prove the last part of the theorem when $\mu(f > \delta) < \infty$ for every $\delta > 0$, simply note that

$$\int_{(0,\infty)} \varphi'(t)\mu(f > t)\,\lambda_{\mathbb{R}}(dt) = \lim_{\substack{\delta \searrow 0 \\ r \nearrow \infty}} \int_{(\delta,r]} \varphi'(t)\mu(f > t)\,\lambda_{\mathbb{R}}(dt),$$

and apply Theorem 5.1.2. □

Exercises

5.1.6 Exercise: Let everything be as in Exercise 5.0.4, and apply the result proved there plus Theorem 5.1.2 to see that every bounded, μ_ψ-almost everywhere continuous $\varphi : [a,b] \longrightarrow \mathbb{R}$ is ψ-Riemann integrable and that

$$\int_{[a,b]} \varphi\, d\bar{\mu}_\psi = (\mathrm{R}) \int_{[a,b]} \varphi(s)\, d\psi(s).$$

In particular, when ψ is continuous, conclude that

$$(5.1.7) \qquad (\mathrm{R}) \int_{[a,b]} f \circ \psi(s)\, d\psi(s) = (\mathrm{R}) \int_{[\psi(a),\psi(b)]} f(t)\, dt, \quad f \in C\big([a,b]\big),$$

which is the classical *change of variables formula*

5.1.8 Exercise: A particularly important case of Theorem 5.1.4 is when $\varphi(t) = t^p$ for some $p \in (0,\infty)$, in which case (5.1.5) yields

$$(5.1.9) \qquad \int_E |f(x)|^p\, \mu(dx) = p \int_{(0,\infty)} t^{p-1} \mu(|f| > t)\, \lambda_{\mathbb{R}}(dt).$$

Use (5.1.9) to show that $|f|^p$ is μ-integrable if and only if

$$\sum_{n=1}^{\infty} \frac{1}{n^{p+1}} \mu\left(|f| > \tfrac{1}{n}\right) + \sum_{n=1}^{\infty} n^{p-1} \mu\big(|f| > n\big) < \infty.$$

Compare this result to the one obtained in the last part of Exercise 3.3.24.

5.2 Polar Coordinates

From now on, at least whenever the meaning is clear from the context, we will use the notation "dx" instead of the more cumbersome "$\lambda_{\mathbb{R}^N}(dx)$" when doing Lebesgue integration with respect to Lebesgue measure on $\mathbb{R}^N$.

In this section, we examine a change of variables which plays an extremely important role in the evaluation of many Lebesgue integrals over $\mathbb{R}^N$. Let $\mathbf{S}^{N-1}$ denote the unit $(N-1)$-**sphere** $\{x \in \mathbb{R}^N : |x| = 1\}$ in $\mathbb{R}^N$, and define $\Phi : \mathbb{R}^N \backslash \{0\} \longrightarrow \mathbf{S}^{N-1}$ by $\Phi(x) = \frac{x}{|x|}$. Clearly Φ is continuous. Next, define the **surface measure** $\lambda_{\mathbf{S}^{N-1}}$ on $\mathbf{S}^{N-1}$ to be the image under Φ of $N\lambda_{\mathbb{R}^N}$ restricted to $\mathcal{B}_{B_{\mathbb{R}^N}(0,1)\backslash\{0\}}$. Noting that $\Phi(rx) = \Phi(x)$ for all $r > 0$ and $x \in \mathbb{R}^N \setminus \{0\}$, we conclude from Exercise 5.0.3 that

$$\int_{B_{\mathbb{R}^N}(0,r)\backslash\{0\}} f \circ \Phi(x)\, dx = r^N \int_{B_{\mathbb{R}^N}(0,1)\backslash\{0\}} f \circ \Phi(x)\, dx$$

and therefore that

$$\int_{B_{\mathbb{R}^N}(0,r)\backslash\{0\}} f \circ \Phi(x)\, dx = \frac{r^N}{N} \int_{\mathbf{S}^{N-1}} f(\boldsymbol{\omega})\, \lambda_{\mathbf{S}^{N-1}}(d\boldsymbol{\omega}) \tag{5.2.1}$$

for every non-negative measurable f on $(\mathbf{S}^{N-1}, \mathcal{B}_{\mathbf{S}^{N-1}})$. In particular, using $\omega_{N-1} \equiv \lambda_{\mathbf{S}^{N-1}}(\mathbf{S}^{N-1})$ to denote the **surface area** of $\mathbf{S}^{N-1}$, we have that $\left|B_{\mathbb{R}^N}(\mathbf{0}, r)\right| = \Omega_N r^N$ where $\Omega_N = \frac{\omega_{N-1}}{N}$ is the **volume** of the unit ball $B_{\mathbb{R}^N}(0,1)$ in $\mathbb{R}^N$.

Next, define $\Psi : (0,\infty) \times \mathbf{S}^{N-1} \longrightarrow \mathbb{R}^N \setminus \{0\}$ by $\Psi(r, \boldsymbol{\omega}) = r\boldsymbol{\omega}$. Note that Ψ is one-to-one and onto; the pair $(r, \boldsymbol{\omega}) \equiv \Psi^{-1}(x) = (|x|, \Phi(x))$ are called the **polar coordinates** of the point $x \in \mathbb{R}^N \setminus \{0\}$. Finally, define the measure R_N on $\big((0,\infty), \mathcal{B}_{(0,\infty)}\big)$ by $R_N(\Gamma) = \int_\Gamma r^{N-1}\, dr$.

The importance of these considerations is contained in the following result.

5.2.2 Theorem. *Referring to the preceding, one has that*

$$\lambda_{\mathbb{R}^N} = \Psi_*\big(R_N \times \lambda_{\mathbf{S}^{N-1}}\big) \text{ on } \mathcal{B}_{\mathbb{R}^N\backslash\{0\}}.$$

In particular, if f is a non-negative, measurable function on $(\mathbb{R}^N, \mathcal{B}_{\mathbb{R}^N})$, then

$$\begin{aligned} \int_{\mathbb{R}^N} f(x)\, dx &= \int_{(0,\infty)} r^{N-1} \left(\int_{\mathbf{S}^{N-1}} f(r\boldsymbol{\omega})\, \lambda_{\mathbf{S}^{N-1}}(d\boldsymbol{\omega}) \right) dr \\ &= \int_{\mathbf{S}^{N-1}} \left(\int_{(0,\infty)} f(r\boldsymbol{\omega}) r^{N-1}\, dr \right) \lambda_{\mathbf{S}^{N-1}}(d\boldsymbol{\omega}). \end{aligned} \tag{5.2.3}$$

PROOF: By Exercise 3.3.23 and Theorem 4.1.5, all that we have to do is check that the first equation in (5.2.3) holds for every

$$f \in C_c(\mathbb{R}^N) \equiv \big\{f \in C(\mathbb{R}^N) : f \equiv 0 \text{ off of some compact set}\big\}.$$

To this end, let $f \in C_c(\mathbb{R}^N)$ be given and set $F(r) = \int_{B_{\mathbb{R}^N}(0,r)} f(x)\,dx$ for $r > 0$. Then, by (5.2.1), for all $r, h > 0$:

$$\begin{aligned} F(r+h) - F(r) &= \int_{B_{\mathbb{R}^N}(0,r+h)\setminus B_{\mathbb{R}^N}(0,r)} f(x)\,dx \\ &= \int\limits_{B_{\mathbb{R}^N}(0,r+h)\setminus B_{\mathbb{R}^N}(0,r)} f\circ\Psi(r,\Phi(x))\,dx \\ &\qquad + \int\limits_{B_{\mathbb{R}^N}(0,r+h)\setminus B_{\mathbb{R}^N}(0,r)} \big(f(x) - f\circ\Psi(r,\Phi(x))\big)\,dx \\ &= \frac{(r+h)^N - r^N}{N} \int_{\mathbf{S}^{N-1}} f\circ\Psi(r,\omega)\,\lambda_{\mathbf{S}^{N-1}}(d\omega) + o(h), \end{aligned}$$

where "$o(h)$" denotes a function which tends to 0 faster than h. Hence, F is continuously differentiable on $(0,\infty)$ and its derivative at $r \in (0,\infty)$ is given by $r^{N-1}\int_{\mathbf{S}^{N-1}} f\circ\Psi(r,\omega)\,\lambda_{\mathbf{S}^{N-1}}(d\omega)$. Since $F(r) \longrightarrow 0$ as $r \searrow 0$, the desired result now follows from Theorem 5.1.2 and the Fundamental Theorem of Calculus. □

Exercises

5.2.4 Exercise: In this exercise we discuss a few elementary properties of $\lambda_{\mathbf{S}^{N-1}}$.

(i) Show that if $\Gamma \neq \emptyset$ is an open subset of $\mathbf{S}^{N-1}$, then $\lambda_{\mathbf{S}^{N-1}}(\Gamma) > 0$. Next, show that $\lambda_{\mathbf{S}^{N-1}}$ is **rotation invariant**. That is, show that if $\mathcal{O}$ is an $N \times N$-orthogonal matrix and $T_{\mathcal{O}}$ is the associated transformation on $\mathbb{R}^N$ (cf. the paragraph preceding Theorem 2.2.2), then $(T_{\mathcal{O}})_*\lambda_{\mathbf{S}^{N-1}} = \lambda_{\mathbf{S}^{N-1}}$. Finally, use this fact to show that

$$\int\limits_{\mathbf{S}^{N-1}} (\boldsymbol{\xi},\boldsymbol{\omega})_{\mathbb{R}^N}\,\lambda_{\mathbf{S}^{N-1}}(d\boldsymbol{\omega}) = 0 \tag{5.2.5 (i)}$$

and

$$\int\limits_{\mathbf{S}^{N-1}} (\boldsymbol{\xi},\boldsymbol{\omega})_{\mathbb{R}^N}(\boldsymbol{\eta},\boldsymbol{\omega})_{\mathbb{R}^N}\,\lambda_{\mathbf{S}^{N-1}}(d\boldsymbol{\omega}) = \Omega_N\,(\boldsymbol{\xi},\boldsymbol{\eta})_{\mathbb{R}^N} \tag{5.2.5 (ii)}$$

for any $\boldsymbol{\xi}$, $\boldsymbol{\eta} \in \mathbb{R}^N$.

Hint: In proving these, let $\boldsymbol{\xi} \in \mathbb{R}^N\setminus\{0\}$ be given and consider the rotation $\mathcal{O}_{\boldsymbol{\xi}}$ which sends $\boldsymbol{\xi}$ to $-\boldsymbol{\xi}$ but acts as the identity on the orthogonal complement of $\boldsymbol{\xi}$.

(ii) Define $\Phi : [0, 2\pi] \longrightarrow \mathbf{S}^1$ by $\Phi(\theta) = \begin{bmatrix} \cos\theta \\ \sin\theta \end{bmatrix}$, and set $\mu = \Phi_* \lambda_{[0,2\pi]}$. Given any rotation invariant finite measure ν on $(\mathbf{S}^1, \mathcal{B}_{\mathbf{S}^1})$, show that $\nu = \frac{\nu(\mathbf{S}^1)}{2\pi}\mu$. In particular, conclude that $\lambda_{\mathbf{S}^1} = \mu$. (Cf. Exercise 5.3.20 below.)

Hint: Define

$$\mathcal{O}_\theta = \begin{bmatrix} \cos\theta & \sin\theta \\ -\sin\theta & \cos\theta \end{bmatrix} \quad \text{for} \quad \theta \in [0, 2\pi],$$

and note that

$$\int_{\mathbf{S}^1} f \, d\nu = \frac{1}{2\pi} \int_{[0,2\pi]} \left(\int_{\mathbf{S}^1} f \circ T_{\mathcal{O}_\theta}(\boldsymbol{\omega}) \, \nu(d\boldsymbol{\omega}) \right) d\theta.$$

(iii) For $N \in \mathbb{Z}^+$, define $\Xi : [-1, 1] \times \mathbf{S}^{N-1} \longrightarrow \mathbf{S}^N$ by

$$\Xi(\rho, \boldsymbol{\omega}) = \begin{bmatrix} (1 - \rho^2)^{\frac{1}{2}} \boldsymbol{\omega} \\ \rho \end{bmatrix},$$

and let

$$\mu_N(\Gamma) = \int_\Gamma (1 - \rho^2)^{\frac{N}{2} - 1} \lambda_{\mathbb{R}} \times \lambda_{\mathbf{S}^{N-1}}(d\rho \times d\boldsymbol{\omega})$$

for $\Gamma \in \mathcal{B}_{[-1,1]} \times \mathcal{B}_{\mathbf{S}^{N-1}}$. Show that $\lambda_{\mathbf{S}^N} = \Xi_* \mu_N$.

(**Hint**: Consider

$$\int_{(0,\infty)} r^N \left(\int_{[-1,1] \times \mathbf{S}^{N-1}} f\big(r\,\Xi(\rho, \boldsymbol{\omega})\big) \mu_N(d\rho \times d\boldsymbol{\omega}) \right) dr$$

for continuous $f : \mathbb{R}^{N+1} \longrightarrow \mathbb{R}$ with compact support.)

Finally, use this result to show that

$$\int_{\mathbf{S}^N} f\big((\boldsymbol{\theta}, \boldsymbol{\omega})_{\mathbb{R}^N}\big) \, \lambda_{\mathbf{S}^N}(d\boldsymbol{\omega}) = \omega_{N-1} \int_{[-1,1]} f(\rho)(1 - \rho^2)^{\frac{N}{2} - 1} \, d\rho$$

for all $\boldsymbol{\theta} \in \mathbf{S}^N$ and all measurable f on $\big([-1, 1], \mathcal{B}_{[-1,1]}\big)$ which are either bounded or non-negative.

5.2.6 Exercise: Perform the calculations outlined in the following.

(i) Justify Gauss's trick:

$$\left(\int_{\mathbb{R}} e^{-\frac{|x|^2}{2}} \, dx \right)^2 = \int_{\mathbb{R}^2} e^{-\frac{|x|^2}{2}} \, dx = 2\pi \int_{(0,\infty)} r e^{-\frac{r^2}{2}} \, dr = 2\pi$$

and conclude that for any $N \in \mathbb{Z}^+$ and symmetric $N \times N$-matrix A which is strictly positive definite (i.e., all the eigenvalues of A are strictly positive),

$$\int_{\mathbb{R}^N} \exp\left[-\frac{1}{2}(x, A^{-1}x)_{\mathbb{R}^N} \right] dx = (2\pi)^{\frac{N}{2}} \big(\det(A)\big)^{\frac{1}{2}}. \tag{5.2.7}$$

Hint: try the change of variable $\Phi(x) = T_{A^{-\frac{1}{2}}} x$.

(ii) Define $\Gamma(\gamma) = \int_{(0,\infty)} t^{\gamma-1}e^{-t}\,dt$ for $\gamma \in (0,\infty)$. Show that, for any $\gamma \in (0,\infty)$, $\Gamma(\gamma+1) = \gamma\Gamma(\gamma)$. Also, show that $\Gamma\left(\frac{1}{2}\right) = \pi^{\frac{1}{2}}$. The function $\Gamma(\,\cdot\,)$ is called Euler's **Gamma function**. Notice that it provides an extension of the factorial function in the sense that $\Gamma(n+1) = n!$ for integers $n \geq 0$.

(iii) Show that

$$\omega_{N-1} = \frac{2(\pi)^{\frac{N}{2}}}{\Gamma\left(\frac{N}{2}\right)},$$

and conclude that the volume Ω_N of the N-dimensional unit ball is given by

$$\Omega_N = \frac{\pi^{\frac{N}{2}}}{\Gamma\left(\frac{N}{2}+1\right)}.$$

(iv) Given $\alpha, \beta \in (0,\infty)$, show that

$$\int_{(0,\infty)} t^{-\frac{1}{2}} \exp\left[-\alpha^2 t - \frac{\beta^2}{t}\right] dt = \frac{\pi^{\frac{1}{2}} e^{-2\alpha\beta}}{\alpha}.$$

Finally, use the preceding to show that

$$\int_{(0,\infty)} t^{-\frac{3}{2}} \exp\left[-\alpha^2 t - \frac{\beta^2}{t}\right] dt = \frac{\pi^{\frac{1}{2}} e^{-2\alpha\beta}}{\beta}.$$

Hint: Define $\psi(s)$ for $s \in \mathbb{R}$ to be the unique $t \in (0,\infty)$ satisfying $s = \alpha t^{\frac{1}{2}} - \beta t^{-\frac{1}{2}}$, and use Exercise 5.1.6 to show that

$$\int_{(0,\infty)} t^{-\frac{1}{2}} \exp\left[-\alpha^2 t - \frac{\beta^2}{t}\right] dt = \frac{e^{-2\alpha\beta}}{\alpha} \int_{\mathbb{R}} e^{-s^2}\,ds.$$

5.3 Jacobi's Transformation and Surface Measure

We begin this section by deriving Jacobi's famous generalization to non-linear maps of the result in Theorem 2.2.2. We will then apply Jacobi's result to show that *Lebesgue measure can be differentiated across a smooth surface.*

Given an open set $G \subseteq \mathbb{R}^N$ and a continuously differentiable map

$$x \in G \longmapsto \mathbf{\Phi}(x) = \begin{bmatrix} \Phi_1(x) \\ \vdots \\ \Phi_N(x) \end{bmatrix} \in \mathbb{R}^N,$$

we define the **Jacobian matrix** $J\mathbf{\Phi}(x) = \frac{\partial \mathbf{\Phi}}{\partial x}(x)$ **of** $\mathbf{\Phi}$ **at** x to be the $N \times N$-matrix whose jth column is the vector

$$\mathbf{\Phi}_{,j}(x) \equiv \begin{bmatrix} \frac{\partial \Phi_1}{\partial x_j} \\ \vdots \\ \frac{\partial \Phi_N}{\partial x_j} \end{bmatrix}$$

In addition, we call $\delta\mathbf{\Phi}(x) \equiv |\det(J\mathbf{\Phi}(x))|$ the **Jacobian of $\mathbf{\Phi}$ at x.**

5.3.1 Lemma. *Let G be an open set in $\mathbb{R}^N$ and $\mathbf{\Phi}$ an element of $C^1(G;\mathbb{R}^N)$ whose Jacobian never vanishes on G. Then $\mathbf{\Phi}$ maps open (or $\mathcal{B}_{\mathbb{R}^N}$-measurable) subsets of G into open (or $\mathcal{B}_{\mathbb{R}^N}$-measurable) sets in $\mathbb{R}^N$. In addition, if $\Gamma \subseteq G$ with $|\Gamma|_e = 0$, then $\left|\mathbf{\Phi}(\Gamma)\right|_e = 0$; and if $\Gamma \subseteq \mathbf{\Phi}(G)$ with $|\Gamma|_e = 0$, then $\left|\mathbf{\Phi}^{-1}(\Gamma)\right|_e = 0$. In particular,*

$$\Gamma \in \overline{\mathcal{B}}_{\mathbb{R}^N}[G] \quad \text{if and only if} \quad \mathbf{\Phi}(\Gamma) \in \overline{\mathcal{B}}_{\mathbb{R}^N}\left[\mathbf{\Phi}(G)\right].$$

PROOF: By the Inverse Function Theorem,[†] for each $x \in G$ there is an open neighborhood $U \subseteq G$ of x such that $\mathbf{\Phi} \restriction U$ is invertible and its inverse has first derivatives which are bounded and continuous. Hence, G can be written as the union of a countable number of open sets on each of which $\mathbf{\Phi}$ admits an inverse having bounded continuous first order derivatives; and so, without loss of generality, we may and will assume that $\mathbf{\Phi}$ admits such an inverse on G itself. But, in that case, it is obvious that both $\mathbf{\Phi}$ and $\mathbf{\Phi}^{-1}$ are continuous. In addition, by Lemma 2.2.1, both take sets of Lebesgue measure 0 into sets of Lebesgue measure 0. Hence, by that same lemma, we now know that both $\mathbf{\Phi}$ and $\mathbf{\Phi}^{-1}$ take Lebesgue measurable sets into Lebesgue measurable sets. □

A continuously differentiable map $\mathbf{\Phi}$ on an open set $U \subseteq \mathbb{R}^N$ into $\mathbb{R}^N$ is called a **diffeomorphism** if it is **injective** (i.e., one-to-one) and $\delta\mathbf{\Phi}$ never vanishes. If $\mathbf{\Phi}$ is a diffeomorphism on the open set U and if $W = \mathbf{\Phi}(U)$, then we say that $\mathbf{\Phi}$ is **diffeomorphic from U onto W**. In what follows, for any given set $\Gamma \subseteq \mathbb{R}^N$ and $\delta > 0$, we use

$$\Gamma^{(\delta)} \equiv \left\{x \in \mathbb{R}^N : |y - x| < \delta \text{ for some } y \in \Gamma\right\}$$

to denote the **open δ-hull of Γ**.

5.3.2 Theorem (Jacobi's Transformation Formula). *Let G be an open set in $\mathbb{R}^N$ and $\mathbf{\Phi}$ an element of $C^2(G;\mathbb{R}^N)$.*[2] *If the Jacobian of $\mathbf{\Phi}$ never vanishes, then, for every measurable function f on $\left(\mathbf{\Phi}(G), \overline{\mathcal{B}}_{\mathbb{R}^N}[\mathbf{\Phi}(G)]\right)$, $f \circ \mathbf{\Phi}$ is measurable on $\left(G, \overline{\mathcal{B}}_{\mathbb{R}^N}[G]\right)$ and*

$$\int_{\mathbf{\Phi}(G)} f(y)\, dy \le \int_G f \circ \mathbf{\Phi}(x)\, \delta\mathbf{\Phi}(x)\, dx \tag{5.3.3}$$

whenever f is non-negative. Moreover, if $\mathbf{\Phi}$ is a diffeomorphism on G, then (5.3.3) can be replaced by

$$\int_{\mathbf{\Phi}(G)} f(y)\, dy = \int_G f \circ \mathbf{\Phi}(x)\, \delta\mathbf{\Phi}(x)\, dx. \tag{5.3.4}$$

[†] See, for example, W. Rudin's *Principles of Mathematical Analysis*, McGraw Hill (1976).
[2] By being a little more careful, one can get the same conclusion for $\Phi \in C^1(G,\mathbb{R}^N)$.

PROOF: We first note that (5.3.4) is a consequence of (5.3.3) when $\mathbf{\Phi}$ is one-to-one. Indeed, if $\mathbf{\Phi}$ is one-to-one, then the Inverse Function Theorem guarantees that $\mathbf{\Phi}^{-1} \in C^2(\mathbf{\Phi}(G);\mathbb{R}^N)$. In addition,

$$J\mathbf{\Phi}^{-1}(y) = \Big(J\mathbf{\Phi}\big(\mathbf{\Phi}^{-1}(y)\big)\Big)^{-1} \quad \text{for} \quad y \in \mathbf{\Phi}(G).$$

Hence we can apply (5.3.3) to $\mathbf{\Phi}^{-1}$ and thereby obtain

$$\int_G f\circ\mathbf{\Phi}(x)\,\delta\mathbf{\Phi}(x)\,dx \le \int_{\mathbf{\Phi}(G)} f(y)\,(\delta\mathbf{\Phi})\circ\mathbf{\Phi}^{-1}(y)\,\delta\mathbf{\Phi}^{-1}(y)\,dy = \int_{\mathbf{\Phi}(G)} f(y)\,dy;$$

which, in conjunction with (5.3.3), yields (5.3.4).

We next note that it suffices to prove (5.3.3) under the assumptions that G is bounded, $\mathbf{\Phi}$ on G has bounded first and second order derivatives, and $\delta\mathbf{\Phi}$ is uniformly positive on G. In fact, if this is not already the case, then we can choose a non-decreasing sequence of bounded open sets G_n so that $\mathbf{\Phi} \restriction G_n$ has these properties for each $n \ge 1$ and $G_n \nearrow G$. Clearly, the result for $\mathbf{\Phi}$ on G follows from the result for $\mathbf{\Phi} \restriction G_n$ on G_n for every $n \ge 1$. Thus, from now on, we assume that G is bounded, the first and second derivatives of $\mathbf{\Phi}$ are bounded, and $\delta\mathbf{\Phi}$ is uniformly positive on G.

Let $Q = Q(c;r) = \prod_1^N [c_i - r, c_i + r] \subseteq G$. Then, by Taylor's Theorem, there is an $L \in [0,\infty)$ (depending only on the bound on the second derivatives of $\mathbf{\Phi}$) such that[3]

$$\mathbf{\Phi}\big(Q(c;r)\big) \subseteq \mathbf{\Phi}(c) + \big(T_{J\mathbf{\Phi}(c)}Q(0;r)\big)^{(Lr^2)}.$$

(Cf. Section 2.2 for the notation here.) At the same time, there is an $M < \infty$ (depending only on L, the lower bound on $\delta\mathbf{\Phi}$, and the upper bounds on the first derivatives of $\mathbf{\Phi}$) such that

$$\big(T_{J\mathbf{\Phi}(c)}Q(0;r)\big)^{(Lr^2)} \subseteq T_{J\mathbf{\Phi}(c)}Q\big(0, r + Mr^2\big).$$

Hence, by Theorem 2.2.2,

$$\big|\mathbf{\Phi}(Q)\big| \le \delta\mathbf{\Phi}(c)\,\big|Q\big(0, r+Mr^2\big)\big| = (1+Mr)^N \delta\mathbf{\Phi}(c)|Q|.$$

Now define $\mu(\Gamma) = \int_\Gamma \delta\mathbf{\Phi}(x)\,dx$ for $\Gamma \in \overline{\mathcal{B}}_{\mathbb{R}^N}[G]$, and set $\nu = \mathbf{\Phi}_*\mu$. Given an open set $H \subseteq \mathbf{\Phi}(G)$, use Lemma 2.1.9 to choose, for each $m \in \mathbb{Z}^+$, a countable, non-overlapping, exact cover $\mathcal{C}_m$ of $\mathbf{\Phi}^{-1}(H)$ by cubes Q with $\operatorname{diam}(Q) < \frac{1}{m}$. Then, by the preceding paragraph,

$$|H| \le \sum_{Q\in\mathcal{C}_m} \big|\mathbf{\Phi}(Q)\big| \le \left(1 + \frac{M}{m}\right)^N \sum_{Q\in\mathcal{C}_m} \delta\mathbf{\Phi}(c_Q)|Q|,$$

[3] It is at this point that the argument has to be modified when Φ is only once continuously differentiable. Namely, the estimate which follows must be replaced by one involving the modulus of continuity of the Φ's first derivatives.

where c_Q denotes the center of the cube Q. After letting $m \to \infty$ in the preceding, we conclude that $|H| \le \nu(H)$ for open $H \subseteq \mathbf{\Phi}(G)$; and so, by Exercise 3.1.9, it follows that

$$|\Gamma| \le \nu(\Gamma) \quad \text{for all} \quad \Gamma \in \overline{\mathcal{B}}_{\mathbb{R}^N}[\mathbf{\Phi}(G)]. \tag{5.3.5}$$

Starting from (5.3.5), working first with simple functions, and then passing to monotone limits, we now conclude that (5.3.3) holds for all non-negative, measurable functions f on $(\mathbf{\Phi}(G), \overline{\mathcal{B}}_{\mathbb{R}^N}[\mathbf{\Phi}(G)])$. □

As an essentially immediate consequence of Theorem 5.3.2, we have the following.

5.3.6 Corollary. *Let G be an open set in $\mathbb{R}^N$ and $\mathbf{\Phi} \in C^2(G;\mathbb{R}^N)$ a diffeomorphism. Set*

$$\mu_{\mathbf{\Phi}}(\Gamma) = \int_\Gamma \delta\mathbf{\Phi}(x)\,dx \quad \textit{for} \quad \Gamma \in \overline{\mathcal{B}}_{\mathbb{R}^N}[G].$$

Then $\mathbf{\Phi}_*\mu_{\mathbf{\Phi}}$ coincides with the restriction $\lambda_{\mathbf{\Phi}(G)}$ of $\lambda_{\mathbb{R}^N}$ to $\overline{\mathcal{B}}_{\mathbb{R}^N}[\mathbf{\Phi}(G)]$. In particular,

$$f \in L^1\big(\mathbf{\Phi}(G), \mathcal{B}_{\mathbb{R}^N}[\mathbf{\Phi}(G)], \lambda_{\mathbf{\Phi}(G)}\big) \iff f \circ \mathbf{\Phi} \in L^1\big(G, \mathcal{B}_{\mathbb{R}^N}[G], \mu_{\mathbf{\Phi}}\big),$$

in which case (5.3.4) *holds.*

As a mnemonic device, it is useful to represent the conclusion of Corollary 5.3.6 as the statement

$$f(y)\,dy = f \circ \mathbf{\Phi}(x)\,\delta\mathbf{\Phi}(x)\,dx \quad \text{when} \quad y = \mathbf{\Phi}(x).$$

We now want to apply Jacobi's formula to show how to *differentiate Lebesgue's measure across a smooth surface.* To be precise, assume that $N \ge 2$ and say that $M \subset \mathbb{R}^N$ is a **hypersurface** if, for each $p \in M$, there exists an $r > 0$ and a three times continuously differentiable $F : B_{\mathbb{R}^N}(p,r) \longrightarrow \mathbb{R}$* with the properties that

$$\begin{gathered} B_{\mathbb{R}^N}(p,r) \cap M = \{y \in B_{\mathbb{R}^N}(p,r) : F(y) = 0\} \\ \text{and } |\nabla F(y)| \ne 0 \text{ for any } y \in B_{\mathbb{R}^N}(p,r), \end{gathered} \tag{5.3.7}$$

where

$$\nabla F(y) \equiv \big[F_{,1}(y), \dots, F_{,N}(y)\big]$$

is the **gradient** of F at y and, once again, we have used the notation $F_{,j}$ to denote the partial derivative in the direction $\mathbf{e}_j$. Given $p \in M$, the **tangent space** $\mathbf{T}_p(M)$ to M at p is the set of $\mathbf{v} \in \mathbb{R}^N$ for which there exists an $\epsilon > 0$ and a twice continuously differentiable curve γ on $(-\epsilon, \epsilon)$ into M such that $\gamma(0) = p$ and $\dot\gamma(0) = \mathbf{v}$. (If γ is a differentiable curve, we use $\dot\gamma(t)$ to denote its derivative $\frac{d\gamma}{dt}$ at t.)

---

* Because we will be dealing with balls in different dimensional Euclidean spaces here, we will use the notation $B_{\mathbb{R}^N}(a,r)$ to emphasize that the ball is in $\mathbb{R}^N$.

Canonical Example: The unit sphere $\mathbf{S}^{N-1}$ is a hypersurface in $\mathbb{R}^N$. In fact, at every point $p \in \mathbf{S}^{N-1}$ we can use the function $F(y) = |y|^2 - 1$ and can identify $\mathbf{T}_p(S^{N-1})$ with the subspace of $\mathbf{v} \in \mathbb{R}^N$ for which $\mathbf{v} - p$ is orthogonal to p.

5.3.8 Lemma. *Every hypersurface M can be written as the countable union of compact sets and is therefore Borel measurable. In addition, for each $p \in M$, $\mathbf{T}_p(M)$ is an $(N-1)$-dimensional subspace of $\mathbb{R}^N$. In fact, if $r > 0$ and $F \in C^3(B_{\mathbb{R}^N}(p,r);\mathbb{R})$ satisfy (5.3.7), then, for every $y \in B_{\mathbb{R}^N}(p,r) \cap M$, $\mathbf{T}_y(M)$ coincides with the space of the vectors $\mathbf{v} \in \mathbb{R}^N$ which are orthogonal to $\nabla F(y)$. Finally, if, for $\Gamma \subseteq M$ and $\rho > 0$,*

$$\Gamma(\rho) \equiv \{y \in \mathbb{R}^N : \exists p \in \Gamma \ (y-p) \perp \mathbf{T}_p(M) \text{ and } |y-p| < \rho\}, \tag{5.3.9}$$

then $\Gamma(\rho) \in \overline{\mathcal{B}}_{\mathbb{R}^N}$ whenever $\Gamma \in \mathcal{B}_{\mathbb{R}^N}[M]$.

PROOF: To see that M is the countable union of compacts, choose, for each $p \in M$, an $r(p) > 0$ and a function F_p so that (5.3.7) holds. Next, select a countable subset $\{p_n\}_1^\infty$ from M so that

$$M \subseteq \bigcup_1^\infty B_{\mathbb{R}^N}\left(p_n, \tfrac{r_n}{2}\right), \quad \text{with } r_n = r(p_n).$$

Clearly, for each $n \in \mathbb{Z}^+$, $K_n \equiv \left\{y \in \overline{B_{\mathbb{R}^N}\left(p_n, \tfrac{r_n}{2}\right)} : F(y) = 0\right\}$ is compact; and $M = \bigcup_1^\infty K_n$.

Next, let $p \in M$ be given, choose associated r and F, and let $y \in B_{\mathbb{R}^N}(p,r) \cap M$. To see that $\nabla F(y) \perp \mathbf{T}_y(M)$, let $\mathbf{v} \in \mathbf{T}_y(M)$ be given and choose $\epsilon > 0$ and γ accordingly. Then, because $\gamma : (-\epsilon, \epsilon) \longrightarrow M$,

$$\left(\nabla F(y), \mathbf{v}\right)_{\mathbb{R}^N} = \frac{d}{dt} F \circ \gamma(t)\big|_{t=0} = 0.$$

Conversely, if $\mathbf{v} \in \mathbb{R}^N$ satisfying $\left(\mathbf{v}, \nabla F(y)\right)_{\mathbb{R}^N} = 0$ is given, set

$$\mathbf{V}(z) = \mathbf{v} - \frac{\left(\mathbf{v}, \nabla F(z)\right)_{\mathbb{R}^N}}{|\nabla F(z)|^2} \nabla F(z)$$

for $z \in B_{\mathbb{R}^N}(p,r)$. By the basic existence theory for ordinary differential equations,* we can then find an $\epsilon > 0$ and a twice continuously differentiable curve $\gamma : (-\epsilon, \epsilon) \longrightarrow B_{\mathbb{R}^N}(p,r)$ such that $\gamma(0) = y$ and $\dot{\gamma}(t) = \mathbf{V}(\gamma(t))$ for all $t \in (-\epsilon, \epsilon)$. Clearly $\dot{\gamma}(0) = \mathbf{v}$, and it is an easy matter to check that

$$\frac{d}{dt}\left(F \circ \gamma(t)\right) = \left(\nabla F(\gamma(t)), \dot{\gamma}(t)\right)_{\mathbb{R}^N} = 0 \quad \text{for} \quad t \in (-\epsilon, \epsilon).$$

---

* See Chapter 1 of E. Coddington and N. Levinson's *Theory of Ordinary Differential Equations*, McGraw Hill (1955).

Hence, $\mathbf{v} \in \mathbf{T}_y(M)$.

To prove the final assertion, note that a covering argument, just like the one given at the beginning of this proof, allows us to reduce to the case when there is an $r > 0$ and a three times continuously differentiable $F : M^{(2r)} \longrightarrow \mathbb{R}$ such that $|\nabla F|$ is uniformly positive and $M = \{x \in M^{(2r)} : F(x) = 0\}$. But in that case, we see that

$$\Gamma(\rho) = \left\{ x + t\frac{\nabla F(x)}{|\nabla F(x)|} : x \in \Gamma \text{ and } |t| < \rho \right\},$$

and so the desired measurability follows as an application of Lemma 2.2.1 to the Lipschitz function

$$(x,t) \in \overline{M^{(r)}} \times \mathbb{R} \longmapsto x + t\frac{\nabla F}{|\nabla F|}(x). \quad \square$$

We are, at last, in a position to say where we are going. Namely, we want to show that there is a unique measure λ_M on $\left(M, \mathcal{B}_{\mathbb{R}^N}[M]\right)$ with the property that (cf. (5.3.9))

$$\lambda_M(\Gamma) = \lim_{\rho \searrow 0} \frac{1}{2\rho} \lambda_{\mathbb{R}^N}\left(\Gamma(\rho)\right) \quad \text{for bounded } \Gamma \in \mathcal{B}_{\mathbb{R}^N} \text{ with } \overline{\Gamma} \subseteq M. \tag{5.3.10}$$

Notice that, aside from the obvious question about whether the limit exists at all, there is a serious question about the additivity of the resulting map $\Gamma \longmapsto \lambda_M(\Gamma)$. Indeed, just because Γ_1 and Γ_2 are disjoint subsets of M, it will not be true, in general, that $\Gamma_1(\rho)$ and $\Gamma_2(\rho)$ will be disjoint. For example, when $M = \mathbf{S}^{N-1}$ and $\rho > 1$, $\Gamma_1(\rho)$ and $\Gamma_2(\rho)$ will intersect as soon as both are non-empty. On the other hand, at least in this example, everything will be all right when $\rho \le 1$; and, in fact, we already know that (5.3.10) defines a measure when $M = \mathbf{S}^{N-1}$. To see this, observe that, when $\rho \in (0,1)$ and $M = \mathbf{S}^{N-1}$,

$$\Gamma(\rho) = \left\{ y : 1 - \rho < |y| < 1 + \rho \text{ and } \frac{y}{|y|} \in \Gamma \right\},$$

and apply (5.2.3) to see that

$$\lambda_{\mathbb{R}^N}\left(\Gamma(\rho)\right) = \lambda_{\mathbf{S}^{N-1}}(\Gamma)\frac{(1+\rho)^N - (1-\rho)^N}{N},$$

where the measure $\lambda_{\mathbf{S}^{N-1}}$ is the one described in Section 5.2. Hence, after letting $\rho \searrow 0$, we see not only that the required limit exists but also gives the measure $\lambda_{\mathbf{S}^{N-1}}$. In other words, the program works in the case $M = \mathbf{S}^{N-1}$, and, perhaps less important, the notation used here is consistent with the notation used in Section 4.2.

In order to handle the problems raised in the preceding paragraph for general hypersurfaces, we are going to have to reduce, at least locally, to the essentially trivial case when $M = \mathbb{R}^{N-1} \times \{0\}$. In this case, it is clear that we can identify M with $\mathbb{R}^{N-1}$ and $\Gamma(\rho)$ with $\Gamma \times (-\rho, \rho)$. Hence, even before passing to a limit, we see that

$$\frac{1}{2\rho}\lambda_{\mathbb{R}^N}(\Gamma(\rho)) = \lambda_{\mathbb{R}^{N-1}}(\Gamma).$$

In the lemmata which follow, we will develop the requisite machinery with which to make the reduction.

5.3.11 Lemma. *For each $p \in M$ there is an open neighborhood U of $\mathbf{0}$ in $\mathbb{R}^{N-1}$ and a three times continuously differentiable injection (i.e., one-to-one) $\mathbf{\Psi} : U \longrightarrow M$ with the properties that $p = \mathbf{\Psi}(\mathbf{0})$ and, for each $u \in U$, the set $\{\mathbf{\Psi}_{,1}(u), \dots, \mathbf{\Psi}_{,N-1}(u)\}$ forms a basis in $\mathbf{T}_{\mathbf{\Psi}(u)}(M)$.*

PROOF: Choose r and F so that (5.3.7) holds. After renumbering the coordinates if necessary, we may and will assume that $F_{,N}(p) \neq 0$. Now consider the map

$$y \in B_{\mathbb{R}^N}(p, r) \longmapsto \mathbf{\Phi}(y) \equiv \begin{bmatrix} y_1 - p_1 \\ \vdots \\ y_{N-1} - p_{N-1} \\ F(y) \end{bmatrix} \in \mathbb{R}^N.$$

Clearly, $\mathbf{\Phi}$ is three times continuously differentiable. In addition,

$$J\mathbf{\Phi}(p) = \begin{bmatrix} \mathbf{I}_{\mathbb{R}^{N-1}} & 0 \\ \mathbf{v} & F_{,N}(p) \end{bmatrix},$$

where $\mathbf{v} = \big[F_{,1}(p), \dots, F_{,N-1}(p)\big]$. In particular, $\delta\mathbf{\Phi}(p) \neq 0$, and so the Inverse Function Theorem guarantees the existence of a $\rho \in (0, r]$ such that $\mathbf{\Phi} \restriction B_{\mathbb{R}^N}(p, \rho)$ is diffeomorphic and $\mathbf{\Phi}^{-1}$ has three continuous derivatives on the open set $W \equiv \mathbf{\Phi}\big(B_{\mathbb{R}^N}(p, \rho)\big)$. Thus, if

$$U \equiv \big\{u \in \mathbb{R}^{N-1} : (u, 0) \in W\big\},$$

then U is an open neighborhood of $\mathbf{0}$ in $\mathbb{R}^{N-1}$, and

$$u \in U \longmapsto \mathbf{\Psi}(u) \equiv \mathbf{\Phi}^{-1}(u, 0) \in M$$

is one-to-one and has three continuous derivatives. Finally, because $\mathbf{\Psi}$ takes its values in M, it is obvious that

$$\mathbf{\Psi}_{,j}(u) = \frac{d}{dt}\mathbf{\Psi}(u + t\mathbf{e}_j)\big|_{t=0} \in \mathbf{T}_{\mathbf{\Psi}(u)}(M) \quad \text{for each } 1 \leq j \leq N-1.$$

At the same time, as the first $(N-1)$ columns in the non-degenerate matrix $J\mathbf{\Phi}^{-1}(u)$, the $\mathbf{\Psi}_{,j}(u)$'s must be linearly independent. Hence, they form a basis in $\mathbf{T}_{\mathbf{\Psi}(u)}(M)$. $\square$

Given a non-empty, open set U in $\mathbb{R}^{N-1}$ and a twice continuously differentiable injection $\mathbf{\Psi} : U \longrightarrow M$ with the property that

$$\{\mathbf{\Psi}_{,1}, \ldots, \mathbf{\Psi}_{,N-1}\} \text{ forms a basis in } \mathbf{T}_{\mathbf{\Psi}(u)}(M)$$

for every $u \in U$, we say that the pair $(\mathbf{\Psi}, U)$ is a **coordinate chart** for M.

5.3.12 Lemma. *Suppose that $(\mathbf{\Psi}, U)$ is a coordinate chart for M, and define*

$$\delta\mathbf{\Psi} = \left[\det\Big(\Big(\big(\mathbf{\Psi}_{,i}, \mathbf{\Psi}_{,j}\big)_{\mathbb{R}^N} \Big)\Big)_{1 \le i,j \le N-1}\right]^{\frac{1}{2}}. \tag{5.3.13}$$

*Then $\delta\mathbf{\Psi}$ never vanishes, and there exists a unique twice continuously differentiable $\mathbf{n} : U \longrightarrow \mathbf{S}^{N-1}$ with the properties that $\mathbf{n}(u) \perp \mathbf{T}_{\mathbf{\Psi}(u)}(M)$ and**

$$\det\Big[\mathbf{\Psi}_{,1}(u) \;\ldots\; \mathbf{\Psi}_{,N-1}(u)\; \mathbf{n}(u)^{\mathrm{T}}\Big] = \delta\mathbf{\Psi}(u)$$

for every $u \in U$. Finally, define

$$\tilde{\mathbf{\Psi}}(u,t) = \mathbf{\Psi}(u) + t\mathbf{n}(u)^{\mathrm{T}} \quad \text{for } (u,t) \in U \times \mathbb{R}.$$

Then

$$\delta\tilde{\mathbf{\Psi}}(u,0) = \delta\mathbf{\Psi}(u), \quad u \in U, \tag{5.3.14}$$

and there exists an open set $\tilde{U}$ in $\mathbb{R}^N$ such that $\tilde{\mathbf{\Psi}} \restriction \tilde{U}$ is a diffeomorphism, $U = \{u \in \mathbb{R}^{N-1} : (u,0) \in \tilde{U}\}$, and $\mathbf{\Psi}(U) = \tilde{\mathbf{\Psi}}(\tilde{U}) \cap M$. In particular, if x and y are distinct elements of $\mathbf{\Psi}(U)$, then $\{x\}(\rho) \cap \tilde{\mathbf{\Psi}}(\tilde{U})$ is disjoint from $\{y\}(\rho) \cap \tilde{\mathbf{\Psi}}(\tilde{U})$ for all $\rho > 0$.

PROOF: Given a $u \in U$ and an $\mathbf{n} \in \mathbf{S}^{N-1}$ which is orthogonal to $\mathbf{T}_{\mathbf{\Psi}(u)}(M)$, $\{\mathbf{\Psi}_{,1}(u), \ldots, \mathbf{\Psi}_{,N-1}(u), \mathbf{n}^{\mathrm{T}}\}$ is a basis for $\mathbb{R}^N$ and therefore

$$\det\Big[\mathbf{\Psi}_{,1}(u) \;\ldots\; \mathbf{\Psi}_{,N-1}(u)\; \mathbf{n}^{\mathrm{T}}\Big] \neq 0.$$

Hence, for each $u \in U$ there is precisely one $\mathbf{n}(u) \in \mathbf{S}^{N-1} \cap \mathbf{T}_{\mathbf{\Psi}(u)}(M)^{\perp}$ with the property that

$$\det\Big[\mathbf{\Psi}_{,1}(u) \;\ldots\; \mathbf{\Psi}_{,N-1}(u)\; \mathbf{n}(u)^{\mathrm{T}}\Big] > 0.$$

To see that $u \in U \longmapsto \mathbf{n}(u) \in \mathbf{S}^{N-1}$ is twice continuously differentiable, set $p = \mathbf{\Psi}(u)$ and choose r and F for p so that (5.3.7) holds. Then $\mathbf{n}(u) = \pm\frac{\nabla F(p)}{|\nabla F(p)|}$, and so, by continuity, we know that, with the same sign throughout,

$$\mathbf{n}(w) = \pm\frac{\nabla F(\mathbf{\Psi}(w))}{|\nabla F(\mathbf{\Psi}(w))|}$$

* Below, and throughout, we use A^{T} to denote the transpose of a matrix A. In particular, if A is a row vector, then A^{T} is the corresponding column vector, and vice versa.

for every w in a neighborhood of u.

Turning to the function $\tilde{\mathbf{\Psi}}$, note that

$$\begin{aligned}
\delta\tilde{\mathbf{\Psi}}(u,0)^2 &= \left(\det\left[\mathbf{\Psi}_{,1}(u)\ \dots\ \mathbf{\Psi}_{,N-1}(u)\ \mathbf{n}(u)^{\mathrm{T}}\right]\right)^2 \\
&= \det\begin{bmatrix} \mathbf{\Psi}_{,1}(u)^{\mathrm{T}} \\ \vdots \\ \mathbf{\Psi}_{,N-1}(u)^{\mathrm{T}} \\ \mathbf{n}(u) \end{bmatrix}\left[\mathbf{\Psi}_{,1}(u)\ \dots\ \mathbf{\Psi}_{,N-1}(u)\ \mathbf{n}(u)^{\mathrm{T}}\right] \\
&= \det\begin{bmatrix} \left(\left(\left(\mathbf{\Psi}_{,i}(u), \mathbf{\Psi}_{,j}(u)\right)_{\mathbb{R}^N}\right)\right) & \mathbf{0} \\ \mathbf{0} & 1 \end{bmatrix} = \delta\mathbf{\Psi}(u)^2.
\end{aligned}$$

Hence, (5.3.14) is proved. In particular, this means, by the Inverse Function Theorem, that, for each $u \in U$, there is a neighborhood of $(u,0)$ in $\mathbb{R}^{N+1}$ on which $\tilde{\mathbf{\Psi}}$ is a diffeomorphism. In fact, given $u \in U$, choose r and F for $p = \mathbf{\Psi}(u)$, and take $\rho > 0$ so that

$$B_{\mathbb{R}^{N-1}}(u, 2\rho) \subseteq U \quad \text{and} \quad \mathbf{\Psi}\big(B_{\mathbb{R}^{N-1}}(u,2\rho)\big) \subseteq B_{\mathbb{R}^N}\left(p, \tfrac{r}{2}\right).$$

Then, because $\mathbf{n}(w) = \pm\frac{\nabla F(\mathbf{\Psi}(w))}{|\nabla F(\mathbf{\Psi}(w))|}$ for all $w \in B_{\mathbb{R}^{N-1}}(u,\rho)$,

$$F\big(\tilde{\mathbf{\Psi}}(w,t)\big) = \pm\, t\big|\nabla F(\mathbf{\Psi}(w))\big| + E(w,t) \qquad \text{for } (w,t) \in B_{\mathbb{R}^{N-1}}(u,\rho) \times \left(-\tfrac{r}{2}, \tfrac{r}{2}\right),$$

where $|E(w,t)| \le Ct^2$ for some $C \in (0,\infty)$. Hence, by readjusting the choice of $\rho > 0$, we can guarantee that $\tilde{\mathbf{\Psi}} \restriction B_{\mathbb{R}^{N-1}}(u,\rho) \times (-\rho,\rho)$ is both diffeomorphic and satisfies

$$\tilde{\mathbf{\Psi}}\Big(B_{\mathbb{R}^{N-1}}(u,\rho) \times (-\rho,\rho)\Big) \cap M = \mathbf{\Psi}\big(B_{\mathbb{R}^{N-1}}(u,\rho)\big).$$

In order to prove the final assertion, we must find an open $\tilde{U}$ in $\mathbb{R}^N$ such that $\mathbf{\Psi}(U) = \tilde{\mathbf{\Psi}}(\tilde{U}) \cap M$ and $\tilde{\mathbf{\Psi}} \restriction \tilde{U}$ is a diffeomorphism. To this end, for each $u \in U$, we use the preceding to choose $\rho(u) > 0$ so that: $B_{\mathbb{R}^{N-1}}\big(u, \rho(u)\big) \subseteq U$ and $\tilde{\mathbf{\Psi}} \restriction B_{\mathbb{R}^{N-1}}\big(u,\rho(u)\big) \times \big(-\rho(u), \rho(u)\big)$ is both a diffeomorphism and satisfies

$$\tilde{\mathbf{\Psi}}\Big(B_{\mathbb{R}^{N-1}}\big(u,\rho(u)\big) \times \big(-\rho(u),\rho(u)\big)\Big) \cap M = \mathbf{\Psi}\Big(B_{\mathbb{R}^{N-1}}\big(u,\rho(u)\big)\Big).$$

Next, choose a countable set $\{u_n\}_1^\infty \subseteq U$ so that

$$U = \bigcup_1^\infty B_{\mathbb{R}^{N-1}}\left(u_n, \tfrac{\rho_n}{3}\right) \quad \text{where} \quad \rho_n = \rho(u_n);$$

and set

$$U_n = \bigcup_1^n B_{\mathbb{R}^{N-1}}\left(u_m, \tfrac{\rho_m}{3}\right) \quad \text{and} \quad R_n = \rho_1 \wedge \cdots \wedge \rho_n.$$

To construct the open set $\tilde{U}$, we proceed inductively as follows. Namely, set $\epsilon_1 = \frac{\rho_1}{3}$ and $\tilde{K}_1 = \overline{U}_1 \times [-\epsilon_1, \epsilon_1]$. Next given $\tilde{K}_n$, define ϵ_{n+1} by

$$2\epsilon_{n+1} = R_{n+1} \wedge \epsilon_n \wedge \left(\inf\left\{\left|\Psi(u) - \tilde{\Psi}(w,t)\right| : u \in \overline{U}_{n+1} \setminus U_n,\ (w,t) \in \tilde{K}_n, \right.\right.$$
$$\left.\left.\text{and } |u-w| \geq \tfrac{R_{n+1}}{3}\right\}\right),$$

and take

$$\tilde{K}_{n+1} = \tilde{K}_n \cup \left[\left(\overline{U}_{n+1} \setminus U_n\right) \times \left[-\epsilon_{n+1}, \epsilon_{n+1}\right]\right].$$

Clearly, for each $n \in \mathbb{Z}^+$, $\tilde{K}_n$ is compact, $\tilde{K}_n \subseteq \tilde{K}_{n+1}$, $U_n \times (-\epsilon_n, \epsilon_n) \subseteq \tilde{K}_n$, and $\delta\tilde{\Psi}$ never vanishes on $\tilde{K}_n$. Thus, if we can show that $\epsilon_n > 0$ and that $\tilde{\Psi} \restriction \tilde{K}_n$ is one-to-one for each $n \in \mathbb{Z}^+$, then we can take $\tilde{U}$ to be the interior of $\bigcup_1^\infty \tilde{K}_n$. To this end, first observe that there is nothing to do when $n = 1$. Furthermore, if $\epsilon_n > 0$ and $\tilde{\Psi} \restriction \tilde{K}_n$ is one-to-one, then $\epsilon_{n+1} = 0$ is possible only if there exists a $u \in B_{\mathbb{R}^{N-1}}(u_{n+1}, \rho_{n+1})$ and a $(w,t) \in B_{\mathbb{R}^{N-1}}(u_m, \rho_m) \times (-\rho_m, \rho_m)$ for some $1 \leq m \leq n$ such that $\Psi(u) = \tilde{\Psi}(w,t)$ and $|u - w| \geq \frac{R_{n+1}}{3}$. But, because

$$(w,t) \in B_{\mathbb{R}^{N-1}}(u_m, \rho_m) \times (-\rho_m, \rho_m) \text{ and } \tilde{\Psi}(w,t) \in M \implies t = 0,$$

this would mean that $\Psi(w) = \Psi(u)$ and therefore, since Ψ is one-to-one, it would lead to the contradiction that $0 = |w - u| \geq \frac{R_{n+1}}{3}$. Hence, $\epsilon_{n+1} > 0$. Finally, to see that $\tilde{\Psi} \restriction \tilde{K}_{n+1}$ is one-to-one, we need only check that

$$(u,s) \in (\overline{U}_{n+1} \setminus U_n) \times [-\epsilon_{n+1}, \epsilon_{n+1}] \text{ and } (w,t) \in \tilde{K}_n \implies \tilde{\Psi}(u,s) \neq \tilde{\Psi}(w,t).$$

But, if $|u - w| < \frac{R_{n+1}}{3}$, then both (u,s) and (w,t) are in $B_{\mathbb{R}^{N-1}}(u_{n+1}, \rho_{n+1}) \times (-\rho_{n+1}, \rho_{n+1})$ and $\tilde{\Psi}$ is one to one there. On the other hand, if $|u-w| \geq \frac{R_{n+1}}{3}$, then

$$\left|\tilde{\Psi}(u,s) - \tilde{\Psi}(w,t)\right| \geq \left|\Psi(u) - \tilde{\Psi}(w,t)\right| - |s| \geq 2\epsilon_{n+1} - \epsilon_{n+1} = \epsilon_{n+1} > 0. \quad \square$$

5.3.15 Lemma. *If (Ψ, U) is a coordinate chart for M and Γ a bounded element of $\mathcal{B}_{\mathbb{R}^N}$ with $\overline{\Gamma} \subseteq \Psi(U)$, then (cf. (5.3.13))*

$$\lim_{\rho \searrow 0} \frac{1}{2\rho} \lambda_{\mathbb{R}^N}(\Gamma(\rho)) = \int_{\Psi^{-1}(\Gamma)} \delta\Psi(u)\, \lambda_{\mathbb{R}^{N-1}}(du). \tag{5.3.16}$$

PROOF: Choose $\tilde{U}$ and $\tilde{\Psi}$ as in Lemma 5.3.12. Since $\overline{\Gamma}$ is a compact subset of the open set $G = \tilde{\Psi}(\tilde{U})$, $\epsilon = \text{dist}(\Gamma, G\complement) > 0$. Hence, for $\rho \in (0, \epsilon)$,

$$\Gamma(\rho) = \tilde{\Psi}\left(\Psi^{-1}(\Gamma) \times (-\rho, \rho)\right) \in \mathcal{B}_{\mathbb{R}^N},$$

which, by (5.3.4) and Tonelli's Theorem, means that

$$\lambda_{\mathbb{R}^N}(\Gamma(\rho)) = \int_{\Psi^{-1}(\Gamma)} \left(\int_{(-\rho,\rho)} \delta\tilde{\Psi}(u,t)\, dt \right) \lambda_{\mathbb{R}^{N-1}}(du).$$

Since $\Psi^{-1}(\overline{\Gamma}) = \tilde{\Psi}^{-1}(\overline{\Gamma})$ is compact, $\delta\tilde{\Psi} \upharpoonright \Psi^{-1}(\Gamma)$ is uniformly bounded and continuous. In particular, by Lebesgue's Dominated Convergence Theorem and (5.3.14),

$$\frac{1}{2\rho}\int_{(-\rho,\rho)} \delta\tilde{\Psi}(u,t)\, dt \longrightarrow \delta\tilde{\Psi}(u,0) = \delta\Psi(u)$$

boundedly and pointwise for $u \in U$. Hence, after a second application of Lebesgue's Dominated Convergence Theorem, we arrive at (5.3.16). □

5.3.17 Theorem. *Let M be a hypersurface in $\mathbb{R}^N$. Then there exists a unique measure λ_M on $(M, \mathcal{B}_{\mathbb{R}^N}[M])$ for which (5.3.10) holds. In fact, $\lambda_M(K) < \infty$ for every compact subset of M. Finally, if (Ψ, U) is a coordinate system for M and f is a non-negative, $\mathcal{B}_{\mathbb{R}^N}[M]$-measurable function, then (cf. (5.4.13))*

$$\int_{\Psi(U)} f(x)\, \lambda_M(dx) = \int_U f \circ \Psi(u) \delta\Psi(u)\, \lambda_{\mathbb{R}^{N-1}}(du). \tag{5.3.18}$$

PROOF: For each $p \in M$, use Lemma 5.3.11 to produce an $r(p) > 0$ and a coordinate chart (Ψ_p, U_p) for M such that $B_{\mathbb{R}^N}(p, 3r(p))$ is contained in (cf. Lemma 5.3.12) $\tilde{\Psi}(\tilde{U}_p)$. Next, select a countable set $\{p_n\}_1^\infty \subseteq M$ so that

$$M \subseteq \bigcup_1^\infty B_{\mathbb{R}^N}(p_n, r_n), \quad \text{where } r_n \equiv r(p_n),$$

set $M_1 = B_{\mathbb{R}^N}(p_1, r_1) \cap M$ and

$$M_n = \left(B_{\mathbb{R}^N}(p_n, r_n) \cap M\right) \setminus \bigcup_1^{n-1} M_m \quad \text{for } n \geq 2.$$

Finally, for each $n \in \mathbb{Z}^+$, define the finite measure μ_n on $(M, \mathcal{B}_{\mathbb{R}^N}[M])$ by

$$\mu_n(\Gamma) = \int_{\Psi_n^{-1}(\Gamma \cap M_n)} \delta\Psi_n(u)\, \lambda_{\mathbb{R}^{N-1}}(du), \quad \text{where } \Psi_n \equiv \Psi_{p_n},$$

and set

$$\lambda_M(\Gamma) = \sum_1^\infty \mu_n. \tag{5.3.19}$$

Given a compact $K \subseteq M$, choose $n \in \mathbb{Z}^+$ so that $K \subseteq \bigcup_1^n B_{\mathbb{R}^N}(p_m, r_m)$, and set $r = r_1 \wedge \cdots \wedge r_n$ and $\epsilon = \frac{r}{3}$. It is then an easy matter to check that, for any pair of distinct elements x and y from K, either $|x - y| \geq r$, in which case it is obvious that $\{x\}(\epsilon) \cap \{y\}(\epsilon) = \emptyset$, or $|x - y| < r$, in which case both $\{x\}(\epsilon)$ and $\{y\}(\epsilon)$ lie in $\tilde{\mathbf{\Psi}}_m(\tilde{U}_{p_m})$ for some $1 \leq m \leq n$ and, therefore, the last part of Lemma 5.3.12 applies and says that $\{x\}(\epsilon) \cap \{y\}(\epsilon) = \emptyset$. In particular, if $\Gamma \in \mathcal{B}_{\mathbb{R}^N}$ is a subset of K and $\Gamma_m = \Gamma \cap M_m$, then, for each $\rho \in [0, \epsilon)$, $\{\Gamma_1(\rho), \ldots, \Gamma_n(\rho)\}$ is a cover of $\Gamma(\rho)$ by mutually disjoint measurable sets. Hence, for each $0 < \rho < \epsilon$,

$$\lambda_{\mathbb{R}^N}\big(\Gamma(\rho)\big) = \sum_1^n \lambda_{\mathbb{R}^N}\big(\Gamma_m(\rho)\big).$$

At the same time, by Lemma 5.3.15,

$$\frac{1}{2\rho}\lambda_{\mathbb{R}^N}\big(\Gamma_m(\rho)\big) \longrightarrow \mu_m(\Gamma_m) \quad \text{for each } 1 \leq m \leq n.$$

In particular, we have now proved that the measure defined in (5.3.19) satisfies (5.3.10) and that λ_M is finite on compacts. Moreover, since (cf. Lemma 5.3.8) M is a countable union of compacts, it is clear that there can be only one measure satisfying (5.3.10).

Finally, if $(\mathbf{\Psi}, U)$ is a coordinate chart for M and $\Gamma \in \mathcal{B}_{\mathbb{R}^N}$ with $\overline{\Gamma} \subseteq \mathbf{\Psi}(U)$ is bounded, then (5.3.18) with $f = \mathbf{1}_\Gamma$ is an immediate consequence of (5.3.16). Hence, (5.3.18) follows in general by taking linear combinations and monotone limits. □

The measure λ_M produced in Theorem 5.3.17 is called the **surface measure** on M.

Exercises

5.3.20 Exercise: In the final assertion of part (**ii**) in Exercise 5.2.4 and again in (**i**) of Exercise 5.2.6, we tacitly accepted the equality of π, the volume Ω_2 of the unit ball $B_{\mathbb{R}^2}(0,1)$ in $\mathbb{R}^2$, with π, the half-period of the sin and cos functions. We are now in a position to justify this identification. To this end, define $\mathbf{\Phi} : G \equiv (0,1) \times (0, 2\pi) \longrightarrow \mathbb{R}^2$ (the π here is the half-period of sin and cos) by $\mathbf{\Phi}(r, \theta) = (r\cos\theta, r\sin\theta)^{\mathrm{T}}$. Note that $\mathbf{\Phi}(G) \subseteq B_{\mathbb{R}^N}(0,1)$ and $B_{\mathbb{R}^N}(\mathbf{0},1) \setminus G \subseteq \{(x_1, x_2) : x_2 = 0\}$ and therefore that $|\mathbf{\Phi}(G)| = \Omega_2$. Now use Jacobi's Transformation Formula to compute $|\mathbf{\Phi}(G)|$.

5.3.21 Exercise: Let M a hypersurface in $\mathbb{R}^N$. Show that, for each $p \in M$, the tangent space $\mathbf{T}_p(M)$ coincides with the set of $\mathbf{v} \in \mathbb{R}^N$ such that

$$\varlimsup_{\xi \to 0} \frac{\operatorname{dist}(p + \xi\mathbf{v}, M)}{\xi^2} < \infty.$$

Hint: Given $\mathbf{v} \in \mathbf{T}_p(M)$, choose a twice continuously differentiable associated curve γ, and consider $\xi \longmapsto \big|p + \xi\mathbf{v} - \gamma(\xi)\big|$.

5.3.22 Exercise: Given $r > 0$, set $\mathbf{S}^{N-1}(r) = \{x \in \mathbb{R}^N : |x| = r\}$, and observe that $\mathbf{S}^{N-1}(r)$ is a smooth region. Next, define $\Phi_r : \mathbf{S}^{N-1} \longrightarrow \mathbf{S}^{N-1}(r)$ by $\Phi_r(\boldsymbol{\omega}) = r\boldsymbol{\omega}$; and show that $\lambda_{\mathbf{S}^{N-1}(r)} = r^{N-1}(\Phi_r)_*\lambda_{\mathbf{S}^{N-1}}$.

5.3.23 Exercise: Show that if $\Gamma \neq \emptyset$ is an open subset of a hypersurface M, then $\lambda_M(\Gamma) > 0$.

5.3.24 Exercise: In this exercise we introduce a function which is intimately related to Euler's Gamma function.

(i) For $(\alpha, \beta) \in (0, \infty)^2$, define

$$B(\alpha, \beta) = \int_{(0,1)} u^{\alpha-1}(1-u)^{\beta-1}\, du.$$

Show that

$$B(\alpha, \beta) = \frac{\Gamma(\alpha)\Gamma(\beta)}{\Gamma(\alpha+\beta)}$$

where $\Gamma(\,\cdot\,)$ is the Gamma function described in **(ii)** of Exercise 5.2.6. (See part **(iv)** of Exercise 6.3.18 for another derivation.) The function B is called the **Beta function**. Clearly it provides an extension of the binomial coefficients in the sense that

$$\frac{m+n+1}{B(m+1, n+1)} = \binom{m+n}{m}$$

for all non-negative integers m and n.

Hint: Think of $\Gamma(\alpha)\,\Gamma(\beta)$ as an integral in (s,t) over $(0,\infty)^2$, and consider the map

$$(u, v) \in (0, \infty) \times (0, 1) \longmapsto \begin{bmatrix} uv \\ u(1-v) \end{bmatrix} \in (0, \infty)^2.$$

(ii) For $\lambda > \frac{N}{2}$ show that

$$\int_{\mathbb{R}^N} \frac{1}{(1+|x|^2)^\lambda}\, dx = \frac{\omega_{N-1}}{2} B\left(\frac{N}{2}, \lambda - \frac{N}{2}\right) = \frac{\pi^{\frac{N}{2}}\Gamma(\lambda - \frac{N}{2})}{\Gamma(\lambda)},$$

where (cf. part **(ii)** of Exercise 5.2.6) ω_{N-1} is the surface area of $\mathbf{S}^{N-1}$. In particular, conclude that

$$\int_{\mathbb{R}^N} \frac{1}{(1+|x|^2)^{\frac{N+1}{2}}}\, dx = \frac{\omega_N}{2}.$$

Hint: Use polar coordinates and then try the change of variable $\mathbf{\Phi}(r) = \frac{r^2}{1+r^2}$.

(iii) For $\lambda \in (0,\infty)$, show that

$$\int_{(-1,1)} \left(1-\xi^2\right)^{\lambda-1} d\xi = \frac{\pi^{\frac{1}{2}}\Gamma(\lambda)}{\Gamma\left(\lambda+\frac{1}{2}\right)};$$

and conclude that, for any $N \in \mathbb{Z}^+$,

$$\int_{(-1,1)} \left(1-\xi^2\right)^{\frac{N}{2}-1} d\xi = \frac{\omega_N}{\omega_{N-1}}.$$

Finally, check that this last result is consistent with part **(iii)** of Exercise 5.2.4.

5.3.25 Exercise: Let U be a non-empty, open subset of $\mathbb{R}^{N-1}$ and $f \in C^3(U;\mathbb{R})$ be given, take

$$M = \left\{\left(u, f(u)\right) : u \in U\right\} \quad \text{and} \quad \mathbf{\Psi}(u) = \begin{bmatrix} u \\ f(u) \end{bmatrix}, \quad u \in U.$$

That is, M is the *graph* of f.

(i) Check that M is a hypersurface and that $(\mathbf{\Psi}, U)$ is coordinate chart for M which is **global** in the sense that $M = \mathbf{\Psi}(U)$.

(ii) Show that

$$\left(\left(\left(\mathbf{\Psi}_{,i}, \mathbf{\Psi}_{,j}\right)_{\mathbb{R}^N}\right)\right)_{1\le i,j\le N-1} = \mathbf{I} + \nabla f^{\mathrm{T}} \nabla f,$$

and conclude that $\delta\mathbf{\Psi} = \sqrt{1+|\nabla f|^2}$.

Hint: Given a non-zero, row vector $\mathbf{v}$, set $A = \mathbf{I} + \mathbf{v}^{\mathrm{T}}\mathbf{v}$ and $\mathbf{e} = \frac{\mathbf{v}^{\mathrm{T}}}{|\mathbf{v}|}$. Note that $A\mathbf{e} = \left(1+|\mathbf{v}|^2\right)\mathbf{e}$ and that $A\mathbf{w} = \mathbf{w}$ for $\mathbf{w} \perp \mathbf{e}$. Thus, A has one eigenvalue equal to $1+|\mathbf{v}|^2$ and all its other eigenvalues equal 1.

(iii) From the preceding, arrive at

$$\int_M \varphi \, d\lambda_M = \int_U \varphi\left(u, f(u)\right)\sqrt{1+|\nabla f(u)|^2}\, du, \quad \varphi \in C_c(M),$$

a formula which should be familiar from elementary calculus.

5.3.26 Exercise: Let G be a non-empty, open set in $\mathbb{R}^N$, $F \in C^3\left(G;\mathbb{R}^3\right)$, and assume that $M \equiv \{x \in G : F(x) = 0\}$ is a non-empty, connected set on which $F_{,N}$ never vanishes.

(i) Set $U = \{u \in \mathbb{R}^{N-1} : (u,t) \in M \text{ for some } t \in \mathbb{R}\}$, and show that there exists an $f \in C^3(U;\mathbb{R})$ with the property that $F\left(u, f(u)\right) = 0$ for all $u \in U$. In particular, M is the graph of f.

(ii) Define $\mathbf{\Psi}$ from f as in Exercise 5.3.25, and conclude that

$$\int_M \varphi\, d\lambda_M = \int_U \frac{\varphi|\nabla F|}{|F_{,N}|} \circ \mathbf{\Psi}(u)\, du.$$

5.3.27 Exercise: Again let G be a non-empty, open set in $\mathbb{R}^N$ and $F \in C^3(G;\mathbb{R})$, but this time assume that $F_{,N}$ vanishes nowhere on G. Set $T = \text{Range}(F)$ and, for $t \in T$, $M_t = \{x \in G : F(x) = t\}$ and $U_t = \{u \in \mathbb{R}^{N-1} : (u,t) \in M_t\}$.

(i) Define $\mathbf{\Phi}$ on G by

$$\mathbf{\Phi}(x) = \begin{bmatrix} x_1 \\ \vdots \\ x_{N-1} \\ F(x) \end{bmatrix},$$

and show that $\mathbf{\Phi}$ is a diffeomorphism. As an application of Exercise 5.3.26, show that, for each $t \in T$, M_t is a hypersurface and that, for each $\varphi \in C_c(G)$:

$$\int_{M_t} \varphi\, d\lambda_{M_t} = \int_{\mathbb{R}^{N-1}} \mathbf{1}_{\mathbf{\Phi}(G)}(u,t) \frac{\varphi|\nabla F|}{|F_{,N}|} \circ \mathbf{\Phi}^{-1}(u,t)\, du, \quad t \in T.$$

In particular, conclude that

$$t \in T \longmapsto \int_{M_t} \varphi\, d\lambda_{M_t} \in \mathbb{R}$$

is bounded and $\mathcal{B}_T$-measurable.

(ii) Using Theorems 4.1.6 and 5.3.2, note that, for $\varphi \in C_c(G)$,

$$\begin{aligned}\int_G \varphi(x)\, dx &= \int_{\mathbf{\Phi}(G)} \frac{\varphi}{|F_{,N}|} \circ \mathbf{\Phi}^{-1}(y)\, dy \\ &= \int_T \left(\int_{\mathbb{R}^{N-1}} \mathbf{1}_{\mathbf{\Phi}(G)}(u,t) \frac{\varphi}{|F_{,N}|} \circ \mathbf{\Phi}^{-1}(u,t)\, du \right) dt.\end{aligned}$$

After combining this with part **(i)**, arrive at the following (somewhat primitive) version of the **co-area formula**

$$\int_G \varphi(x)\, dx = \int_T \left(\int_{M_t} \frac{\varphi}{|\nabla F|}\, d\lambda_{M_t} \right) dt, \quad \varphi \in C_c(G). \tag{5.3.28}$$

(iii) Take $G = \{x \in \mathbb{R}^N : x_N \neq 0\}$, $F = |x|$ for $x \in G$, and show that (5.2.3) can be easily derived from (5.3.28).

5.4 The Divergence Theorem

Again let $N \geq 2$. Perhaps the single most striking application of the construction made in the second part of Section 5.3 is to multidimensional integration by parts formulae. Namely, given a non-empty open G in $\mathbb{R}^N$, we will say that the region G is **smooth** if its boundary ∂G is a hypersurface in $\mathbb{R}^N$ and, for each $p \in \partial G$, the number $r > 0$ and $F : B_{\mathbb{R}^N}(p,r) \longrightarrow \mathbb{R}$ in (5.3.7) can be chosen so that, in addition to (5.3.7),

$$B_{\mathbb{R}^N}(p,r) \cap G = \big\{y \in B_{\mathbb{R}^N}(p,r) : F(y) < 0\big\}. \tag{5.4.1}$$

Notice that if G is a smooth region, then, for each $p \in \partial G$, there is a unique $\mathbf{n}(p) \in \mathbf{S}^{N-1} \cap \mathbf{T}_p(\partial G)^\perp$ with the property that, for some $\epsilon > 0$:

$$p + \xi\mathbf{n}(p) \in \begin{cases} \overline{G}\complement & \text{if } \xi \in (0,\epsilon) \\ G & \text{if } \xi \in (-\epsilon, 0). \end{cases} \tag{5.4.2}$$

For obvious reasons, $\mathbf{n}(p)$ is called the **outer normal** to ∂G at p. Notice that $x \in \partial G \longmapsto \mathbf{n}(x) \in \mathbf{S}^{N-1}$ is locally (i.e., on each compact) Lipschitz continuous, since if r and F are chosen for p so that (5.3.7) and (5.4.1) hold, then (cf. Lemma 5.3.8)

$$\mathbf{n}(x) = \frac{\nabla F(x)}{|\nabla F(x)|}, \quad x \in B_{\mathbb{R}^N}(p,r) \cap \partial G.$$

Our main goal in this section will be to prove that if $f \in C_c(\mathbb{R}^N;\mathbb{R})$ (i.e., vanishes off of some compact subset) and G is a smooth region, then for every $\omega \in \mathbf{S}^{N-1}$:

$$\frac{d}{d\xi} \int_G f(x + \xi\omega)\, dx \bigg|_{\xi=0} = \int_{\partial G} f(x) \big(\omega, \mathbf{n}(x)\big)_{\mathbb{R}^N} \lambda_{\partial G}(dx). \tag{5.4.3}$$

Observe that (5.4.3) is another way of seeing that surface measure is in truth the derivative of $\lambda_{\mathbb{R}^N}$ across the surface. Indeed, since there is no requirement that f be differentiable, it must be Lebesgue measure which is absorbing the derivative on the left hand side of (5.4.3).

The key to (5.4.3) is contained in the following lemma.

5.4.4 Lemma. *Let G be a smooth region. Then, for each $p \in \partial G$, there is a coordinate chart (Ψ, U) for ∂G and an $\epsilon > 0$ with the properties that U is bounded, $p \in \Psi(U)$, and the associated map*

$$(u,t) \in \tilde{U} \equiv U \times (-\epsilon,\epsilon) \longmapsto \tilde{\Psi}(u,t) \equiv \Psi(u) + t\mathbf{n}\big(\Psi(u)\big)^{\mathrm{T}} \in \mathbb{R}^N \tag{5.4.5}$$

is a diffeomorphism such that both $\tilde{\Psi}$ and $\tilde{\Psi}^{-1}$ have bounded, continuous first and second order derivatives, and, for each $u \in U$,

$$\tilde{\Psi}(u,t) \in G \text{ if and only if } t \in (-\epsilon, 0). \tag{5.4.6}$$

Furthermore, given such a coordinate chart, (5.4.3) holds for every continuous $f : \mathbb{R}^N \longrightarrow \mathbb{R}$ with compact support in (i.e., vanishing off a compact subset of) $\tilde{\Psi}(\tilde{U})$.

PROOF: To prove the first part, use Lemma 5.3.11 to choose some coordinate chart $(\mathbf{\Phi}, W)$ for ∂G with $p \in \mathbf{\Phi}(W)$, assume that W is connected, and define associated $\mathbf{n}(w)$, $w \in W$, $\tilde{W}$, and $\tilde{\mathbf{\Phi}}$ as in Lemma 5.3.12. Clearly, $\mathbf{n}(w) = \pm\mathbf{n}(\mathbf{\Phi}(w))$, with the same choice of sign for every $w \in W$. Hence, if necessary after replacing W by $W' = \{w : (w_1, \dots, -w_{N-1}) \in W\}$ and $\mathbf{\Phi}$ by

$$w \in W' \longmapsto \mathbf{\Phi}(w_1, \dots, -w_{N-1}),$$

we may and will assume that $\mathbf{n}(w) = \mathbf{n}(\mathbf{\Phi}(w))$ for all $w \in W$. In particular, because $\tilde{\mathbf{\Phi}}(\tilde{W}) \cap \partial G = \mathbf{\Phi}(W)$ and $\tilde{\mathbf{\Phi}}(w,t) \in G$ for sufficiently small strictly negative t's, we know that, for each $(w,t) \in \tilde{W}$, $\tilde{\mathbf{\Phi}}(w,t) \in G$ if and only if $t < 0$. Finally, choose $\epsilon > 0$ so that $B_{\mathbb{R}^N}(p, 2\epsilon) \subseteq \tilde{\mathbf{\Phi}}(\tilde{U})$, set $U = \mathbf{\Phi}^{-1}\big(B_{\mathbb{R}^N}(p,\epsilon) \cap \partial G\big)$, and take $\mathbf{\Psi}$ and $\tilde{\mathbf{\Psi}}$ to be, respectively, the restrictions of $\mathbf{\Phi}$ and $\tilde{\mathbf{\Phi}}$ to U and $\tilde{U} = U \times (-\epsilon, \epsilon)$.

Turning to the second part of the lemma, set $\Omega = \tilde{\mathbf{\Psi}}(\tilde{U})$, and define

$$p(y) = \mathbf{\Psi}\Big(\tilde{\mathbf{\Psi}}^{-1}(y)_1, \dots, \tilde{\mathbf{\Psi}}^{-1}(y)_{N-1}\Big) \quad \text{and} \quad \mathbf{n}(y) = \mathbf{n}\big(p(y)\big)$$

for $y \in \Omega$. Given $\boldsymbol{\omega} \in \mathbf{S}^{N-1}$, set

$$\omega_{\mathbf{n}}(y) = \big(\boldsymbol{\omega}, \mathbf{n}(y)\big)_{\mathbb{R}^N} \quad \text{and} \quad \boldsymbol{\omega}_{\mathbf{n}}(y) = \omega_{\mathbf{n}}(y)\mathbf{n}(y), \quad y \in \Omega;$$

and, given $f \in C(\mathbb{R}^N; \mathbb{R})$ with compact support $K \subseteq \Omega$, set $r = \operatorname{dist}\big(K, \Omega\complement\big)$. Then, by the translation invariance of Lebesgue measure, for every $\xi \in \mathbb{R}$,

$$\Delta(\xi) \equiv \int_G f(x + \xi\boldsymbol{\omega})\,dx - \int_G f(x)\,dx = \Delta_1(\xi) + \Delta_2(\xi),$$

where

$$\Delta_1(\xi) \equiv \int_K \Big(\mathbf{1}_G(x - \xi\boldsymbol{\omega}) - \mathbf{1}_G\big(x - \xi\boldsymbol{\omega}_{\mathbf{n}}(x)\big)\Big) f(x)\,dx$$

and

$$\Delta_2(\xi) \equiv \int_K \Big(\mathbf{1}_G\big(x - \xi\boldsymbol{\omega}_{\mathbf{n}}(x)\big) - \mathbf{1}_G(x)\Big) f(x)\,dx.$$

In order to prove that (5.4.3) holds, we will show that

$$\text{(5.4.7)} \qquad \begin{gathered} \lim_{\xi \to 0} \frac{\Delta_1(\xi)}{\xi} = 0 \\ \lim_{\xi \to 0} \frac{\Delta_2(\xi)}{\xi} = \int_G f(x)\omega_{\mathbf{n}}(x)\,\lambda_{\partial G}(dx); \end{gathered}$$

and we will begin with the second, and easier part of (5.4.7). To this end, we use Theorem 5.3.2 and Fubini's Theorem to write

$$\Delta_2(\xi) = \int_U g(\xi, u)\,du, \qquad |\xi| < r,$$

where

$$g(\xi,u) \equiv \int_{(-\epsilon,\epsilon)} \Big[\mathbf{1}_{(-\infty,0)}\Big(t-\xi\omega_{\mathbf{n}}\big(\Psi(u)\big)\Big) - \mathbf{1}_{(-\infty,0)}(t)\Big] f\big(\tilde{\Psi}(u,t)\big)\delta\tilde{\Psi}(u,t)\,dt.$$

By elementary reasoning, one sees that, as $\xi \to 0$,

$$\frac{g(\xi,u)}{\xi} \longrightarrow f\big(\Psi(u)\big)\omega_{\mathbf{n}}\big(\Psi(u)\big)\delta\Psi(u)$$

uniformly and boundedly for $u \in U$; and so (cf. (5.3.18))

$$\begin{aligned}\lim_{\xi\to 0}\frac{\Delta_2(x)}{\xi} &= \int_U f\big(\Psi(u)\big)\omega_{\mathbf{n}}\big(\Psi(u)\big)\delta\Psi(u)\,\lambda_{\mathbb{R}^{N-1}}(du)\\ &= \int_{\partial G} f(x)\omega_{\mathbf{n}}(x)\,\lambda_{\partial G}(dx).\end{aligned}$$

The first part of (5.4.7) is more involved. To handle it, we introduce the notation

$$D(y) = \tilde{\Psi}^{-1}(y)_N = \Big(y-p(y),\mathbf{n}(y)\Big)_{\mathbb{R}^N}, \quad y\in\Omega.$$

Observe that if $p \in \Omega\cap\partial G$, then (cf. Exercise 5.3.21)

$$\lim_{t\to 0}\frac{D(p+t\mathbf{v})-D(p)}{t} = \begin{cases} 0 & \text{if } \mathbf{v}\in\mathbf{T}_p(\partial G)\\ 1 & \text{if } \mathbf{v}=\mathbf{n}(p).\end{cases}$$

Hence,

$$\nabla D(p) = \mathbf{n}(p) \quad \text{for } p\in\Omega\cap\partial G. \tag{5.4.8}$$

Next, set

$$E(y,\xi) = D\big(y-\xi\boldsymbol{\omega}_{\mathbf{n}}(y)\big) - D(y-\xi\boldsymbol{\omega}) \quad \text{for } (y,\xi)\in K\times(-r,r).$$

Clearly,

$$\begin{aligned} x-\xi\boldsymbol{\omega}\in G \text{ but } x-\xi\boldsymbol{\omega}_{\mathbf{n}}(y)\notin G &\implies 0\le D\big(x-\xi\boldsymbol{\omega}_{\mathbf{n}}(x)\big)\le E(x,\xi)\\ x-\xi\boldsymbol{\omega}\notin G \text{ but } x-\xi\boldsymbol{\omega}_{\mathbf{n}}(y)\in G &\implies E(x,\xi)\le D\big(x-\xi\boldsymbol{\omega}_{\mathbf{n}}(x)\big)\le 0.\end{aligned}$$

Thus,

$$\begin{gathered}\big|\Delta_1(\xi)\big| \le \|f\|_{\mathrm{u}}\lambda_{\mathbb{R}^N}\big(\Gamma(\xi)\big)\\ \text{where } \Gamma(\xi)\equiv\Big\{x\in K:\ \big|D\big(x-\xi\boldsymbol{\omega}_{\mathbf{n}}(x)\big)\big|\le\big|E(x,\xi)\big|\Big\}.\end{gathered}$$

In order to estimate $\lambda_{\mathbb{R}^N}\big(\Gamma(\xi)\big)$, first observe that, for some $C_1<\infty$,

$$|E(y,\xi)|\le C_1|\xi|, \quad (y,\xi)\in K\times(-r,r).$$

Hence, since $p(x - \xi\boldsymbol{\omega}_{\mathbf{n}}(x)) = p(x)$,

$$x \in \Gamma(\xi) \implies \big|x - \xi\boldsymbol{\omega}_{\mathbf{n}}(x) - p(x)\big| = \big|D(x - \xi\boldsymbol{\omega}_{\mathbf{n}})\big| \le C_1|\xi|;$$

which, together with (5.4.8), leads to the existence of a $C_2 < \infty$ for which

$$x \in \Gamma(\xi) \implies \big|\nabla D(x - \xi\boldsymbol{\omega}_{\mathbf{n}}(x)) - \mathbf{n}(x)\big| \le C_2|\xi|.$$

But, since $\boldsymbol{\omega} - \boldsymbol{\omega}_{\mathbf{n}}(x) \perp \mathbf{n}(x)$, this, in conjunction with Taylor's Theorem, says that

$$x \in \Gamma(\xi) \implies \big|E(x,\xi)\big| \le C\xi^2$$

for some $C < \infty$. In other words, we have now shown that

$$\begin{aligned}\Gamma(\xi) &\subseteq \Big\{x \in K : \big|D\big(x - \xi\boldsymbol{\omega}_{\mathbf{n}}(x)\big)\big| \le C\xi^2\Big\} \\ &\subseteq \tilde{\boldsymbol{\Psi}}\Big(\Big\{(u,t) \in \tilde{U} : \big|t - \xi\boldsymbol{\omega}_{\mathbf{n}}\big(\boldsymbol{\Psi}(u)\big)\big| \le C\xi^2\Big\}\Big).\end{aligned}$$

Finally, set

$$I(\xi) = (-\epsilon,\epsilon) \cap \big[\xi\boldsymbol{\omega}_{\mathbf{n}}\big(\boldsymbol{\Psi}(u)\big) - C\xi^2, \xi\boldsymbol{\omega}_{\mathbf{n}}\big(\boldsymbol{\Psi}(u)\big) + C\xi^2\big],$$

and use Theorem 5.3.2 to conclude that

$$\begin{aligned}\lambda_{\mathbb{R}^N}\big(\Gamma(\xi)\big) &\le \int_U \left(\int_{I(\xi)} \delta\tilde{\boldsymbol{\Psi}}(u,t)\,dt\right)\lambda_{\mathbb{R}^{N-1}}(dw) \\ &\le C\xi^2 \sup_{(u,t)\in\tilde{U}} \delta\tilde{\boldsymbol{\Psi}}(u,t)\lambda_{\mathbb{R}^{N-1}}(U),\end{aligned}$$

which is more than enough to prove the first part of (5.4.7). □

5.4.9 Theorem. *Let G be a smooth region in $\mathbb{R}^N$ and U an open neighborhood of $\overline{G}$. Then* (5.4.3) *holds for every $f \in C_c(U;\mathbb{R})$. In particular, if G is bounded, then* (5.4.3) *holds for every $f \in C(U;\mathbb{R})$.*

PROOF: We first observe that, without loss of generality, we may assume that $U = \mathbb{R}^N$ and that $f \in C_c(\mathbb{R}^N;\mathbb{R})$. Indeed, when G is unbounded but $f \in C_c(U;\mathbb{R})$, choose a compact set $K \subseteq U$ so that f vanishes off K; and when G is bounded, choose K so that $\overline{G} \subseteq \mathring{K}$ and $K \subseteq U$. Next, in either case, let ϵ denote the distance between K and $U\complement$ and define $\tilde{f} : \mathbb{R}^N \longrightarrow \mathbb{R}$ so that

$$\tilde{f}(x) = \Big[\Big(\tfrac{2}{\epsilon}\mathrm{dist}(x, U\complement)\Big) \wedge 1\Big] f(x) \quad \text{for } x \in U$$

and $\tilde{f} \equiv 0$ off U. It is then clear that $\tilde{f} \in C_c(\mathbb{R}^N;\mathbb{R})$. Moreover, since

$$\int_G \tilde{f}(x + \xi\boldsymbol{\omega})\,dx = \int_G f(x + \xi\boldsymbol{\omega})\,dx \quad \text{for } |\xi| < \frac{\epsilon}{2},$$

it is obvious that (5.4.3) for $\tilde{f}$ is equivalent to (5.4.3) for f.

In view of the preceding, we now assume that $f \in C_c(\mathbb{R}^N;\mathbb{R})$; and we start the proof of (5.4.3) by observing that when f vanishes off the compact set K and $K \cap \partial G = \emptyset$, then $\epsilon = \operatorname{dist}(K, \partial G) > 0$ and

$$\int_G f(x+\xi\boldsymbol{\omega})\,dx = \int_K \mathbf{1}_G(x-\xi\boldsymbol{\omega})f(x)\,dx = \int_G f(x)\,dx \quad \text{for all } |\xi| < \epsilon.$$

Hence, in this case, (5.4.3) is essentially trivial.

To complete the proof, we must show how to reduce to the situations for which the result is already proved. Thus, let $f \in C_c(\mathbb{R}^N;\mathbb{R})$ be given and choose a compact K off which f vanishes. As we noted above, there is nothing more to do if $K \cap \partial G = \emptyset$. Thus, we assume that $K \cap \partial G \neq \emptyset$. Using the compactness of $K \cap \partial G$ and Lemma 5.4.4, we now choose coordinate charts $\{(\boldsymbol{\Psi}_m, U_m)\}_1^n$ of the sort described in that lemma, points $\{p_m\}_1^n \subseteq K \cap \partial G$, and numbers $\{r_m\}_1^n \subseteq (0,\infty)$ so that $B_{\mathbb{R}^N}(p_m, 4r_m) \subseteq \tilde{\boldsymbol{\Psi}}_m(\tilde{U}_m)$ for each $1 \le m \le n$ and

$$K \cap \partial G \subseteq \bigcup_1^n B_{\mathbb{R}^N}(p_m, r_m).$$

Set

$$H = \mathbb{R}^N \setminus \bigcup_1^n B_{\mathbb{R}^N}(p_m, r_m),$$

and define

$$\psi_0(x) = \operatorname{dist}\big(x, H\complement\big) \text{ and } \psi_m(x) = \operatorname{dist}\big(x, B_{\mathbb{R}^N}(p_m, 3r_m)\complement\big) \text{ for } 1 \le m \le n.$$

It is then clear that each ψ_m is a non-negative, continuous function and that

$$s(x) \equiv \sum_0^n \psi_m(x) \ge r_1 \wedge \cdots \wedge r_n > 0 \quad \text{for all } x \in \mathbb{R}^N.$$

Hence, if

$$f_m(x) = \frac{\psi_m(x) f(x)}{s(x)}, \quad x \in \mathbb{R}^N \text{ and } 0 \le m \le n,$$

then each f_m is continuous, $f = \sum_0^n f_m$, f_0 vanishes off a compact subset of $\partial G\complement$, and, for $1 \le m \le n$, f_m vanishes off of a compact subset of $\tilde{\boldsymbol{\Psi}}_{p_m}(\tilde{U}_{p_m})$. In particular, this means that

$$\frac{d}{d\xi}\int_G f(x+\xi\boldsymbol{\omega})\,dx\bigg|_{\xi=0} = \sum_0^n \frac{d}{d\xi}\int_G f_m(x+\xi\boldsymbol{\omega})\,dx\bigg|_{\xi=0}$$

$$= \sum_0^n \int_{\partial G} f_m(x)\big(\boldsymbol{\omega}, \mathbf{n}(x)\big)_{\mathbb{R}^N}\lambda_{\partial G}(dx) = \int_{\partial G} f(x)\big(\boldsymbol{\omega}, \mathbf{n}(x)\big)_{\mathbb{R}^N}\lambda_{\partial G}(dx). \quad \square$$

Although (5.4.3) is the basic fact which we wanted to prove in this section, it does not present the result in the form which is most frequently encountered in applications. To see how to pass from (5.4.3) to the expression which we are after, assume that f is continuously differentiable in a neighborhood of $\overline{G}$, and (cf. Exercise 4.1.13) note that (5.4.3) becomes

$$\int_G \big(\nabla f(x), \boldsymbol{\omega}\big)_{\mathbb{R}^N}\, dx = \int_{\partial G} f(x)\big(\boldsymbol{\omega}, \mathbf{n}(x)\big)_{\mathbb{R}^N}\, \lambda_{\partial G}(dx).$$

Next, suppose that $\mathbf{F} : \mathbb{R}^N \longrightarrow \mathbb{R}^N$ is once continuously differentiable and that $\mathbf{F} \equiv \mathbf{0}$ off a compact set. By applying the preceding with $f = F_k$ and $\boldsymbol{\omega} = \mathbf{e}_k$ and summing over $1 \le k \le N$, we obtain

$$\int_G \operatorname{div} \mathbf{F}(x)\, dx = \int_{\partial G} \big(\mathbf{F}(x), \mathbf{n}(x)\big)_{\mathbb{R}^N}\, \lambda_{\partial G}(dx), \tag{5.4.10}$$

where $\operatorname{div} \mathbf{F} \equiv \sum_1^N F_{k,k}$ is the **divergence** of $\mathbf{F}$. The formula in (5.4.10) is sufficiently important to warrant our stating it as a theorem.

5.4.11 Theorem (Divergence Theorem). *Again let G be a smooth region in $\mathbb{R}^N$ and U an open neighborhood of $\overline{G}$. If $\mathbf{F} : U \longrightarrow \mathbb{R}^N$ is continuously differentiable and either G is bounded or $\mathbf{F} \equiv \mathbf{0}$ off of a compact subset of U, then*

$$\int_G \operatorname{div} \mathbf{F}(x)\, dx = \int_{\partial G} \big(\mathbf{F}(x), \mathbf{n}(x)\big)_{\mathbb{R}^N}\, \lambda_{\partial G}(dx).$$

Before dropping this topic, we will give some examples of the way in which The Divergence Theorem is used in the analysis of partial differential equations.

Let $\Delta = \sum_{i=1}^N \frac{\partial^2}{\partial x_i{}^2}$ denote the standard **Laplacian**. The following variant on The Divergence Theorem provides one of the keys to the analysis of equations in which Δ appears.

5.4.12 Theorem (Green's Identity). *Let G be a smooth region in $\mathbb{R}^N$ and U an open neighborhood of $\overline{G}$. If u and v are twice continuously differentiable $\mathbb{R}$-valued functions on U and either G is bounded or u has compact support in U, then*

$$\begin{aligned} &\int_G u\, \Delta v\, dx - \int_G v\, \Delta u\, dx \\ &\qquad = \int_{\partial G} u\, \frac{\partial v}{\partial \mathbf{n}}\, \lambda_{\partial G}(dx) - \int_{\partial G} v\, \frac{\partial u}{\partial \mathbf{n}}\, \lambda_{\partial G}(dx), \end{aligned} \tag{5.4.13}$$

where

$$\frac{\partial f}{\partial \mathbf{n}}(x) \equiv \frac{d}{d\xi} f(x + \xi \mathbf{n}(x)) \bigg|_{\xi=0} = \big(\nabla f(x), \mathbf{n}(x)\big)_{\mathbb{R}^N}$$

denotes differentiation in the direction (cf. (5.4.2)*)* $\mathbf{n}$.

PROOF: Simply note that $u\,\Delta v - v\,\Delta u = \operatorname{div}\big(u\,\nabla v - v\,\nabla u\big)$, and apply The Divergence Theorem. □

In order to extract information from Green's Identity, one must make judicious choices of v for a given u. For example, one often wants to take v to be the **fundamental solution** g given by $g(0) = \infty$ and

$$(5.4.14)\qquad g(x) = \begin{cases} -\log|x| & \text{if } N = 2 \\ |x|^{2-N} & \text{if } N \ge 3 \end{cases} \quad \text{for} \quad x \in \mathbb{R}^N \setminus \{0\}.$$

Note that g and $|\nabla g|$ are integrable on every compact subset of $\mathbb{R}^N$. In addition, the following facts about g are easy to verify:

$$(5.4.15)\qquad \Delta g(x) = 0 \quad \text{and} \quad |x|^N\,\nabla g(x) = \begin{cases} -x & \text{if } N = 2 \\ (2-N)x & \text{if } N \ge 3 \end{cases}$$

on $\mathbb{R}^N \setminus \{0\}$.

Our first application allows us to solve the **Poisson equation** $\Delta u = -f$.

5.4.16 Theorem. *Set $c_N = 2\pi$ or $(N-2)\,\omega_{N-1}$ (cf. part (iii) of Exercise 5.2.6) depending on whether $N = 2$ or $N \ge 3$. Given $f \in C^2_c\big(\mathbb{R}^N;\mathbb{R}\big)$, define u_f on $\mathbb{R}^N$ by (cf. (5.4.14))*

$$u_f(x) = \frac{1}{c_N}\int_{\mathbb{R}^N} g(x-y)f(y)\,dy.$$

Then $u_f \in C^2\big(\mathbb{R}^N;\mathbb{R}\big)$ and $\Delta u_f = -f$.

PROOF: Since another expression for $c_N u_f(\cdot)$ is $\int_{\mathbb{R}^N} g(y)f(\cdot\,-y)\,dy$, it is clear (cf. Exercise 4.1.13) both that $u_f \in C^2(\mathbb{R}^N;\mathbb{R})$ and that $c_N\Delta u_f(x) = \int_{\mathbb{R}^N} g(y)\Delta f(x-y)\,dy$. Thus, all that we need to do is check that

$$\int_{\mathbb{R}^N} g(y)\Delta f(x-y)\,dy = -c_N f(x).$$

Fix x and choose $R > 1$ so that $f \equiv 0$ off of $B_{\mathbb{R}^N}(x, R-1)$. For $0 < r < R$, set $G_r = B_{\mathbb{R}^N}(0,R) \setminus \overline{B_{\mathbb{R}^N}(0,r)}$. Then

$$\int_{\mathbb{R}^N} g(y)\Delta f(x-y)\,dy = \lim_{r\searrow 0}\int_{G_r} g(y)\Delta f(x-y)\,dy;$$

and, by Green's Identity and (5.4.15), for each $0 < r < R$ (cf. Exercise 5.3.22):

$$\begin{aligned}
&\int_{G_r} g(y)\Delta f(x-y)\,dy \\
&= \int_{\partial G_r} g(y)\frac{\partial f}{\partial \mathbf{n}}(x-y)\,\lambda_{\mathbf{S}^{N-1}(r)}(dy) - \int_{\partial G_r} f(x-y)\frac{\partial g}{\partial \mathbf{n}}(y)\,\lambda_{\partial G_r}(dy) \\
&= -\int_{\partial B_{\mathbb{R}^N}(0,r)} g(y)\frac{\partial f}{\partial \rho}(x-y)\,\lambda_{\partial B_{\mathbb{R}^N}(0,r)}(dy) \\
&\qquad\qquad + \int_{\mathbf{S}^{N-1}(r)} f(x-y)\frac{\partial g}{\partial \rho}(y)\,\lambda_{\partial B_{\mathbb{R}^N}(0,r)}(dy) \\
&= -r^{N-1}\int_{\mathbf{S}^{N-1}} g(r\omega)\frac{\partial f}{\partial \rho}(x+r\omega)\lambda_{\mathbf{S}^{N-1}}(d\omega) \\
&\qquad\qquad + r^{N-1}\int_{\mathbf{S}^{N-1}} f(x+r\omega)\frac{\partial g}{\partial \rho}(r\omega)\lambda_{\mathbf{S}^{N-1}}(d\omega),
\end{aligned}$$

where $\frac{\partial}{\partial \rho}$ denotes differentiation in the outward radial direction and we have used Exercise 5.3.22 together with the fact that, for G_r, $\frac{\partial}{\partial \mathbf{n}} = -\frac{\partial}{\partial \rho}$ on $\mathbf{S}^{N-1}(r)$. But $r^{N-1}g(r\omega) \longrightarrow 0$ uniformly as $r \searrow 0$ and (cf. (5.4.15))

$$r^{N-1}\frac{\partial g}{\partial \rho}(r\omega) = \begin{cases} -1 & \text{if } \; N = 2 \\ -(N-2) & \text{if } \; N \geq 3. \end{cases}$$

After combining this with the preceding, we now see that

$$\begin{aligned}
&\lim_{r\searrow 0}\int_{G_r} g(y)\Delta f(x-y)\,dy \\
&= -\frac{c_N}{\omega_{N-1}}\lim_{r\searrow 0}\int_{\mathbf{S}^{N-1}} f(x+r\omega)\,\lambda_{\mathbf{S}^{N-1}}(d\omega) = -c_N f(x). \quad \square
\end{aligned}$$

Our second application of Green's Identity will be to harmonic functions. A function $u \in C^2(G;\mathbb{R})$ is said to be **harmonic** in G if $\Delta u = 0$. Notice that if $N = 1$ and u is harmonic on (a,b) and continuous on $[a,b]$, then $u(x) = \frac{(b-x)}{(b-a)}u(a) + \frac{(x-a)}{(b-a)}u(b)$ for $x \in [a,b]$. In particular, $u\left(\frac{a+b}{2}\right)$ is precisely the mean of the values that u takes on $\partial(a,b)$. We will now use Green's Identity to derive the analogous fact about harmonic functions in higher dimensions.

5.4.17 Theorem (The Mean Value Property). *Suppose that u is a harmonic element of $C^2(G;\mathbb{R})$. Then, for each $x \in G$ and $R > 0$ satisfying $\overline{B_{\mathbb{R}^N}(x,R)} \subseteq G$,*

$$u(x) = \frac{1}{\omega_{N-1}}\int_{\mathbf{S}^{N-1}} u(x+R\omega)\,\lambda_{\mathbf{S}^{N-1}}(d\omega). \tag{5.4.18}$$

PROOF: Without loss of generality, we will assume that $x = 0$.

Set (cf. (5.4.14)) $g_R(x) = g(x) - g(R\mathbf{e}_1)$ where $\mathbf{e}_1 = (1, 0, \dots, 0) \in \mathbf{S}^{N-1}$. Then, by Green's Identity applied to the functions u and g_R in the region (cf. the proof of the preceding) G_r

$$\begin{aligned}
0 &= \int_{G_r} \big(g_R(y)\,\Delta u(y) - u(y)\,\Delta g_R(y)\big)\,dy \\
&= -R^{N-1} \int_{\mathbf{S}^{N-1}} u(R\omega) \frac{\partial g_R}{\partial \rho}(R\omega) \lambda_{\mathbf{S}^{N-1}}(d\omega) \\
&\quad - r^{N-1} \int_{\mathbf{S}^{N-1}} g_R(r\omega) \frac{\partial u}{\partial \rho}(r\omega) \lambda_{\mathbf{S}^{N-1}}(d\omega) \\
&\quad + r^{N-1} \int_{\mathbf{S}^{N-1}} u(r\omega) \frac{\partial g_R}{\partial \rho}(r\omega) \lambda_{\mathbf{S}^{N-1}}(d\omega),
\end{aligned}$$

where we have used the same notation as in the preceding proof. Note that the first term on the right equals (cf. (5.4.15))

$$\frac{c_N}{\omega_{N-1}} \int_{\mathbf{S}^{N-1}} u(R\omega)\,\lambda_{\mathbf{S}^{N-1}}(d\omega),$$

the second term tends to 0 as $r \searrow 0$, while the third term tends to $-c_N u(0)$. □

Exercises

5.4.19 Exercise: Let u be a twice continuously differentiable function in a neighborhood of the closed ball $\overline{B_{\mathbb{R}^N}(x, r)}$, and assume that $\Delta u \le 0$ in $B_{\mathbb{R}^N}(x, r)$. Generalize the Mean Value Property by showing that

$$u(x) \ge \frac{1}{\omega_{N-1}} \int_{\mathbf{S}^{N-1}} u(x + r\omega)\,\lambda_{\mathbf{S}^{N-1}}(d\omega). \tag{5.4.20}$$

Next, show that (5.4.18) and (5.4.20) yield, respectively,

$$u(x) = \frac{1}{\Omega_N r^N} \int_{B_{\mathbb{R}^N}(x,r)} u(x + y)\,dy \tag{5.4.21 (i)}$$

and

$$u(x) \ge \frac{1}{\Omega_N r^N} \int_{B_{\mathbb{R}^N}(x,r)} u(x + y)\,dy. \tag{5.4.21 (ii)}$$

Using (5.4.21 (ii)), argue that if G is a connected open set in $\mathbb{R}^N$ and $u \in C^2(G;\mathbb{R})$ satisfies $\Delta u \le 0$, then u achieves its minimum value at an $x \in G$ if and only if u is constant on G. This fact is known as the **strong minimum principle**.

5.4.22 Exercise: Let G be a bounded, smooth region in the plain $\mathbb{R}^2$. In addition, assume ∂G is a closed curve in the sense that there is a $\gamma \in C^2([0,1);\mathbb{R}^2)$ with the properties that

$$t \in [0,1) \longmapsto \gamma(t) \in \partial G \text{ is an injective surjection,}$$
$$\gamma(0) = \lim_{t \nearrow 1} \gamma(t), \quad \dot\gamma(0) = \lim_{t \nearrow 1} \dot\gamma(t),$$
$$\ddot\gamma(0) = \lim_{t \nearrow 1} \ddot\gamma(t), \quad \text{and } |\dot\gamma(t)| > 0 \text{ for } t \in [0,1).$$

(**i**) Show that

$$\int_{\partial G} \varphi(\zeta)\, \lambda_{\partial G}(d\zeta) = \int_{[0,1]} \varphi \circ \gamma(t)\, |\dot\gamma(t)|\, dt$$

for all bounded measurable φ on ∂G.

(**ii**) Let $\mathbf{n}(t)$ denote the outer normal to G at $\gamma(t)$, check that

$$\mathbf{n}(t) = \pm |\dot\gamma(t)|^{-1} \big(\dot\gamma_2(t), -\dot\gamma_1(t)\big),$$

with the same sign for all $t \in [0,1)$, and assume that γ has been parameterized so that the plus sign is the correct one. Next, suppose that $h \in C^2(G^{(\rho)};\mathbb{R})$ for some $\rho > 0$, define $u = \frac{\partial h}{\partial x}$ and $v = -\frac{\partial h}{\partial y}$, and set $f = u + \sqrt{-1}v$ and $\Gamma(t) = \gamma_1(t) + \sqrt{-1}\gamma_2(t)$. Show that

$$\int_{[0,1]} f(\gamma(t))\, \dot\Gamma(t)\, dt = \sqrt{-1} \int_G [\Delta h](\zeta)\, d\zeta. \tag{5.4.23}$$

A particularly important case of (5.4.23) is the one when $\Delta h = 0$, in which case f is a complex analytic function and (5.4.23) leads to the famous **Cauchy integral theorem**.

5.4.24 Exercise: Let G, a non-empty, open set in $\mathbb{R}^N$, and $F \in C^3(G;\mathbb{R})$ be given, and assume that $|\nabla F|$ never vanishes in G. Next, set $T = \text{Range}(F)$ and, for $t \in T$, define $M_t = \{x \in \mathbb{R}^N : F(x) = t\}$. Check that, for each $t \in T$, M_t is a hypersurface. Further, by combining the result obtained in Exercise 5.3.27 with the localization technique used in the proof of Theorem 5.4.9, show that, for each $\varphi \in C_c(G)$, $t \in T \longmapsto \int_{M_t} \varphi\, d\lambda_{M_t} \in \mathbb{R}$ is bounded and measurable and that the co-area formula (5.3.28) continues to hold.

Chapter VI

Some Basic Inequalities

6.1 Jensen, Minkowski, and Hölder

There are a few general inequalities which play a central role in measure theory and its applications. The ones dealt with in this section are all consequences of convexity considerations.

A subset $C \subseteq \mathbb{R}^N$ is said to be **convex** if $(1-t)x + ty \in C$ whenever $x, y \in C$ and $t \in [0,1]$. Given a convex set $C \subseteq \mathbb{R}^N$, we say that $g : C \longrightarrow \mathbb{R}$ is a **concave function** on C if

$$g\big((1-t)x + ty\big) \geq (1-t)g(x) + tg(y) \quad \text{for all} \quad x, y \in C \text{ and } t \in [0,1].$$

Note that g is concave on C if and only if $\big\{(x,t) \in C \times \mathbb{R} : t \leq g(x)\big\}$ is a convex subset of $\mathbb{R}^{N+1}$. In addition, one can use induction on $n \geq 2$ to see that

$$\sum_1^n \alpha_k y_k \in C \quad \text{and} \quad g\left(\sum_1^n \alpha_k y_k\right) \geq \sum_1^n \alpha_k g(y_k)$$

for all $n \geq 2$, $\{y_1, \dots, y_n\} \subseteq C$ and $\{\alpha_1, \dots, \alpha_n\} \subseteq [0,1]$ with $\sum_1^n \alpha_k = 1$. Namely, if $n = 2$ or $\alpha_n \in \{0,1\}$, then there is nothing to do. On the other hand, if $n \geq 3$ and $\alpha_n \in (0,1)$, set $y = (1-\alpha_n)^{-1}\sum_{k=1}^{n-1} \alpha_k$, and conclude that

$$\begin{aligned} g\left(\sum_1^n \alpha_k y_k\right) &= g\big((1-\alpha_n)y + \alpha_n y_n\big) \\ &\geq (1-\alpha_n) g\left(\sum_1^{n-1} \alpha_k (1-\alpha_n)^{-1} y_k\right) + \alpha_n g(y_n) \geq \sum_{k=1}^n \alpha_m g(y_k). \end{aligned}$$

The essence of the relationship between these notions and measure theory is contained in the following.

6.1.1 Theorem (Jensen's Inequality). *Let C be a closed, convex subset of $\mathbb{R}^N$, and suppose that g is a continuous, concave, non-negative function on C. Let $(E, \mathcal{B}, \mu)$ be a probability space and $\mathbf{F} : E \longrightarrow C$ a measurable function on $(E, \mathcal{B})$ with the property that $|\mathbf{F}| \in L^1(\mu)$. Then*

$$\int_E \mathbf{F}\, d\mu \equiv \begin{bmatrix} \int_E F_1\, d\mu \\ \vdots \\ \int_E F_N\, d\mu \end{bmatrix} \in C$$

and

$$\int_E g \circ \mathbf{F}\, d\mu \le g\Big(\int_E \mathbf{F}\, d\mu\Big).$$

(See Exercise *6.1.9 for another derivation.)*

PROOF: First assume that $\mathbf{F}$ is simple. Then $\mathbf{F} = \sum_{k=0}^n y_k \mathbf{1}_{\Gamma_k}$ for some $n \in \mathbb{Z}^+$, $y_0, \dots, y_n \in C$, and cover $\{\Gamma_0, \dots, \Gamma_n\}$ of E by mutually disjoints elements of $\mathcal{B}$. Hence, since $\sum_0^n \mu(\Gamma_k) = 1$ and C is convex, $\int_E \mathbf{F}\, d\mu = \sum_0^n y_k \mu(\Gamma_k) \in C$ and, because g is concave,

$$g\Big(\int_E \mathbf{F}\, d\mu\Big) = g\Big(\sum_0^n y_k\mu(\Gamma_k)\Big) \ge \sum_0^n g(y_k)\mu(\Gamma_k) = \int_E g \circ \mathbf{F}\, d\mu.$$

Now let $\mathbf{F}$ be general. We want to approximate $\mathbf{F}$ by C-valued simple functions. For this purpose, choose and fix some element y_0 of C, and let $\{y_k\}_1^\infty$ be a dense sequence in C. Given $m \in \mathbb{Z}^+$, choose $R_m > 0$ so that

$$\int_{\{|\mathbf{F}| \ge R_m\}} (|\mathbf{F}||y_0|)\, d\mu \le \frac{1}{m}$$

and $n_m \in \mathbb{Z}^+$ so that $C \cap \overline{B_{\mathbb{R}^N}(0, R_m)} \subseteq \bigcup_{k=1}^{n_m} B\left(y_k, \frac{1}{m}\right)$. Set $\Gamma_{m,0} = \{\xi \in E : |\mathbf{F}(\xi)| \ge R_m\}$, and use induction to define

$$\Gamma_{m,\ell} = \left\{\xi \in E \setminus \bigcup_{k=0}^{\ell-1} \Gamma_{m,k} : \mathbf{F}(\xi) \in B\left(y_\ell, \tfrac{1}{m}\right)\right\}$$

for $1 \le \ell \le n_m$. Finally, set $\mathbf{F}_m = \sum_{k=0}^{n_m} y_k \mathbf{1}_{\Gamma_{m,k}}$.

By construction, the $\mathbf{F}_m$'s are simple and C-valued. Hence, by the preceding, $\int_E \mathbf{F}_m\, d\mu \in C$ and

$$g\Big(\int_E \mathbf{F}_m\, d\mu\Big) \ge \int_E g \circ \mathbf{F}_m\, d\mu$$

for each $m \in \mathbb{Z}^+$. Moreover, since $|\mathbf{F} - \mathbf{F}_m| \le \frac{1}{m}$ on $\bigcup_1^{n_m} \Gamma_{m,\ell} = E \setminus \Gamma_{0,m}$,

$$\int_E |\mathbf{F} - \mathbf{F}_m|\, d\mu = \sum_{\ell=0}^{n_m} \int_{\Gamma_{m,\ell}} |\mathbf{F} - \mathbf{F}_m|\, d\mu \le \frac{1}{m} + \int_{\Gamma_{m,0}} (|\mathbf{F}| + |y_0|)\, d\mu \le \frac{2}{m}.$$

Thus, $\big\| |\mathbf{F}_m - \mathbf{F}| \big\|_{L^1(\mu)} \longrightarrow 0$ as $m \to \infty$; and so, because C is closed, we now see that $\int_E \mathbf{F}\, d\mu \in C$. At the same time, because g is continuous, $g \circ \mathbf{F}_m \longrightarrow g \circ \mathbf{F}$ in μ-measure as $m \to \infty$. Hence, by the version of Fatou's Lemma in Lemma 3.3.12,

$$\int_E g \circ \mathbf{F}\, d\mu \le \varliminf_{m\to\infty} \int_E g \circ \mathbf{F}_m\, d\mu \le \varliminf_{m\to\infty} g\Big(\int_E \mathbf{F}_m\, d\mu\Big) = g\Big(\int_E \mathbf{F}\, d\mu\Big). \quad \square$$

We now need to develop a criterion for recognizing when a function is concave. Such a criterion is contained in the next theorem. Recall that the **Hessian matrix** $H_g(x)$ of a function g which is twice continuously differentiable at x is the symmetric matrix given by

$$H_g(x) \equiv \left[\frac{\partial^2 g}{\partial x_i \partial x_j}(x)\right]_{1\le i,j\le N}.$$

6.1.2 Lemma. *Suppose that U is an open, convex subset of $\mathbb{R}^N$, and set $C = \overline{U}$. Then C is also convex. Moreover, if $g : C \longrightarrow \mathbb{R}$ is continuous and $g \restriction U$ is concave, then g is concave on all of C. Finally, if $g : C \longrightarrow \mathbb{R}$ is continuous and $g \restriction U$ is twice continuously differentiable, then g is concave on C if and only if its Hessian matrix is non-positive definite[1] for each $x \in U$.*

PROOF: The convexity of C is obvious. In addition, if $g \restriction U$ is concave, the concavity of g on C follows trivially by continuity. Thus, what remains to show is that if $g : U \longrightarrow \mathbb{R}$ is twice continuously differentiable, then g is concave on U if and only if its Hessian is non-positive definite at each $x \in U$.

In order to prove that g is concave on C if $H_g(x)$ is non-positive definite at every $x \in U$, we will use the following simple result about functions on the interval [0,1]. Namely, suppose that u is continuous on $[0, 1]$ and that u has two continuous derivatives on $(0, 1)$. Then $u(t) \ge 0$ for every $t \in [0, 1]$ if $u(0) = u(1) = 0$ and $u''(t) \le 0$ for every $t \in (0, 1)$. To see this, let $\epsilon > 0$ be given and consider the function $u_\epsilon \equiv u - \epsilon t(t - 1)$. Clearly it is enough for us to show that $u_\epsilon \ge 0$ on $[0, 1]$ for every $\epsilon > 0$. Note that $u_\epsilon(0) = u_\epsilon(1) = 0$ and $u_\epsilon''(t) < 0$ for every $t \in (0, 1)$. In particular, if $u_\epsilon(t) < 0$ for some $t \in [0, 1]$, then there is an $s \in (0, 1)$ at which u_ϵ achieves its absolute minimum. But this is impossible, since then we would have that $u_\epsilon''(s) \ge 0$. (The astute reader will undoubtedly see that this result could have been derived as a consequence of the strong minimum principle in Exercise 5.4.19 for $N = 1$.)

Now assume that $H_g(x)$ is non-positive definite for every $x \in U$. Given $x, y \in U$, define $u(t) = g((1-t)x + ty) - (1-t)g(x) - tg(y)$ for $t \in [0, 1]$. Then $u(0) = u(1) = 0$ and

$$u''(t) = \Big(y - x, H_g((1-t)x + ty)(y - x)\Big)_{\mathbb{R}^N} \le 0$$

for every $t \in (0, 1)$. Hence, by the preceding paragraph, $u \ge 0$ on $[0, 1]$; and so $g((1-t)x + ty) \ge (1-t)g(x) + tg(y)$ for all $t \in [0, 1]$. In other words, g is concave on U.

To complete the proof, suppose that $H_g(x)$ has a positive eigenvalue for some $x \in U$. We can then find an $\omega \in \mathbf{S}^{N-1}$ and an $\epsilon > 0$ such that

[1] That is, $(\boldsymbol{\xi}, H_g(x)\boldsymbol{\xi})_{\mathbb{R}^N} \le 0$ for all $\boldsymbol{\xi} \in \mathbb{R}^N$ or, equivalently, all of the eigenvalues of $H_g(x)$ are non-positive.

$\big(\boldsymbol{\omega}, H_g(x)\boldsymbol{\omega}\big)_{\mathbb{R}^N} > 0$ and $x + t\boldsymbol{\omega} \in U$ for all $t \in (-\epsilon, \epsilon)$. Set $u(t) = g(x + t\boldsymbol{\omega})$ for $t \in (-\epsilon, \epsilon)$. Then $u''(0) = \big(\boldsymbol{\omega}, H_g(x)\boldsymbol{\omega}\big)_{\mathbb{R}^N} > 0$. On the other hand,

$$u''(0) = \lim_{t \to 0} \frac{u(t) + u(-t) - 2u(0)}{t^2},$$

and, if g were concave,

$$\begin{aligned} 2u(0) = 2u\left(\frac{t-t}{2}\right) &= 2g\left(\frac{1}{2}(x + t\boldsymbol{\omega}) + \frac{1}{2}(x - t\boldsymbol{\omega})\right) \\ &\geq g(x + t\boldsymbol{\omega}) + g(x - t\boldsymbol{\omega}) = u(t) + u(-t), \end{aligned}$$

from which would we would get the contradictory conclusion that $u''(0) \leq 0$. □

6.1.3 Lemma. *Let $A = \begin{bmatrix} a & b \\ b & c \end{bmatrix}$ be a real symmetric matrix. Then A is non-positive if and only if both $a + c \leq 0$ and $ac \geq b^2$. In particular, for each $\alpha \in (0,1)$, the functions $(x,y) \in [0,\infty)^2 \longmapsto x^\alpha y^{1-\alpha}$ and $(x,y) \in [0,\infty)^2 \longmapsto (x^\alpha + y^\alpha)^{\frac{1}{\alpha}}$ are continuous and concave.*

PROOF: In view of Lemma 6.1.2, it suffices for us to check the first assertion. To this end, let $T = a + c$ be the trace and $D = ac - b^2$ the determinant of A. Also, let λ and μ denote the eigenvalues of A. Then, $T = \lambda + \mu$ and $D = \lambda\mu$.

If A is non-positive and therefore $\lambda \vee \mu \leq 0$, then it is obvious that $T \leq 0$ and that $D \geq 0$. If $D > 0$, then either both λ and μ are positive or both are negative. Hence if, in addition, $T \leq 0$, then λ and μ are negative. Finally, if $D = 0$ and $T \leq 0$, then either $\lambda = 0$ and $\mu = T \leq 0$ or $\mu = 0$ and $\lambda = T \leq 0$. □

6.1.4 Theorem (Minkowski's Inequality). *Let f_1 and f_2 be non-negative, measurable functions on the measure space $(E, \mathcal{B}, \mu)$. Then, for every $p \in [1, \infty)$,*

$$\left(\int_E (f_1 + f_2)^p \, d\mu\right)^{\frac{1}{p}} \leq \left(\int_E f_1^p \, d\mu\right)^{\frac{1}{p}} + \left(\int_E f_2^p \, \mu\right)^{\frac{1}{p}}.$$

PROOF: The case when $p = 1$ follows from (3.2.13). Also, without loss of generality, we assume that f_1^p and f_2^p are μ-integrable and that f_1 and f_2 are $[0,\infty)$-valued.

Let $p \in (1, \infty)$ be given. If we assume that $\mu(E) = 1$ and we take $\alpha = \frac{1}{p}$, then, by Lemma 6.1.3 and Jensen's inequality,

$$\begin{aligned} \int_E (f_1 + f_2)^p \, d\mu &= \int_E \left[(f_1^p)^\alpha + (f_2^p)^\alpha\right]^{\frac{1}{\alpha}} d\mu \\ &\leq \left[\left(\int_E f_1^p \, d\mu\right)^\alpha + \left(\int_E f_2^p \, d\mu\right)^\alpha\right]^{\frac{1}{\alpha}} \\ &= \left[\left(\int_E f_1^p \, d\mu\right)^{\frac{1}{p}} + \left(\int_E f_2^p \, d\mu\right)^{\frac{1}{p}}\right]^p. \end{aligned}$$

More generally, if $\mu(E) = 0$ there is nothing to do, and if $0 < \mu(E) < \infty$ we can replace μ by $\frac{\mu}{\mu(E)}$ and apply the preceding. Hence, all that remains is the case when $\mu(E) = \infty$. But if $\mu(E) = \infty$, take $E_n = \{f_1 \vee f_2 \geq \frac{1}{n}\}$, note that $\mu(E_n) \leq n^p \int f_1^p \, d\mu + n^p \int f_2^p \, d\mu < \infty$, apply the preceding to f_1, f_2, and μ all restricted to E_n, and let $n \to \infty$. $\square$

6.1.5 Theorem (Hölder's Inequality). *Given $p \in (1,\infty)$, define the* **Hölder conjugate** *p' of p by the equation $\frac{1}{p} + \frac{1}{p'} = 1$. Then, for every pair of non-negative, measurable functions f_1 and f_2 on the measure space $(E, \mathcal{B}, \mu)$,*

$$\int_E f_1 f_2 \, d\mu \leq \left(\int_E f_1^p \, d\mu\right)^{\frac{1}{p}} \left(\int_E f_2^{p'} \, d\mu\right)^{\frac{1}{p'}}$$

for every $p \in (1,\infty)$.

PROOF: First note that if either factor on the right hand side of the above inequality is 0, then $f_1 f_2 = 0$ (a.e., μ), and so the left hand side is also 0. Thus we will assume that both factors on the right are strictly positive, in which case, we may and will assume in addition that both f_1^p and $f_2^{p'}$ are μ-integrable and that f_1 and f_2 are both $[0,\infty)$-valued. Also, just as in the proof of Minkowski's inequality, we can reduce everything to the case when $\mu(E) = 1$. But then we can use apply Jensen's Inequality and Lemma 6.1.3 with $\alpha = \frac{1}{p}$ to see that

$$\begin{aligned}\int_E f_1 f_2 \, d\mu &= \int_E \left(f_1^p\right)^\alpha \left(f_2^{p'}\right)^{1-\alpha} d\mu \leq \left(\int_E f_1^p \, d\mu\right)^\alpha \left(\int_E f_2^{p'} \, d\mu\right)^{1-\alpha} \\ &= \left(\int_E f_1^p \, d\mu\right)^{\frac{1}{p}} \left(\int_E f_2^{p'} \, d\mu\right)^{\frac{1}{p'}}. \quad \square\end{aligned}$$

Exercises

6.1.6 Exercise: Here are a few easy applications of the preceding.

(**i**) Show that log is continuous and concave on every interval $[\epsilon, \infty)$ with $\epsilon > 0$. Use this together with Jensen's inequality to show that for any $n \in \mathbb{Z}^+$, $\mu_1, \dots, \mu_n \in (0,1)$ satisfying $\sum_{m=1}^n \mu_m = 1$, and $a_1, \dots, a_n \in (0,\infty)$,

$$\prod_{m=1}^n a_m^{\mu_m} \leq \sum_{m=1}^n \mu_m a_m.$$

In particular, when $\mu_m = \frac{1}{n}$ for every $1 \leq m \leq n$, this yields $(a_1 \cdots a_n)^{\frac{1}{n}} \leq \frac{1}{n} \sum_{m=1}^n a_m$, which is the statement that *the arithmetic mean dominates the geometric mean.*

(ii) Let $n \in \mathbb{Z}^+$, and suppose that $f_1, \dots, f_n$ are non-negative, measurable functions on the measure space $(E, \mathcal{B}, \mu)$. Given $p_1, \dots, p_n \in (1, \infty)$ satisfying $\sum_{m=1}^n \frac{1}{p_m} = 1$, show that

$$\int_E f_1 \cdots f_n \, d\mu \le \prod_{m=1}^n \left(\int_E f_m^{p_m} \, d\mu \right)^{\frac{1}{p_m}}.$$

6.1.7 Exercise: When $p = 2$, Minkowski's and Hölder's inequalities are intimately related and are both very simple to prove. Indeed, let f_1 and f_2 be bounded, measurable functions on the finite measure space $(E, \mathcal{B}, \mu)$. Given any $\alpha \neq 0$, observe that

$$0 \le \int_E \left(\alpha f_1 \pm \frac{1}{\alpha} f_2 \right)^2 d\mu = \alpha^2 \int_E f_1^2 \, d\mu \pm 2 \int_E f_1 f_2 \, d\mu + \frac{1}{\alpha^2} \int_E f_2^2 \, d\mu,$$

from which it follows that

$$2 \left| \int_E f_1 f_2 \, d\mu \right| \le t \int_E f_1^2 \, d\mu + \frac{1}{t} \int_E f_2^2 \, d\mu$$

for every $t > 0$. If either integral on the right vanishes, show from the preceding that $\int_E f_1 f_2 \, d\mu \le 0$. On the other hand, if neither integral vanishes, choose $t > 0$ so that the preceding yields

$$\left| \int_E f_1 f_2 \, d\mu \right| \le \left(\int_E f_1^2 \, d\mu \right)^{\frac{1}{2}} \left(\int_E f_2^2 \, d\mu \right)^{\frac{1}{2}}. \tag{6.1.8}$$

Hence, in any case, (6.1.8) holds. Finally, argue that one can remove the restriction that f_1 and f_2 be bounded, and then remove the condition that $\mu(E) < \infty$. In particular, even if they are not bounded, so long as f_1^2 and f_2^2 are μ-integrable, conclude that $f_1 f_2$ must be μ-integrable and that (6.1.8) continues to hold.

Clearly (6.1.8) is the special case of Hölder's inequality when $p = 2$. Because it is a particularly significant case, it is often referred to by a different name and is called **Schwarz's inequality**. Assuming that both f_1^2 and f_2^2 are μ-integrable, show that the inequality in Schwarz's inequality is an equality if and only if there exist $(\alpha, \beta) \in \mathbb{R}^2 \setminus \{\mathbf{0}\}$ such that $\alpha f_1 + \beta f_2 = 0$ (a.e., μ).

Finally, use Schwarz's inequality to obtain Minkowski's inequality for the case when $p = 2$. Notice the similarity between the development here and that of the classical *triangle inequality* for the Euclidean metric on $\mathbb{R}^N$.

6.1.9 Exercise: A geometric proof of Jensen's inequality can be based on the following. Given a closed, convex subset C of $\mathbb{R}^N$, show that $q \notin C$ if and only if there is a $\mathbf{w}_q \in \mathbf{S}^{N-1}$ such that $(\mathbf{w}_q, q - x)_{\mathbb{R}^N} > 0$ for all $x \in C$. Next, given a probability space $(E, \mathcal{B}, \mu)$ and a μ-integrable $\mathbf{F} : E \longrightarrow C$, use the preceding to show that $p \equiv \int \mathbf{F} \, d\mu \in C$. Finally, let $g : C \longrightarrow [0, \infty)$ be a continuous, concave function, show that $\hat{C} = \{(x,t) \in \mathbb{R}^{N+1} : x \in C \text{ and } t \leq g(x)\}$ is a convex subset of $\mathbb{R}^{N+1}$, and, for each $\epsilon > 0$, apply the first part to find $\hat{\mathbf{w}}_\epsilon \equiv (\mathbf{w}_\epsilon, \rho_\epsilon) \in \mathbb{R}^N \times \mathbb{R}$ such that $\rho_\epsilon(t - g(p) - \epsilon) < (\mathbf{w}_\epsilon, x - p)_{\mathbb{R}^N}$ for all $(x,t) \in \hat{C}$. Note that $\rho_\epsilon > 0$, set $\mathbf{v}_\epsilon = \frac{\mathbf{w}_\epsilon}{\rho_\epsilon}$, and conclude that $g(x) < g(p) + \epsilon + (\mathbf{v}_\epsilon, x - p)_{\mathbb{R}^N}$ for all $x \in C$. Now replace x in the preceding by $\mathbf{F}$, integrate with respect to μ, and get $\int g \circ \mathbf{F}\, d\mu < g(p) + \epsilon$ for every $\epsilon > 0$.

6.2 The Lebesgue Spaces

In Section 3.2 we introduced $\|\cdot\|_{L^1(\mu)}$ and the space $L^1(\mu)$. We are now ready to embed $L^1(\mu)$ into a one-parameter family of spaces.

Given a measure space $(E, \mathcal{B}, \mu)$ and a $p \in [1, \infty)$, define

$$\|f\|_{L^p(\mu)} = \left(\int_E |f|^p \, d\mu \right)^{\frac{1}{p}}$$

for measurable functions f on $(E, \mathcal{B})$. Also, if f is a measurable function on $(E, \mathcal{B})$ define

$$\|f\|_{L^\infty(\mu)} = \inf\Big\{ M \in [0, \infty] : |f| \leq M \text{ (a.e., } \mu) \Big\}.$$

Obviously, as p varies $\|f\|_{L^p(\mu)}$ provides different estimates on the size of f as it is "seen" by the measure μ.

Although information about f can be gleaned from a study of $\|f\|_{L^p(\mu)}$ as p changes (for example, *spikes* in f will be emphasized by taking p to be large), all these quantities share the same flaw as $\|f\|_{L^1(\mu)}$: they cannot detect properties of f which occur on sets having μ-measure 0. Thus, before we can hope to use any of them to get a metric on measurable functions, we must invoke the same subterfuge which we introduced at the end of Section 3.2 in connection with the space $L^1(\mu)$. Namely, for $p \in [1, \infty]$, we denote by $L^p(\mu) = L^p(E, \mathcal{B}, \mu)$ the collection of equivalence classes $[f]_{\sim}^{\mu}$ (cf. Remark 3.2.14) of $\mathbb{R}$-valued, measurable functions f satisfying $\|f\|_{L^p(\mu)} < \infty$, and, once again, we will abuse notation by using f to denote its own equivalence class $[f]_{\sim}^{\mu}$.

Note that, by (3.2.13) and Minkowski's inequality,

$$\|\alpha f_1 + \beta f_2\|_{L^p(\mu)} \leq |\alpha| \, \|f_1\|_{L^p(\mu)} + |\beta| \, \|f_2\|_{L^p(\mu)} \tag{6.2.1}$$

for all $p \in [1, \infty)$, $f_1, f_2 \in L^p(\mu)$, and $\alpha, \beta \in \mathbb{R}$. Moreover, it is a simple matter to check that (6.2.1) continues to hold when $p = \infty$. Thus, each of the spaces $L^p(\mu)$ is a vector space. In addition, because of our convention and Markov's inequality (Theorem 3.2.8), $\|f\|_{L^p(\mu)} = 0$ if and only if $f = 0$ as an element of $L^p(\mu)$. Finally, (6.2.1) allows us to check that $\|f_2 - f_1\|_{L^p(\mu)}$ satisfies the triangle inequality and, together with the preceding, this shows that it determines a metric on $L^p(\mu)$. Thus, when $\{f_n\}_1^\infty \cup \{f\} \subseteq L^p(\mu)$, we often write $f_n \longrightarrow f$ in $L^p(\mu)$ when we mean $\|f_n - f\|_{L^p(\mu)} \longrightarrow 0$.

The following theorem simply summarizes obvious applications of the results in Sections 3.2 and 3.3 to the present context. The reader should check that he sees how each of the assertions here follows from the relevant result there.

6.2.2 Theorem. *Let $(E, \mathcal{B}, \mu)$ be a measure space. Then, for any $p \in [1, \infty]$ and $f,\ g \in L^p(\mu)$,*

$$\big|\, \|g\|_{L^p(\mu)} - \|f\|_{L^p(\mu)} \big| \leq \|g - f\|_{L^p(\mu)}.$$

Next suppose that $\{f_n\}_1^\infty \subseteq L^p(\mu)$ for some $p \in [1, \infty]$ and that f is an $\mathbb{R}$-valued measurable function on $(E, \mathcal{B})$.

(i) *If $p \in [1, \infty)$ and $f_n \longrightarrow f$ in $L^p(\mu)$, then $f_n \longrightarrow f$ in μ-measure. If $f_n \longrightarrow f$ in $L^\infty(\mu)$, then $f_n \longrightarrow f$ uniformly off of a set of μ-measure 0.*

(ii) *If $p \in [1, \infty]$ and $f_n \longrightarrow f$ in μ-measure or (a.e., μ), then $\|f\|_{L^p(\mu)} \leq \varliminf_{n\to\infty} \|f_n\|_{L^p(\mu)}$. Moreover, if $p \in [1, \infty)$ and, in addition, there is a $g \in L^p(\mu)$ such that $|f_n| \leq g$ (a.e., μ) for each $n \in \mathbb{Z}^+$, then $f_n \longrightarrow f$ in $L^p(\mu)$.*

(iii) *If $p \in [1, \infty]$ and $\lim_{m\to\infty} \sup_{n\geq m} \|f_n - f_m\|_{L^p(\mu)} = 0$, then there is an $f \in L^p(\mu)$ such that $f_n \longrightarrow f$ in $L^p(\mu)$. In other words, the space $L^p(\mu)$ is complete with respect to the metric determined by $\|\cdot\|_{L^p(\mu)}$.*

Finally, we have the following variants of Theorem 3.3.14 and Corollary 3.3.15.

(iv) *Assume that $\mu(E) < \infty$ and that $p, q \in [1, \infty)$. Referring to Theorem 3.3.14, define $\mathcal{S}$ as in that theorem. Then, for each $f \in L^p(\mu) \cap L^q(\mu)$, there is a sequence $\{\varphi_n\}_1^\infty \subseteq \mathcal{S}$ such that $\varphi_n \longrightarrow f$ both in $L^p(\mu)$ and in $L^q(\mu)$. In particular, if $\mathcal{B}$ is generated by a countable collection $\mathcal{C}$, then each of the spaces $L^p(\mu)$, $p \in [1, \infty)$, is separable.*

(v) *Let (E, ρ) be a metric space, and suppose that μ is a measure on $(E, \mathcal{B}_E)$ for which there exists a non-decreasing sequence of open sets $E_n \nearrow E$ satisfying $\mu(E_n) < \infty$ for each $n \geq 1$. Then, for each pair $p, q \in [1, \infty)$ and $f \in L^p(\mu) \cap L^q(\mu)$, there is a sequence $\{\varphi_n\}_1^\infty$ of bounded ρ-uniformly continuous functions such that $\varphi_n \equiv 0$ off of E_n and $\varphi_n \longrightarrow f$ both in $L^p(\mu)$ and in $L^q(\mu)$.*

The version of Lieb's variation on Fatou's Lemma for L^p-spaces with $p \neq 1$ is not so easy as the assertions in Theorem 6.2.2. To prove it we will need the following lemma.

6.2.3 Lemma. *Let $p \in (1, \infty)$, and suppose that $\{f_n\}_1^\infty \subseteq L^p(\mu)$ satisfies $\sup_{n\geq 1} \|f_n\|_{L^p(\mu)} < \infty$ and that $f_n \longrightarrow 0$ either in μ-measure or (a.e., μ). Then, for every $g \in L^p(\mu)$,*

$$\lim_{n\to\infty} \int |f_n|^{p-1} |g|\, d\mu = 0 = \lim_{n\to\infty} \int |f_n|\, |g|^{p-1}\, d\mu.$$

PROOF: Without loss of generality, we assume that all of the f_n's as well as g are non-negative. Given $\delta > 0$, we have that

$$\begin{aligned}\int f_n^{p-1} g\, d\mu &= \int_{\{f_n \le \delta g\}} f_n^{p-1} g\, d\mu + \int_{\{f_n > \delta g\}} f_n^{p-1} g\, d\mu \\ &\le \delta^{p-1} \|g\|_{L^p(\mu)}^p + \int_{\{f_n \ge \delta^2\}} f_n^{p-1} g\, d\mu + \int_{\{g \le \delta\}} f_n^{p-1} g\, d\mu.\end{aligned}$$

Applying Hölder's inequality to each of the last two terms, we now see that

$$\begin{aligned}\int f_n^{p-1} g\, d\mu \le & \delta^{p-1} \|g\|_{L^p(\mu)}^p \\ &+ \|f_n\|_{L^p(\mu)}^{p-1} \left[\left(\int_{\{f_n \ge \delta^2\}} g^p\, d\mu \right)^{\frac{1}{p}} + \left(\int_{\{g \le \delta\}} g^p\, d\mu \right)^{\frac{1}{p}} \right].\end{aligned}$$

Since, by Lebesgue's Dominated Convergence Theorem, the first term in the final brackets tends to 0 as $n \to 0$, we conclude that

$$\overline{\lim_{n\to\infty}} \int f_n^{p-1} g\, d\mu \le \delta^{p-1} \|g\|_{L^p(\mu)}^p + \sup_{n\ge 1} \|f_n\|_{L^p(\mu)}^{p-1} \, \|\mathbf{1}_{\{g\le\delta\}} g\|_{L^p(\mu)}$$

for every $\delta > 0$. Thus, after another application of Lebesgue's Dominated Convergence Theorem, we get the first result upon letting $\delta \searrow 0$.

To treat the other case, apply the preceding with f_n^{p-1} and g^{p-1} replacing f_n and g, respectively, and with p' in place of p. □

6.2.4 Theorem (Lieb). *Let $(E, \mathcal{B}, \mu)$ be a measure space, $p \in [1, \infty)$, and $\{f_n\}_1^\infty \cup \{f\} \subseteq L^p(\mu)$. If $\sup_{n\ge 1} \|f_n\|_{L^p(\mu)} < \infty$ and $f_n \longrightarrow f$ in μ-measure or (a.e., μ), then*

$$\lim_{n\to\infty} \int \big|\, |f_n|^p - |f|^p - |f_n - f|^p \,\big|\, d\mu = 0; \tag{6.2.5}$$

and therefore $\|f_n - f\|_{L^p(\mu)} \longrightarrow 0$ if $\|f_n\|_{L^p(\mu)} \longrightarrow \|f\|_{L^p(\mu)}$.

PROOF: The case when $p = 1$ is covered by Theorems 3.3.5 and 3.3.12, and so we will assume that $p \in (1, \infty)$. Given such a p, we first check that there is a $K_p < \infty$ such that

$$\big|\, |b|^p - |a|^p - |b-a|^p \,\big| \le K_p \left(|b-a|^{p-1}|a| + |a|^{p-1}|b-a| \right), \quad a, b \in \mathbb{R}. \tag{6.2.6}$$

Since (6.2.6) clearly holds for all $a, b \in \mathbb{R}$ if it does for all $a \in \mathbb{R} \setminus \{0\}$ and $b \in \mathbb{R}$, we can divide both sides of (6.2.6) by $|a|^p$ and thereby show that (6.2.6) is equivalent to

$$\big|\, |c|^p - 1 - |c-1|^p \,\big| \le K_p \left(|c-1|^{p-1} + |c-1| \right), \quad c \in \mathbb{R}.$$

Finally, the existence of a $K_p < \infty$ for which this inequality holds can be easily verified with elementary consideration of what happens when c is near 1 and when $|c|$ is near infinity.

Applying (6.2.6) with $a = f_n(x)$ and $b = f(x)$, we see that

$$\big|\,|f_n|^p - |f|^p - |f_n - f|^p\,\big| \le K_p\left(|f_n - f|^{p-1}|f| + |f_n - f||f|^{p-1}\right)$$

pointwise. Thus, by Lemma 6.2.3 with f_n and g there replaced by $f_n - f$ and f, respectively, our result follows. □

We now turn to the application of Hölder's inequality to the L^p-spaces. In order to do so, we first complete the definition of the Hölder conjugate p' which, thus far, has only been defined (cf. Theorem 6.1.5) for $p \in (1,\infty)$. Thus, we define $p' = \infty$ or 1 according to whether $p = 1$ or ∞. Notice that this is completely consistent with the equation $\frac{1}{p} + \frac{1}{p'} = 1$ used before.

6.2.7 Theorem. *Let $(E,\mathcal{B},\mu)$ be a measure space.*

(i) *If f and g are measurable functions on $(E,\mathcal{B})$, then for every $p \in [1,\infty]$*

$$\|fg\|_{L^1(\mu)} \le \|f\|_{L^p(\mu)}\,\|g\|_{L^{p'}(\mu)}. \tag{6.2.8}$$

In particular, if $f \in L^p(\mu)$ and $g \in L^{p'}(\mu)$, then $fg \in L^1(\mu)$.

(ii) *If $p \in [1,\infty)$ and $f \in L^p(\mu)$, then*

$$\|f\|_{L^p(\mu)} = \sup\left\{\int fg\,d\mu : g \in L^{p'}(\mu) \text{ and } \|g\|_{L^{p'}(\mu)} \le 1\right\}. \tag{6.2.9}$$

In fact, if $\|f\|_{L^p(\mu)} > 0$, then the supremum in (6.2.9) is achieved by the function

$$g = \frac{|f|^{p-1}\,\mathrm{sgn}\circ f}{\|f\|^{p-1}_{L^p(\mu)}}.$$

(iii) *More generally, for any f which is measurable on $(E,\mathcal{B})$,*

$$\|f\|_{L^p(\mu)} = \sup\left\{\|fg\|_{L^1(\mu)} : g \in L^{p'}(\mu) \text{ and } \|g\|_{L^{p'}(\mu)} \le 1\right\} \tag{6.2.10}$$

if $p = 1$ or if $p \in (1,\infty)$ and either $\mu(|f| \ge \delta) < \infty$ for every $\delta > 0$ or μ is σ-finite.

PROOF: Part **(i)** is an immediate consequence of Hölder's inequality when $p \in (1,\infty)$. At the same time, when $p \in \{1,\infty\}$, the conclusion is clear without any further comment. Given **(i)**, **(ii)** is easy.

When $p = 1$, **(iii)** is obvious; and, in view of **(ii)**, the proof of **(iii)** for $p \in (1,\infty)$ reduces to showing that, under either one of the stated conditions, $\|f\|_{L^p(\mu)} = \infty$ implies that the right hand side of (6.2.10) is infinite. To this end, first suppose that $\mu(|f| \geq \delta) < \infty$ for every $\delta > 0$. Then, for each $n \geq 1$, the function

$$\psi_n \equiv |f|^{p-1} \left(\mathbf{1}_{[\frac{1}{n},n]} \circ |f| \right) + n\mathbf{1}_{\{\infty\}} \circ f \in L^{p'}(\mu).$$

Moreover, if $\|f\|_{L^p(\mu)} = \infty$, then, by the Monotone Convergence Theorem, $\|\psi_n\|_{L^{p'}(\mu)} \longrightarrow \infty$. Thus, since $\|f\psi_n\|_{L^1(\mu)} \geq \|\psi_n\|^{p'}_{L^{p'}(\mu)}$, we see that

$$\|fg_n\|_{L^1(\mu)} \longrightarrow \infty \text{ if } \|f\|_{L^p(\mu)} = \infty \text{ and } g_n \equiv \frac{\psi_n}{1 + \|\psi_n\|_{L^{p'}(\mu)}}.$$

Finally, suppose that μ is σ-finite and that $\mu(|f| \geq \delta) = \infty$ for some $\delta > 0$. Choose $\{E_n\}_1^\infty \subseteq \mathcal{B}$ so that $E_n \nearrow E$ and $\mu(E_n) < \infty$ for every $n \geq 1$. Then it is easy to see that $\lim_{n\to\infty} \|fg_n\|_{L^1(\mu)} = \infty$ when

$$g_n \equiv \frac{\mathbf{1}_{\Gamma_n}}{\left(1 + \mu(\Gamma_n)\right)^{1-\frac{1}{p}}} \quad \text{with} \quad \Gamma_n = E_n \cap \{|f| \geq \delta\}.$$

Since $\|g_n\|_{L^{p'}(\mu)} \leq 1$, this completes the proof. □

For reasons which will become clearer in the next section, it is sometimes useful to consider the following slight variation on the basic L^p-spaces. Namely, let $(E_1, \mathcal{B}_1, \mu_1)$ and $(E_2, \mathcal{B}_2, \mu_2)$ be a pair of σ-finite measure spaces and let $p_1, p_2 \in [1, \infty)$. Given a measurable function f on $(E_1 \times E_2, \mathcal{B}_1 \times \mathcal{B}_2)$, define

$$\|f\|_{L^{(p_1,p_2)}(\mu_1,\mu_2)} \equiv \left[\int_{E_2} \left(\int_{E_1} |f(x_1,x_2)|^{p_1} \, \mu_1(dx_1) \right)^{\frac{p_2}{p_1}} \mu_2(dx_2) \right]^{\frac{1}{p_2}},$$

and let $L^{(p_1,p_2)}(\mu_1, \mu_2)$ denote the **mixed Lebesgue space** of $\mathbb{R}$-valued, $\mathcal{B}_1 \times \mathcal{B}_2$-measurable f's for which $\|f\|_{L^{(p_1,p_2)}(\mu_1,\mu_2)} < \infty$. Obviously, when $p_1 = p = p_2$, $\|f\|_{L^{(p_1,p_2)}(\mu_1,\mu_2)} = \|f\|_{L^p(\mu_1 \times \mu_2)}$ and $L^{(p_1,p_2)}(\mu_1, \mu_2) = L^p(\mu_1 \times \mu_2)$.

6.2.11 Lemma. *For all f and g which are measurable on $(E_1 \times E_2, \mathcal{B}_1 \times \mathcal{B}_2)$ and all α, $\beta \in \mathbb{R}$,*

$$\|\alpha f + \beta g\|_{L^{(p_1,p_2)}(\mu_1,\mu_2)} \leq |\alpha| \, \|f\|_{L^{(p_1,p_2)}(\mu_1,\mu_2)} + |\beta| \, \|g\|_{L^{(p_1,p_2)}(\mu_1,\mu_2)}$$

(6.2.12)

$$\|fg\|_{L^1(\mu_1\times\mu_2)} \leq \|f\|_{L^{(p_1,p_2)}(\mu_1,\mu_2)} \, \|g\|_{L^{(p_1',p_2')}(\mu_1,\mu_2)}.$$

Moreover, if $\{f_n\}_1^\infty \cup \{f\} \subseteq L^{(p_1,p_2)}(\mu_1, \mu_2)$, $f_n \longrightarrow f$ (a.e., $\mu_1 \times \mu_2$), and $|f_n| \leq g$ (a.e., $\mu_1 \times \mu_2$) for each $n \geq 1$ and some $g \in L^{(p_1,p_2)}(\mu_1, \mu_2)$, then $\|f_n - f\|_{L^{(p_1,p_2)}(\mu_1,\mu_2)} \longrightarrow 0$. Finally, if μ_1 and μ_2 are finite and $\mathcal{G}$ denotes the class of all ψ's on $E_1 \times E_2$ having the form $\sum_{m=1}^n \mathbf{1}_{\Gamma_{1,m}}(\,\cdot_1)\varphi_m(\,\cdot_2)$ for some $n \geq 1$, $\{\varphi_m\}_1^n \subseteq L^\infty(\mu_2)$, and mutually disjoint $\Gamma_{1,1}, \ldots, \Gamma_{1,n} \in \mathcal{B}_1$, then, for every measurable $f \in L^{(p_1,p_2)}(\mu_1, \mu_2)$ and $\epsilon > 0$, there is a $\psi \in \mathcal{G}$ such that $\|f - \psi\|_{L^{(p_1,p_2)}(\mu_1,\mu_2)} < \epsilon$.

PROOF: Note that

$$\|f\|_{L^{(p_1,p_2)}(\mu_1,\mu_2)} = \Big\| \, \|f(\,\cdot_1,\,\cdot_2)\|_{L^{p_1}(\mu_1)} \Big\|_{L^{p_2}(\mu_2)}. \tag{6.2.13}$$

Hence the assertions in (6.2.12) are consequences of repeated application of Minkowski's and Hölder's inequalities, respectively. Moreover, to prove the second statement, observe (cf. Exercise 4.1.11) that for μ_2-almost every $x_2 \in E_2$, $f_n(\,\cdot\,,x_2) \longrightarrow f(\,\cdot\,,x_2)$ (a.e., μ_1), $|f_n(\,\cdot\,,x_2)| \le g(\,\cdot\,,x_2)$ (a.e., μ_1), and $g(\,\cdot\,,x_2) \in L^{p_1}(\mu_1)$. Thus, by part **(ii)** of Theorem 6.2.2,

$$\|f_n(\,\cdot\,,x_2) - f(\,\cdot\,,x_2)\|_{L^{p_1}(\mu_1)} \longrightarrow 0$$

for μ_2-almost every $x_2 \in E_2$. In addition,

$$\|f_n(\,\cdot\,,x_2) - f(\,\cdot\,,x_2)\|_{L^{p_1}(\mu_1)} \le 2\|g(\,\cdot\,,x_2)\|_{L^{p_1}(\mu_1)}$$

for μ_2-almost every $x_2 \in E_2$ and, by (6.2.13) with g replacing f,

$$\Big\| \, \|g(\,\cdot_1,\,\cdot_2)\|_{L^{p_1}(\mu_1)} \Big\|_{L^{p_2}(\mu_2)} < \infty.$$

Hence the required result follows after a second application of **(ii)** in Theorem 6.2.2.

We turn now to the final part of the lemma, in which the measures μ_1 and μ_2 are assumed to be finite. In fact, without loss of generality, we will assume that they are probability measures. In addition, by the preceding, it is clear that, for each $f \in L^{(p_1,p_2)}(\mu_1,\mu_2)$,

$$\|f - f_n\|_{L^{(p_1,p_2)}(\mu_1,\mu_2)} \longrightarrow 0 \quad \text{where } f_n \equiv f\mathbf{1}_{[-n,n]} \circ f.$$

Thus, we need only consider f's which are bounded. Finally, because $\mu_1 \times \mu_2$ is also a probability measure, Jensen's inequality and (6.2.13) imply that

$$\|f - \psi\|_{L^{(p_1,p_2)}(\mu_1,\mu_2)} \le \|f - \psi\|_{L^q(\mu_1 \times \mu_2)} \quad \text{where} \quad q = p_1 \vee p_2.$$

Hence, it suffices to show that, for every bounded measurable f on $(E_1 \times E_2, \mathcal{B}_1 \times \mathcal{B}_2)$ and $\epsilon > 0$, there is a $\psi \in \mathcal{G}$ for which $\|f - \psi\|_{L^q(\mu_1 \times \mu_2)} < \epsilon$. But, by part **(iv)** of Theorem 6.2.2, the class of simple functions having the form

$$\psi = \sum_{m=1}^{n} a_m \mathbf{1}_{\Gamma_{1,m} \times \Gamma_{2,m}}$$

with $\Gamma_{i,m} \in \mathcal{B}_i$ is dense in $L^q(\mu_1 \times \mu_2)$. Thus, we will be done once we check that such a ψ is an element of $\mathcal{G}$. To this end, we use the same technique as we did in the final part of the proof of Lemma 3.2.3. That is, set $\mathcal{I} = \big(\{0,1\}\big)^n$ and, for $\boldsymbol{\eta} \in \mathcal{I}$, define $\Gamma_{1,\boldsymbol{\eta}} = \bigcap_{m=1}^n \Gamma_{1,m}^{(\eta_m)}$ where $\Gamma^{(0)} \equiv \Gamma\complement$ and $\Gamma^{(1)} \equiv \Gamma$. Then

$$\psi(x_1,x_2) = \sum_{m=1}^{n} a_m \left(\sum_{\boldsymbol{\eta} \in \mathcal{I}} \eta_m \mathbf{1}_{\Gamma_{1,\boldsymbol{\eta}}}(x_1) \right) \mathbf{1}_{\Gamma_{2,m}}(x_2) = \sum_{\boldsymbol{\eta} \in \mathcal{I}} \mathbf{1}_{\Gamma_{1,\boldsymbol{\eta}}}(x_1)\, \varphi_{\boldsymbol{\eta}}(x_2),$$

where

$$\varphi_{\boldsymbol{\eta}} = \sum_{m=1}^{n} \eta_m a_m \mathbf{1}_{\Gamma_{2,m}}.$$

Since the $\Gamma_{1,\boldsymbol{\eta}}$'s are mutually disjoint, this completes the proof. □

For our purposes, the most important fact that comes out of these considerations is the following **continuous version of Minkowski's inequality**.

6.2.14 Theorem. *Let $(E_i, \mathcal{B}_i, \mu_i)$, $i \in \{1, 2\}$, be σ-finite measure spaces. Then, for any $1 \le p_1 \le p_2 < \infty$ and any measurable function f on $(E_1 \times E_2, \mathcal{B}_1 \times \mathcal{B}_2)$, $\|f\|_{L^{(p_1,p_2)}(\mu_1,\mu_2)} \le \|f\|_{L^{(p_2,p_1)}(\mu_2,\mu_1)}$.*

PROOF: Since it is easy to reduce the general case to the one in which both μ_1 and μ_2 are finite, we may take them to be finite. In fact, without loss of generality, we will assume, from the outset, that they are probability measures.

Let $\mathcal{G}$ be the class described in the last part of Lemma 6.2.11. Given $\psi = \sum_1^n \mathbf{1}_{\Gamma_{1,m}}(\,\cdot\,_1)\,\varphi_m(\,\cdot\,_2)$ which is an element of $\mathcal{G}$, note that, since the $\Gamma_{1,m}$'s are mutually disjoint, $|\sum_1^n a_m \mathbf{1}_{\Gamma_{1,m}}|^r = \sum_1^n |a_m|^r \mathbf{1}_{\Gamma_{1,m}}$ for any $r \in [0, \infty)$ and $a_1, \dots, a_n \in \mathbb{R}$. Hence, by Minkowski's inequality for $p = \frac{p_2}{p_1}$,

$$\begin{aligned}
\|\psi\|_{L^{(p_1,p_2)}(\mu_1,\mu_2)} &= \left[\int_{E_2} \left(\sum_1^n \mu_1(\Gamma_{1,m})|\varphi_m(x_2)|^{p_1}\right)^{\frac{p_2}{p_1}} \mu_2(dx_2)\right]^{\frac{1}{p_2}} \\
&= \left\| \sum_1^n \mu_1(\Gamma_{1,m})|\varphi_m(\,\cdot\,_2)|^{p_1} \right\|_{L^{\frac{p_2}{p_1}}(\mu_2)}^{\frac{1}{p_1}} \\
&\le \left[\sum_1^n \mu_1(\Gamma_{1,m}) \|\, |\varphi_m|^{p_1}\, \|_{L^{\frac{p_2}{p_1}}(\mu_2)}\right]^{\frac{1}{p_1}} \\
&= \left[\sum_1^n \mu_1(\Gamma_{1,m}) \|\varphi_m\|_{L^{p_2}(\mu_2)}^{p_1}\right]^{\frac{1}{p_1}} \\
&= \left[\int_{E_1} \sum_1^n \mathbf{1}_{\Gamma_{1,m}}(x_1) \|\varphi_m\|_{L^{p_2}(\mu_2)}^{p_1} \mu_1(dx_1)\right]^{\frac{1}{p_1}} \\
&= \left[\int_{E_1} \left(\sum_1^n \mathbf{1}_{\Gamma_{1,m}}(x_1) \|\varphi_m\|_{L^{p_2}(\mu_2)}^{p_2}\right)^{\frac{p_1}{p_2}} \mu_1(dx_1)\right]^{\frac{1}{p_1}} \\
&= \left[\int_{E_1} \left(\int_{E_2} \left|\sum_1^n \mathbf{1}_{\Gamma_{1,m}}(x_1)\varphi_m(x_2)\right|^{p_2} \mu_2(dx_2)\right)^{\frac{p_1}{p_2}} \mu_1(dx_1)\right]^{\frac{1}{p_1}} \\
&= \|\psi\|_{L^{(p_2,p_1)}(\mu_2,\mu_1)}.
\end{aligned}$$

Hence, we are done when the function f is an element of $\mathcal{G}$.

To complete the proof, let f be a measurable function on $(E_1 \times E_2, \mathcal{B}_1 \times \mathcal{B}_2)$. Clearly we may assume that $\|f\|_{L^{(p_2,p_1)}(\mu_2,\mu_1)} < \infty$. Using the last part of Lemma 6.2.11, choose $\{\psi_n\}_1^\infty \subseteq \mathcal{G}$ so that $\|\psi_n - f\|_{L^{(p_2,p_1)}(\mu_2,\mu_1)} \longrightarrow 0$. Then, by Jensen's inequality, it is easy to check that $\|\psi_n - f\|_{L^1(\mu_1\times\mu_2)} \longrightarrow 0$, and therefore that $\psi_n \longrightarrow f$ in $\mu_1 \times \mu_2$-measure. Hence, without loss of generality, we will assume that $\psi_n \longrightarrow f$ (a.e., $\mu_1 \times \mu_2$). In particular, by Fatou's Lemma and Exercise 4.1.11, this means that

$$\int_{E_1} |f(x_1,x_2)|^{p_1}\, \mu_1(dx_1) \le \varliminf_{n\to\infty} \int_{E_1} |\psi_n(x_1,x_2)|^{p_1}\, \mu_1(dx_1)$$

for μ_2-almost every $x_2 \in E_2$; and so, by the result for $\mathcal{G}$ and another application of Fatou's Lemma, the required result follows for f. $\square$

The following result is typical of the way in which one applies Theorem 6.2.14.

6.2.15 Theorem. *Let $(E_1, \mathcal{B}_1, \mu_1)$ and $(E_2, \mathcal{B}_2, \mu_2)$ be a pair of σ-finite measure spaces, and suppose that K is a measurable function on $(E_1 \times E_2, \mathcal{B}_1 \times \mathcal{B}_2)$ which satisfies*

$$M_1 \equiv \sup_{x_2\in E_2} \|K(\,\cdot\,,x_2)\|_{L^q(\mu_1)} < \infty \quad \text{and} \quad M_2 \equiv \sup_{x_1\in E_1} \|K(x_1,\,\cdot\,)\|_{L^q(\mu_2)} < \infty$$

for some $q \in [1,\infty)$. Define

$$\mathcal{K}f(x_1) = \int_{E_2} K(x_1,x_2) f(x_2)\, \mu_2(dx_2) \tag{6.2.16}$$

for $f \in L^{q'}(\mu_2)$. Then for each $p \in [1,\infty]$ satisfying $\frac{1}{r} \equiv \frac{1}{p} + \frac{1}{q} - 1 \ge 0$,

$$\|\mathcal{K}f\|_{L^r(\mu_1)} \le M_1^{\frac{q}{r}} M_2^{1-\frac{q}{r}} \|f\|_{L^p(\mu_2)}. \tag{6.2.17}$$

PROOF: First, suppose that $r = \infty$ and therefore that $p = q'$. Then, by part **(i)** of Theorem 6.2.7,

$$\big|\mathcal{K}f(x_1)\big| \le \big\|K(x_1,\cdot)\big\|_{L^q(\mu_2)} \|f\|_{L^p(\mu_2)} \quad \text{for all } x_1 \in E_1;$$

and so (6.2.17) is trivial in this case.

Next, suppose that $p = 1$ and therefore that $q = r$. Noting that $\|\mathcal{K}f\|_{L^r(\mu_1)} \le \|Kf\|_{L^{(1,r)}(\mu_2,\mu_1)}$, we can apply Theorem 6.2.14 to obtain

$$\begin{aligned}\|\mathcal{K}f\|_{L^r(\mu_1)} &\le \|Kf\|_{L^{(1,r)}(\mu_2,\mu_1)} \le \|Kf\|_{L^{(r,1)}(\mu_1,\mu_2)} \\ &= \int_{E_2} \left(\int_{E_1} |K(x_1,x_2)f(x_2)|^r\, \mu_1(dx_1) \right)^{\frac{1}{r}} \mu_2(dx) \\ &= \int_{E_2} \|K(\,\cdot\,,x_2)\|_{L^r(\mu_1)} |f(x_2)|\, \mu_2(dx_2) \le M_1 \|f\|_{L^1(\mu_2)}.\end{aligned}$$

Finally, the only case remaining is when $r \in [1,\infty)$ and $p \in (1,\infty)$. Noting that $r \in (q,\infty)$, set $\alpha = \frac{q}{r}$. Then, $\alpha \in (0,1)$ and $(1-\alpha)p' = q$. Given $g \in L^{r'}(\mu_1)$, we have, by the second inequality in (6.2.12), that

$$\begin{aligned}\|g\,\mathcal{K}f\|_{L^1(\mu_1)} &\le \|g\,Kf\|_{L^1(\mu_1\times\mu_2)}\\ &\le \big\|\,|K|^\alpha f\big\|_{L^{(r,p)}(\mu_1,\mu_2)}\,\big\|g\,|K|^{1-\alpha}\big\|_{L^{(r',p')}(\mu_1,\mu_2)}.\end{aligned}$$

Next, observe that

$$\begin{aligned}&\big\|\,|K|^\alpha f\big\|_{L^{(r,p)}(\mu_1,\mu_2)}\\ &= \left[\int_{E_2}\left(\int_{E_1}|K(x_1,x_2)|^{\alpha r}|f(x_2)|^r\,\mu_1(dx_1)\right)^{\frac{p}{r}}\mu_2(dx_2)\right]^{\frac{1}{p}} \le M_1^\alpha\|f\|_{L^p(\mu_2)}.\end{aligned}$$

At the same time, since $p \le r$ and therefore $r' \le p'$, we can apply Theorem 6.2.14 to see that $\|g\,|K|^{1-\alpha}\|_{L^{(r',p')}(\mu_1,\mu_2)} \le \|g\,|K|^{1-\alpha}\|_{L^{(p',r')}(\mu_2,\mu_1)}$. Hence, by the same reasoning as we just applied to $\||K|^\alpha f\|_{L^{(r,p)}(\mu_1,\mu_2)}$, we find that $\|g\,|K|^{1-\alpha}\|_{L^{(r',p')}(\mu_1,\mu_2)} \le M_2^{1-\alpha}\|g\|_{L^{r'}(\mu_2)}$. Combining these two, we arrive at $\|g\,\mathcal{K}f\|_{L^1(\mu_1)} \le M_1^\alpha M_2^{1-\alpha}\|f\|_{L^p(\mu_1)}$ for all $g \in L^{r'}(\mu_1)$ with $\|g\|_{L^{r'}(\mu_1)} \le 1$; and so (6.2.17) now follows from part **(iii)** of Theorem 6.2.7. □

6.2.18 Corollary. *Let everything be as in Theorem 6.2.15, and, for measurable $f : E_2 \longrightarrow \mathbb{R}$, define*

$$\Lambda_K(f) = \left\{x_1 \in E_1 : \int_{E_2}|K(x_1,x_2)|\,|f(x_2)|\,\mu_2(dx_2) < \infty\right\}$$

and

$$\overline{\mathcal{K}}f(x_1) = \begin{cases}\int_{E_2}K(x_1,x_2)\,f(x_2)\,\mu_2(dx_2) & \text{if } x_1 \in \Lambda_K(f)\\ 0 & \text{otherwise.}\end{cases}$$

Next, let $p \in [1,\infty]$ satisfying $\frac{1}{p}+\frac{1}{q} \ge 1$ be given, and define $r \in [1,\infty]$ by $\frac{1}{r} = \frac{1}{p}+\frac{1}{q}-1$. Then

$$\mu_1\left(\Lambda_K(f)\complement\right) = 0 \quad\text{and}\quad \big\|\overline{\mathcal{K}}f\big\|_{L^r(\mu_1)} \le M_1^{\frac{q}{r}}M_2^{1-\frac{q}{r}}\|f\|_{L^p(\mu_2)} \tag{6.2.19}$$

for $f \in L^p(\mu_2)$. In particular, as a map from $L^p(\mu_2)$ into $L^r(\mu_1)$, $\overline{\mathcal{K}}$ is linear. In fact, $f \in L^p(\mu_2) \longmapsto \overline{\mathcal{K}}f \in L^r(\mu_1)$ is the unique continuous mapping from $L^p(\mu_2)$ into $L^r(\mu_1)$ whose restriction to $L^p(\mu_2)\cap L^{q'}(\mu_2)$ is given by the map $\mathcal{K}$ in (6.2.16).

PROOF: If $r = \infty$, and therefore $p = q'$, there is nothing to do. Thus, we will assume that r and therefore p are finite.

Let $f \in L^p(\mu_2)$ be given, and set $f_n = f\mathbf{1}_{[-n,n]} \circ f$ for $n \in \mathbb{Z}^+$. Because $p < q'$ and $f_n \in L^p(\mu_2) \cap L^\infty(\mu_2)$ (cf. Exercise 6.2.20), $f_n \in L^p(\mu_2) \cap L^{q'}(\mu_2)$. Hence, by (6.2.17) applied to $|K|$ and $|f_n|$,

$$\int_{E_1} \left(\int_{\{x_2:|f(x_2)|\le n\}} \big|K(x_1,x_2)\big|\,\big|f(x_2)\big|\,\mu_2(dx_2) \right)^r \mu_1(dx_1)$$
$$\le M_1^q\, M_2^{r-q} \|f_n\|^r_{L^p(\mu_2)} \le M_1^q\, M_2^{r-q} \|f\|^r_{L^p(\mu_2)}.$$

In particular, by the Monotone Convergence Theorem, this proves both parts of (6.2.19). Furthermore, if $f,\ g \in L^p(\mu_2)$ and $\alpha,\ \beta \in \mathbb{R}$, then

$$\overline{\mathcal{K}}(\alpha f + \beta g) = \alpha \overline{\mathcal{K}} f + \beta \overline{\mathcal{K}} g \quad \text{on} \quad \Lambda_K(f) \cap \Lambda_K(g).$$

Thus, since both $\Lambda_K(f)\complement$ and $\Lambda_K(g)\complement$ have μ_1-measure 0, we now see that, as a mapping into $L^r(\mu_1)$, $\overline{\mathcal{K}}$ is linear. Finally, it is obvious that $\overline{\mathcal{K}} f = \mathcal{K} f$ for $f \in L^p(\mu_2) \cap L^{q'}(\mu_2)$. Hence, if $\mathcal{K}'$ is any extension of $\mathcal{K} \restriction L^p(\mu_2) \cap L^{q'}(\mu_2)$ as a continuous, linear mapping from $L^p(\mu_2)$ to $L^r(\mu_1)$, then (with the same choice of $\{f_n\}_1^\infty$ as above)

$$\begin{aligned}\big\|\overline{\mathcal{K}} f - \mathcal{K}' f\big\|_{L^r(\mu_1)} &\le \overline{\lim_{n\to\infty}} \big\|\overline{\mathcal{K}} f - \mathcal{K} f_n\big\|_{L^r(\mu_1)} \\ &= \overline{\lim_{n\to\infty}} \big\|\overline{\mathcal{K}}(f - f_n)\big\|_{L^r(\mu_1)} \le M_1^{\frac{q}{r}}\, M_2^{1-\frac{q}{r}} \overline{\lim_{n\to\infty}} \|f - f_n\|_{L^p(\mu_1)} = 0. \quad \square\end{aligned}$$

Exercises

6.2.20 Exercise: Let $(E, \mathcal{B}, \mu)$ be a measure space and $1 \le q_1 \le q_2 \le \infty$ be given. If $f \in L^{q_1}(\mu) \cap L^{q_2}(\mu)$, show that for any $t \in (0,1)$

$$\|f\|_{L^{p_t}(\mu)} \le \|f\|^t_{L^{q_1}(\mu)} \|f\|^{1-t}_{L^{q_2}(\mu)} \quad \text{where } \frac{1}{p_t} = \frac{t}{q_1} + \frac{1-t}{q_2}. \tag{6.2.21}$$

Note that (6.2.21) says that $p \longmapsto -\log \|f\|_{L^p(\mu)}$ is a concave function of $\frac{1}{p}$.

6.2.22 Exercise: The following exercises give some insight into the L^p-spaces in various situations.

(i) If $(E, \mathcal{B}, \mu)$ is a probability space, show that $p \in [1,\infty] \longmapsto \|f\|_{L^p(\mu)}$ is a non-decreasing function for any measurable f on $(E, \mathcal{B})$.

(ii) Let $E = \mathbb{Z}^+$ and define μ on $\mathcal{B} = \mathcal{P}(E)$ by $\mu(\{n\}) = 1$ for all $n \in \mathbb{Z}^+$. In this case show that $p \in [1,\infty] \longmapsto \|f\|_{L^p(\mu)}$ is non-increasing for every f on E.

(iii) Let $(E, \mathcal{B}, \mu)$ be a measure space and $f : E \longrightarrow \overline{\mathbb{R}}$ is a $\mathcal{B}$-measurable function. Show that $\|f\|_{L^\infty(\mu)} \le \underline{\lim}_{p\nearrow\infty} \|f\|_{L^p(\mu)}$. Further, assuming either that $\mu(E) < \infty$ or that $\|f\|_{L^1(\mu)} < \infty$, show that $\|f\|_{L^\infty(\mu)} = \lim_{p\nearrow\infty} \|f\|_{L^p(\mu)}$.

(iv) Let $(E_i, \mathcal{B}_i)$, $i \in \{1,2\}$, be measurable spaces, and suppose that μ_2 is a σ-finite measure on $(E_2, \mathcal{B}_2)$. Using part **(iii)**, show that, for every measurable function f on $(E_1 \times E_2, \mathcal{B}_1 \times \mathcal{B}_2)$, the function $x_1 \longmapsto \|f(x_1, \cdot_2)\|_{L^\infty(\mu_2)}$ is measurable on $(E_1, \mathcal{B}_1)$. In particular, we could have defined $\|f\|_{L^{(p_1,p_2)}(\mu_1,\mu_2)}$ for all $p_1, p_2 \in [1, \infty]$.

6.2.23 Exercise: Let $(E, \mathcal{B}, \mu)$ be a measure space, g a non-negative element of $L^p(\mu)$ for some $p \in (1, \infty)$, and f is non-negative, $\mathcal{B}$-measurable function for which there exists a $C \in (0, \infty)$ such that

$$\mu(f \geq t) \leq \frac{C}{t} \int_{\{f \geq t\}} g \, d\mu, \quad t \in (0, \infty). \tag{6.2.24}$$

(i) Set $\nu(\Gamma) = \int_\Gamma g \, d\mu$ for $\Gamma \in \mathcal{B}$, note that (6.2.24) is equivalent to

$$\mu(f > t) \leq \frac{C}{t} \nu(f > t), \quad t \in (0, \infty),$$

and use (5.1.4) (cf. Exercise 5.1.8) to justify

$$\begin{aligned} \|f\|^p_{L^p(\mu)} &= p \int_{(0,\infty)} t^{p-1} \mu(f > t) \, dt \\ &\leq Cp \int_{(0,\infty)} t^{p-2} \nu(f > t) \, dt = \frac{Cp}{p-1} \int f^{p-1} \, d\nu. \end{aligned}$$

Finally, note that $\int f^{p-1} \, d\nu = \int f^{p-1} g \, d\mu$, and apply Hölder's inequality to conclude that

$$\|f\|^p_{L^p(\mu)} \leq \frac{Cp}{p-1} \|f\|^{p-1}_{L^p(\mu)} \|g\|_{L^p(\mu)}. \tag{6.2.25}$$

(ii) Under the condition that $\|f\|_{L^p(\mu)} < \infty$, it is clear that (6.2.25) implies

$$\|f\|_{L^p(\mu)} \leq \frac{Cp}{p-1} \|g\|_{L^p(\mu)}. \tag{6.2.26}$$

Now suppose that $\mu(E) < \infty$. After checking that (6.2.24) for f implies (6.2.24) for $f_R \equiv f \wedge R$, conclude that (6.2.26) holds first with f_R replacing f and then, after $R \nearrow \infty$, for f itself. In other words, when μ is finite, (6.2.24) always implies (6.2.26).

(iii) Even if μ is not finite, show that (6.2.24) implies (6.2.26) as long as $\mu(f > \epsilon) < \infty$ for every $\epsilon > 0$.

Hint: Given $\epsilon > 0$, consider $\mu_\epsilon = \mu \upharpoonright \mathcal{B}[\{f > \epsilon\}]$, note that (6.2.24) with μ implies itself with μ_ϵ, and use **(ii)** to conclude that (6.2.26) holds with μ_ϵ in place of μ. Finally, let $\epsilon \searrow 0$.

6.2.27 Exercise: Recall the Hardy–Littlewood maximal function Mf defined in (3.4.2), and observe that the definition extends, without change, to any measurable f on $\mathbb{R}$ which is integrable on compacts. Show that

$$\|Mf\|_{L^p(\mathbb{R})} \le \frac{2p}{p-1}\|f\|_{L^p(\mathbb{R})}, \quad p \in (1,\infty) \text{ and } f \in L^p(\mathbb{R}). \tag{6.2.28}$$

Hint: When $f \in C_c(\mathbb{R})$, use (3.4.7) together with **(iii)** of Exercise 6.2.23. To handle general $f \in L^p(\mathbb{R})$, use **(v)** in Theorem 6.2.2 to find $\{f_n\}_1^\infty \subseteq C_c(\mathbb{R})$ so that $f_n \longrightarrow f$ in $L^p(\mathbb{R})$, and apply Fatou's Lemma to see that

$$\|Mf\|_{L^p(\mathbb{R})} \le \varliminf_{n\to\infty} \|Mf_n\|_{L^p(\mathbb{R})}.$$

6.3 Convolution and Approximate Identities

We will use the ideas of the last section to develop in this section an important notion of *multiplication* for functions on $\mathbb{R}^N$; and, because the only measure involved will be Lebesgue's, we will use the notation $L^p(\mathbb{R}^N)$ instead of the more cumbersome $L^p(\lambda_{\mathbb{R}^N})$.

6.3.1 Theorem (Young's Inequality). *Let p and q from $[1,\infty]$ satisfying $\frac{1}{p}+\frac{1}{q}\ge 1$ be given, and define $r \in [1,\infty]$ by $\frac{1}{r}=\frac{1}{p}+\frac{1}{q}-1$. Then, for each $f \in L^p(\mathbb{R}^N)$ and $g \in L^q(\mathbb{R}^N)$, the complement of the set*

$$\Lambda(f,g) \equiv \left\{x \in \mathbb{R}^N : \int_{\mathbb{R}^N} |f(x-y)|\,|g(y)|\,dy < \infty\right\} \tag{6.3.2}$$

has Lebesgue measure 0. Furthermore, if

$$f * g(x) \equiv \begin{cases} \int_{\mathbb{R}^N} f(x-y)\,g(y)\,dy & \text{when } x \in \Lambda(f,g) \\ 0 & \text{otherwise,} \end{cases} \tag{6.3.3}$$

then $f*g = g*f$ and

$$\|f*g\|_{L^r(\mathbb{R}^N)} \le \|f\|_{L^p(\mathbb{R}^N)}\|g\|_{L^q(\mathbb{R}^N)}. \tag{6.3.4}$$

Finally, the mapping $(f,g) \in L^p(\mathbb{R}^N)\times L^q(\mathbb{R}^N) \longmapsto f*g \in L^r(\mathbb{R}^N)$ is bilinear.

PROOF: We begin with the observation that there is nothing to do when $r=\infty$. Thus, we will assume throughout that r and therefore also p and q are all finite. Next, using the translation invariance of Lebesgue measure, first note that $\Lambda(f,g)=\Lambda(g,f)$ and then conclude that $f*g=g*f$. Finally, given $q\in[1,\infty)$ and $g \in L^q(\mathbb{R}^N)$, set $K(x,y)=g(x-y)$ for $x,\ y \in \mathbb{R}^N$. Obviously,

$$\sup_{y\in\mathbb{R}^N}\|K(\,\cdot\,,y)\|_{L^q(\mathbb{R}^N)} = \sup_{x\in\mathbb{R}^N}\|K(x,\cdot\,)\|_{L^q(\mathbb{R}^N)} = \|g\|_{L^q(\mathbb{R}^N)} < \infty;$$

and, in the notation of Corollary 6.2.18, $\Lambda(f,g) = \Lambda_K(f)$ and $f * g = \overline{\mathcal{K}}f$. In particular, for each $f \in L^p(\mathbb{R}^N)$, $\Lambda(f,g)\complement$ has Lebesgue measure 0 and (6.3.4) holds. In addition, $f \in L^p(\mathbb{R}^N) \longmapsto f * g \in L^r(\mathbb{R}^N)$ is linear for each $g \in L^q(\mathbb{R}^N)$; and therefore the bilinearity assertion follows after one reverses the roles of f and g. □

The quantity $f * g$ described in (6.3.3) is called the **convolution** of f with g. In applications, the most useful cases are those when $f \in L^p(\mathbb{R}^N)$ and $g \in L^q(\mathbb{R}^N)$ where either $p = q'$ (and therefore $r = \infty$) or $p = 1$ (and therefore $r = q$). To get more information about the case when $p = q'$, we will need the following.

6.3.5 Lemma. *Given $h \in \mathbb{R}^N$, define $\tau_h f$ for functions f on $\mathbb{R}^N$ by $\tau_h f(x) = f(x+h)$. Then τ_h is an isometry on $L^p(\mathbb{R}^N)$ for every $h \in \mathbb{R}^N$ and $p \in [1,\infty]$. Moreover, if $p \in [1,\infty)$ and $f \in L^p(\mathbb{R}^N)$, then*

$$\lim_{h \to 0} \|\tau_h f - f\|_{L^p(\mathbb{R}^N)} = 0. \tag{6.3.6}$$

PROOF: The first assertion is an immediate consequence of the translation invariance of Lebesgue measure.

Next, suppose that $p \in [1,\infty)$ is given. If $\mathcal{G}$ denotes the class of $f \in L^p(\mathbb{R}^N)$ for which (6.3.6) holds, it is clear that $C_c(\mathbb{R}^N) \subseteq \mathcal{G}$. Hence, by (**v**) in Theorem 6.2.2, we will know that $\mathcal{G} = L^p(\mathbb{R}^N)$ as soon as we show that $\mathcal{G}$ is closed in $L^p(\mathbb{R}^N)$. To this end, let $\{f_n\}_1^\infty \subseteq \mathcal{G}$ and suppose that $f_n \longrightarrow f$ in $L^p(\mathbb{R}^N)$. Then

$$\begin{aligned}
&\overline{\lim_{h\to 0}} \|\tau_h f - f\|_{L^p(\mathbb{R}^N)} \\
&\quad \leq \overline{\lim_{h\to 0}} \|\tau_h(f - f_n)\|_{L^p(\mathbb{R}^N)} + \overline{\lim_{h\to 0}} \|\tau_h f_n - f_n\|_{L^p(\mathbb{R}^N)} + \|f_n - f\|_{L^p(\mathbb{R}^N)} \\
&\quad = 2\|f_n - f\|_{L^p(\mathbb{R}^N)} \longrightarrow 0
\end{aligned}$$

as $n \to \infty$. □

6.3.7 Theorem. *Let $p \in [1,\infty]$, $f \in L^p(\mathbb{R}^N)$, and $g \in L^{p'}(\mathbb{R}^N)$. Then*

$$\tau_h(f * g) = (\tau_h f) * g = f * (\tau_h g) \quad \text{for all } h \in \mathbb{R}^N.$$

Moreover, $f * g$ is uniformly continuous on $\mathbb{R}^N$ and

$$\|f * g\|_u \leq \|f\|_{L^p(\mathbb{R}^N)} \|g\|_{L^{p'}(\mathbb{R}^N)}. \tag{6.3.8}$$

Finally, if $p \in (1,\infty)$, then

$$\lim_{|x| \to \infty} f * g(x) = 0. \tag{6.3.9}$$

PROOF: The first assertion is again just an expression of translation invariance for $\lambda_{\mathbb{R}^N}$. Further, (6.3.8) is a simple application of Hölder's inequality. To see that $f * g$ is uniformly continuous, first suppose that $p \in [1,\infty)$. Then, by (6.3.8) and (6.3.6),

$$\|\tau_h(f * g) - f * g\|_{\mathrm{u}} = \|(\tau_h f - f) * g\|_{\mathrm{u}} \leq \|\tau_h f - f\|_{L^p(\mathbb{R}^N)} \|g\|_{L^{p'}(\mathbb{R}^N)} \longrightarrow 0$$

as $|h| \to 0$; and when $p = \infty$, simply reverse the roles of f and g in this argument.

To prove the final assertion, first let $f \in C_{\mathrm{c}}(\mathbb{R}^N)$ be given, and define $\mathcal{G}_f$ to be the class of $g \in L^{p'}(\mathbb{R}^N)$ for which (6.3.9) holds. Then it is easy to check that $C_{\mathrm{c}}(\mathbb{R}^N) \subseteq \mathcal{G}_f$. Moreover, by (6.3.8), one sees that $\mathcal{G}_f$ is closed in $L^{p'}(\mathbb{R}^N)$. Hence, just as in the final step of the proof of Lemma 6.3.5, we conclude that $\mathcal{G}_f = L^{p'}(\mathbb{R}^N)$. Next, let $g \in L^{p'}(\mathbb{R}^N)$ be given and define $\mathcal{H}_g$ to be the class of $f \in L^p(\mathbb{R}^N)$ for which (6.3.9) is true. By the preceding, we know that $C_{\mathrm{c}}(\mathbb{R}^N) \subseteq \mathcal{H}_g$. Moreover, just as before, $\mathcal{H}_g$ is closed in $L^p(\mathbb{R}^N)$; and therefore $\mathcal{H}_g = L^p(\mathbb{R}^N)$. □

6.3.10 Remark: Both Theorem 6.3.3 and Theorem 6.3.7 tell us that the convolution of two functions is often more regular than either or both of its factors. An application of this fact is given in Exercise 6.3.26 below, where one sees how it leads to an elegant derivation of Lemma 2.1.15.

The next result can be considered as another example of the observation made in the preceding remark.

6.3.11 Lemma. *Let $g \in C^1(\mathbb{R}^N)$, and assume that g as well as $g_{,1}, \dots, g_{,N}$ are elements of $L^{p'}(\mathbb{R}^N)$ for some $p \in [1,\infty]$. (Recall that $g_{,i} \equiv \frac{\partial g}{\partial x_i}$.) Then $f * g \in C^1(\mathbb{R}^N)$ for every $f \in L^p(\mathbb{R}^N)$, and*

$$\frac{\partial (f * g)}{\partial x_i} = f * g_{,i}, \quad 1 \leq i \leq N. \tag{6.3.12}$$

PROOF: Let $\boldsymbol{\omega} \in \mathbf{S}^{N-1}$ be given. If $p' \in [1,\infty)$, then, by Theorem 6.3.7, $\tau_{t\boldsymbol{\omega}}(f * g) - f * g = f * \big(\tau_{t\boldsymbol{\omega}} g - g\big)$ for every $t \in \mathbb{R}$. Since

$$\frac{\tau_{t\boldsymbol{\omega}} g(y) - g(y)}{t} = \int_{[0,1]} \big(\boldsymbol{\omega}, \nabla g(y + st\boldsymbol{\omega})\big)_{\mathbb{R}^N}\, ds$$

and, by Theorem 6.2.14,

$$\begin{aligned} &\left\| \int_{[0,1]} \big(\boldsymbol{\omega}, \nabla g(\,\cdot\, + st\boldsymbol{\omega}) - \nabla g(\,\cdot\,)\big)_{\mathbb{R}^N} ds \right\|_{L^{p'}(\mathbb{R}^N)} \\ &\qquad \leq \int_{[0,1]} \left\| \tau_{st\boldsymbol{\omega}} \big(\boldsymbol{\omega}, \nabla g\big)_{\mathbb{R}^N} - \big(\boldsymbol{\omega}, \nabla g\big)_{\mathbb{R}^N} \right\|_{L^{p'}(\mathbb{R}^N)} ds \longrightarrow 0 \end{aligned}$$

as $t \to 0$, the required result follows from (6.3.4). On the other hand, if $p' = \infty$, then

$$\frac{\tau_{t\omega}g(y) - g(y)}{t} = \int_{[0,1]} \big(\omega, \nabla g(y + st\omega)\big)_{\mathbb{R}^N}\, ds \longrightarrow \big(\omega, \nabla g(y)\big)_{\mathbb{R}^N}$$

boundedly and pointwise, and therefore the result follows, in this case, from Lebesgue's Dominated Convergence Theorem. □

The preceding result leads immediately to the conclusion that the smoother g is the smoother is $f*g$. More precisely, given a multi-index $\boldsymbol{\alpha} = (\alpha_1, \dots, \alpha_N)$, where the α_i 's are non-negative integers, define $\|\boldsymbol{\alpha}\| = \sum_1^N \alpha_i$ and

$$\partial^{\boldsymbol{\alpha}} = \frac{\partial^{\|\boldsymbol{\alpha}\|}}{\partial x^{\boldsymbol{\alpha}}} \equiv \frac{\partial^{\alpha_1}}{\partial x_1{}^{\alpha_1}} \cdots \frac{\partial^{\alpha_N}}{\partial x_1{}^{\alpha_N}} (\equiv \text{Identity when } \|\boldsymbol{\alpha}\| = 0).$$

Then, as an immediate corollary to Lemma 6.3.11, we see that if $g \in C^\infty(\mathbb{R}^N)$ and $\partial^{\boldsymbol{\alpha}} g \in L^{p'}(\mathbb{R}^N)$ for some $p \in [1, \infty]$ and all $\boldsymbol{\alpha}$'s, then

(6.3.13) $$f * g \in C^\infty(\mathbb{R}^N) \quad \text{and} \quad \partial^{\boldsymbol{\alpha}}(f * g) = f * (\partial^{\boldsymbol{\alpha}} g)$$

for every $f \in L^p(\mathbb{R}^N)$.

We next turn our attention to the case when $g \in L^1(\mathbb{R}^N)$. The main result here is the one which follows.

6.3.14 Theorem. *Given $g \in L^1(\mathbb{R}^N)$ and $t > 0$, define $g_t(\,\cdot\,) = t^{-N} g(t^{-1}\,\cdot\,)$. Then $g_t \in L^1(\mathbb{R}^N)$ and $\int g_t\, dx = \int g\, dx$. In addition, if $\int g\, dx = 1$, then for every $p \in [1, \infty)$ and $f \in L^p(\mathbb{R}^N)$:*

(6.3.15) $$\lim_{t \searrow 0} \|f * g_t - f\|_{L^p(\mathbb{R}^N)} = 0.$$

PROOF: We need only deal with the last statement.

Assume that $\int g\, dx = 1$. Given $f \in L^p(\mathbb{R}^N)$, note that, for almost every $x \in \mathbb{R}^N$,

$$f * g_t(x) - f(x) = \int_{\mathbb{R}^N} \big(f(x-y) - f(x)\big) g_t(y)\, dy = \int_{\mathbb{R}^N} \big(f(x-ty) - f(x)\big) g(y)\, dy.$$

Hence, if $\Psi^t(x, y) = \big(f(x - ty) - f(x)\big) g(y)$, then, by Theorem 6.2.14,

$$\begin{aligned} \|f * g_t - f\|_{L^p(\mathbb{R}^N)} &\le \|\Psi^t\|_{L^{(1,p)}(\lambda_{\mathbb{R}^N}, \lambda_{\mathbb{R}^N})} \\ &\le \|\Psi^t\|_{L^{(p,1)}(\lambda_{\mathbb{R}^N}, \lambda_{\mathbb{R}^N})} = \int_{\mathbb{R}^N} \|\tau_{-ty} f - f\|_{L^p(\mathbb{R}^N)} |g(y)|\, dy. \end{aligned}$$

Since $\|\tau_{-ty} f - f\|_{L^p(\mathbb{R}^N)} \le 2\|f\|_{L^p(\mathbb{R}^N)}$, we now see that the result follows from the above by Lebesgue's Dominated Convergence Theorem and (6.3.6). □

For reasons which ought to be made clear by Theorem 6.3.14, if $g \in L^1(\mathbb{R}^N)$ and $\int g\,dx = 1$, the corresponding family $\{g_t : t > 0\}$ is called an **approximate identity**. To understand how an approximate identity actually carries out *an approximation of the identity*, consider the case when g is non-negative and vanishes off of $B_{\mathbb{R}^N}(0,1)$. Then *the volume under the graph* of g_t continues to be 1 as $t \searrow 0$ while the base of the graph is restricted to $B_{\mathbb{R}^N}(0,t)$. Hence, all the *mass* is getting concentrated over the origin.

Combining Theorem 6.3.14 and (6.3.13), we get the following important approximation procedure.

6.3.16 Corollary. *Let $g \in C^\infty(\mathbb{R}^N) \cap L^1(\mathbb{R}^N)$ with $\int_{\mathbb{R}^N} g(x)\,dx = 1$ be given. In addition, let $p \in [1,\infty)$ and assume that $\partial^{\boldsymbol{\alpha}} g \in L^{p'}(\mathbb{R}^N)$ for all $\boldsymbol{\alpha} \in \mathbb{N}^N$. Then, for each $f \in L^p(\mathbb{R}^N)$, $f * g_t \longrightarrow f$ in $L^p(\mathbb{R}^N)$ as $t \searrow 0$; and, for all $t > 0$, $f * g_t$ has bounded, continuous derivatives of all orders and $\partial^{\boldsymbol{\alpha}}(f * g_t) = f * (\partial^{\boldsymbol{\alpha}} g_t)$, $\alpha \in \mathbb{N}^N$.*

Exercises

6.3.17 Exercise: Given $f,\ g \in L^1(\mathbb{R}^N)$, show that

$$\int f * g(x)\,dx = \int f(x)\,dx \int g(x)\,dx.$$

6.3.18 Exercise: Given a family $\{f_t : t \in (0,\infty)\} \subseteq L^1(\mathbb{R}^N)$, we say that the family is a **convolution semigroup** if $f_{s+t} = f_s * f_t$ for all $s,\ t \in (0,\infty)$. Here are four famous examples of convolution semigroups, three of which are also approximate identities.

(i) Define the **Gauss kernel** $\gamma(x) = (2\pi)^{-\frac{N}{2}} \exp\left(-\frac{|x|^2}{2}\right)$ for $x \in \mathbb{R}^N$. Using the result in part **(i)** Exercise 5.2.6, show that $\int_{\mathbb{R}^N} \gamma(x)\,dx = 1$ and that

$$\gamma_{\sqrt{s}} * \gamma_{\sqrt{t}} = \gamma_{\sqrt{s+t}} \quad \text{for } s,\ t \in (0,\infty).$$

Clearly this says that the approximate identity $\{\gamma_{\sqrt{t}} : t \in (0,\infty)\}$ is a convolution semigroup of functions. It is known as either the **heat flow semigroup** or the **Weierstrass semigroup**.

(ii) Define ν on $\mathbb{R}$ by

$$\nu(\xi) = \frac{\mathbf{1}_{(0,\infty)}(\xi)\, e^{-\frac{1}{\xi}}}{\pi^{\frac{1}{2}} \xi^{\frac{3}{2}}}.$$

Show that $\int_{\mathbb{R}} \nu(\xi)\,d\xi = 1$ and that

$$\nu_{s^2} * \nu_{t^2} = \nu_{(s+t)^2} \quad \text{for } s,\ t \in (0,\infty). \tag{6.3.19}$$

Hence, here again, we have an approximate identity which is a convolution semigroup. The family $\{\nu_{t^2} : t > 0\}$, or, more precisely, the probability measures $A \longmapsto \int_A \nu_{t^2}(\xi)\,d\xi$, play a role in probability theory, where they are called the **one-sided stable laws of order** $\frac{1}{2}$.

Hint: Note that for $\eta \in (0, \infty)$

$$\nu_{s^2} * \nu_{t^2}(\eta) = \frac{st}{\pi} \int_{(0,\eta)} \frac{1}{\left(\xi(\eta - \xi)\right)^{\frac{3}{2}}} \exp\left[-\frac{s^2}{\xi} - \frac{t^2}{\eta - \xi}\right] d\xi,$$

try the change of variable $\Phi(\xi) = \frac{\xi}{\eta - \xi}$, and use part (**iv**) of Exercise 5.2.6.

(**iii**) Using part (**ii**) of Exercise 5.3.24, check that the function P on $\mathbb{R}^N$ given by

$$P(x) = \frac{2}{\omega_N}\left(1 + |x|^2\right)^{-\frac{N+1}{2}}, \quad x \in \mathbb{R}^N,$$

has Lebesgue integral 1. Next prove the representation

$$P_t(x) = \int_{(0,\infty)} \gamma_{\sqrt{\frac{\xi}{2}}}(x)\nu_{t^2}(\xi)\,d\xi. \tag{6.3.20}$$

Finally, using (6.3.20) together with the preceding parts of this exercise, show that

$$P_s * P_t = P_{s+t}, \quad s, t \in (0, \infty); \tag{6.3.21}$$

and therefore that $\{P_t : t > 0\}$ is a convolution semigroup. This semigroup is known as the **Poisson semigroup** among harmonic analysts and as the **Cauchy semigroup** in probability theory; the representation (6.3.20) is an example of how to obtain one semigroup from another by the method of **subordination**.

(**iv**) For each $\alpha > 0$, define $g_\alpha : \mathbb{R} \longrightarrow [0, \infty)$ so that $g_\alpha(x) = 0$ if $x \leq 0$ and (cf. (**ii**) in Exercise 5.2.6) $g_\alpha(x) = \left(\Gamma(\alpha)\right)^{-1} x^{\alpha-1}e^{-x}$ for $x > 0$. Clearly, $\int_{\mathbb{R}} g_\alpha(x)\,dx = 1$. Next, check that

$$g_\alpha * g_\beta(x) = \left(\frac{\Gamma(\alpha + \beta)}{\Gamma(\alpha)\Gamma(\beta)} \int_{(0,1)} t^{\alpha-1}(1-t)^{\beta-1}\,dt\right) g_{\alpha+\beta}(x),$$

and use this, together with Exercise 6.3.17, to give another derivation of the formula, in (**i**) of Exercise 5.3.24, for the Beta function. Clearly, it is also shows that $\{g_\alpha : \alpha > 0\}$ is yet another convolution semigroup, although this one is not an approximate identity.

6.3.22 Exercise: Show that if μ is a finite measure on $\mathbb{R}^N$ and $p \in [1, \infty]$, then for all $f \in C_c(\mathbb{R}^N)$ the function $f * \mu$ given by

$$f * \mu(x) = \int_{\mathbb{R}^N} f(x - y)\, \mu(dy), \quad x \in \mathbb{R}^N$$

is continuous and satisfies

$$\|f * \mu\|_{L^p(\mathbb{R}^N)} \leq \mu(\mathbb{R}^N) \|f\|_{L^p(\mathbb{R}^N)}. \tag{6.3.23}$$

Next, use (6.3.23) to show that for each $p \in [1, \infty)$ there is a unique continuous map $\overline{\mathcal{K}}_\mu : L^p(\mathbb{R}^N) \longrightarrow L^p(\mathbb{R}^N)$ such that $\overline{\mathcal{K}}_\mu f = f * \mu$ for $f \in C_c(\mathbb{R}^N)$. Finally, note that (6.3.23) continues to hold when $f * \mu$ is replaced by $\overline{\mathcal{K}}_\mu f$, but that $\overline{K}_\mu f$ need not be continuous for every $f \in L^p(\mathbb{R}^N)$.

6.3.24 Exercise: In many applications it is extremely important to have compactly supported approximate identities.

(i) Set

$$c_N = \left(\int_{B_{\mathbb{R}^N}(0,1)} \exp\left[-\frac{1}{1 - |x|^2} \right] dx \right)^{-1}$$

and define

$$\rho(x) = \begin{cases} c_N \exp\left[-\frac{1}{1-|x|^2} \right] & \text{if } x \in B_{\mathbb{R}^N}(0,1) \\ 0 & \text{if } x \notin B_{\mathbb{R}^N}(0,1) \end{cases}.$$

Show that $\rho \in C^\infty(\mathbb{R}^N)$.

(ii) Use the preceding to show that if F is a closed subset of $\mathbb{R}^N$ and G is an open subset satisfying $\operatorname{dist}(F, G\complement) > 0$, then there is an $\eta \in C^\infty(\mathbb{R}^N)$ such that $\mathbf{1}_F \leq \eta \leq \mathbf{1}_G$. Such a function η is sometimes called a **bump** function.

(iii) Show that for each pair $p, q \in [1, \infty)$ and every $f \in L^p(\mathbb{R}^N) \cap L^q(\mathbb{R}^N)$, there exists a sequence $\psi_n \in C_c^\infty(\mathbb{R}^N)$ such that $\psi_n \longrightarrow f$ both in $L^p(\mathbb{R}^N)$ and in $L^q(\mathbb{R}^N)$. In particular, $C_c^\infty(\mathbb{R}^N)$ is dense in $L^p(\mathbb{R}^N)$ for every $p \in [1, \infty)$.

(iv) Let G be an open subset of $\mathbb{R}^N$. When $N \geq 2$, we showed in Theorem 5.4.17 that every $u \in C^2(G)$ which is harmonic on G satisfies the Mean Value Property (5.4.18) for balls $B_{\mathbb{R}^N}(x, R)$ whose closures are in G. Moreover, as pointed out in the paragraph preceding that theorem, the Mean Value Property is a triviality when $N = 1$. In this exercise, prove the converse of the Mean Value Property. Namely, show that if $u \in C(G)$ satisfies (5.4.18) whenever $\overline{B_{\mathbb{R}^N}(x, R)} \subseteq G$, then $u \in C^\infty(G)$ and u is harmonic on G. The proof can be accomplished in two steps. First show that if $\overline{B_{\mathbb{R}^N}(x, 2t)} \subseteq G$, then the Mean Value Property implies that (cf. part **(i)**)

$$u(\xi) = \left[\left(\mathbf{1}_{B_{\mathbb{R}^N}(x,2t)}\, u \right) * \rho_t \right](\xi) \quad \text{for} \quad \xi \in B_{\mathbb{R}^N}(x, t);$$

and conclude from this that $u \in C^\infty(G)$. Second, show that for any $f \in C^\infty(G)$ and $x \in G$,

$$\Delta f(x) = \frac{2}{\Omega_N} \lim_{t \searrow 0} \frac{1}{t^2} \int_{\mathbb{S}^{N-1}} \big(f(x + t\omega) - f(x)\big)\, \lambda_{\mathbb{S}^{N-1}}(d\omega). \tag{6.3.25}$$

Hint: To prove (6.3.25), expand f in a two place Taylor expansion around x and use the relations in (5.2.5).

6.3.26 Exercise: Let $\Gamma \in \overline{\mathcal{B}}_{\mathbb{R}^N}$ have finite Lebesgue measure and set $u(x) = \mathbf{1}_{-\Gamma} * \mathbf{1}_\Gamma$. Show that $u(x) \le |\Gamma| \mathbf{1}_\Delta$, where $\Delta = \Gamma - \Gamma \equiv \{y - x : x, y \in \Gamma\}$, and that $u(0) = |\Gamma|$. Use these observations, together with Theorem 6.3.7, to give another proof of Lemma 2.1.15.

6.3.27 Exercise: Define the σ-finite measure μ on $\big((0,\infty), \mathcal{B}_{(0,\infty)}\big)$ by

$$\mu(\Gamma) = \int_\Gamma \frac{1}{x}\, dx \quad \text{for} \quad \Gamma \in \mathcal{B}_{(0,\infty)},$$

and show that μ is **invariant under the multiplicative group** in the sense that

$$\int_{(0,\infty)} f(\alpha x)\, \mu(dx) = \int_{(0,\infty)} f(x)\, \mu(dx), \qquad \alpha \in (0,\infty),$$

and

$$\int_{(0,\infty)} f\left(\frac{1}{x}\right) \mu(dx) = \int_{(0,\infty)} f(x)\, \mu(dx)$$

for every $\mathcal{B}_{(0,\infty)}$-measurable $f : (0,\infty) \longrightarrow [0,\infty]$. Next, for $\mathcal{B}_{(0,\infty)}$-measurable, $\mathbb{R}$-valued functions f and g, set

$$\Lambda_\mu(f,g) = \left\{ x \in (0,\infty) : \int_{(0,\infty)} \left| f\left(\frac{x}{y}\right) \right| |g(y)|\, \mu(dy) < \infty \right\},$$

$$f \bullet g(x) = \begin{cases} \int_{(0,\infty)} f\left(\frac{x}{y}\right) g(y)\, \mu(dy) & \text{when} \quad x \in \Lambda_\mu(f,g) \\ 0 & \text{otherwise,} \end{cases}$$

and show that $f \bullet g = g \bullet f$. In addition, show that if $p,\, q \in [1,\infty]$ satisfy $\frac{1}{r} \equiv \frac{1}{p} + \frac{1}{q} - 1 \ge 0$, then

$$\mu\left(\Lambda_\mu(f,g)\complement\right) = 0 \quad \text{and} \quad \|f \bullet g\|_{L^r(\mu)} \le \|f\|_{L^p(\mu)} \|g\|_{L^q(\mu)}$$

for all $f \in L^p(\mu)$ and $g \in L^q(\mu)$. Finally, use these considerations to prove the following one of G.H. Hardy's many inequalities:

$$\left[\int_{(0,\infty)} \frac{1}{x^{1+\alpha}} \left(\int_{(0,x)} \varphi(y)\, dy \right)^p dx \right]^{\frac{1}{p}} \le \frac{p}{\alpha} \left(\int_{(0,\infty)} \frac{\big(y\varphi(y)\big)^p}{y^{1+\alpha}}\, dy \right)^{\frac{1}{p}}$$

for all $\alpha \in (0,\infty)$, $p \in [1,\infty)$, and non-negative, $\mathcal{B}_{(0,\infty)}$-measurable φ.

Hint: To prove everything except Hardy's inequality, simply repeat the argument used in the proof of Young's Inequality. To prove Hardy's result, take

$$f(x) = \left(\frac{1}{x}\right)^{\frac{\alpha}{p}} \mathbf{1}_{[1,\infty)}(x) \quad \text{and} \quad g(x) = x^{1-\frac{\alpha}{p}} \varphi(x),$$

and use $\|f \bullet g\|_{L^p(\mu)} \leq \|f\|_{L^1(\mu)} \, \|g\|_{L^p(\mu)}$.

Chapter VII

Elements of Fourier Analysis

7.1 Hilbert Space

In Exercise 6.1.7 we saw evidence that, among the L^p-spaces, the space L^2 is the most closely related to familiar Euclidean geometry. In the present section, we will expand on this observation and give an application of it.

Throughout $(E, \mathcal{B}, \mu)$ will be a measure space. By part (**iii**) of Theorem 6.2.2, the space $L^2(\mu)$ is a vector space which becomes a complete metric space when we use $\|f - f'\|_{L^2(\mu)}$ to measure the distance between f and f'. In addition, if we define

$$(f, f') \in \left(L^2(\mu)\right)^2 \longmapsto \left(f, f'\right)_{L^2(\mu)} \equiv \int_E f f' \, d\mu \in \mathbb{R},$$

then $\left(f, f'\right)_{L^2(\mu)}$ is bilinear (i.e., it is linear as a function of each its entries) and (cf. (6.2.9)), for $f \in L^2(\mu)$,

$$\begin{aligned} \|f\|_{L^2(\mu)} &= \sqrt{(f, f)_{L^2(\mu)}} \\ &= \sup\left\{ \left(f, f'\right)_{L^2(\mu)} : f' \in L^2(\mu) \text{ with } \|f'\|_{L^2(\mu)} \leq 1 \right\}. \end{aligned}$$

Note that $\left(f, f'\right)_{L^2(\mu)}$ plays the same role for $L^2(\mu)$ that the Euclidean inner product plays in $\mathbb{R}^N$. That is, by Schwarz's inequality (cf. Exercise 6.1.7),

$$\left|(f, f')_{L^2(\mu)}\right| \leq \|f\|_{L^2(\mu)} \|f'\|_{L^2(\mu)} \quad \text{and} \quad \arccos\left(\frac{(f, f')_{L^2(\mu)}}{\|f\|_{L^2(\mu)} \|f'\|_{L^2(\mu)}} \right)$$

can be thought of as the *angle between f and f'* in the plane spanned by f and f'. Thus, we say that $f \in L^2(\mu)$ is **orthogonal** or **perpendicular** to $S \subseteq L^2(\mu)$ and write $f \perp S$ if $\left(f, f'\right)_{L^2(\mu)} = 0$ for every $f' \in S$. When $S = \{f'\}$, we write $f \perp f'$ instead of $f \perp \{f'\}$.

For ease of presentation as well as conceptual clarity, it is helpful to abstract these properties. Thus, we call a topological vector space with this kind of structure a **real Hilbert space**. More precisely, we say that $\mathbf{H}$ is a **Hilbert space over the field** $\mathbb{R}$ if it is a vector space over $\mathbb{R}$ which possesses a symmetric bilinear map $(\mathbf{x}, \mathbf{x}') \in \mathbf{H}^2 \longmapsto (x, x')_{\mathbf{H}} \in \mathbb{R}$ with the properties that:

(**a**) $(\mathbf{x}, \mathbf{x}')_{\mathbf{H}} \geq 0$ for all $x \in \mathbf{H}$ and $\|\mathbf{x}\|_{\mathbf{H}} \equiv \sqrt{(\mathbf{x}, \mathbf{x})_{\mathbf{H}}} = 0$ if and only if $\mathbf{x} = \mathbf{0}$ in $\mathbf{H}$.

(b) $\mathbf{H}$ is complete as a metric space when the metric assigns $\|\mathbf{x}'-\mathbf{x}\|_{\mathbf{H}}$ to be the distance between $\mathbf{x}$ and $\mathbf{x}'$.

A look at the reasoning suggested in Exercise 6.1.7 should suffice to convince one that (**a**) implies the abstract *Schwarz's inequality*

$$\left|(\mathbf{x},\mathbf{x}')_{\mathbf{H}}\right| \le \|\mathbf{x}\|_{\mathbf{H}}\|\mathbf{x}'\|_{\mathbf{H}}.$$

Thus, once again, it is natural to interpret

$$\arccos\left(\frac{(\mathbf{x},\mathbf{x}')_{\mathbf{H}}}{\|\mathbf{x}\|_{\mathbf{H}}\|\mathbf{x}'\|_{\mathbf{H}}}\right)$$

as the angle between $\mathbf{x}$ and $\mathbf{x}'$. In particular, if $S \subseteq \mathbf{H}$ and $\mathbf{x} \in \mathbf{H}$, then we say that $\mathbf{x}$ is **orthogonal** to S and will write $\mathbf{x} \perp S$ if $(\mathbf{x},\mathbf{x}')_{\mathbf{H}} = 0$ for all $\mathbf{x}' \in S$.

A further abstraction, and one which will prove important to us, entails replacing the field $\mathbb{R}$ of real numbers by the field $\mathbb{C}$ of complex numbers and considering **complex Hilbert space**. That is, $\mathbf{H}$ is a Hilbert space over $\mathbb{C}$ if it is a vector space over $\mathbb{C}$ which possesses a **Hermition inner product** (property (**a**) below) $(\mathbf{z},\mathbf{z}') \in \mathbf{H}^2 \longmapsto (\mathbf{z},\mathbf{z}')_{\mathbf{H}} \in \mathbb{C}$ with the properties that

(a) $\overline{(\mathbf{z},\mathbf{z}')_{\mathbf{H}}} = (\mathbf{z}',\mathbf{z})_{\mathbf{H}}$,
(b) $(\mathbf{z},\mathbf{z})_{\mathbf{H}} \ge 0$ and $\|\mathbf{z}\|_{\mathbf{H}} \equiv \sqrt{(\mathbf{z},\mathbf{z})_{\mathbf{H}}} = 0$ if and only if $\mathbf{z} = \mathbf{0}$ in $\mathbf{H}$.
(c) $\mathbf{H}$ is complete as a metric space when the metric assigns $\|\mathbf{z}'-\mathbf{z}\|_{\mathbf{H}}$ to be the distance between $\mathbf{z}$ and $\mathbf{z}'$.

Once again, the reasoning in Exercise 6.1.7 allows us to pass from (**a**) and (**b**) to Schwarz's inequality $|(\mathbf{z},\mathbf{z}')_{\mathbf{H}}| \le \|\mathbf{z}\|_{\mathbf{H}}\|\mathbf{z}'\|_{\mathbf{H}}$. Perhaps the easiest way to see that the argument survives the introduction of complex numbers is to begin with the observation that it suffices to handle the case in which $(\mathbf{z},\mathbf{z}')_{\mathbf{H}} \ge 0$. Indeed, if this is not already the case, one can achieve it after multiplying $\mathbf{z}$ by a $\theta \in \mathbb{C}$ of modulus (i.e., absolute value) 1. Clearly, none of the quantities entering the inequality is altered by this multiplication. On the other hand, when $(\mathbf{z},\mathbf{z}')_{\mathbf{H}} \ge 0$, then the reasoning given in Exercise 6.1.7 works without change. Having verified Schwarz's inequality, there is some reason to continue interpreting

$$\arccos\left(\frac{\mathfrak{Re}\big((\mathbf{z},\mathbf{z}')_{\mathbf{H}}\big)}{\|\mathbf{z}\|_{\mathbf{H}}\|\mathbf{z}'\|_{\mathbf{H}}}\right)$$

as giving the angle between $\mathbf{z}$ and $\mathbf{z}'$. In particular, we will continue to say that $\mathbf{z}$ is orthogonal to S and write $\mathbf{z} \perp S$ if $(\mathbf{z},\mathbf{z}')_{\mathbf{H}} = 0$ for all $\mathbf{z}' \in S$.

Notice that if $\mathbf{H}$ is a Hilbert space over $\mathbb{C}$, then its *real part* $\mathfrak{Re}(\mathbf{H}) \equiv \{\mathfrak{Re}(\mathbf{x}) : \mathbf{x} \in \mathbf{H}\}$ together with the restriction of its inner product to $\mathfrak{Re}(\mathbf{H})$ is a Hilbert space over $\mathbb{R}$. Conversely, if $\mathbf{H}$ is a Hilbert space over $\mathbb{R}$, then it can be *complexified.* Namely, set

$$\widetilde{\mathbf{H}} \equiv \{\mathbf{x} + \sqrt{-1}\,\mathbf{y} : (\mathbf{x},\mathbf{y}) \in \mathbf{H}^2\},$$

take

$$\left(\mathbf{x}+\sqrt{-1}\,\mathbf{y},\mathbf{x}'+\sqrt{-1}\,\mathbf{y}'\right)_{\widetilde{\mathbf{H}}} \equiv (\mathbf{x},\mathbf{x}')_{\mathbf{H}} - (\mathbf{y},\mathbf{y}')_{\mathbf{H}} + \sqrt{-1}\left((\mathbf{x},\mathbf{y}')_{\mathbf{H}} + (\mathbf{x}',\mathbf{y})_{\mathbf{H}}\right),$$

and check that $\widetilde{\mathbf{H}}$ together with this inner product is a Hilbert space over $\mathbb{C}$ and that $\mathbf{H} = \mathfrak{Re}(\mathbf{H})$.

Of course, when $\mathbf{H} = \mathbb{R}$, $\widetilde{\mathbf{H}} = \mathbb{C}$ with its standard Hermitian inner product $(z, z')_{\mathbb{C}} = z\bar{z}$. To understand the meaning of $\widetilde{L^2(\mu)}$, first, for each $p \in [1, \infty]$, let $L^p(\mu; \mathbb{C})$ denote the space of $\mathcal{B}$-measurable $f : E \longrightarrow \mathbb{C}$ for which $|f| \in L^p(\mu)$. Obviously, $L^p(\mu; \mathbb{C})$ is a vector space over $\mathbb{C}$. Second, note that

$$f \in L^p(\mu;\mathbb{C}) \implies \left(\mathfrak{Re}(f), \mathfrak{Im}(f)\right) \in L^p(\mu)^2.$$

In particular, we can define the integral of $f \in L^1(\mu; \mathbb{C})$ by

$$\int f\,d\mu = \int \mathfrak{Re}(f)\,d\mu + \sqrt{-1}\int \mathfrak{Im}(f)\,d\mu. \tag{7.1.1}$$

In fact, this definition makes integration of $f \in L^1(\mu; \mathbb{C})$ a complex linear operation. Finally, note that

$$(f, f') \in L^2(\mu;\mathbb{C})^2 \implies f\overline{f'} \in L^1(\mu;\mathbb{C}),$$

and take

$$\left(f, f'\right)_{L^2(\mu;\mathbb{C})} = \int f\,\overline{f'}\,d\mu.$$

It is then an easy matter to check that $L^2(\mu; \mathbb{C})$ with this inner product is the complexification of $L^2(\mu)$.

Anyone who has studied linear algebra knows that all N-dimensional vector spaces over $\mathbb{R}$ or $\mathbb{C}$ are algebraically isomorphic to, respectively, $\mathbb{R}^N$ or $\mathbb{C}^N$. Moreover, this purely algebraic statement can be strengthened when the vector space comes equipped with a Hilbert structure. More precisely, if $\mathbf{H}$ is an N-dimensional Hilbert space over either $\mathbb{R}$ or $\mathbb{C}$ and if $\{\mathbf{x}_m : 1 \le m \le N\}$ is a basis for $\mathbf{H}$, then one can first apply the Gram–Schmidt orthogonalization procedure to transform $\{\mathbf{x}_m : 1 \le m \le N\}$ into a basis $\{\mathbf{e}_m : 1 \le m \le N\}$ which is **orthonormal** in the sense that

$$(\mathbf{e}_m, \mathbf{e}_{m'})_{\mathbf{H}} = \delta_{m,m'} \equiv \begin{cases} 1 & \text{if } m = m' \\ 0 & \text{otherwise,} \end{cases} \tag{7.1.2}$$

and one can then construct the isomorphism

$$\mathbf{x} \in \mathbf{H} \longmapsto \Phi(\mathbf{x}) \equiv \begin{bmatrix} (\mathbf{x}, \mathbf{e}_1)_{\mathbf{H}} \\ \vdots \\ (\mathbf{x}, \mathbf{e}_N)_{\mathbf{H}} \end{bmatrix} \in \mathbb{R}^N \text{ or } \mathbb{C}^N$$

$$\begin{bmatrix} \alpha_1 \\ \vdots \\ \alpha_N \end{bmatrix} \in \mathbb{R}^N \text{ or } \mathbb{C}^N \longmapsto \sum_{m=1} N \longmapsto \Phi^{-1}\left(\begin{bmatrix} \alpha_1 \\ \vdots \\ \alpha_N \end{bmatrix}\right) \equiv \sum_{m=1}^{N} \alpha_m \mathbf{e}_m \in \mathbf{H}.$$

In fact, this isomorphism is an **isometry** in the sense that it *preserves the inner product*: the inner product of $\Phi(\mathbf{x})$ with $\Phi(\mathbf{x}')$ in $\mathbb{R}^N$ or $\mathbb{C}^N$ is precisely $(\mathbf{x}, \mathbf{x}')_{\mathbf{H}}$.

We are now going to prove the analogous statement for infinite dimensional Hilbert spaces, at least when they are separable. Namely, the analog of $\mathbb{R}^N$ and $\mathbb{C}^N$ will be $\ell^2(\mathbb{Z}^+)$ and $\ell^2(\mathbb{Z}^+;\mathbb{C})$, which are, respectively, $L^2(\mu)$ and $L^2(\mu;\mathbb{C})$ when $E = \mathbb{Z}^+$, $\mathcal{B} = \mathcal{P}(\mathbb{Z}^+)$, and μ is counting measure. Obviously, $\ell^2(\mathbb{Z}^+)$ or $\ell^2(\mathbb{Z}^+;\mathbb{C})$ can be thought of as the subset of $\mathbb{R}^{\mathbb{Z}^+}$ or $\mathbb{C}^{\mathbb{Z}^+}$ whose coordinates are square summable.

A key step in this program is provided by the following.

7.1.3 Theorem. *Let* $\mathbf{F}$ *be a closed linear subspace of a Hilbert space* $\mathbf{H}$ *over either* $\mathbb{R}$ *or* $\mathbb{C}$. *Then, for each* $\mathbf{y} \in \mathbf{H}$, *there is a unique* $\mathbf{x} \in \mathbf{F}$ *for which*

$$\|\mathbf{y} - \mathbf{x}\|_{\mathbf{H}} = \min\Big\{\|\mathbf{y} - \mathbf{x}'\|_{\mathbf{H}} : \mathbf{x}' \in F\Big\}. \tag{7.1.4}$$

In fact, the unique $\mathbf{x} \in \mathbf{F}$ *which satisfies* (7.1.4) *is the unique* $\mathbf{x} \in \mathbf{F}$ *such that* $(\mathbf{y} - \mathbf{x}) \perp F$.

PROOF: In view of the fact that all real Hilbert spaces can be realized as the real part of a complex Hilbert space, we may and will carry out the proof only in the complex case.

We first check that $\mathbf{x} \in \mathbf{F}$ satisfies (7.1.4) if and only if $\mathbf{y} - \mathbf{x} \perp \mathbf{F}$. To this end, suppose that $\mathbf{x} \in \mathbf{F}$ satisfies (7.1.4), and observe that, for any $\mathbf{x}' \in \mathbf{F}$, the function

$$t \in \mathbb{R} \longmapsto \big\|\mathbf{y} - \mathbf{x} - t\mathbf{x}'\big\|_{\mathbf{H}}^2 = \|\mathbf{y} - \mathbf{x}\|_{\mathbf{H}}^2 - 2t\,\mathfrak{Re}\big((\mathbf{y} - \mathbf{x}, \mathbf{x}')_{\mathbf{H}}\big) + t^2\|\mathbf{x}'\|_{\mathbf{H}}^2$$

has a minimum at $t = 0$. Hence, by the first derivative test, we see that $\mathfrak{Re}\big((\mathbf{y} - \mathbf{x}, \mathbf{x}')_{\mathbf{H}}\big) = 0$ for every $\mathbf{x}' \in \mathbf{F}$. But for any $\mathbf{x}' \in \mathbf{F}$, we can choose $\theta \in \mathbb{C}$ with modulus 1 so that

$$(\mathbf{y} - \mathbf{x}, \theta\mathbf{x}')_{\mathbf{H}} = \bar{\theta}(\mathbf{y} - \mathbf{x}, \mathbf{x}')_{\mathbf{H}} \geq 0,$$

and therefore, since $\theta\mathbf{x}'$ is again in $\mathbf{F}$, we conclude that

$$\big|(\mathbf{y} - \mathbf{x}, \mathbf{x}')_{\mathbf{H}}\big| = (\mathbf{y} - \mathbf{x}, \theta\mathbf{x}')_{\mathbf{H}} = 0.$$

Conversely, if $\mathbf{x} \in \mathbf{F}$ and $(\mathbf{y} - \mathbf{x}) \perp \mathbf{F}$, then, for any $\mathbf{x}' \in \mathbf{F}$,

$$\begin{aligned}\|\mathbf{y} - \mathbf{x}'\|_{\mathbf{H}}^2 &= \|\mathbf{y} - \mathbf{x}\|_{\mathbf{H}}^2 + 2\mathfrak{Re}\big((\mathbf{y} - \mathbf{x}, \mathbf{x} - \mathbf{x}')_{\mathbf{H}}\big) + \|\mathbf{x} - \mathbf{x}'\|_{\mathbf{H}}^2 \\ &= \|\mathbf{y} - \mathbf{x}\|_{\mathbf{H}}^2 + \|\mathbf{x} - \mathbf{x}'\|_{\mathbf{H}}^2 \geq \|\mathbf{y} - \mathbf{x}\|_{\mathbf{H}}^2.\end{aligned}$$

Thus, we have now proved the equivalence of the two characterizations of $\mathbf{x}$; and, as a consequence of the second characterization, uniqueness is easy.

Indeed, if $\mathbf{x}_i$, $\mathbf{x}_2 \in \mathbf{F}$ and $(\mathbf{y}-\mathbf{x}_i) \perp \mathbf{F}$, $i \in \{1,2\}$, then $\mathbf{x}_2-\mathbf{x}_1 \in \mathbf{F}$, $(\mathbf{x}_2-\mathbf{x}_1) \perp \mathbf{F}$, and therefore $\|\mathbf{x}_2 - \mathbf{x}_1\|_{\mathbf{H}} = 0$.

In view of the preceding, it remains only to prove that there is an $\mathbf{x} \in \mathbf{F}$ for which (7.1.4) holds. To this end, choose $\{\mathbf{x}_n\}_1^\infty \subseteq \mathbf{F}$ so that

$$\|\mathbf{y} - \mathbf{x}_n\|_{\mathbf{H}} \longrightarrow \alpha \equiv \inf\Big\{\|\mathbf{y} - \mathbf{x}'\|_{\mathbf{H}} : \mathbf{x}' \in \mathbf{F}\Big\}.$$

Since $\mathbf{H}$ and therefore $\mathbf{F}$ are complete, all that we have to do is show that $\{\mathbf{x}_n\}_1^\infty$ is Cauchy convergent. In order to do this, note that for any $\mathbf{a}$, $\mathbf{b} \in \mathbf{H}$, we have the **law of the parallelagram**

$$\|\mathbf{a}+\mathbf{b}\|_{\mathbf{H}}^2 + \|\mathbf{a}-\mathbf{b}\|_{\mathbf{H}}^2 = 2\|\mathbf{a}\|_{\mathbf{H}}^2 + 2\|\mathbf{b}\|_{\mathbf{H}}^2. \tag{7.1.5}$$

Taking $\mathbf{a} = \mathbf{y} - \mathbf{x}_m$ and $\mathbf{b} = \mathbf{y} - \mathbf{x}_n$ in (7.1.5), we obtain

$$\begin{aligned}\|\mathbf{x}_n - \mathbf{x}_m\|_{\mathbf{H}}^2 &= 2\|\mathbf{y} - \mathbf{x}_n\|_{\mathbf{H}}^2 + 2\|\mathbf{y} - \mathbf{x}_m\|_{\mathbf{H}}^2 - 4\left\|\mathbf{y} - \frac{\mathbf{x}_n + \mathbf{x}_m}{2}\right\|_{\mathbf{H}}^2 \\ &\le 2\Big(\|\mathbf{y} - \mathbf{x}_n\|_{L^2(\mu)}^2 - \alpha^2\Big) + 2\Big(\|\mathbf{y} - \mathbf{x}_m\|_{\mathbf{H}}^2 - \alpha^2\Big),\end{aligned}$$

where we have used the fact that $\frac{\mathbf{x}_n+\mathbf{x}_m}{2} \in \mathbf{F}$ in order to get the last inequality. Now let $\epsilon > 0$ be given, choose $N \in \mathbb{Z}^+$ so that $\|\mathbf{y}-\mathbf{x}_n\|_{\mathbf{H}}^2 < \alpha^2 + \frac{\epsilon^2}{4}$ for $n \ge N$, and conclude that $\|\mathbf{x}_n - \mathbf{x}_m\|_{\mathbf{H}} < \epsilon$ for m, $n \ge N$. $\square$

The reader would do well to compare the preceding proof to what one would be inclined to do when $\mathbf{F}$ is finite dimensional. Namely, when $\mathbf{F}$ is finite dimensional, one could produce $\mathbf{x}$ by arguing that $\{\mathbf{x}_m : m \ge 1\}$ is a bounded subset of $\mathbf{F}$ and is therefore relatively compact. When $\mathbf{F}$ is infinite dimensional, it is no longer true that bounded subsets must be relatively compact. For example, consider the sequence $\{\mathbf{e}_m : m \ge 1\} \subseteq \ell^2(\mathbb{Z}^+)$ in which (cf. (7.1.2)) $(\mathbf{e}_m)_j = \delta_{j,m}$. Nonetheless, completeness can survive the transition to infinite dimensions, and it is this fact on which the preceding proof relies. In any case, we will call $\mathbf{x}$ the **perpendicular or orthogonal projection** of $\mathbf{y}$ onto $\mathbf{F}$ and will write $\mathbf{x} = \Pi_{\mathbf{F}}\mathbf{y}$. The mapping $\Pi_{\mathbf{F}} : \mathbf{H} \longrightarrow \mathbf{F}$ is called the **orthogonal projection operator** from $\mathbf{H}$ onto $\mathbf{F}$. Obviously, $\Pi_{\mathbf{F}}$ is *idempotent* in the sense that $\Pi_{\mathbf{F}} = \Pi_{\mathbf{F}}^2$. In addition, for any $\mathbf{x} \in \mathbf{H}$,

$$\|\mathbf{x}\|_{\mathbf{H}}^2 = \|\Pi_{\mathbf{F}}\mathbf{x}\|_{\mathbf{H}}^2 + \|\mathbf{x} - \Pi_{\mathbf{F}}\mathbf{x}\|_{\mathbf{H}}^2 \ge \|\Pi_{\mathbf{F}}\mathbf{x}\|_{\mathbf{H}}^2. \tag{7.1.6}$$

Given a sequence $\{\mathbf{x}_m\}_1^\infty$ in a Hilbert space $\mathbf{H}$, we say that *the series* $\sum_{m=1}^\infty \mathbf{x}_m$ *converges to* $\mathbf{x} \in \mathbf{H}$ and will write $\mathbf{x} = \sum_{m=1}^\infty \mathbf{x}_m$ if

$$\lim_{M\to\infty} \left\|\mathbf{x} - \sum_{m=1}^{M} \mathbf{x}_m\right\|_{\mathbf{H}} = 0.$$

In this connection, we say simply that $\sum_{m=1}^{\infty} \mathbf{x}_m$ *converges* when we mean that there exists an $\mathbf{x}$ to which it converges.

7.1.7 Theorem. *Assume that* $\mathbf{H}$ *is an infinite dimensional real or complex Hilbert space, and suppose that* $\{\mathbf{e}_m : m \in \mathbb{Z}^+\}$ *is an orthonormal (i.e., (7.1.2) holds) sequence in* $\mathbf{H}$. *Then, for any* $\{\alpha_m\}_1^\infty$ *from* $\mathbb{R}^{\mathbb{Z}^+}$ $(\mathbb{C}^{\mathbb{Z}^+})$, $\sum_{m=1}^{\infty} \alpha_m \mathbf{e}_m$ *converges if and only if* $\{\alpha_m\}_1^\infty$ *is an element of* $\ell^2(\mathbb{Z}^+)$ $(\ell^2(\mathbb{Z}^+;\mathbb{C}))$. *Moreover, if* $\mathbf{F}$ *denotes the closure of the subspace* $\operatorname{span}\big(\{\mathbf{e}_m : m \geq 1\}\big)$ *spanned*[1] *by* $\{\mathbf{e}_m : m \in \mathbb{Z}^+\}$, *then*

$$\mathbf{x} \rightsquigarrow \Phi(\mathbf{x}) \equiv \big\{(\mathbf{x}, \mathbf{e}_m)_{\mathbf{H}} : m \in \mathbb{Z}^+\big\}$$

is an isometric, isomorphism from $\mathbf{F}$ *onto* $\ell^2(\mathbb{Z}^+)$ *or* $\ell(\mathbb{Z}^+;\mathbb{C})$ *whose inverse is*

$$\{\alpha_m : m \in \mathbb{Z}^+\} \rightsquigarrow \sum_{m=1}^{\infty} \alpha_m \mathbf{e}_m.$$

Finally, for any $\mathbf{x} \in \mathbf{H}$, $\sum_{m=1}^{\infty} (\mathbf{x}, \mathbf{e}_m)_{\mathbf{H}} \mathbf{e}_m$ *converges to* $\Pi_{\mathbf{F}} \mathbf{x}$. *In particular,*

$$\sum_{m=1}^{\infty} \big|(\mathbf{x}, \mathbf{e}_m)_{\mathbf{H}}\big|^2 = \|\Pi_{\mathbf{F}} \mathbf{x}\|_{\mathbf{H}}^2 \leq \|\mathbf{x}\|_{\mathbf{H}}^2. \tag{7.1.8}$$

PROOF: For each $M \geq 1$, let $\mathbf{F}_M = \operatorname{span}\big(\{\mathbf{e}_1, \dots, \mathbf{e}_M\}\big)$. Because it is finite dimensional, $\mathbf{F}_M$ is closed. In fact,

$$\Pi_{\mathbf{F}_M} \mathbf{x} = \sum_{m=1}^{M} (\mathbf{x}_m, \mathbf{e}_m)_{\mathbf{H}} \mathbf{e}_m \quad \text{for all } \mathbf{x} \in \mathbf{H}. \tag{7.1.9}$$

To see this, let $\mathbf{x}_M$ denote the right hand side of (7.1.9). Obviously, $\mathbf{x}_M \in \mathbf{F}_M$. In addition, $(\mathbf{x} - \mathbf{x}_M, \mathbf{e}_\ell)_{\mathbf{H}} = 0$ for each $1 \leq \ell \leq M$; and so $\mathbf{x} - \mathbf{x}_M \perp \mathbf{F}_M$.

Next, suppose that $\sum_{m=1}^{\infty} \alpha_m \mathbf{x}_m$ converges. Clearly, the $\mathbf{x}$ to which it converges must be in $\mathbf{F}$. In addition,

$$(\mathbf{x}, \mathbf{e}_\ell)_{\mathbf{H}} = \lim_{M \to \infty} \left(\sum_{m=1}^{M} \alpha_m \mathbf{e}_m, \mathbf{e}_\ell \right)_{\mathbf{H}} = \alpha_\ell.$$

Hence, by (7.1.9), we have proved that $\Pi_{\mathbf{F}_M} \mathbf{x} = \sum_{m=1}^{M} \alpha_m \mathbf{x}_m$. In particular, by (7.1.6), we now know that

$$\|\mathbf{x}\|_{\mathbf{H}}^2 \geq \|\Pi_{\mathbf{F}_M} \mathbf{x}\|_{\mathbf{H}}^2 = \sum_{m=1}^{M} |\alpha_m|^2,$$

[1] The subspace spanned by a set S is the set of all vectors which can be written as a finite linear combination of elements of S.

which certainly means that the sequence $\{\alpha_m\}_1^\infty$ is square summable. In fact, because $\sum_{m=1}^M \alpha_m \mathbf{x}_m \longrightarrow \mathbf{x}$, we have also proved that

$$\mathbf{x} = \sum_{m=1}^\infty \alpha_m \mathbf{x}_m \implies \alpha_m = (\mathbf{x}, \mathbf{e}_m)_{\mathbf{H}} \text{ for all } m \in \mathbb{Z}^+ \tag{7.1.10}$$
$$\text{and } \|\mathbf{x}\|_{\mathbf{H}}^2 = \sum_{m=1}^\infty |\alpha_m|^2.$$

Next suppose that $\{\alpha_m\}_1^\infty$ is square summable, and set $\mathbf{x}_M = \sum_{m=1}^M \alpha_m \mathbf{x}_m$. Then, because

$$\sup_{n>M} \|\mathbf{x}_n - \mathbf{x}_M\|_{\mathbf{H}}^2 = \sup_{n>M} \sum_{m=M+1}^n |\alpha_m|^2 = \sum_{m>M} |\alpha_m|^2 \longrightarrow 0$$

as $M \to \infty$, the completeness of $\mathbf{F}$ implies the existence of an $\mathbf{x} \in \mathbf{F}$ to which $\mathbf{x}_M$ converges. In other words, $\sum_{m=1}^\infty \alpha_m \mathbf{e}_m$ converges to an element of $\mathbf{F}$.

So far, we have proved the first assertion as well as (7.1.9) and (7.1.10). Notice that, as a consequence of (7.1.10), we know that the map

$$\{\alpha_m\}_1^\infty \in \ell^2(\mathbb{Z}^+) \text{ or } \ell^2(\mathbb{Z}^+;\mathbb{C}) \longmapsto \sum_{m=1}^\infty \alpha_m \mathbf{e}_m \in \mathbf{F}$$

is length preserving and therefore one-to-one. Since, in addition, it is linear, we will know that Φ is an isomorphism once we show that

$$\Pi_{\mathbf{F}} \mathbf{x} = \sum_{m=1}^\infty (\mathbf{x}, \mathbf{e}_m)_{\mathbf{H}} \mathbf{e}_m, \quad \mathbf{x} \in \mathbf{H}. \tag{7.1.11}$$

To this end, first observe that $(\Pi_{\mathbf{F}}\mathbf{x}, \mathbf{e}_m)_{\mathbf{H}} = (\mathbf{x}, \mathbf{e}_m)_{\mathbf{H}}$ for all $m \in \mathbb{Z}^+$. Hence, it suffices to prove (7.1.11) when $\mathbf{x} \in \mathbf{F}$. But if $\mathbf{x} \in \mathbf{F}$, then $\Pi_{\mathbf{F}_M}\mathbf{x} \longrightarrow \Pi_{\mathbf{F}}\mathbf{x}$. To check this, note that, because $\mathbf{F}$ is the closure of $\bigcup_{M=1}^\infty \mathbf{F}_M$, we can choose $\{\mathbf{y}_M\}_1^\infty$ so that $\mathbf{y}_M \in \mathbf{F}_M$ and $\mathbf{y}_M \longrightarrow \mathbf{x}$; and therefore $\Pi_{\mathbf{F}_M}\mathbf{x} \longrightarrow \mathbf{x}$ follows from $\|\mathbf{x} - \Pi_{\mathbf{F}_M}\mathbf{x}\|_{\mathbf{H}} \le \|\mathbf{x} - \mathbf{y}_M\|_{\mathbf{H}}$. Knowing this, it should be clear that (7.1.11) for $\mathbf{x} \in \mathbf{F}$ is a corollary of (7.1.9).

Finally, we must verify that the isomorphism Φ is an isometry. However, because we already know that it is length preserving, the fact that it must also preserve inner products can be seen from the easily verified identity

$$4\mathfrak{Re}\big((\mathbf{x}, \mathbf{x}')_{\mathbf{H}}\big) = \|\mathbf{x} + \mathbf{x}'\|_{\mathbf{H}}^2 - \|\mathbf{x} - \mathbf{x}'\|_{\mathbf{H}}^2, \tag{7.1.12}$$

which holds for all Hilbert spaces. Clearly (7.1.12) finishes the job when $\mathbf{H}$ is real. When $\mathbf{H}$ is complex, it is necessary to reduce to the case when $(\mathbf{x}, \mathbf{x}')_{\mathbf{H}} \ge 0$ by replacing $\mathbf{x}$ with $\theta\mathbf{x}$ for an appropriately chosen $\theta \in \mathbb{C}$ of modulus 1. □

The inequality (7.1.9) is known as **Bessel's inequality**. In addition, we can now prove the isomorphism result alluded to just before Theorem 7.1.3.

7.1.13 Corollary. *If* $\mathbf{H}$ *is a real or complex, infinite dimensional (i.e., not finite dimensional) Hilbert space which is separable, then there exists an orthonormal sequence which is complete in the sense that*

$$\mathbf{x} = \sum_{m=1}^{\infty} (\mathbf{x}, \mathbf{e}_m)_{\mathbf{H}} \mathbf{e}_m \quad \textit{for all } \mathbf{x} \in \mathbf{H}. \tag{7.1.14}$$

In particular,

$$\mathbf{x} \in \mathbf{H} \longmapsto \{(\mathbf{x}, \mathbf{e}_m)_{\mathbf{H}} : m \in \mathbb{Z}^+\} \in \ell^2(\mathbb{Z}^+) \text{ or } \ell^2(\mathbb{Z}^+; \mathbb{C})$$

is an isometric isomorphism whose inverse is

$$\{\alpha_m : m \in \mathbb{Z}^+\} \in \ell^2(\mathbb{Z}^+) \text{ or } \ell^2(\mathbb{Z}^+; \mathbb{C}) \longmapsto \sum_{m=1}^{\infty} \alpha_m \mathbf{e}_m \in \mathbf{H}.$$

PROOF: In view of the results in Theorem 7.1.7, the only thing that still must be proved is the existence of $\{\mathbf{e}_m : m \in \mathbb{Z}^+\}$. To do this, we start with a countable, dense sequence $\{\mathbf{x}_n\}_1^\infty$ in $\mathbf{H}$. Second, we choose a subsequence $\{\mathbf{x}_{n_m} : m \in \mathbb{Z}^+\}$ inductively by the algorithm: $n_1 = 1$ and

$$n_{m+1} = \min\{n > n_m : \mathbf{x}_n \notin \operatorname{span}(\{\mathbf{x}_1, \dots, \mathbf{x}_{n_m}\})\}.$$

By construction, $\mathbf{x}_{n_1}, \dots, \mathbf{x}_{n_M}$ is a basis in $\mathbf{F}_M \equiv \operatorname{span}(\{\mathbf{x}_1, \dots, \mathbf{x}_{n_M}\})$ for each $M \in \mathbb{Z}^+$, and $\bigcup_{M=1}^\infty \mathbf{F}_M$ is dense in $\mathbf{H}$. Finally, we apply the Gram–Schmidt orthogonalization procedure to $\{\mathbf{x}_{n_m} : m \in \mathbb{Z}^+\}$ and thereby produce an orthonormal sequence $\{\mathbf{e}_m : m \in \mathbb{Z}^+\}$ with the property that $\mathbf{F}_M = \operatorname{span}(\{\mathbf{e}_1, \dots, \mathbf{e}_M\})$ for each $M \in \mathbb{Z}^+$. In particular, $\operatorname{span}(\{\mathbf{e}_m : m \in \mathbb{Z}^+\})$ is dense in $\mathbf{H}$, and therefore Theorem 7.1.7 applies with $\mathbf{F} = \mathbf{H}$. □

An orthonormal sequence for which (7.1.14) holds is called an **orthonormal basis** in $\mathbf{H}$.

Warning: The danger in results like Corollary 7.1.13 is that they incline one to think that, to paraphrase ex-president Reagen, "if you've seen one separable Hilbert space, you've seen them all." Of course, what it is really doing is validating that sentiment if and only if one is willing to ignore all but the most rudimentary properties of any particular separable Hilbert space.

Exercises

7.1.15 Exercise: Let $\mathbf{F}$ be a not necessarily closed subspace of the Hilbert space $\mathbf{H}$. Show that $\overline{\mathbf{F}} \neq \mathbf{H}$ if and only if there exists an $\mathbf{x} \in \mathbf{H} \setminus \{\mathbf{0}\}$ such that $\mathbf{x} \perp \mathbf{F}$. In this connection, show that the orthonormal sequence $\{\mathbf{e}_m : m \in \mathbb{Z}^+\}$ is complete (i.e., (7.1.14) obtains) if and only if equality holds in Bessel's inequality for each $\mathbf{x} \in \mathbf{H}$.

7.1.16 Exercise: Let $\mathbf{H}$ be a real or complex Hilbert space, and note that every closed subspace of $\mathbf{H}$ becomes a Hilbert space with the inner product obtained by restriction. Also, show that every finite dimensional subspace is necessarily closed. Finally, show that $C([0,1];\mathbb{R})$ is a non-closed subspace of $L^2(\lambda_{[0,1]};\mathbb{R})$. Hence, when dealing with infinite dimensional subspaces, closedness is something that requires checking.

7.1.17 Exercise: Suppose that Π is a linear map from the Hilbert space $\mathbf{H}$ into itself. Show that Π is the orthogonal projection operator onto the closed subspace $\mathbf{F}$ if and only if $\mathbf{F} = \text{Range}(\Pi)$, $\Pi^2 = \Pi$, and $(\Pi\mathbf{x}, \mathbf{x}')_{\mathbf{H}} = (\mathbf{x}, \Pi\mathbf{x}')_{\mathbf{H}}$ for all $(\mathbf{x}, \mathbf{x}') \in \mathbf{H}^2$. In particular, if $\mathbf{F}$ is a subspace and $\mathbf{F}^\perp \equiv \{\mathbf{x} \in \mathbf{H} : \mathbf{x} \perp \mathbf{F}\}$ is its **perpendicular complement**, show that $\mathbf{F}^\perp$ is closed (even if $\mathbf{F}$ is not) and that $\Pi_{\mathbf{F}^\perp} = I - \Pi_{\overline{\mathbf{F}}}$, where I denotes the identity operator on $\mathbf{H}$.

7.1.18 Exercise: It may be reassuring to know that, in some sense, the dimension of a separable, infinite dimensional Hilbert space is well-defined and equal to the cardinality of the integers. To see this, show that if E is a subset of $\mathbf{e} \in \mathbf{H}$ with the properties that

$$(\mathbf{e}, \mathbf{e}')_{\mathbf{H}} = \begin{cases} 1 & \text{if } \mathbf{e}' = \mathbf{e} \\ 0 & \text{otherwise,} \end{cases}$$

and if $\text{span}(E)$ is dense in $\mathbf{H}$, then the elements of E are in one-to-one correspondence with the integers.

7.1.19 Exercise: Assume that $\mathbf{H}$ is a separable, infinite dimensional, real or complex Hilbert space, and let $\mathbf{F}$ be a closed subspace of $\mathbf{H}$. Augment the statement of Corollary 7.1.13 by showing that *every orthonormal basis for* $\mathbf{F}$ *can be extended to an orthonormal basis for* $\mathbf{H}$. That is, if E is an orthonormal basis for $\mathbf{F}$, then there is an orthonormal basis $\tilde{E}$ for $\mathbf{H}$ with $E \subseteq \tilde{E}$. In fact, show that $\tilde{E} = E \cup E'$, where E' is an orthonormal basis for $\mathbf{F}^\perp$.

7.2 Fourier Series

Much of analysis rests on a clever selection of the "right" orthonormal basis for a particular task. Unfortunately, the "right" choice is often unavailable in any practical sense, and one has to make do with a choice which represents a compromise between what is ideal and what is available. For example, when dealing with a situation in which it is important to exploit features which derive from translation invariance, a reasonable choice is a basis whose elements transform nicely under translations, namely, exponentials of linear functions. This is the choice made by Fourier, and it is still the basis of choice in the largest variety of applications.

The basic fact on which the theory of Fourier series depends is the content of the following theorem. Throughout this section, if I is an interval, $L^2(I;\mathbb{R})$ and $L^2(I;\mathbb{C})$ will denote, respectively, $L^2(\lambda_I)$ and $L^2(\lambda_I;\mathbb{C})$.

7.2.1 Theorem. *Let $L \in (0,\infty)$ and $a \in \mathbb{R}$ be given, and set*

$$(7.2.2)\quad (\mathfrak{e}_{a,L})_m(x) = L^{-\frac{1}{2}} \exp\left(\sqrt{-1}\,\frac{2\pi m(x-a)}{L}\right),\quad m \in \mathbb{Z}\ \&\ x \in [a, a+L].$$

Then the family $\{(\mathfrak{e}_{a,L})_m : m \in \mathbb{Z}\}$ forms an orthonormal basis in $L^2\big([a,a+L];\mathbb{C}\big)$. Equivalently, for every $f \in L^2\big([a,a+L];\mathbb{C}\big)$,

$$(7.2.3)\quad \begin{aligned} f &= \sum_{m\in\mathbb{Z}} \big(f,(\mathfrak{e}_{a,L})_m\big)_{L^2([a,a+L];\mathbb{C})}(\mathfrak{e}_{a,L})_m \\ &= \sum_{m\in\mathbb{Z}} \left(\frac{1}{L}\int_{[a,a+L]} f(y)\, e^{-\sqrt{-1}\frac{2\pi m(y-a)}{L}}\, dy\right) e^{\sqrt{-1}\frac{2\pi m(x-a)}{L}}, \end{aligned}$$

where the convergence is in $L^2\big([a,a+L];\mathbb{C}\big)$.

PROOF: We begin by noting that it suffices to handle the case when $a=0$ and $L=1$. In fact, for any $a \in \mathbb{R}$ and $L>0$, the map $\Phi_{a,L} : L^2\big([0,1];\mathbb{C}\big) \longrightarrow L^2\big([a,a+L];\mathbb{C}\big)$ given by

$$\big[\Phi_{a,L} f\big](x) = L^{-\frac{1}{2}} f\left(\frac{x-a}{L}\right),\quad x \in [a,a+L]$$

is an isometric isomorphism. Thus, we will assume that $a=0$ and $L=1$, and we will use $\mathfrak{e}_m$ to denote $(\mathfrak{e}_{0,1})_m$.

Clearly, $(\mathfrak{e}_m,\mathfrak{e}_{m'})_{L^2([0,1];\mathbb{C})} = \delta_{m,m'}$. Thus, the only question is whether $\mathbf{F} \equiv \operatorname{span}\big(\{\mathfrak{e}_m : m \in \mathbb{Z}\}\big)$ is dense in $L^2\big([0,1];\mathbb{C}\big)$, and, in view of Exercise 7.1.15, this comes down to showing that $f = \mathbf{0}$ is the only $f \in L^2\big([0,1];\mathbb{C}\big)$ satisfying $(f,\mathfrak{e}_m)_{L^2([0,1];\mathbb{C})} = 0$ for all $m \in \mathbb{Z}$. Thus, suppose that f is such an element of $L^2\big([0,1];\mathbb{C}\big)$. By part (**v**) of Theorem 6.2.2 applied to the real and imaginary parts of functions, we know that the space $C\big([0,1];\mathbb{C}\big)$ of continuous $\psi : [0,1] \longrightarrow \mathbb{C}$ is dense in $L^2\big([0,1];\mathbb{C}\big)$. In particular, we will know that $f = \mathbf{0}$ once we show that $f \perp C\big([0,1];\mathbb{C}\big)$.

Let $\psi \in C\big([0,1];\mathbb{C}\big)$ be given, and, for $r \in (0,1)$, set

$$\psi_r = \sum_{m\in\mathbb{Z}} r^{|m|}\big(\psi,\mathfrak{e}_m\big)_{L^2([0,1];\mathbb{C})}\mathfrak{e}_m.$$

Because, by Bessel's inequality,

$$\|\psi\|^2_{L^2([0,1];\mathbb{C})} \geq \sum_{m\in\mathbb{Z}} \left|\left(\psi, \mathfrak{e}_m\right)_{L^2([0,1];\mathbb{C})}\right|^2,$$

it is clear that the series defining ψ_r converges uniformly on $[0,1]$ for each $r \in (0,1)$. In particular, ψ_r is continuous and $f \perp \psi_r$ for each $r \in (0,1)$. Hence, we will know that $f \perp \psi$ once we show that $\psi_r \longrightarrow \psi$ in $L^2([0,1];\mathbb{C})$ as $r \nearrow 1$.

The key to proving that $\psi_r \longrightarrow \psi$ is the observation that an alternative expression for ψ_r is

$$\psi_r(x) = \int_{[0,1]} \psi(y)\, P_r(x,y)\, dy, \quad x \in [0,1], \tag{7.2.4}$$

where

$$P_r(x,y) \equiv \frac{1-r^2}{\left|r\mathfrak{e}_1(x) - \mathfrak{e}_1(y)\right|^2}.$$

The proof of (7.2.4) is easy. Namely, one simply has to observe that, for each $r \in (0,\infty)$,

$$\begin{aligned} P_r(x,y) &= \frac{1}{1 - r\mathfrak{e}_1(x)\mathfrak{e}_{-1}(y)} + \frac{r\mathfrak{e}_{-1}(x)\mathfrak{e}_1(y)}{1 - r\mathfrak{e}_{-1}(x)\mathfrak{e}_1(y)} \\ &= \sum_{m=0}^{\infty} r^m \mathfrak{e}_m(x)\mathfrak{e}_{-m}(y) + \sum_{m=1}^{\infty} r^m \mathfrak{e}_{-m}(x)\mathfrak{e}_m(y) \\ &= \sum_{m\in\mathbb{Z}} r^{|m|}\mathfrak{e}_m(x)\mathfrak{e}_{-m}(y), \end{aligned}$$

where the convergence is uniform in $(x,y) \in [0,1]^2$. Hence, since

$$\psi_r(x) = \sum_{m\in\mathbb{Z}} r^{|m|} \left(\int_{[0,1]} \psi(y)\mathfrak{e}_{-m}(y)\, dy\right) \mathfrak{e}_m(x),$$

(7.2.4) is proved.

Obviously, $P_r(x,y) > 0$. Secondly, by applying (7.2.4) when $\psi = \mathfrak{e}_0$, we see that, for each $x \in [0,1]$, $\int_{[0,1]} P_r(x,y)\, dy = 1$. Indeed, $(\mathfrak{e}_0, \mathfrak{e}_m)_{L^2([0,1];\mathbb{C})} = \delta_{0,m}$ and so $(\mathfrak{e}_0)_r \equiv 1$. In particular, these two combine to prove that $\|\psi_r\|_{\mathrm{u}} \leq \|\psi\|_{\mathrm{u}}$ for $r \in (0,1)$. Finally, if, for $0 < \epsilon < \frac{1}{2}$ and $x \in [0,1]$, $I_\epsilon(x) = \{y \in [0,1] : |x-y| \wedge |1-x-y| < \epsilon\}$, then

$$P_r(x,y) \leq 2\frac{1-r}{(1-\cos 2\pi\epsilon)^2}, \quad r \in (0,1) \text{ and } y \in [0,1]\setminus I_\epsilon(x).$$

Hence,

$$\begin{aligned} \left|\psi_r(x) - \psi(x)\right| &= \left|\int_{[0,1]} (\psi(y) - \psi(x)) P_r(x,y)\, dy\right| \\ &\leq \sup_{y\in I_\epsilon(x)} |\psi(y) - \psi(x)| + 4\|\psi\|_{\mathrm{u}} \frac{1-r}{(1-\cos 2\pi\epsilon)^2}, \end{aligned}$$

and so

$$\varlimsup_{r\nearrow 1}\big|\psi_r(x)-\psi(x)\big| \le \sup_{y\in I_\epsilon(x)} |\psi(y)-\psi(x)| \longrightarrow 0 \text{ as } \epsilon\searrow 0 \quad \text{for } x\in(0,1).$$

In other words, we have now proved that $\{\psi_r : r\in(0,1)\}$ is uniformly bounded in $C\big([0,1];\mathbb{C}\big)$ and that $\psi_r \longrightarrow \psi$ pointwise on $(0,1)$. Since this is more than enough to assure that $\psi_r \longrightarrow \psi$ in $L^2\big([0,1];\mathbb{C}\big)$, we are done. □

The astute reader will have noticed that the preceding proof of $\psi_r \longrightarrow \psi$ would have been smoother had we assumed that ψ is periodic on $[0,1]$. Indeed, in that case, our argument would have shown that $\psi_r \longrightarrow \psi$ uniformly on the whole of $[0,1]$. The point is, of course, that Theorem 7.2.1 involves the representation of general functions on an interval in terms of periodic ones. Thus, it is not surprising that something goes slightly awry at the endpoints. A more geometric explanation would be that we are identifying $[0,1]$ with the circle S of radius $(2\pi)^{-1}$, and we should not make the mistake of identifying $C\big([0,1];\mathbb{C}\big)$ with $C(S;\mathbb{C})$.

7.2.5 Corollary. *For each $L\in(0,\infty)$ and $a\in\mathbb{R}$,*

$$\left\{L^{-\frac12}\cos\frac{2\pi m(x-a)}{L} : m\in\mathbb{N}\right\}\cup\left\{L^{-\frac12}\sin\frac{2\pi m(x-a)}{L} : m\in\mathbb{Z}^+\right\}$$

forms an orthonormal basis in $L^2\big([a,a+L];\mathbb{R}\big)$ and in $L^2\big([a,a+L];\mathbb{C}\big)$.

PROOF: Clearly these functions span the same subspace of $L^2\big([a,a+L];\mathbb{C}\big)$ as $\{(\mathfrak{e}_{a,L})_m : m\in\mathbb{Z}\}$. Furthermore, they are all real-valued. Hence, the only question is whether they are orthonormal. But to check this, one simply uses

$$L^{-\frac12}\cos\frac{2\pi m(x-a)}{L} = \frac{(\mathfrak{e}_{a,L})_m(x)+(\mathfrak{e}_{a,L})_{-m}(x)}{2}, \quad m\in\mathbb{Z}^+$$
$$L^{-\frac12}\sin\frac{2\pi m(x-a)}{L} = \frac{(\mathfrak{e}_{a,L})_m(x)-(\mathfrak{e}_{a,L})_{-m}(x)}{2\sqrt{-1}}, \quad m\in\mathbb{Z}^+. \quad □$$

Exercises

7.2.6 Exercise: As we mentioned after proof of Theorem 7.2.1, Fourier series are sensitive to boundary conditions, and for this reason it is often useful to know how to adjust the boundary conditions. For example, show that each of

$$\left\{L^{-\frac12}\cos\frac{\pi m x}{L} : m\in\mathbb{N}\right\}$$

and

$$\left\{L^{-\frac12}\sin\frac{\pi m x}{L} : m\in\mathbb{Z}^+\right\}$$

is an orthonormal basis in both $L^2\big([0,L];\mathbb{R}\big)$ and $L^2\big([0,L];\mathbb{C}\big)$. The first of these corresponds to the boundary condition that the function vanish at both end points, whereas the second corresponds to the boundary condition that the function's first derivative vanish at both end points.

Hint: First show that

$$\left\{(2L)^{-\frac{1}{2}}e^{\frac{\sqrt{-1}\,\pi m x}{L}} : m \in \mathbb{Z}\right\}$$

is an orthonormal basis in for $L^2([-L,L];\mathbb{C})$. Next, conclude that

$$\left\{(2L)^{-\frac{1}{2}}\cos\frac{\pi m x}{L} : m \in \mathbb{N}\right\} \cup \left\{(2L)^{-\frac{1}{2}}\sin\frac{\pi m x}{L} : m \in \mathbb{Z}^+\right\}$$

is an orthonormal basis in both $L^2([-L,L];\mathbb{R})$ and in $L^2([-L,L];\mathbb{C})$. Now say that $f \in L^2([-L,L];\mathbb{C})$ is even or odd depending on whether $f(x) = f(-x)$ or $f(x) = -f(-x)$ for Lebesgue almost every $x \in [0,L]$, and show that $\mathbf{E} \equiv \{f \in L^2([-L,L];\mathbb{C}) : f \text{ is even}\}$ and $\mathbf{O} \equiv \{f \in L^2([-L,L];\mathbb{C}) : f \text{ is odd}\}$ are both closed subspaces. Finally, show that

$$\left\{(2L)^{-\frac{1}{2}}\cos\frac{\pi m x}{L} : m \in \mathbb{N}\right\}$$

is an orthonormal basis for $\mathbf{E}$ and that

$$\left\{(2L)^{-\frac{1}{2}}\sin\frac{\pi m x}{L} : m \in \mathbb{Z}^+\right\}$$

is an orthonormal basis for $\mathbf{O}$. Since $L^2([0,L];\mathbb{C})$ can be identified with either $\mathbf{E}$ or $\mathbf{O}$, this completes the proof.

7.2.7 Exercise: Suppose that $f : \mathbb{R} \longrightarrow \mathbb{C}$ is a smooth function which is periodic with period $2L$. Show that

$$\begin{aligned} f(x) &= \frac{1}{2L}\sum_{m\in\mathbb{Z}} \big(f, (\mathfrak{e}_{L,2L})_m\big)_{L^2([-L,L];\mathbb{C})} (\mathfrak{e}_{L,2L})_m(x) \\ &= \frac{1}{2L}\sum_{m\in\mathbb{Z}} \int_{[0,L]} f(y)\, e^{\sqrt{-1}\,\frac{\pi m(x-y)}{L}}\, dy, \end{aligned}$$

where the convergence of the series is absolute and uniform.

Hint: Integrate by parts to see that

$$\sup_{m\in\mathbb{Z}} |m|^n \left| \big(f, \mathfrak{e}_{-L,L})_m\big)_{L^2([-L,2L];\mathbb{C})} \right| < \infty$$

for all $n \in \mathbb{Z}^+$.

7.3 The Fourier Transform, L^1-theory

Let f be a smooth, compactly supported, complex-valued function on $\mathbb{R}$. Then as soon as $L > 0$ is large enough so that $[-L,L]$ contains its support, we know (cf. Exercise 7.2.7) that

$$f(x) = \frac{1}{2L}\sum_{m\in\mathbb{Z}} \int_{[-L,L]} f(y) e^{\frac{\sqrt{-1}\,2\pi m(x-y)}{2L}}\, dy.$$

Thus, if we close our eyes, suspend our disbelief, let $L \to \infty$, interpret the sum as a Riemann approximation, and indulge in a certain amount of re-arrangement, we are led to guess that

$$f(x) = \int_{\mathbb{R}} \mathfrak{e}(\xi x)\, d\xi \left(\int_{\mathbb{R}} f(y)\, \mathfrak{e}(-\xi y)\, dy \right) dx \tag{7.3.1}$$

where

$$\mathfrak{e}(\eta) \equiv e^{\sqrt{-1}(2\pi)\eta}, \quad \eta \in \mathbb{R}. \tag{7.3.2}$$

The purpose of this section is to show that, in a sense to be made precise below, (7.3.1) holds not only for smooth, compactly supported f but even for any $f \in L^1(\mathbb{R};\mathbb{C})$.

7.3.3 Lemma. *Given $f \in L^1(\mathbb{R}^N;\mathbb{C})$, set*[2]

$$\begin{aligned} \hat{f}(\xi) &= \int_{\mathbb{R}^N} f(x)\, \mathfrak{e}\big((\boldsymbol{\xi}, \mathbf{x})_{\mathbb{R}^N}\big)\, d\mathbf{x}, \qquad \boldsymbol{\xi} \in \mathbb{R}^N, \\ \check{f}(\mathbf{x}) &= \int_{\mathbb{R}^N} f(\mathbf{x})\mathfrak{e}\big(-(\boldsymbol{\xi}, \mathbf{x})_{\mathbb{R}^N}\big)\, d\mathbf{x}, \quad \mathbf{x} \in \mathbb{R}^N. \end{aligned} \tag{7.3.4}$$

Then $\check{f}(\mathbf{x}) = \hat{f}(-\mathbf{x})$, $\mathbf{x} \in \mathbb{R}^N$,

$$\hat{f} \in C_0(\mathbb{R}^N;\mathbb{C}) \equiv \left\{ \varphi \in C_{\mathrm{b}}(\mathbb{R}^N;\mathbb{C}) : \lim_{|\mathbf{x}|\to\infty} \varphi(\mathbf{x}) = 0 \right\},$$

and the map $f \in L^1(\mathbb{R}^N;\mathbb{C}) \longmapsto \hat{f} \in C_0(\mathbb{R}^N;\mathbb{C})$ is both linear and continuous. Furthermore, if $f,\ g \in L^1(\mathbb{R}^N;\mathbb{C})$, then

$$\int_{\mathbb{R}^N} f(\mathbf{x})\, \overline{\check{g}(\mathbf{x})}\, dx = \int_{\mathbb{R}^N} \hat{f}(\boldsymbol{\xi})\, \overline{g(\boldsymbol{\xi})}\, d\boldsymbol{\xi}. \tag{7.3.5}$$

PROOF: Since it is obvious that $\hat{f} \in C_{\mathrm{b}}(\mathbb{R}^N;\mathbb{C})$ with $\|\hat{f}\|_{\mathrm{u}} \le \|f\|_{L^1(\mathbb{R}^N;\mathbb{C})}$ and that $f \rightsquigarrow \hat{f}$ is linear, the only part of the first assertion which requires comment is the statement that $f \in C_0(\mathbb{R}^N;\mathbb{C})$. However, $C_0(\mathbb{R}^N;\mathbb{C})$ is a closed subspace of $C_{\mathrm{b}}(\mathbb{R}^N;\mathbb{C})$, and so, by continuity, part **(iii)** of Exercise 6.3.24 tells us it is enough to check $\hat{f} \in C_0(\mathbb{R}^N;\mathbb{C})$ when $f \in C_{\mathrm{c}}^\infty(\mathbb{R}^N;\mathbb{C})$. But if $f \in C_{\mathrm{c}}^\infty(\mathbb{R}^N;\mathbb{C})$, then we can use Green's Identity (cf. Theorem 5.4.12) to see that, because f has compact support,

$$\begin{aligned} -|2\pi\boldsymbol{\xi}|^2 \hat{f}(\boldsymbol{\xi}) &= \int_{\mathbb{R}^N} f(\mathbf{x})\Delta_{\mathbf{x}}\mathfrak{e}\big((\boldsymbol{\xi}, \cdot\,)_{\mathbb{R}^N}\big)\, d\mathbf{x} \\ &= \int_{\mathbb{R}^N} \Delta f(\mathbf{x})\mathfrak{e}\big((\boldsymbol{\xi}, \mathbf{x})_{\mathbb{R}^N}\big)\, d\mathbf{x} = \widehat{\Delta f}(\boldsymbol{\xi}), \end{aligned}$$

[2] Throughout the rest of this chapter, we will use bold letters like $\mathbf{x}$ or $\boldsymbol{\xi}$ to denote elements of $\mathbb{R}^N$ and will use x_j or ξ_j to denote the jth coordinate of $\mathbf{x}$ or $\boldsymbol{\xi}$.

where $\Delta_{\mathbf{x}}\mathfrak{e}\big((\boldsymbol{\xi},\,\cdot\,)_{\mathbb{R}^N}\big)$ is used to indicate that we are applying the Laplacian Δ to $\mathbf{x} \rightsquigarrow \mathfrak{e}\big((\boldsymbol{\xi},\mathbf{x})_{\mathbb{R}^N}\big)$. Thus $|\hat{f}(\mathbf{x})| \le |2\pi\boldsymbol{\xi}|^{-2}\|\Delta f\|_{\mathrm{u}} \longrightarrow 0$ as $|\boldsymbol{\xi}| \to \infty$.

Finally, to prove (7.3.5) we apply Fubini's Theorem to justify

$$\int_{\mathbb{R}^N} \hat{f}(\boldsymbol{\xi})\overline{g(\boldsymbol{\xi})}\,d\boldsymbol{\xi} = \iint\limits_{\mathbb{R}^N\times\mathbb{R}^N} f(x)\mathfrak{e}\big((\boldsymbol{\xi},\mathbf{x})_{\mathbb{R}^N}\big)\overline{g(\boldsymbol{\xi})}\,d\mathbf{x}\,d\boldsymbol{\xi}$$
$$= \iint\limits_{\mathbb{R}^N\times\mathbb{R}^N} f(x)\overline{\mathfrak{e}\big(-(\boldsymbol{\xi},\mathbf{x})_{\mathbb{R}^N}\big)g(\boldsymbol{\xi})}\,d\mathbf{x}\,d\boldsymbol{\xi} = \int_{\mathbb{R}^N} f(\mathbf{x})\overline{\check{g}(\mathbf{x})}\,d\mathbf{x}. \qquad \square$$

The function $\hat{f}$ is called the **Fourier transform** of the function f. The fact that $\hat{f}$ tends to 0 at infinity was observed originally (in the context of Fourier series) by Riemann and is usually called the **Riemann–Lebesgue Lemma**. In this connection, the reader should not be deluded into thinking that $\{\hat{f} : f \in L^1(\mathbb{R}^N;\mathbb{C})\} = C_0(\mathbb{R}^N;\mathbb{C})$; it is not! In fact, there is no simple characterization of $\{\hat{f} : f \in L^1(\mathbb{R}^N;\mathbb{C})\}$.

To complete our justification of (7.3.1) for $f \in L^1(\mathbb{R}^N;\mathbb{C})$, we need the computation contained in the next lemma.

7.3.6 Lemma. *Given $t \in (0,\infty)$, define $\gamma_t : \mathbb{R}^N \longrightarrow (0,\infty)$ so that*

$$\gamma_t(\mathbf{x}) = t^{-\frac{N}{2}} \exp\left(-\frac{\pi|\mathbf{x}|^2}{t}\right), \quad \mathbf{x} \in \mathbb{R}^N. \tag{7.3.7}$$

Then, for all $t > 0$, $\int_{\mathbb{R}^N} \gamma_t(\mathbf{x})\,\mathbf{x} = 1$. In fact, for all $\boldsymbol{\zeta} \in \mathbb{C}^N$,

$$\int_{\mathbb{R}^N} e^{2\pi(\boldsymbol{\zeta},\mathbf{x})_{\mathbb{C}^N}}\gamma_t(\mathbf{x})\,d\mathbf{x} = \exp\left(t\pi\sum_{j=1}^N \zeta_j^2\right). \tag{7.3.8}$$

In particular,

$$\widehat{\gamma_t}(\boldsymbol{\xi}) = g_t(\boldsymbol{\xi}) \equiv e^{-t\pi|\boldsymbol{\xi}|^2} \quad \textit{and} \quad \check{g}_t = \gamma_t. \tag{7.3.9}$$

PROOF: The first part of (7.3.9) is an easy application of (7.3.8), and, given the first part, the second part is another application of (7.3.8). Thus we need only prove (7.3.8). To this end, first note that, by Fubini's Theorem, it is enough to handle the case when $N = 1$. That is, we have to prove that

$$t^{-\frac{1}{2}} \int_{\mathbb{R}} \exp\left(-\frac{\pi x^2}{t} + 2\pi\zeta x\right)\,dx = e^{t\pi\zeta^2} \tag{*}$$

for all $t > 0$ and $\zeta \in \mathbb{C}$. Next, note that, for any given $t > 0$, both sides of (*) are holomorphic functions of ζ in the entire complex plane $\mathbb{C}$. Hence, for each $t > 0$, we need only prove (*) for $\zeta \in \mathbb{R}$. Furthermore, given $\zeta \in \mathbb{R}$, a change of variables shows that (*) for some $t > 0$ implies (*) for all $t > 0$. Thus, we need only prove (*) for $\zeta \in \mathbb{R}$ and $t = 2\pi$. But

$$\begin{aligned}\frac{1}{(2\pi)^{\frac{1}{2}}}\int_{\mathbb{R}} e^{-\frac{x^2}{2}+2\pi\zeta x}\,dx &= \frac{1}{(2\pi)^{\frac{1}{2}}}e^{2\pi^2\zeta^2}\int_{\mathbb{R}} e^{-\frac{1}{2}(x-2\pi\zeta)^2}\,dx \\ &= \frac{1}{(2\pi)^{\frac{1}{2}}}e^{2\pi^2\zeta^2}\int_{\mathbb{R}} e^{-\frac{x^2}{2}}\,dx = e^{2\pi^2\zeta^2},\end{aligned}$$

where, in the last equation, we have used the computation made in part (**i**) of Exercise 5.2.6. □

7.3.10 Theorem. *Given $f \in L^1(\mathbb{R}^N;\mathbb{C})$ and $t \in (0,\infty)$, (cf. (7.3.7) and (7.3.9))*

$$f * \gamma_t(x) = \int_{\mathbb{R}^N} g_t(\boldsymbol{\xi})\mathfrak{e}\big(-(\boldsymbol{\xi},\mathbf{x})_{\mathbb{R}^N}\big)\hat{f}(\boldsymbol{\xi})\,d\boldsymbol{\xi}. \tag{7.3.11}$$

In particular,

$$\int_{\mathbb{R}^N} g_t(\boldsymbol{\xi})\mathfrak{e}\big(-(\boldsymbol{\xi},\mathbf{x})_{\mathbb{R}^N}\big)\hat{f}(\boldsymbol{\xi})\,d\boldsymbol{\xi} \longrightarrow f(\mathbf{x}) \quad \textit{in } L^1(\mathbb{R}^N;\mathbb{C}) \textit{ as } t \searrow 0.$$

Thus, if $f \in L^1(\mathbb{R}^N;\mathbb{C})$ and $\hat{f} = 0$ Lebesgue almost everywhere, then $f = 0$ Lebesgue almost everywhere.

PROOF: By Theorem 6.3.14, it suffices for us to prove (7.3.11). To this end, set (cf. Lemma 6.3.5) $f_{\mathbf{x}} = \tau_{\mathbf{x}} f$ for $\mathbf{x} \in \mathbb{R}^N$, and observe that $\widehat{f_{\mathbf{x}}}(\boldsymbol{\xi}) = \mathfrak{e}\big(-(\boldsymbol{\xi},\mathbf{x})_{\mathbb{R}^N}\big)\hat{f}(\boldsymbol{\xi})$. Hence, because $\gamma_t(-\mathbf{y}) = \gamma_t(\mathbf{y})$, (7.3.11) follows from (7.3.5), applied to f_x and g_t, combined with (7.3.9). □

The preceding proof turns on a general principle which, because of its interpretation in quantum mechanics, is known as the **uncertainty principle**. Crudely stated, this principle says that the more *localized* is a function f, the more *delocalized* is its Fourier transform $\hat{f}$. Thus, because γ_t gets more and more concentrated near $\mathbf{0}$ as $t \searrow 0$, $\widehat{\gamma_t}$ gets more and more evenly spread. See Exercise 9.3.14 below to see this phenomenon in a more general context.

Exercises

7.3.12 Exercise: Suppose that $f, g \in L^1(\mathbb{R}^N;\mathbb{C})$, and recall (cf. Theorem 6.3.1 and make the obvious extension to $\mathbb{C}$-valued functions) that the convolution $f * g$ is again an element of $L^1(\mathbb{R}^N;\mathbb{C})$. By an application of Fubini's Theorem, show that $\widehat{f * g} = \hat{f}\,\hat{g}$. Conclude, in particular, that $\{\hat{f} : f \in L^1(\mathbb{R}^N;\mathbb{C})\}$ is an algebra.

7.3.13 Exercise: Using part (**iii**) of Exercise 5.2.6, show that, for $\mathbf{x} \in \mathbb{R}^N$, (cf. part (**iii**) of Exercise 6.3.18)

$$P(\mathbf{x}) \equiv \frac{2}{\omega_N (1+|\mathbf{x}|^2)^{\frac{N+1}{2}}} = \frac{1}{(2\pi)^{\frac{N+1}{2}}} \int_{(0,\infty)} t^{-\frac{N+3}{2}} e^{-\frac{1}{2t}} e^{-\frac{|\mathbf{x}|^2}{2t}}\, dt.$$

Next, using (7.3.9), conclude that

$$\hat{P}(\boldsymbol{\xi}) = \frac{1}{(2\pi)^{\frac{1}{2}}} \int_{(0,\infty)} t^{-\frac{3}{2}} e^{-\frac{1}{2t}} e^{-t|2\pi\boldsymbol{\xi}|^2}\, dt.$$

Finally, use part (**iv**) of Exercise 5.2.6 to conclude that

$$\hat{P}(\boldsymbol{\xi}) = e^{-2\pi|\boldsymbol{\xi}|}.$$

Use this computation to give another derivation of (6.3.21).

7.3.14 Exercise: Let $g \in L^1(\mathbb{R}^N)$ with $\int_{\mathbb{R}} g(\mathbf{x})\, d\mathbf{x} = 1$ be given, and define $g_t(\mathbf{x}) = t^{-N} g(t^{-1}\mathbf{x})$. Clearly, as $t \searrow 0$, g_t gets more and more localized at $\mathbf{0}$ in a sense made precise by Theorem 6.3.14. Show that, at the same time, $\widehat{g_t}$ is becoming delocalized in the sense that $\widehat{g_t}(\boldsymbol{\xi}) = \hat{g}(t\boldsymbol{\xi}) \longrightarrow 1$ uniformly on compacts.

7.4 Hermite Functions

In this section we will be making preparations to extend the Fourier transform to give meaning to $\hat{f}$ when $f \in L^2(\mathbb{R}^N;\mathbb{C})$. Our approach will be to construct an orthonormal basis $\{h_{\mathbf{n}} : n \in \mathbb{N}^N\} \subseteq L^1(\mathbb{R}^N;\mathbb{R})$ for $L^2(\mathbb{R}^N;\mathbb{C})$ with the property that

$$\hat{h}_{\mathbf{n}} = (\sqrt{-1})^{\|\mathbf{n}\|} h_{\mathbf{n}}, \quad \text{where } \|\mathbf{n}\| \equiv \sum_{j=1}^{N} n_j \tag{7.4.1}$$

when $\mathbf{n} = (n_1, \dots, n_N)$. We will then define the Fourier transform of $f \in L^2(\mathbb{R}^N;\mathbb{C})$ to be given by the series

$$\sum_{\mathbf{n}\in\mathbb{N}^N} (\sqrt{-1})^{\|\mathbf{n}\|} \big(f, h_{\mathbf{n}}\big)_{L^2(\mathbb{R}^N;\mathbb{C})} h_{\mathbf{n}}.$$

We begin with the observation that it suffices to carry out our construction when $N = 1$. Indeed, Fubini's Theorem shows that

$$\hat{f}(\boldsymbol{\xi}) = \prod_{j=1}^{N} \hat{f}_j(\xi_j) \quad \text{if } f(\mathbf{x}) = \prod_{j=1}^{N} f_j(x_j) \text{ where } f_1, \dots, f_N \in L^1(\mathbb{R}^N; \mathbb{C}),$$

and so there is no problem about the computation of $\hat{h}_{\mathbf{n}}$ if

$$h_{\mathbf{n}}(\mathbf{x}) \equiv \prod_{j=1}^{N} h_{n_j}(x_j) \quad \text{for } \mathbf{x} = (x_1, \dots, x_N) \in \mathbb{R}^N \tag{7.4.2}$$

and we know how to compute $\hat{h}_n$ for $n \in \mathbb{N}$. At the same time, the following simple lemma tells us that we can construct a basis in $L^2(\mathbb{R}^N; \mathbb{C})$ by taking products of elements of a basis in $L^2(\mathbb{R}; \mathbb{C})$.

7.4.3 Lemma. *Suppose that $\{h_n : n \in \mathbb{N}\}$ is an orthonormal basis in $L^2(\mathbb{R}; \mathbb{C})$, and define $h_{\mathbf{n}}$ for $\mathbf{n} \in \mathbb{N}^N$ by (7.4.2). Then $\{h_{\mathbf{n}} : \mathbf{n} \in \mathbb{N}^N\}$ is an orthonormal basis in $L^2(\mathbb{R}^N; \mathbb{C})$.*

PROOF: By Fubini's Theorem, it is easy to verify that $\{h_{\mathbf{n}} : \mathbf{n} \in \mathbb{N}^N\}$ is orthonormal. Thus, by Exercise 7.1.15, it suffices for us to assume that $f \in L^2(\mathbb{R}^N; \mathbb{C})$ is perpendicular to $\operatorname{span}(\{h_{\mathbf{n}} : \mathbf{n} \in \mathbb{N}^N\})$ and deduce that $f = \mathbf{0}$. To this end, first note that if $g(\mathbf{x}) = \prod_{j=1}^{N} g_j(x_j)$ for some $g_1, \dots, g_N \in L^2(\mathbb{R}; \mathbb{C})$, then

$$g = \sum_{\mathbf{n} \in \mathbb{N}^N} \left(g, h_{\mathbf{n}}\right)_{L^2(\mathbb{R}^N;\mathbb{C})} h_{\mathbf{n}}$$

follows from Fubini's Theorem and the fact that $\{h_n : n \in \mathbb{N}\}$ is a basis in $L^2(\mathbb{R}; \mathbb{C})$. Hence, $(f, g)_{L^2(\mathbb{R}^N;\mathbb{C})} = 0$ for all such g's; and so f is perpendicular $\mathbf{1}_\Gamma$ for all $\Gamma \in \mathcal{B}_{\mathbb{R}^N}$ which can be expressed as the product of bounded, Borel subsets of $\mathbb{R}$. Next, given $R > 0$, consider the set $\mathcal{S}_R$ of $\Gamma \in \mathcal{B}_{\mathbb{R}^N}$ such that $\Gamma \subseteq [-R, R]^N$ and $f \perp \mathbf{1}_\Gamma$. It is then an easy matter to check that $\mathcal{S}_R$ is a λ-system, and because $\mathcal{S}_R$ contains the π-system

$$\mathcal{C}_R \equiv \left\{ \prod_{j=1}^{N} \Gamma_j : \Gamma_j \in \mathcal{B}_{[-R,R]} \text{ for each } 1 \le j \le N \right\},$$

it follows that $f \perp \mathbf{1}_\Gamma$ for all $\Gamma \in \mathcal{B}_{[-R,R]^N}$. Since this is true for every $R > 0$, an obvious limit procedure now shows that f is orthogonal to $\mathbf{1}_\Gamma$ for all $\Gamma \in \mathcal{B}_{\mathbb{R}^N}$ with finite Lebesgue measure and therefore to all $\mathbb{C}$-valued simple functions $g \in L^2(\mathbb{R}^N; \mathbb{C})$. But we already know that the simple, $\mathbb{R}$-valued elements of $L^2(\mathbb{R}^N; \mathbb{R})$ are dense in $L^2(\mathbb{R}^N; \mathbb{R})$, and one can easily pass from this to the analogous $\mathbb{C}$-valued statement by considering real and imaginary parts separately. In other words, we have now shown that f is perpendicular to a dense subspace of $L^2(\mathbb{R}^N; \mathbb{C})$ and therefore must be 0 almost everywhere. □

In view of the preceding considerations, we need only find an orthonormal basis $\{h_n : n \in \mathbb{N}\} \subseteq L^1(\mathbb{R};\mathbb{R})$ in $L^2(\mathbb{R};\mathbb{C})$ with the property that

$$\hat{h}_n = (\sqrt{-1})^n h_n, \quad n \in \mathbb{N}. \tag{7.4.4}$$

Although it hardly explains why the construction works, we begin by observing that we already know one candidate for h_0, namely, by (7.3.9), we know that $\hat{h}_0 = h_0$ when $h_0(x) = 2^{\frac{1}{4}} e^{-\pi x^2}$. Unfortunely, it is not so easy to explain why we should expect that the other h_n's are given by

$$h_n = \frac{2^{\frac{1}{4}}}{\beta_n} g_n \text{ where } g_n(x) \equiv (-1)^n e^{\pi x^2} \frac{d^n}{dx^n} e^{-2\pi x^2} \;\&\; \beta_n \equiv \left((4\pi)^n n!\right)^{\frac{1}{2}}. \tag{7.4.5}$$

The function h_n is called the nth **Hermite function**, and, as we now show, $\{h_n : n \in \mathbb{N}\}$ serves our purposes.

Obviously,

$$g_n(x) = G_n(x) e^{-\pi x^2} \quad \text{where } G_n(x) \equiv (-1)^n e^{2\pi x^2} \frac{d^n}{dx^n} e^{-2\pi x^2}. \tag{7.4.6}$$

Next, starting from (7.4.6), observe that

$$G_n = A G_{n-1}, \; n \in \mathbb{Z}^+, \quad \text{where } A \equiv -\frac{d}{dx} + 4\pi x. \tag{7.4.7}$$

For this reason, A is called the **raising operator**. Because $G_0 = \mathbf{1}$, an easy inductive argument shows that G_n is an n-order, real polynomial with $(4\pi)^n$ as the coefficient of x^n. In particular,

$$\begin{aligned} &\operatorname{span}\left(\{G_m : 0 \le m \le n\}\right) = \operatorname{span}\left(\{x^m : 0 \le m \le n\}\right) \\ &\frac{d^n G_n}{dx^n} = (4\pi)^n n! = \beta_n^2 \end{aligned} \quad \text{for all } n \in \mathbb{N} \tag{7.4.8}$$

7.4.9 Lemma. *Define the Borel, probability measure $\Gamma(dx) = 2^{\frac{1}{2}} e^{-2\pi x^2}\,dx$ on $\mathbb{R}$, and let $C^\infty_\nearrow(\mathbb{R};\mathbb{C})$ denote the space of $f \in C^\infty(\mathbb{R};\mathbb{C})$ with the property that, for each $n \in \mathbb{N}$, $\left|\frac{d^n f}{dx^n}(x)\right|$ is dominated by $C_n(1+|x|^2)^{\nu_n}$ for some choice of $C_n < \infty$ and $\nu_n \in [0,\infty)$. Then $C^\infty_\nearrow(\mathbb{R};\mathbb{C})$ is an algebra which is contained in $L^2(\Gamma;\mathbb{C})$ and is preserved by differentiation. Moreover (cf. (7.4.7))*

$$\left(Af, g\right)_{L^2(\Gamma;\mathbb{C})} = \left(f, Dg\right)_{L^2(\Gamma;\mathbb{C})} \quad \textit{for all } f, g \in C^\infty_\nearrow(\mathbb{R};\mathbb{C}), \tag{7.4.10}$$

where $Dg \equiv \frac{dg}{dx}$. In particular, (cf. (7.4.5))

$$\left(G_m, G_n\right)_{L^2(\Gamma;\mathbb{C})} = \delta_{m,n}\beta_n^2 \quad \textit{for all } m, n \in \mathbb{N}, \tag{7.4.11}$$

and

$$DG_n = 4\pi n G_{n-1} \quad \text{for all } n \in \mathbb{Z}^+. \tag{7.4.12}$$

PROOF: The first assertions are trivial. To see (7.4.10), we use integration by parts to obtain

$$\begin{aligned}(f, Dg)_{L^2(\Gamma;\mathbb{C})} &= \lim_{R\to\infty} (\mathrm{R})\int_{[-R,R]} f(x)\frac{d\bar g}{dx}(x)e^{-2\pi x^2}\,dx \\ &= \lim_{R\to\infty}\big(f(R)\bar g(R) - f(-R)g(-R)\big)e^{-2\pi R^2} \\ &\quad + \lim_{R\to\infty} (\mathrm{R})\int_{[-R,R]} Af(x)\bar g(x)e^{-2\pi x^2}\,dx = (Af, g)_{L^2(\Gamma;\mathbb{C})}.\end{aligned}$$

Given (7.4.10), we prove (7.4.11) by assuming that $m \geq n$, writing (cf. (7.4.7)) $G_m = A^m G_0$, and concluding that

$$(G_m, G_n)_{L^2(\Gamma;\mathbb{C})} = (G_0, D^m G_n)_{L^2(\Gamma;\mathbb{C})} = \begin{cases} 0 & \text{if } m > n \\ \beta_n^2 & \text{if } m = n, \end{cases}$$

where, in the second case, we have used the second line of (7.4.8) and the fact that Γ is a probability measure.

Finally, to prove (7.4.12), first note (cf. (7.4.8)) that DG_n is in the span of $\{G_m : 0 \leq m \leq n-1\}$. Thus,

$$\begin{aligned}DG_n &= \sum_{m=0}^{n-1} \beta_m^{-2}(DG_n, G_m)_{L^2(\Gamma;\mathbb{C})}G_m = \sum_{m=0}^{n-1} \beta_m^{-2}(G_n, AG_m)_{L^2(\Gamma;\mathbb{C})}G_m \\ &= \sum_{m=0}^{n-1} \beta_m^{-2}(G_n, G_{m+1})_{L^2(\Gamma;\mathbb{C})}G_m = \frac{\beta_n^2}{\beta_{n-1}^2}G_{n-1}. \quad \square\end{aligned}$$

Notice that one conclusion which can be drawn from the preceding is that, for each $n \in \mathbb{N}$, $\{\beta_m^{-1}G_m : 0 \leq m \leq n\}$ is an orthonormalization in $L^2(\Gamma;\mathbb{C})$ of $\{x^m : 0 \leq m \leq n\}$. Of course, we do not know yet that the polynomials are dense in $L^2(\mathbb{R};\mathbb{C})$, and we therefore cannot assert yet that $\{\beta_m^{-1}G_m : m \in \mathbb{N}\}$ is an orthonormal basis in $L^2(\Gamma;\mathbb{C})$. However, the density of the polynomials in $L^2(\Gamma;\mathbb{C})$ is equivalent (cf. Exercise 7.4.23 below) to the first part of the following lemma.

7.4.13 Lemma. *The set $\{h_n : n \in \mathbb{N}\}$ is an orthonormal basis in $L^2(\mathbb{R};\mathbb{C})$. Moreover, for each $n \in \mathbb{N}$,*

$$\begin{aligned} h_n' &= \pi^{\frac12}\big(-(n+1)^{\frac12}h_{n+1} + n^{\frac12}h_{n-1}\big) \\ 2\pi x h_n &= \pi^{\frac12}\big((n+1)^{\frac12}h_{n+1} + n^{\frac12}h_{n-1}\big) \\ -\frac{d^2h_n}{dx^2}(x) &+ (2\pi x)^2 h_n = 4\pi\big(n + \tfrac12\big)h_n,\end{aligned} \tag{7.4.14}$$

where, in the first two lines, $h_{-1} \equiv 0$. In particular,

$$\|h_n'\|^2_{L^2(\mathbb{R};\mathbb{C})} = \|2\pi x h_n\|^2_{L^2(\mathbb{R};\mathbb{C}} = 2\pi\big(n+\tfrac{1}{2}\big). \tag{7.4.15}$$

PROOF: Because of the (7.4.11), we know that $\{h_n : n \in \mathbb{N}\}$ is an orthonormal sequence. To prove that it is a basis, suppose that $f \in L^2(\mathbb{R};\mathbb{C})$ satisfies $\big(f, h_n\big)_{L^2(\mathbb{R};\mathbb{C})} = 0$ for all $n \in \mathbb{N}$. We want to show that $f = \mathbf{0}$ in $L^2(\mathbb{R};\mathbb{C})$. To this end, set $\varphi(x) = e^{\pi x^2} f(x)$. Clearly, $\varphi \in L^2(\Gamma;\mathbb{C})$, $\big(\varphi, G_n\big)_{L^2(\Gamma;\mathbb{C})} = 0$ for all $n \in \mathbb{N}$, and it suffices for us to show that $\psi(x) \equiv \varphi(x)e^{-2\pi x^2} = 0$ for $\lambda_{\mathbb{R}}$-almost everywhere $x \in \mathbb{R}$.

Our proof that $\psi = 0$ $\lambda_{\mathbb{R}}$-almost everywhere has two steps. First, we note that, by Schwarz's inequality,

$$\|\psi\|_{L^1(\mathbb{R};\mathbb{C})} = \int_{\mathbb{R}} \big|f(x)\big| e^{-\pi x^2}\,dx \le 2^{-\frac{1}{4}}\|f\|_{L^2(\mathbb{R};\mathbb{C})} < \infty,$$

which means that $\psi \in L^1(\mathbb{R};\mathbb{C})$. Second, we want to show that $\hat{\psi} = \mathbf{0}$. But, because of the orthogonality relation satisfied by φ together with the first part of (7.4.8), we know that, for every $n \in \mathbb{Z}^+$,

$$\int_{\mathbb{R}} \psi(x)\mathfrak{e}_n(\xi\, x)\,dx = \sum_{m=0}^{n-1} \frac{\big(\sqrt{-1}\,2\pi x\big)^m}{m!} \int_{\mathbb{R}} \varphi(x) x^m e^{-2\pi x^2}\,dx = 0,$$

where $\mathfrak{e}_n(\eta) \equiv \sum_{m=0}^{n-1} \frac{(\sqrt{-1}\,2\pi\eta)^m}{m!}$. At the same time, by Taylor's Theorem,

$$\big|\mathfrak{e}(\eta) - \mathfrak{e}_n(\eta)\big| \le \frac{|2\pi\eta|^n}{n!} e^{2\pi|\eta|}, \quad \eta \in \mathbb{C},$$

and so

$$B_n(\xi) \equiv \left|\hat{\psi}(\xi) - \int_{\mathbb{R}} \psi(x)\mathfrak{e}_n(\xi x)\,dx\right| \le \int_{\mathbb{R}} \big|\psi(x)\big| \frac{(2\pi|\xi x|)^n}{n!} e^{2\pi|\xi x|}\,dx.$$

Hence,

$$\begin{aligned}\sum_{n=1}^{\infty} B_n(\xi) &\le \int_{\mathbb{R}} \big|\psi(x)\big| e^{2\pi|\xi x|}\,dx = 2^{-\frac{1}{2}} \int_{\mathbb{R}} \big|\varphi(x)\big| e^{2\pi|\xi x|}\,\Gamma(dx) \\ &\le 2^{-\frac{1}{2}}\|\varphi\|_{L^2(\Gamma;\mathbb{C})} \left(2^{\frac{1}{2}} \int_{\mathbb{R}} e^{4\pi|\xi x| - 2\pi x^2}\,dx\right)^{\frac{1}{2}} < \infty,\end{aligned}$$

and so $B_n(\xi) \longrightarrow 0$ as $n \to \infty$. Thus, we have now proved that $\{h_n : n \in \mathbb{N}\}$ is an orthonormal basis in $L^2(\mathbb{R};\mathbb{C})$.

Turning to (7.4.14), note that the third line results from substracting two applications of the first line from two applications of the second line. To prove the first line, note that, by (7.4.7) and (7.4.12),

$$\begin{aligned}g_n'(x) &= -e^{-\pi x^2} 2\pi x G_n(x) + e^{-\pi x^2} G_n'(x) \\ &= -\tfrac{1}{2} e^{-\pi x^2} G_{n+1}(x) + e^{-\pi x^2} 2\pi n G_{n-1}(x) = -\tfrac{1}{2} g_{n+1}(x) + 2\pi n g_{n-1}(x),\end{aligned}$$

where the last term vanishes when $n = 0$. Hence,

$$h_n' = -\frac{\beta_{n+1}}{2\beta_n}h_{n+1} + \frac{2\pi n\beta_{n-1}}{\beta_n}h_{n-1} = \pi^{\frac{1}{2}}\left(-(n+1)^{\frac{1}{2}}h_{n+1} + n^{\frac{1}{2}}h_{n-1}\right)$$

when $h_{-1} \equiv 0$. The proof of second line is similar:

$$\begin{aligned}2\pi x g_n &= e^{-\pi x^2}2\pi x G_n(x) = \tfrac{1}{2}e^{-\pi x^2}G_{n+1}(x) + e^{-\pi x^2}2\pi n G_{n-1}(x)\\ &= \tfrac{1}{2}g_{n+1}(x) + 2\pi n g_{n-1}(x),\end{aligned}$$

which, just as before, yields the second line after multiplication by β_n^{-1}.

Finally, given the first two lines of (7.4.14), (7.4.15) is a trivial application of the orthonormality of the h_n's. □

In quantum mechanics, the operator $-\frac{d^2}{dx^2} + (2\pi x)^2$ which appears in (7.4.14) is the Hamiltonian corresponding to the *harmonic oscillator*, and Lemma 7.4.13 says that the Hermite functions diagonalize this operator. For our immediate purposes, the most important application of (7.4.14) is (7.4.15), which, after an application of Schwarz's inequality, leads in turn to:

$$\begin{aligned}\|h_n\|_{L^1(\mathbb{R};\mathbb{C})}^2 &= \left(\int_{\mathbb{R}} (1+x^2)^{-\frac{1}{2}}(1+x^2)^{\frac{1}{2}}|h_n(x)|\,dx\right)^2\\ &\le \int_{\mathbb{R}} \frac{1}{1+x^2}\,dx \int_{\mathbb{R}} (1+x^2)h_n(x)^2\,dx \le \pi\left(1 + (2\pi)^{-1}(n+\tfrac{1}{2})\right).\end{aligned}$$

Hence,

$$\|h_n\|_{L^1(\mathbb{R};\mathbb{C})} \le n^{\frac{1}{2}} + 2, \quad n \in \mathbb{N}. \tag{7.4.16}$$

7.4.17 Theorem. *Let $\{h_n : n \in \mathbb{N}\}$ be given by (7.4.5), define $\{h_{\mathbf{n}} : \mathbf{n} \in \mathbb{N}^N\}$ accordingly, as in (7.4.2). Then $\{h_{\mathbf{n}} : \mathbf{n} \in \mathbb{N}^N\}$ is an orthonormal basis in $L^2(\mathbb{R}^N;\mathbb{C})$,*

$$\|h_{\mathbf{n}}\|_{L^1(\mathbb{R}^N;\mathbb{C})} \vee \|h_{\mathbf{n}}\|_{\mathrm{u}} \le \prod_{j=1}^N \left(n_j^{\frac{1}{2}} + 2\right), \quad \mathbf{n} \in \mathbb{N}^N, \tag{7.4.18}$$

and (7.4.1) holds.

PROOF: By Lemmas 7.4.3 and 7.4.13, we know that $\{h_{\mathbf{n}} : \mathbf{n} \in \mathbb{N}^N\}$ is an orthonormal basis in $L^2(\mathbb{R}^N;\mathbb{C})$. Further, the L^1-estimate (7.4.18) is an immediate corollary of (7.4.16) and Tonelli's Theorem, and the uniform estimate will follow from the L^1-estimate once we verify (7.4.1). Finally, by the remark preceding (7.4.2), we will know that (7.4.1) holds in general once we show that it holds when $N = 1$. To this end, observe that, for any $\zeta \in \mathbb{C}$, Taylor's Theorem says that

$$\begin{aligned}\sum_{m=0}^{\infty} \frac{(2\pi^{\frac{1}{2}}\zeta)^m}{2^{\frac{1}{4}}(m!)^{\frac{1}{2}}} h_m(x) &= \sum_{m=0}^{\infty} \frac{\zeta^m}{m!} g_m(x) \\ &= e^{\pi x^2} e^{-2\pi(x-\zeta)^2} = \exp\bigl(-\pi x^2 + 4\pi\zeta x - 2\pi\zeta^2\bigr),\end{aligned}$$

where the convergence is uniform as (x, ζ) range over compact subsets of $\mathbb{R}\times\mathbb{C}$. By combining this with (7.4.16), we conclude that, for each $\zeta \in \mathbb{C}$,

$$\sum_{m=0}^{n} \frac{(2\pi^{\frac{1}{2}}\zeta)^m}{2^{\frac{1}{4}}(m!)^{\frac{1}{2}}} h_m(x) \longrightarrow \exp\bigl(-\pi x^2 + 4\pi\zeta x - 2\pi\zeta^2\bigr)$$

in $L^1(\mathbb{R};\mathbb{C})$. In particular, by (7.3.8) with $N = 1$ and ζ replaced by $2\zeta + \sqrt{-1}\,\xi$,

$$\begin{aligned}\sum_{m=0}^{\infty} \frac{(2\pi^{\frac{1}{2}}\zeta)^m}{2^{\frac{1}{4}}(m!)^{\frac{1}{2}}} \hat{h}_m(\xi) &= e^{-2\pi\zeta^2} \int_{\mathbb{R}} e^{2\pi(2\zeta+\sqrt{-1}\,\xi)} \gamma_1(x)\,dx \\ &= e^{-2\pi\zeta^2} \exp\Bigl(\pi\bigl(2\zeta + \sqrt{-1}\,\xi\bigr)^2\Bigr) = \exp\bigl(-\pi\xi^2 + 4\pi\sqrt{-1}\,\zeta\xi + 2\pi\zeta^2\bigr).\end{aligned}$$

But, by our earlier calculation,

$$\exp\bigl(-\pi\xi^2 + 4\pi\sqrt{-1}\,\zeta\xi + 2\pi\zeta^2\bigr) = \sum_{m=0}^{\infty} \frac{(2\pi^{\frac{1}{2}}\sqrt{-1}\,\zeta)^m}{2^{\frac{1}{4}}(m!)^{\frac{1}{2}}} h_m(\xi),$$

and so $\hat{h}_m = (\sqrt{-1})^m h_m$ follows from equating coefficients of ζ^m in these two series. $\square$

7.4.19 Corollary. *Assume that $f \in C_c^\infty(\mathbb{R}^N;\mathbb{C})$. Then, for each $L \in \mathbb{Z}^+$,*

$$\sup_{\mathbf{n}\in\mathbb{N}^N} \bigl(1 + \|\mathbf{n}\|\bigr)^L \bigl|(f, h_{\mathbf{n}})_{L^2(\mathbb{R}^N;\mathbb{C})}\bigr| < \infty, \tag{7.4.20}$$

and therefore

$$\sum_{\|\mathbf{m}\|<M} (f, h_{\mathbf{m}})_{L^2(\mathbb{R}^N;\mathbb{C})} \longrightarrow f$$

both in $L^1(\mathbb{R}^N;\mathbb{C})$ as well as uniformly on compacts.

Proof: Obviously, it is enough to prove (7.4.20). To this end, note that, by the third line of (7.4.14),

$$-\Delta h_{\mathbf{n}} + |2\pi\mathbf{x}|^2 h_{\mathbf{n}} = 4\pi\bigl(\|\mathbf{n}\| + \tfrac{1}{2}\bigr). \tag{7.4.21}$$

Hence, if $f \in C_c^\infty(\mathbb{R}^N;\mathbb{C})$, then, by Green's Identity (cf. Theorem 5.4.12), we know that

$$4\pi\bigl(\|\mathbf{n}\| + \tfrac{1}{2}\bigr)(f, h_{\mathbf{n}})_{L^2(\mathbb{R}^N;\mathbb{C})} = \Bigl(-\Delta f + |2\pi\mathbf{x}|^2 f, h_{\mathbf{n}}\Bigr)_{L^2(\mathbb{R}^N;\mathbb{C})}.$$

By iterating this identity, we obtain

$$\Bigl(4\pi\bigl(\|\mathbf{n}\| + \tfrac{1}{2}\bigr)\Bigr)^L (f, h_{\mathbf{n}})_{L^2(\mathbb{R}^N;\mathbb{C})} = \Bigl(\bigl(-\Delta + |2\pi\mathbf{x}|^2\bigr)^L f, h_{\mathbf{n}}\Bigr)_{L^2(\mathbb{R}^N;\mathbb{C})}$$

for any $L \in \mathbb{Z}^+$, and clearly (7.4.20) is an immediate consequence of this. $\square$

Exercises

7.4.22 Exercise: Starting from the first two lines in (7.4.14), use induction on $k \in \mathbb{N}$ and $\ell \in \mathbb{N}$ to show that, for each $(k,\ell) \in \mathbb{N}^2$ and $m \in \mathbb{N}$,

$$(2\pi x)^k \frac{d^\ell h_m}{dx^\ell}(x) = \sum_{n=0}^{m+k+\ell} c_n(m,k,\ell) h_{m-k-\ell+2n}(x),$$

where $h_n \equiv 0$ if $n < 0$ and there exists $\{C_{k,\ell} : (k,\ell) \in \mathbb{N}^2\} \subseteq [0,\infty)$ such that

$$\left|c_n(m,k,\ell)\right| \le C_{k,\ell}(1+m)^{\frac{k+\ell}{2}} \quad \text{for all } m \in \mathbb{N}.$$

7.4.23 Exercise: Set $H_n = \beta_n^{-1} G_n$ for $n \in \mathbb{N}$.

(i) Using the isometric, isomorphism $f \in L^2(\Gamma;\mathbb{C}) \longmapsto 2^{\frac{1}{4}} e^{-\pi x^2} f \in L^1(\mathbb{R};\mathbb{C})$, show that $\{h_n : n \in \mathbb{N}\}$ being an orthonormal basis in $L^2(\mathbb{R};\mathbb{C})$ is equivalent to $\{H_n : n \in \mathbb{N}\}$ being an orthonormal basis in $L^2(\Gamma;\mathbb{C})$.

(ii) Define $Lf = \frac{d^2 f}{dx^2}(x) - 4\pi x f(x)$ for $f \in C^2(\mathbb{R};\mathbb{C})$, and notice that $L = -A \circ D$. By combining (7.4.7) and (7.4.12) as we did in the proof of (7.4.14), show that, for any (cf. Lemma 7.4.9) $f \in C^\infty_{\nearrow}(\mathbb{R};\mathbb{C})$,

$$\big(Lf, H_n\big)_{L^2(\Gamma;\mathbb{C})} = -4\pi n (f, H_n)_{L^2(\Gamma;\mathbb{C})}.$$

(iii) Take $N = 1$ in (7.3.7), and define

$$u_f(t,x) = \int_{\mathbb{R}} f(y) \gamma_{2^{-1}(1-e^{-8\pi t})}\big(y - e^{-4\pi t} x\big)\, dy \quad \text{on } (0,\infty) \times \mathbb{R}$$

for $f \in C^\infty_{\nearrow}(\mathbb{R};\mathbb{C})$. Show that $u_f \in C^\infty\big([0,\infty) \times \mathbb{R};\mathbb{C}\big)$ and that, for each $\ell \in \mathbb{N}$,

$$\sup_{t \ge 0} \left|\frac{\partial^\ell u_f}{\partial x^\ell}(t,x)\right| \le C_\ell \big(1 + x^2\big)^{\nu_\ell}$$

for some $C_\ell < \infty$ and $\nu_\ell \in \mathbb{N}$.

(iv) For each $f \in C^\infty_{\nearrow}(\mathbb{R};\mathbb{C})$, show first that $\frac{\partial u_f}{\partial t} = L u_f$ and then that

$$\lim_{t \searrow s} \left\| \frac{u_f(t,\,\cdot\,) - u_f(s,\,\cdot\,)}{t-s} - L u_f(s,\,\cdot\,) \right\|_{L^2(\Gamma;\mathbb{C})} = 0 \quad \text{for } s \in (0,\infty)$$

$$\lim_{t \searrow 0} \|u_f(t,\,\cdot\,) - f\|_{L^2(\Gamma;\mathbb{C})} = 0.$$

By combining this with part **(ii)**, conclude that

$$\big(u_f(t,\,\cdot\,), H_n\big)_{L^2(\Gamma;\mathbb{C})} = e^{-4\pi n t}(f, H_n)_{L^2(\Gamma;\mathbb{C})} \text{ for } t \in (0,\infty) \text{ and } f \in C_{\nearrow}(\mathbb{R};\mathbb{C}).$$

Deduce, in particular, that

$$u_{H_n}(t,\,\cdot\,) = e^{-4\pi n t} H_n, \quad n \in \mathbb{N} \text{ and } t > 0.$$

Hint: Begin by verifying the fact that

$$\frac{\partial \gamma_t}{\partial t} = \frac{1}{4\pi}\frac{\partial^2 \gamma_t}{\partial x^2}.$$

(**v**) Define the **Mehler kernel**

$$M(t,x,y) = \left(\frac{1}{1-e^{-8\pi t}}\right)^{\frac{1}{2}} \exp\left(-2\pi \frac{e^{-8\pi t}x^2 - 2e^{-4\pi t}xy + e^{-8\pi t}y^2}{1-e^{-8\pi t}}\right)$$

for $(t,x,y) \in (0,\infty)\times\mathbb{R}\times\mathbb{R}$. Using part (**iii**) together with the second estimate in (7.4.18), show that

$$M(t,x,y) = \sum_{n=0}^{\infty} e^{-4\pi n t} H_n(x) H_n(y),$$

where the convergence on the right hand side is uniform on compact subsets of $(0,\infty)\times\mathbb{R}\times\mathbb{R}$.

Hint: First prove that, for each $(t,x) \in (0,\infty)\times\mathbb{R}$, the preceding holds in $L^2(\Gamma;\mathbb{R})$ with respect to y, and then apply the estimate in (7.4.18) to get the final conclusion.

7.5 The Fourier Transform, L^2-theory

Let $\{h_{\mathbf{n}} : \mathbf{n} \in \mathbb{N}^N\}$ be the Hermite functions described in § 7.4. Because $\{h_{\mathbf{n}} : \mathbf{n} \in \mathbb{N}^N\}$ is an orthonormal basis in $L^2(\mathbb{R}^N;\mathbb{C})$, we can define a linear isometry $\mathfrak{F} : L^2(\mathbb{R}^N;\mathbb{C}) \longrightarrow L^2(\mathbb{R}^N;\mathbb{C})$ by the prescription (cf. (7.4.1))

$$\mathfrak{F}f = \sum_{\mathbf{n}\in\mathbb{N}^N} (\sqrt{-1})^{\|\mathbf{n}\|}(f,h_{\mathbf{n}})_{L^2(\mathbb{R}^N;\mathbb{C})} h_{\mathbf{n}}. \tag{7.5.1}$$

In fact, because the $h_{\mathbf{n}}$'s are real and therefore

$$\overline{\mathfrak{F}\bar{f}} = \sum_{\mathbf{n}\in\mathbb{N}^N} (\sqrt{-1})^{-\|\mathbf{n}\|}(f,h_{\mathbf{n}})_{L^2(\mathbb{R}^N;\mathbb{C})} h_{\mathbf{n}},$$

we see that $\mathfrak{F}$ is an invertible isometry and that

$$\mathfrak{F}^{-1}f = \overline{\mathfrak{F}\bar{f}}. \tag{7.5.2}$$

7.5.3 Theorem. *For each $f \in L^1(\mathbb{R}^N;\mathbb{C}) \cap L^2(\mathbb{R}^N;\mathbb{C})$, $\hat{f} \in C_0(\mathbb{R}^N;\mathbb{C}) \cap L^2(\mathbb{R}^N;\mathbb{C})$. Moreover, the map $\mathfrak{F}$ given by (7.5.1) is the one and only extension of $f \in L^1(\mathbb{R}^N;\mathbb{C}) \cap L^2(\mathbb{R}^N;\mathbb{C}) \longmapsto \hat{f} \in L^2(\mathbb{R}^N;\mathbb{C})$ as a continuous map from $L^2(\mathbb{R}^N;\mathbb{C})$ into $L^2(\mathbb{R}^N;\mathbb{C})$. In particular, this extension is an invertible isometry, and, for each $f \in L^2(\mathbb{R}^N;\mathbb{C})$,*

$$\text{(7.5.4)}\qquad \begin{aligned} \int_{|\mathbf{x}|\le R} f(\mathbf{x})\mathfrak{e}\big((\boldsymbol{\xi},\mathbf{x})_{\mathbb{R}^N}\big)\,dx &\longrightarrow \mathfrak{F}f \\ \int_{|\mathbf{x}|\le R} f(\mathbf{x})\mathfrak{e}\big(-(\boldsymbol{\xi},\mathbf{x})_{\mathbb{R}^N}\big)\,dx &\longrightarrow \mathfrak{F}^{-1}f \end{aligned} \qquad \text{in } L^2(\mathbb{R}^N;\mathbb{C}).$$

PROOF: Let $f \in L^1(\mathbb{R}^N;\mathbb{C}) \cap L^2(\mathbb{R}^N;\mathbb{C})$ be given. Using part **(iii)** of Exercise 6.3.24, choose $\{f_k\}_1^\infty \subseteq C_c^\infty(\mathbb{R}^N;\mathbb{C})$ so that $f_k \longrightarrow f$ in both $L^1(\mathbb{R}^N;\mathbb{C})$ and $L^2(\mathbb{R}^N;\mathbb{C})$. By Corollary 7.4.19 and (7.4.1), we know that, for each $k \in \mathbb{Z}^+$,

$$\sum_{\|\mathbf{m}\|\le M} (\sqrt{-1})^{\|\mathbf{m}\|}(f_k, h_{\mathbf{m}})_{L^2(\mathbb{R}^N;\mathbb{C})} h_{\mathbf{m}}(\boldsymbol{\xi}) \longrightarrow \hat{f}_k(\boldsymbol{\xi}) \quad \text{as } M \to \infty$$

uniformly in $\boldsymbol{\xi} \in \mathbb{R}^N$. At the same time, we know that

$$\sum_{\|\mathbf{m}\|\le M} (\sqrt{-1})^{\|\mathbf{m}\|}(f_k, h_{\mathbf{m}})_{L^2(\mathbb{R}^N;\mathbb{C})} h_{\mathbf{m}} \longrightarrow \mathfrak{F}f$$

in $L^2(\mathbb{R}^N;\mathbb{C})$. Hence, for each $k \in \mathbb{Z}^+$, $\hat{f}_k \in L^2(\mathbb{R}^N;\mathbb{C})$ and, as an element of $L^2(\mathbb{R}^N;\mathbb{C})$, $\hat{f}_k = \mathfrak{F}f_k$. Now let $k \to \infty$. Then, $\hat{f}_k \longrightarrow \hat{f}$ uniformly while $\mathfrak{F}f_k \longrightarrow \mathfrak{F}f$ in $L^2(\mathbb{R}^N;\mathbb{C})$. In particular, $\hat{f} \in L^2(\mathbb{R}^N;\mathbb{C})$ and, as an element of $L^2(\mathbb{R}^N;\mathbb{C})$, $\hat{f} = \mathfrak{F}f$.

Having proved that $\hat{f} = \mathfrak{F}f$ when $f \in L^1(\mathbb{R}^N;\mathbb{C}) \cap L^2(\mathbb{R}^N;\mathbb{C})$, it is now clear that $f \rightsquigarrow \mathfrak{F}f$ is the one and only extension of $f \rightsquigarrow \hat{f}$ as a continuous map on $L^2(\mathbb{R}^N;\mathbb{C})$ into itself. Thus, by taking $f_R = \mathbf{1}_{B_{\mathbb{R}^N}(\mathbf{0},R)} f \in L^1(\mathbb{R}^N;\mathbb{C}) \cap L^2(\mathbb{R}^N;\mathbb{C})$ and noting that, as $R \nearrow \infty$, $f_R \longrightarrow f$ in $L^2(\mathbb{R}^N;\mathbb{C})$, we get the first line of (7.5.4). Finally, to get the second line, note that $\hat{f}_R(-\boldsymbol{\xi}) = \widehat{\overline{f_R}}(\boldsymbol{\xi})$. □

Warning: The reader would do well to notice that it is not possible to express $\mathfrak{F}f$ as a Lebesgue integral for general $f \in L^2(\mathbb{R}^N;\mathbb{C})$. Indeed, the situation here is very much like the one treated in Corollary 6.2.18 in the sense that, unless $f \in L^1(\mathbb{R}^N;\mathbb{C}) \cap L^2(\mathbb{R}^N;\mathbb{C})$, $\mathfrak{F}f$ has to be defined by an extension procedure based on *a priori* continuity considerations.

The fact the $\mathfrak{F}$ is an isometry and therefore that

$$\text{(7.5.5)}\qquad \big(f,g\big)_{L^2(\mathbb{R}^N;\mathbb{C})} = \big(\mathfrak{F}f,\mathfrak{F}g\big)_{L^2(\mathbb{R}^N;\mathbb{C})} \quad \text{for } f,g \in L^2(\mathbb{R}^N;\mathbb{C})$$

is usually called **Parseval's identity**, and it is Parseval's identity which makes $L^2(\mathbb{R}^N;\mathbb{C})$ the *natural* place in which to do Fourier analysis. To give a feeling for the sort of applications which have made the Fourier transform the single most powerful tool in modern analysis, we will make use of the following simple but powerful observation.

7.5.6 Lemma. *Given $f \in L^2(\mathbb{R}^N;\mathbb{C})$ and $g \in L^1(\mathbb{R}^N;\mathbb{C})$, $\mathfrak{F}(f * g) = \hat{g}\mathfrak{F}f$. Also, if $f \in C^1(\mathbb{R}^N;\mathbb{C}) \cap L^2(\mathbb{R}^N;\mathbb{C})$ and $\frac{\partial f}{\partial x_j} \in L^2(\mathbb{R}^N;\mathbb{C})$ for some $1 \leq j \leq N$, then $\xi_j \mathfrak{F}f \in L^2(\mathbb{R}^N;\mathbb{C})$ and*

$$\mathfrak{F}\frac{\partial f}{\partial x_j}(\boldsymbol{\xi}) = \frac{2\pi\xi_j}{\sqrt{-1}}\mathfrak{F}f(\boldsymbol{\xi}).$$

PROOF: Choose $\eta \in C^\infty\big(\mathbb{R}^N;[0,1]\big)$ so that $\eta \equiv 1$ on $B_{\mathbb{R}^N}(\mathbf{0};1)$ and $\eta \equiv 0$ off of $B_{\mathbb{R}^N}(\mathbf{0};2)$, and set $\eta_R(\mathbf{x}) = \eta(R^{-1}\mathbf{x})$, $\mathbf{x} \in \mathbb{R}^N$, for $R > 0$.

Now let $f \in L^2(\mathbb{R}^N;\mathbb{C})$ be given, and set $f_R = \eta_R f$. Clearly, $f_R \in L^1(\mathbb{R}^N;\mathbb{C}) \cap L^2(\mathbb{R}^N;\mathbb{C})$ and $f_R \longrightarrow f$ in $L^2(\mathbb{R}^N;\mathbb{C})$. Hence (cf. Theorem 6.3.1), if $g \in L^1(\mathbb{R}^N;\mathbb{C})$, then

$$L^1(\mathbb{R}^N;\mathbb{C}) \cap L^2(\mathbb{R}^N;\mathbb{C}) \ni f_R * g \longrightarrow f * g \text{ in } L^2(\mathbb{R}^N;\mathbb{C}),$$

which, by Exercise 7.3.12 and Theorem 7.5.3, means that

$$\widehat{f_R * g} = \widehat{f_R}\,\hat{g} \longrightarrow \hat{g}\mathfrak{F}f \text{ in } L^2(\mathbb{R}^N;\mathbb{C}).$$

Finally, suppose that f is as in the second part of the statement, and again take $f_R = \eta_R f$. Obviously, both f_R and $\frac{\partial f_R}{\partial x_j}$ are in $L^1(\mathbb{R}^N;\mathbb{C}) \cap L^2(\mathbb{R}^N;\mathbb{C})$. Moreover,

$$\frac{\partial f_R}{\partial x_j}(\mathbf{x}) = R^{-1}\eta'(R^{-1}\mathbf{x})f(\mathbf{x}) + \eta_R(\mathbf{x})\frac{\partial f}{\partial x_j}(\mathbf{x}) \longrightarrow \frac{\partial f}{\partial x_j}(\mathbf{x})$$

in $L^2(\mathbb{R}^N;\mathbb{C})$. Hence, it suffices for us to show that

$$\widehat{\frac{\partial f_R}{\partial x_j}} \longrightarrow \frac{2\pi\xi_j}{\sqrt{-1}}\mathfrak{F}f(\boldsymbol{\xi}) \text{ in } \lambda_{\mathbb{R}^N}\text{-measure.}$$

But f_R has compact support and therefore, either by Fubini's Theorem plus integration by parts or by the Divergence Theorem, we know that

$$\int_{\mathbb{R}^N} \frac{\partial}{\partial x_j}\Big(\mathfrak{e}\big((\boldsymbol{\xi},\mathbf{x})_{\mathbb{R}^N}\big)f(\mathbf{x})\Big)\,dx = 0.$$

Hence,

$$\widehat{\frac{\partial f_R}{\partial x_j}}(\boldsymbol{\xi}) = \frac{2\pi\xi_j}{\sqrt{-1}}\widehat{f_R}(\boldsymbol{\xi}) \longrightarrow \frac{2\pi\xi_j}{\sqrt{-1}}\mathfrak{F}f(\boldsymbol{\xi}) \text{ in } \lambda_{\mathbb{R}^N}\text{-measure}$$

because $\widehat{f_R} \longrightarrow \mathfrak{F}f$ in $L^2(\mathbb{R}^N;\mathbb{C})$. □

Both parts of Lemma 7.5.6 are manifestations of the same basic fact: *translation invariant operations are given by multiplication on the Fourier transform side.* To wit, convolution with g becomes multiplication by $\hat{g}$ and partial differentiation in a coordinate direction becomes multiplication by $-\sqrt{-1}\,2\pi$ times the corresponding coordinate. To see the potential power of these observations, consider the problem of solving the **wave equation**

$$\frac{\partial^2 u}{\partial t^2} = \Delta u \quad \text{with } \lim_{t\searrow 0} u(t,\,\cdot\,) = f_0 \text{ and } \lim_{t\searrow 0} \frac{\partial u}{\partial t} = f_1,$$

where $f_0,\ f_1 \in L^2(\mathbb{R}^N;\mathbb{C})$. Proceeding somewhat formally, suppose that u is a solution, and set $v(t,\,\cdot\,) = \mathfrak{F}u(t,\,\cdot\,)$. On the basis of the second part of Lemma 7.5.6, we would expect that

$$\frac{\partial^2 v}{\partial t^2}(\boldsymbol{\xi}) = \widehat{\Delta u(t,\,\cdot\,)}(\boldsymbol{\xi}) = -(2\pi)^2|\boldsymbol{\xi}|^2 v(t,\boldsymbol{\xi}).$$

Hence, by viewing this as an ordinary differential equation in t for each fixed $\boldsymbol{\xi} \in \mathbb{R}^N$, we find that

$$v(t,\boldsymbol{\xi}) = \cos\bigl(2\pi|\boldsymbol{\xi}|t\bigr)\mathfrak{F}f_0(\boldsymbol{\xi}) + \sin\bigl(2\pi|\boldsymbol{\xi}|t\bigr)\mathfrak{F}f_1(\boldsymbol{\xi}).$$

Notice that, in spite of our having ignored questions of justification, our formal computation has led us to a rigorously defined candidate for a solution. Namely, because

$$\boldsymbol{\xi} \rightsquigarrow \cos\bigl(2\pi|\boldsymbol{\xi}|t\bigr)\mathfrak{F}f_0(\boldsymbol{\xi}) + \sin\bigl(2\pi|\boldsymbol{\xi}|t\bigr)\mathfrak{F}f_1(\boldsymbol{\xi})$$

is square-integrable, we can now take

$$u(t,\mathbf{x}) = \bigl[\mathfrak{F}^{-1}v(t,\,\cdot\,)\bigr](\mathbf{x})$$
$$= \int_{\mathbb{R}^N} \mathfrak{e}\bigl(-(\boldsymbol{\xi},\mathbf{x})_{\mathbb{R}^N}\bigr)\Bigl(\cos\bigl(2\pi|\boldsymbol{\xi}|t\bigr)\mathfrak{F}f_0(\boldsymbol{\xi}) + \sin\bigl(2\pi|\boldsymbol{\xi}|t\bigr)\mathfrak{F}f_1(\boldsymbol{\xi})\Bigr)\,d\boldsymbol{\xi}.$$

It turns out that, in a suitable sense, the preceding expression for u is the unique solution to the wave equation with the given initial data. Of course, although our expression looks "explicit," a great deal of work is required before it can be considered truly useful. For instance, it is not at all clear how one can derives Huygen's principle[3] from this expression. On the other hand, certain questions about u can be answered on the basis of this representation. In particular, an application of a line of reasoning which was systematically developed by Sobolev, one can say something about the smoothness of u. A primative example of Sobolev's thinking is contained in the theorem which follows.

7.5.7 Theorem. *Suppose that $f \in L^2(\mathbb{R}^N;\mathbb{C})$ and that $\bigl(1+|\boldsymbol{\xi}|^2\bigr)^{\frac{n}{2}}\mathfrak{F}f \in L^1(\mathbb{R}^N;\mathbb{C})$ for some $n \in \mathbb{N}$. Then there exists an $\tilde{f} \in C^n_{\mathrm{b}}(\mathbb{R}^N;\mathbb{C})$ such that*

[3] The fact that light passes you in odd dimensions.

$f = \tilde{f}$ Lebesgue almost everywhere. In fact, for each multi-index[4] $\boldsymbol{\alpha} \in \mathbb{N}^N$ with $\|\boldsymbol{\alpha}\| \le n$,

$$\partial^{\boldsymbol{\alpha}} \tilde{f} = \check{\varphi}_{\boldsymbol{\alpha}} = \mathfrak{F}^{-1}\varphi_{\boldsymbol{\alpha}} \quad \text{where } \varphi_{\boldsymbol{\alpha}}(\boldsymbol{\xi}) \equiv \left(\frac{2\pi \mathbf{x}}{\sqrt{-1}}\right)^{\boldsymbol{\alpha}} \mathfrak{F}f(\boldsymbol{\xi}). \tag{7.5.8}$$

PROOF: For $R > 0$, set $\psi_R = \mathbf{1}_{B_{\mathbb{R}^N}(\mathbf{0},R)}\mathfrak{F}f$ and $f_R = \check{\psi}_R$. Because ψ_R is integrable and has compact support, it is easy to see that $f_R \in C^\infty_{\mathrm{b}}(\mathbb{R}^N;\mathbb{C})$ and

$$\partial^{\boldsymbol{\alpha}} f_R(\mathbf{x}) = \int_{\mathbb{R}^N} \left(-\sqrt{-1}\,2\pi\boldsymbol{\xi}\right)^{\boldsymbol{\alpha}} \psi_R(\boldsymbol{\xi}) \mathfrak{e}\left(-(\boldsymbol{\xi},\mathbf{x})_{\mathbb{R}^N}\right) d\boldsymbol{\xi}.$$

Since, as long as $\|\boldsymbol{\alpha}\| \le n$,

$$\left(-\sqrt{-1}\,2\pi\boldsymbol{\xi}\right)^{\boldsymbol{\alpha}} \psi_R(\boldsymbol{\xi}) \longrightarrow \varphi_{\boldsymbol{\alpha}} \quad \text{in } L^1(\mathbb{R}^N;\mathbb{C})$$

as $R \nearrow \infty$, it follows that $\partial^{\boldsymbol{\alpha}} f_R \longrightarrow \check{\psi}_{\boldsymbol{\alpha}}$ uniformly. At the same time, $\check{\psi}_R \longrightarrow f$ in $L^2(\mathbb{R}^N;\mathbb{C})$. Hence, if $\tilde{f} \equiv \check{\varphi}_{\mathbf{0}}$, then $\tilde{f} \in C^n_{\mathrm{b}}(\mathbb{R}^N;\mathbb{C})$ and $\partial^{\boldsymbol{\alpha}} \tilde{f} = \check{\varphi}_{\boldsymbol{\alpha}}$. $\square$

7.5.9 Corollary. *Let $f \in L^2(\mathbb{R}^N;\mathbb{C})$ be given. Then the following are equivalent:*

(1) *there exist $\tilde{f} \in C^\infty(\mathbb{R}^N;\mathbb{C})$ such that $f = \tilde{f}$ Lebesgue almost everywhere and $\mathbf{x}^{\boldsymbol{\alpha}}\partial^{\boldsymbol{\beta}}\tilde{f} \in L^2(\mathbb{R}^N;\mathbb{C})$ for all $(\boldsymbol{\alpha},\boldsymbol{\beta}) \in (\mathbb{N}^N)^2$,*
(2) *there exists a $\varphi \in C^\infty(\mathbb{R}^N;\mathbb{C})$ such that $\mathfrak{F}f = \varphi$ Lebesgue almost everywhere and $\boldsymbol{\xi}^{\boldsymbol{\alpha}}\partial^{\boldsymbol{\beta}}\varphi \in L^2(\mathbb{R}^N;\mathbb{C}) \cap C_{\mathrm{b}}(\mathbb{R}^N;\mathbb{C})$ for all $(\boldsymbol{\alpha},\boldsymbol{\beta}) \in (\mathbb{N}^N)^2$,*
(3) *there exist $\tilde{f} \in C^\infty(\mathbb{R}^N;\mathbb{C})$ such that $f = \tilde{f}$ Lebesgue almost surely and $\mathbf{x}^{\boldsymbol{\alpha}}\partial^{\boldsymbol{\beta}}\tilde{f} \in C_{\mathrm{b}}(\mathbb{R}^N;\mathbb{C})$ for all $(\boldsymbol{\alpha},\boldsymbol{\beta}) \in (\mathbb{N}^N)^2$.*

PROOF: We begin with two observations. First, if $\psi : \mathbb{R}^N \longrightarrow \mathbb{C}$ is measurable, then an application of Schwarz's inequality shows that

$$\mathbf{x}^{\boldsymbol{\alpha}}\psi \in L^2(\mathbb{R}^N;\mathbb{C}) \quad \text{for all } \boldsymbol{\alpha} \in \mathbb{N}^N \implies \mathbf{x}^{\boldsymbol{\alpha}}\psi \in L^1(\mathbb{R}^N;\mathbb{C}) \quad \text{for all } \boldsymbol{\alpha} \in \mathbb{N}^N.$$

Second, if $\psi \in C^\infty(\mathbb{R}^N;\mathbb{C})$, then Leibnitz's rule shows that

$$\begin{aligned} &\mathbf{x}^{\boldsymbol{\alpha}}\partial^{\boldsymbol{\beta}}\psi \in L^2(\mathbb{R}^N;\mathbb{C}) \quad \text{for all } (\boldsymbol{\alpha},\boldsymbol{\beta}) \in (\mathbb{N}^N)^2 \\ \iff &\partial^{\boldsymbol{\alpha}}\left(\mathbf{x}^{\boldsymbol{\beta}}\psi\right) \in L^2(\mathbb{R}^N;\mathbb{C}) \quad \text{for all } (\boldsymbol{\alpha},\boldsymbol{\beta}) \in (\mathbb{N}^N)^2. \end{aligned}$$

Now assume that (1) holds, and set $\varphi_{\boldsymbol{\beta}} = (-\sqrt{-1}\,2\pi\boldsymbol{\xi})^{\boldsymbol{\beta}}\mathfrak{F}f$. By repeated application of Lemma 7.5.6, $\varphi_{\boldsymbol{\beta}} = \mathfrak{F}\partial^{\boldsymbol{\beta}}\tilde{f}$. Equivalently,

$$\overline{\partial^{\boldsymbol{\beta}}\tilde{f}} = \mathfrak{F}\bar{\varphi}_{\boldsymbol{\beta}}.$$

[4] Recall the introduction of multi-indices in § 6.3, and introduce $\boldsymbol{\zeta}^{\boldsymbol{\alpha}}$ to stand for $\prod_{j=1}^N \zeta_j$.

But, by the first observation above, $(1+|\mathbf{x}|^2)^{\frac{n}{2}}\overline{\partial^{\beta}\tilde{f}} \in L^1(\mathbb{R}^N;\mathbb{C})$ for all $n \in \mathbb{N}$. Hence, by Theorem 7.5.7, there exist a $\tilde{\varphi}_{\boldsymbol{\beta}} \in C_{\mathrm{b}}^{\infty}(\mathbb{R}^N;\mathbb{C})$ such that $\varphi_{\boldsymbol{\beta}} = \tilde{\varphi}_{\boldsymbol{\beta}}$ almost everywhere and

$$\overline{\partial^{\boldsymbol{\alpha}}\varphi_{\boldsymbol{\beta}}} = \left((\overline{\partial^{\beta}\tilde{f}})_{\boldsymbol{\alpha}}\right)^{\vee}$$

where $(\partial^{\beta}\tilde{f})_{\boldsymbol{\alpha}} = (\sqrt{-1}\,2\pi\mathbf{x})^{\boldsymbol{\alpha}}\partial^{\beta}\tilde{f}$. In particular, this proves that $\partial^{\boldsymbol{\alpha}}\tilde{\varphi}_{\boldsymbol{\beta}} \in L^2(\mathbb{R}^N;\mathbb{C})\cap C_{\mathrm{b}}(\mathbb{R}^N;\mathbb{C})$. Finally, it is clear the $\tilde{\varphi}_{\boldsymbol{\beta}}$ must be equal to $(-\sqrt{-1}\,2\pi\boldsymbol{\xi})^{\boldsymbol{\beta}}\tilde{\varphi}$ where $\tilde{\varphi} = \tilde{\varphi}_{\mathbf{0}}$, and therefore, by the second observation made above, we are done.

To prove that (2) implies (3), we simply note that, because $\mathfrak{F}^{-1}\varphi = \overline{\mathfrak{F}\bar{\varphi}}$, the implication (2) $\implies$ (3) reduces to the implication (1) $\implies$ (2). Finally, the implication (3) $\implies$ (1) is trivial, since, for each $(\boldsymbol{\alpha},\boldsymbol{\beta}) \in (\mathbb{N}^N)^2$, $\mathbf{x}^{\boldsymbol{\alpha}}\partial^{\boldsymbol{\beta}}f$ is bounded and goes to 0 at infinity faster than $(1+|\mathbf{x}|^2)^{-n}$ for any $n \in \mathbb{N}$. □

The space of $f \in C^{\infty}(\mathbb{R}^N;\mathbb{C})$ with the property that $\mathbf{x}^{\boldsymbol{\alpha}}\partial^{\boldsymbol{\beta}}f$ is bounded for all $(\boldsymbol{\alpha},\boldsymbol{\beta}) \in (\mathbb{N}^N)^2$ is usually denoted by $\mathcal{S}(\mathbb{R}^N;\mathbb{C})$ and called the **Schwartz test function space**, in honor of Laurent Schwartz[5] who introduced it as the test functions class for tempered distributions. As Corollary 7.5.9 demonstrates, $\mathcal{S}(\mathbb{R}^N:\mathbb{C})$ is a subspace of $C^{\infty}(\mathbb{R}^N;\mathbb{C})\cap L^2(\mathbb{R}^N;\mathbb{C})$ which is invariant under the Fourier transform. In fact, a little thought leads to the conclusion that it is the largest such subspace.

Exercises

7.5.10 Exercise Using the estimate obtained in Exercise 7.4.22, show that $f \in \mathcal{S}(\mathbb{R}^N;\mathbb{C})$ if and only if $f \in C(\mathbb{R}^N;\mathbb{C})\cap L^2(\mathbb{R}^N;\mathbb{C})$ and

$$\sup_{\mathbf{n}\in\mathbb{N}^N} \|\mathbf{n}\|^k \left|(f,h_{\mathbf{n}})_{L^2(\mathbb{R}^N;\mathbb{C})}\right| < \infty \quad \text{for each } k \in \mathbb{N}, \tag{7.5.11}$$

in which case

$$\lim_{n\to\infty}\ \max_{\|\boldsymbol{\alpha}\|+\|\boldsymbol{\beta}\|\le k}\ \sup_{\mathbf{x}\in\mathbb{R}^N} |\mathbf{x}|^k \left|\partial^{\boldsymbol{\beta}}f(x) - \sum_{\|\mathbf{n}\|\le n}(f,h_{\mathbf{n}})_{L^2(\mathbb{R}^N;\mathbb{C})}\partial^{\boldsymbol{\beta}}h_{\mathbf{n}}(x)\right| = 0$$

for all $k \in \mathbb{N}$. Further, given $f \in C^{\infty}(\mathbb{R}^N;\mathbb{C})\cap L^2(\mathbb{R}^N;\mathbb{C})$, show that $f \in \mathcal{S}(\mathbb{R}^N;\mathbb{C})$ if and only if $\left(-\Delta + |2\pi\mathbf{x}|^2\right)^k f \in L^2(\mathbb{R}^N;\mathbb{C})$ for each $k \in \mathbb{N}$.

---

[5] Not the "Schwarz" of inequality fame.

Chapter VIII
A Little Abstract Theory

8.1 An Existence Theorem

In Chapter II we constructed Lebesgue measure on $\mathbb{R}^N$, and in ensuing chapters we saw how to construct various other measures from a given measure. However, as yet, we have not discussed any general procedure for the construction of measures *ab initio*; it is the purpose of the present section to provide such a procedure.

The basic idea behind what we will be doing appears to be due to F. Riesz and entails the reversal, in some sense, of the process by which we went in Chapter III from the existence of a measure to the existence of integrals. That is, we will suppose that we have at hand an *integral* and will attempt to show that it must have come from a measure. Thus, we must first describe what we mean by an *integral*.

Let E be a non-empty set. We will say that a subset $\mathbf{L}$ of the functions $f : E \longrightarrow \overline{\mathbb{R}}$ is a **lattice** if $f \wedge g$ and $f \vee g$ are both elements of $\in \mathbf{L}$ whenever f and g are. Given a lattice $\mathbf{L}$ of $\mathbb{R}$-valued functions, we will say that $\mathbf{L}$ is a **vector lattice** if the constant function $\mathbf{0}$ is an element of $\mathbf{L}$ and $\mathbf{L}$ is a vector space over $\mathbb{R}$. Note that if $\mathbf{L}$ is a vector space of $\mathbb{R}$-valued functions on E, then it is a vector lattice if and only if $f^+ \equiv f \vee 0 \in \mathbf{L}$ whenever $f \in \mathbf{L}$. Next, given a vector lattice $\mathbf{L}$, we will say that the mapping $I : \mathbf{L} \longrightarrow \mathbb{R}$ is an **integral on L** if

(**a**) I is linear,
(**b**) I is **non-negative** in the sense that $I(f) \geq 0$ for every non-negative $f \in \mathbf{L}$,
(**c**) $I(f_n) \searrow 0$ whenever $\{f_n\}_1^\infty \subseteq \mathbf{L}$ is a non-increasing sequence which tends (point-wise) to 0.

Finally, we will say that the triple $\mathcal{I} = (E, \mathbf{L}, I)$ is an **integration theory** if $\mathbf{L}$ is a vector lattice of functions $f : E \longrightarrow \mathbb{R}$ and I is an integral on $\mathbf{L}$.

8.1.1 Examples: Here are three situations to which the preceding notions apply.

(**i**) The basic model on which the preceding definitions are based is the one which comes from the integration theory for a measure space $(E, \mathcal{B}, \mu)$. Indeed, in that case, $\mathbf{L} = L^1(\mu)$ and $I(f) = \int f \, d\mu$.

(ii) A second basic source of integration theories is the one which comes from *finitely additive functions on an algebra.* That is, let $\mathcal{A}$ be an algebra of subsets of E and denote by $\mathbf{L}(\mathcal{A})$ the space of simple functions $f : E \longrightarrow \mathbb{R}$ with the property that $\{f = a\} \in \mathcal{A}$ for every $a \in \mathbb{R}$. It is then an easy matter to check that $\mathbf{L}(\mathcal{A})$ is a vector lattice. Now let $\mu : \mathcal{A} \longrightarrow [0, \infty)$ be **finitely additive** in the sense that

$$\mu(\Gamma_1 \cup \Gamma_2) = \mu(\Gamma_1) + \mu(\Gamma_2) \quad \text{for disjoint } \Gamma_1, \Gamma_2 \in \mathcal{A}.$$

Note that, since $\mu(\emptyset) = \mu(\emptyset \cup \emptyset) = 2\mu(\emptyset)$, $\mu(\emptyset)$ must be 0. Also, by proceeding in precisely the same way as we did (via Lemma 3.2.3) in the proof of Lemma 3.2.4 and then in the proof of Lemma 3.2.11, one can show that

$$f \in \mathbf{L}(\mathcal{A}) \longmapsto I(f) \equiv \sum_{a \in \text{Range}(f)} a\, \mu(\{f = a\})$$

is linear and non-negative. Finally, observe that I cannot be an integral unless μ has the property that

$$\mu(\Gamma_n) \searrow 0 \quad \text{whenever } \{\Gamma_n\}_1^\infty \subseteq \mathcal{A} \text{ decreases to } \emptyset. \tag{8.1.2}$$

On the other hand, if (8.1.2) holds and $\{f_n\}_1^\infty \subseteq \mathbf{L}(\mathcal{A})$ is a non-increasing sequence which tends pointwise to 0, then for each $\epsilon > 0$,

$$\varlimsup_{n\to\infty} I(f_n) \le \epsilon I(\mathbf{1}) + \|f_1\|_{\text{u}} \varlimsup_{n\to\infty} \mu(\{f_n > \epsilon\}) = \epsilon I(\mathbf{1}).$$

Thus, in this setting, (8.1.2) is equivalent to I being an integral.

(iii) A third important example of an integration theory is provided by the following abstraction of Riemann's theory. Namely, let E be a compact topological space, and note that $C(E;\mathbb{R})$ is a vector lattice. Next, suppose that $I : C(E;\mathbb{R}) \longrightarrow \mathbb{R}$ is a linear map which is non-negative. It is then clear that $|I(f)| \le C\|f\|_{\text{u},E}$, $f \in C(E;\mathbb{R})$, where $C = I(\mathbf{1})$ and $\|f\|_{\text{u},E} \equiv \sup_{x\in E}|f(x)|$ is the **uniform norm** of f on E. In particular, this means that $|I(f_n) - I(f)| \le C\|f_n - f\|_{\text{u},E} \longrightarrow 0$ if $f_n \longrightarrow f$ uniformly. Thus, to see that I is an integral, all that we have to do is use Dini's Lemma (cf. Lemma 8.1.23 below) which says that $f_n \longrightarrow 0$ uniformly on E if $\{f_n\}_1^\infty \subseteq C(E;\mathbb{R})$ decreases pointwise to 0.

Our main goal will be to show that, at least when $\mathbf{1} \in \mathbf{L}$, every integration theory is the sub-theory of the sort of theory described in **(i)** above. Thus, we must learn how to extract the *measure* μ from the *integral.* At least in case **(ii)** above, it is clear how one might begin such a procedure. Namely, $\mathcal{A} = \{\Gamma \subseteq E : \mathbf{1}_\Gamma \in \mathbf{L}(\mathcal{A})\}$ and $\mu(\Gamma) = I(\mathbf{1}_\Gamma)$ for $\Gamma \in \mathcal{A}$. Hence, what we are attempting to do in this case is tantamount to showing that μ can be extended as a measure to the σ-algebra $\sigma(\mathcal{A})$ generated by $\mathcal{A}$. On the other hand, it is not so immediately clear where to start looking for the measure μ in case **(iii)**;

the procedure which got us started in case (**ii**) does not work here since there will seldom be many $\Gamma \subseteq E$ for which $\mathbf{1}_\Gamma \in C(E;\mathbb{R})$. Thus, in any case, we must learn first how to extend I to a larger class of functions $f : E \longrightarrow \mathbb{R}$ and only then look for μ.

Our extension procedure has two steps, the first of which is nothing but a rerun of what we did in Section 3.2, and the second one is a minor variant on what we did in Section 2.1.

8.1.3 Lemma. *Let $(E, \mathbf{L}, I)$ be an integration theory, and define $\mathbf{L}_u$ to be the class of $f : E \longrightarrow (-\infty, \infty]$ which can be written as the pointwise limit of a non-decreasing sequence $\{\varphi_n\}_1^\infty \subseteq \mathbf{L}$. Then $\mathbf{L}_u$ is a lattice which is closed under non-negative linear operations and non-decreasing sequential limits (i.e., $\{f_n\}_1^\infty \subseteq \mathbf{L}_u$ and $f_n \nearrow f$ implies that $f \in \mathbf{L}_u$). Moreover, I admits a unique extension to $\mathbf{L}_u$ in such a way that $I(f_n) \nearrow I(f)$ whenever f is the limit of a non-decreasing $\{f_n\}_1^\infty \subseteq \mathbf{L}_u$. In particular, for all f, $g \in \mathbf{L}_u$, $-\infty < I(f) \leq I(g)$ if $f \leq g$ and $I(\alpha f + \beta g) = \alpha I(f) + \beta I(g)$ for all α, $\beta \in [0, \infty)$.*

PROOF: The closedness properties of $\mathbf{L}_u$ are obvious. Moreover, given that an extension of I with the stated properties exists, it is clear that that extension is unique, monotone, and linear under non-negative linear operations.

Just as in the development in Section 3.2 which eventually led to The Monotone Convergence Theorem, the proof (cf. Lemma 3.2.6) that I extends to $\mathbf{L}_u$ is simply a matter of checking that the desired extension of I is consistent. Thus, what we must show is that when $\psi \in \mathbf{L}$ and $\{\varphi_n\}_1^\infty \subseteq \mathbf{L}$ is a non-decreasing sequence with the property that $\psi \leq \lim_{n\to\infty} \varphi_n$ point-wise, then $I(\psi) \leq \lim_{n\to\infty} I(\varphi_n)$. To this end, note that $\varphi_n = \psi - (\psi - \varphi_n) \geq \psi - (\psi - \varphi_n)^+$, $L \ni (\psi - \varphi_n)^+ \searrow 0$, and therefore that

$$\lim_{n\to\infty} I(\varphi_n) \geq I(\psi) - \lim_{n\to\infty} I\big((\psi - \varphi_n)^+\big) = I(\psi).$$

As we said before, once one knows that I is consistently defined on $\mathbf{L}_u$, the rest of the proof differs in no way from the proof of The Monotone Convergence Theorem (cf. Theorem 3.3.2). □

Lemma 8.1.3 gives the first step in the extension of I. What it provides is a rich class functions which will be used to play the role that open sets played in our construction of Lebesgue's measure. Thus, given any $f : E \longrightarrow \overline{\mathbb{R}}$, we define

$$\overline{I}(f) = \inf\Big\{I(\varphi) : \varphi \in \mathbf{L}_u \text{ and } f \leq \varphi\Big\}. \tag{8.1.4}$$

(We use the convention that the infimum over the empty set is $+\infty$.) Clearly $\overline{I}(f)$ is the analog here of the outer measure $\Gamma \longmapsto |\Gamma|_e$ in Section 2.1. At the same time as we consider $\overline{I}$, it will be convenient to have

$$\underline{I}(f) = \sup\Big\{-I(\varphi) : \varphi \in \mathbf{L}_u \text{ and } -\varphi \leq f\Big\}; \tag{8.1.5}$$

the analog of which in Section 2.1 would have been the **interior measure**

$$\Gamma \longmapsto |\Gamma|_{\mathrm{i}} \equiv \sup\Big\{|F| : F \text{ is closed and } F \subseteq \Gamma\Big\}.$$

(In keeping with our convention about the infimum, we take the supremum over the empty set to be $-\infty$.)

8.1.6 Lemma. *For any $\overline{\mathbb{R}}$-valued function f on E*

$$\underline{I}(f) \le \overline{I}(f), \tag{8.1.7}$$

and

$$\begin{aligned} &\overline{I}(\alpha f) = \alpha \overline{I}(f) \text{ and } \underline{I}(\alpha f) = \alpha \underline{I}(f) \quad \text{if } \alpha \in [0,\infty) \\ &\overline{I}(\alpha f) = \alpha \underline{I}(f) \text{ and } \underline{I}(\alpha f) = \alpha \overline{I}(f) \quad \text{if } \alpha \in (-\infty, 0]. \end{aligned} \tag{8.1.8}$$

Moreover, if $(f,g) : E \longrightarrow \overline{\mathbb{R}}^2$, then

$$f \le g \implies \overline{I}(f) \le \overline{I}(g) \text{ and } \underline{I}(f) \le \underline{I}(g), \tag{8.1.9}$$

and, when (f,g) takes values in $\widehat{\mathbb{R}^2}$ (cf. Section 3.2),

$$\begin{aligned} &\big(\overline{I}(f), \overline{I}(g)\big) \in \widehat{\mathbb{R}^2} \implies \overline{I}(f+g) \le \overline{I}(f) + \overline{I}(g) \\ &\big(\underline{I}(f), \underline{I}(g)\big) \in \widehat{\mathbb{R}^2} \implies \underline{I}(f+g) \ge \underline{I}(f) + \underline{I}(g). \end{aligned} \tag{8.1.10}$$

Finally,

$$f \in \mathbf{L}_u \implies \overline{I}(f) = I(f) = \underline{I}(f). \tag{8.1.11}$$

PROOF: To prove (8.1.7), note first that it suffices to treat the case in which $\underline{I}(f) > -\infty$ and $\overline{I}(f) < \infty$ and then that, for any $(\varphi, \psi) \in \mathbf{L}_u^2$ with $-\varphi \le f \le \psi$, $\varphi + \psi \ge 0$ and therefore $I(\varphi) + I(\psi) = I(\varphi + \psi) \ge 0$.

Both (8.1.8) and (8.1.9) are obvious, and, because of (8.1.8), it suffices to prove only the first line in (8.1.10). Moreover, when $\overline{I}(f) \wedge \overline{I}(g) > -\infty$, the required result is easy. On the other hand, if $\overline{I}(f) = -\infty$ and $\overline{I}(g) < \infty$, then we can choose $\{\varphi_n\}_1^\infty \cup \{\psi\} \subseteq \mathbf{L}_u$ so that $f \le \varphi_n$ and $I(\varphi_n) \le -n$ for each $n \in \mathbb{Z}^+$, $g \le \psi$, and $I(\psi) < \infty$. In particular, $f + g \le \varphi_n + \psi$ for all $n \in \mathbb{Z}^+$, and so $\overline{I}(f+g) \le \lim_{n\to\infty} I(\varphi_n) + I(\psi) = -\infty$.

Finally, suppose that $f \in \mathbf{L}_u$. Obviously, $\overline{I}(f) \le I(f)$. At the same time, if $\{\varphi_n\}_1^\infty \subseteq \mathbf{L}$ is chosen so that $\varphi_n \nearrow f$, then (because $-\varphi_n \in \mathbf{L} \subseteq \mathbf{L}_u$ and $-(-\varphi_n) = \varphi_n \le f$ for each $n \in \mathbb{Z}^+$)

$$\underline{I}(f) \ge \lim_{n\to\infty} -I(-\varphi_n) = \lim_{n\to\infty} I(\varphi_n) = I(f). \quad \square$$

From now on, we will use $\mathfrak{M}(\mathcal{I})$ to denote the class of those $f : E \longrightarrow \overline{\mathbb{R}}$ for which $\overline{I}(f) = \underline{I}(f)$, and we define $\tilde{I} : \mathfrak{M}(\mathcal{I}) \longrightarrow \overline{\mathbb{R}}$ so that $\underline{I}(f) = \tilde{I}(f) = \overline{I}(f)$. Obviously (cf. (8.1.8))

$$(8.1.12) \qquad f \in \mathfrak{M}(\mathcal{I}) \implies \alpha f \in \mathfrak{M}(\mathcal{I}) \text{ and } \tilde{I}(\alpha f) = \alpha \tilde{I}(f), \quad \alpha \in \mathbb{R}.$$

Finally, let $L^1(\mathcal{I})$ denote the class of $\mathbb{R}$-valued $f \in \mathfrak{M}(\mathcal{I})$ with $\tilde{I}(f) \in \mathbb{R}$.

8.1.13 Lemma. *If $f : E \longrightarrow \mathbb{R}$, then $f \in L^1(\mathcal{I})$ if and only if, for each $\epsilon > 0$, there exist $\varphi, \psi \in \mathbf{L}_u$ such that $-\varphi \le f \le \psi$ and $I(\varphi) + I(\psi) < \epsilon$. In particular, $f \in L^1(\mathcal{I}) \implies f^+ \in L^1(\mathcal{I})$. Moreover, $L^1(\mathcal{I})$ is a vector space and $\tilde{I}$ is linear on $L^1(\mathcal{I})$. Finally, if $\{f_n\}_1^\infty \subseteq L^1(\mathcal{I})$ and $0 \le f_n \nearrow f$, then $f \in \mathfrak{M}(\mathcal{I})$ and $0 \le \tilde{I}(f_n) \nearrow \tilde{I}(f)$.*

PROOF: First suppose that $f \in L^1(\mathcal{I})$. Given $\epsilon > 0$, there exists $(\varphi, \psi) \in \mathbf{L}_u^2$ such that $-\varphi \le f \le \psi$, $\tilde{I}(f) \le -I(\varphi) + \frac{\epsilon}{2}$, and $I(\psi) \le \tilde{I}(f) + \frac{\epsilon}{2}$; from which it is clear that $I(\varphi) + I(\psi) \le \epsilon$. Conversely, suppose that $f : E \longrightarrow \mathbb{R}$ and that, for some $\epsilon \in (0, \infty)$, there exists $(\varphi, \psi) \in \mathbf{L}_u^2$ such that $-\varphi \le f \le \psi$ and $I(\varphi) + I(\psi) < \epsilon$. Because $I(\varphi) \wedge I(\psi) > -\infty$, $-\infty < I(\psi) < \epsilon - I(\varphi) < \infty$ and $-\infty < I(\varphi) < \epsilon - I(\psi) < \infty$. In addition, $-I(\varphi) \le \underline{I}(f) \le \overline{I}(f) \le I(\psi)$. Hence, not only are both $\underline{I}(f)$ and $\overline{I}(f)$ in $\mathbb{R}$, but also $\overline{I}(f) - \underline{I}(f) < \epsilon$, which completes the proof of the first assertion.

Next, suppose that $f \in L^1(\mathcal{I})$. To prove that $f^+ \in L^1(\mathcal{I})$, let $\epsilon \in (0, \infty)$ be given, and choose $(\varphi, \psi) \in \mathbf{L}_u^2$ so that $-\varphi \le f \le \psi$ and $I(\varphi) + I(\psi) < \epsilon$. Note that $-\varphi^- = \varphi \wedge 0 \in \mathbf{L}_u$, $\varphi^+ = \varphi \vee 0 \in \mathbf{L}_u$, and that $\varphi^- \le f^+ \le \psi^+$. Moreover, because $\varphi + \psi \ge 0$, it is easy to see that $-\varphi^- + \psi^+ \le \varphi + \psi$, and therefore we know that

$$I(-\varphi^-) + I(\psi^+) = I(-\varphi^- + \psi^+) \le I(\varphi + \psi) = I(\varphi) + I(\psi) < \epsilon.$$

To see that $L^1(\mathcal{I})$ is a vector space and that $\tilde{I}$ is linear there, simply apply (8.1.8) and (8.1.10). Finally, let $\{f_n\}_1^\infty$ be a non-decreasing sequence of non-negative elements of $L^1(\mathcal{I})$, and set $f = \lim_{n\to\infty} f_n$. Obviously, $\lim_{n\to\infty} \tilde{I}(f_n) \le \underline{I}(f) \le \overline{I}(f)$. Thus, all that we have to do is prove that $\overline{I}(f) \le \lim_{n\to\infty} \tilde{I}(f_n)$. To this end, set $h_1 = f_1$, $h_n = f_n - f_{n-1}$ for $n \ge 2$, and note that each h_n is a non-negative element of $L^1(\mathcal{I})$. Next, given $\epsilon > 0$, choose, for each $m \in \mathbb{Z}^+$, $\psi_m \in \mathbf{L}_u$ so that $h_m \le \psi_m$ and $I(\psi_m) \le \tilde{I}(h_m) + 2^{-m}\epsilon$. Clearly, $\psi \equiv \sum_1^\infty \psi_m \in \mathbf{L}_u$ and $f \le \psi$. Thus, $\overline{I}(f) \le I(\psi)$. Moreover, by Lemma 8.1.3 and the linearity of $\tilde{I}$ on $L^1(\mathcal{I})$,

$$\begin{aligned} I(\psi) = \lim_{n\to\infty} I\left(\sum_1^n \psi_m\right) &= \lim_{n\to\infty} \sum_1^n I(\psi_m) \\ &\le \lim_{n\to\infty} \sum_1^n \tilde{I}(h_m) + \epsilon = \lim_{n\to\infty} \tilde{I}(f_n) + \epsilon. \quad \square \end{aligned}$$

8.1.14 Theorem (Daniell). *Let $\mathcal{I} = (E, \mathbf{L}, I)$ be an integration theory. Then $\tilde{\mathcal{I}} = \big(E, L^1(\mathcal{I}), \tilde{I}\,\big)$ is again an integration theory, $\mathbf{L} \subseteq L^1(\mathcal{I})$, and $\tilde{I}$ agrees with I on $\mathbf{L}$. Moreover, if $\{f_n\}_1^\infty \subseteq L^1(\mathcal{I})$ is non-decreasing and $f_n \nearrow f$, then $f \in L^1(\mathcal{I})$ if and only if $\sup_n f(x) < \infty$ for each $x \in E$ and $\sup_n \tilde{I}(f_n) < \infty$, in which case $\tilde{I}(f_n) \nearrow \tilde{I}(f)$.*

PROOF: In order to prove the first assertion, note that, by Lemma 8.1.13, $L^1(\mathcal{I})$ is a vector lattice and $\tilde{I}$ is linear there. Moreover, by (8.1.11), $\mathbf{L} \subseteq L^1(\mathcal{I})$ and $\tilde{I} \restriction \mathbf{L} = I \restriction \mathbf{L}$; and, by (8.1.9), $\tilde{I}(f) \geq I(\mathbf{0}) = 0$ if $f \geq 0$. Finally, if $\{f_n\}_1^\infty \subseteq L^1(\mathcal{I})$ and $f_n \nearrow f$, then, by the last part of Lemma 8.1.13 applied to $\{f_n - f_1\}_1^\infty$, $f - f_1 \in \mathfrak{M}(\mathcal{I})$,

$$\tilde{I}(f_n) = \tilde{I}(f_1) + \tilde{I}(f_n - f_1) \nearrow \tilde{I}(f_1) + \tilde{I}(f - f_1).$$

But, by (8.1.10), $f = f_1 + (f - f_1) \in \mathfrak{M}(\mathcal{I})$ and $\tilde{I}(f) = \tilde{I}(f_1) + \tilde{I}(f - f_1)$. Hence, the last assertion is now proved. In addition, if $\{f_n\}_1^\infty \subseteq L^1(\mathcal{I})$ and $f_n \searrow 0$, then $\tilde{I}(f_n) \searrow 0$ follows from the preceding applied to $\{-f_n\}_1^\infty$, which completes the proof that $\tilde{\mathcal{I}}$ is an integration theory. □

We are now ready to return to the problem, raised in the discussion following Examples 8.1.1, of identifying the measure underlying a given integration theory $(E, \mathbf{L}, I)$. For this purpose, we introduce the notation $\sigma(\mathbf{L})$ to denote the smallest σ-algebra over E with respect to which all of the functions in the vector lattice $\mathbf{L}$ are measurable. Obviously, $\sigma(\mathbf{L})$ is generated by the sets $\{f > a\}$ as f runs over $\mathbf{L}$ and a runs over $\mathbb{R}$.

8.1.15 Theorem (Stone). *Let $(E, \mathbf{L}, I)$ be an integration theory, and assume that $\mathbf{1} \in \mathbf{L}$. Then*

$$\sigma\big(L^1(\mathcal{I})\big) = \Big\{\Gamma \subseteq E : \mathbf{1}_\Gamma \in L^1(\mathcal{I})\Big\}, \tag{8.1.16}$$

the mapping

$$\Gamma \in \sigma\big(L^1(\mathcal{I})\big) \longmapsto \mu_I(\Gamma) \equiv \tilde{I}\big(\mathbf{1}_\Gamma\big) \in [0, \infty) \tag{8.1.17}$$

is a finite measure on $\big(E, \sigma(L^1(\mathcal{I}))\big)$, $\sigma\big(L^1(\mathcal{I})\big)$ is the completion of $\sigma(\mathbf{L})$ with respect to μ_I, $L^1(\mu_I) = L^1(\mathcal{I})$, and

$$\tilde{I}(f) = \int_E f\, d\mu_I, \qquad f \in L^1(\mathcal{I}). \tag{8.1.18}$$

Finally, if $\big(E, \mathcal{B}, \nu\big)$ is any finite measure space with the properties that $\mathbf{L} \subseteq L^1(E, \mathcal{B}, \nu)$ and $I(f) = \int_E f\, d\nu$ for every $f \in \mathbf{L}$, then $\sigma(\mathbf{L}) \subseteq \mathcal{B}$ and ν coincides with μ_I on $\sigma(\mathbf{L})$.

PROOF: Let $\mathcal{H}$ denote the collection described on the right hand side of equation (8.1.16). Using Theorem 8.1.14, one can easily show that $\mathcal{H}$ is a σ-algebra over E and that

$$\Gamma \in \mathcal{H} \longmapsto \mu_I(\Gamma) \equiv \tilde{I}(\mathbf{1}_\Gamma) \in [0,\infty)$$

defines a finite measure on $(E,\mathcal{H})$. Our first goal is to prove that

$$(8.1.19) \qquad L^1(\mathcal{I}) = L^1(E,\mathcal{H},\mu_I) \quad \text{and} \quad \tilde{I}(f) = \int_E f\, d\mu_I \text{ for all } f \in L^1(\mathcal{I}).$$

To this end, for given $f : E \longrightarrow \mathbb{R}$ and $a \in \mathbb{R}$, consider the functions

$$g_n \equiv \big[n(f - f \wedge a)\big] \wedge \mathbf{1}, \quad n \in \mathbb{Z}^+.$$

If $f \in L^1(\mathcal{I})$, then each g_n is also an element of $L^1(\mathcal{I})$, $g_n \nearrow \mathbf{1}_{\{f>a\}}$ as $n \longrightarrow \infty$, and therefore $\mathbf{1}_{\{f>a\}} \in L^1(\mathcal{I})$. Thus, we see that every $f \in L^1(\mathcal{I})$ is $\mathcal{H}$-measurable. Next, for given $f : E \longrightarrow [0,\infty)$, define

$$f_n = \sum_{k=0}^{4^n} \frac{k}{2^n} \mathbf{1}_{\{k<2^n f \le k+1\}} \quad \text{for} \quad n \in \mathbb{Z}^+.$$

If $f \in L^1(\mathcal{I}) \cup L^1(E,\mathcal{H},\mu_I)$, then (cf. the preceding and use linearity) $f_n \in L^1(\mathcal{I}) \cap L^1(\mu_I)$, $f_n \nearrow f$, and so $f \in L^1(\mathcal{I}) \cap L^1(E,\mathcal{H},\mu_I)$ and $\tilde{I}(f) = \int_E f\, d\mu_I$. Hence, we have now proved (8.1.19).

Our next goal is to show that

$$(8.1.20) \qquad \overline{\sigma(\mathbf{L})}^{\mu_I} = \mathcal{H} = \sigma\big(L^1(\mathcal{I})\big).$$

Since $\mathbf{L} \subseteq L^1(\mathcal{I})$ and every element of $L^1(\mathcal{I})$ is $\mathcal{H}$-measurable, what we know so far is that

$$\sigma(\mathbf{L}) \subseteq \sigma\big(L^1(\mathcal{I})\big) \subseteq \mathcal{H}.$$

Thus, to prove (8.1.20), all we need to do is show that

$$(8.1.21) \qquad \overline{\mathcal{H}}^{\mu_I} \subseteq \overline{\sigma(\mathbf{L})}^{\mu_I}.$$

But if $\Gamma \in \overline{\mathcal{H}}^{\mu_I}$, then there exist $A,\, B \in \mathcal{H}$ such that $A \subseteq \Gamma \subseteq B$ and $\tilde{I}(\mathbf{1}_A) = \mu_I(A) = \mu_I(B) = \tilde{I}(\mathbf{1}_B)$. Hence, we can choose sequences $\{\varphi_n\}_1^\infty$ and $\{\psi_n\}_1^\infty$ from $\mathbf{L}_u \cap L^1(\mathcal{I})$ so that $-\varphi_n \le \mathbf{1}_A \le \mathbf{1}_\Gamma \le \mathbf{1}_B \le \psi_n$, $-I(\varphi_n) \nearrow \tilde{I}(\mathbf{1}_A)$, and $I(\psi_n) \searrow \tilde{I}(\mathbf{1}_B)$. Further, after replacing φ_n and ψ_n by $\varphi_1 \wedge \cdots \wedge \varphi_n$ and $\psi_1 \wedge \cdots \wedge \psi_n$, we may and will assume that each of these sequences is non-increasing. Next, take $\varphi = \lim_{n\to\infty} \varphi_n$ and $\psi = \lim_{n\to\infty} \psi_n$. Note that φ and ψ are in $L^1(\mathcal{I})$, $-\varphi \le \mathbf{1}_\Gamma \le \psi$, and $-\tilde{I}(\varphi) = \tilde{I}(\psi)$. To complete the proof that $\Gamma \in \overline{\sigma(\mathbf{L})}^{\mu_I}$ from here, first note that every element of $\mathbf{L}_u$ is $\sigma(\mathbf{L})$-measurable. Hence, both φ and ψ are $\sigma(\mathbf{L})$-measurable, and so both the sets $C = \{\varphi < 0\}$ and $D = \{\psi \ge 1\}$ are elements of $\sigma(\mathbf{L})$. Finally, from $-\varphi \le \mathbf{1}_\Gamma \le \psi$ and

$-\tilde{I}(\varphi) = \tilde{I}(\psi)$, it is easy to check that $-\varphi \le \mathbf{1}_C \le \mathbf{1}_\Gamma \le \mathbf{1}_D \le \psi$ and therefore that $\mu_I(D \setminus C) = \mu_I(D) - \mu_I(C) \le \tilde{I}(\psi) + \tilde{I}(\varphi) = 0$, which means that $\Gamma \in \overline{\sigma(\mathbf{L})}^{\mu_I}$. Hence, we have now proved (8.1.21) and therefore (8.1.20).

We have now completed the proof of everything except the concluding assertion of uniqueness. But if $\mathbf{L} \subseteq L^1(E, \mathcal{B}, \nu)$, then obviously $\sigma(\mathbf{L}) \subseteq \mathcal{B}$. Moreover, if $\int f\, d\nu = I(f)$ for all $f \in \mathbf{L}$, then we can prove that $\nu \restriction \sigma(\mathbf{L}) = \mu_I \restriction \sigma(\mathbf{L})$ as follows. Namely, it is clear that $\sigma(\mathbf{L})$ is generated by the π-system of sets Γ of the form

$$\Gamma = \{f_1 > a_1, \ldots, f_\ell > a_\ell\},$$

where $\ell \in \mathbb{Z}^+$, $\{a_1, \ldots, a_\ell\} \subseteq \mathbb{R}$, and $\{f_1, \ldots, f_\ell\} \subseteq L$. Thus, by Exercise 3.1.8, we need only check that ν agrees with μ_I on such sets Γ. To this end, define

$$g_n = \left[n \min_{1 \le k \le \ell} (f_k - f_k \wedge a_k) \right] \wedge 1,$$

note that $\{g_n\}_1^\infty \subseteq \mathbf{L}$ and $0 \le g_n \nearrow \mathbf{1}_\Gamma$, and conclude that

$$\nu(\Gamma) = \lim_{n \to \infty} \int g_n\, d\nu = \lim_{n \to \infty} \int g_n\, d\mu_I = \mu_I(\Gamma). \quad \square$$

With the preceding result, we are now ready to handle the situations described in **(ii)** and **(iii)** of Examples 8.1.1.

8.1.22 Corollary (Carathéodory Extension). *Let $\mathcal{A}$ be an algebra of subsets of E, and suppose that $\mu : \mathcal{A} \longrightarrow [0, \infty)$ is a finitely additive function (cf.* **(ii)** *in* Examples *8.1.1) with the property that* (8.1.2) *holds. Then there is a unique finite measure $\tilde{\mu}$ on $(E, \sigma(\mathcal{A}))$ with the property that $\tilde{\mu}$ coincides with μ on $\mathcal{A}$.*

PROOF: Define $\mathbf{L}(\mathcal{A})$ and I on $\mathbf{L}(\mathcal{A})$ as in **(ii)** of Examples 8.1.1. It is then an easy matter to see that $\sigma(\mathcal{A}) = \sigma(\mathbf{L}(\mathcal{A}))$. In addition, as was shown in **(ii)** of Examples 8.1.1, I is an integral on $\mathbf{L}(\mathcal{A})$. Hence the desired existence and uniqueness statements follow immediately from Theorem 8.1.15. $\square$

Before we can complete **(iii)** in Examples 8.1.1, we must first prove the lemma alluded to there.

8.1.23 Lemma (Dini's Lemma). *Let $\{f_n\}_1^\infty$ be a non-increasing sequence of non-negative, continuous functions on the topological space E. If $f_n \searrow 0$, then $f_n \longrightarrow 0$ uniformly on each compact subset $K \subseteq E$ (i.e., $\|f_n\|_{\mathrm{u},K} \longrightarrow 0$.)*

PROOF: Without loss of generality, we assume that E itself is compact.

Let $\epsilon > 0$ be given. By assumption, we can find for each $x \in E$ an $n(x) \in \mathbb{Z}^+$ and an open neighborhood $U(x)$ of x such that $f_{n(x)}(y) \le \epsilon$ for all $y \in U(x)$. Moreover, by the Heine-Borel Theorem, we can choose a finite set $\{x_1, \ldots, x_L\} \subseteq E$ so that $E = \bigcup_{\ell=1}^L U(x_\ell)$. Thus, if $N(\epsilon) = n(x_1) \vee \cdots \vee n(x_L)$, then $f_n \le \epsilon$ as long as $n \ge N(\epsilon)$. $\square$

Given a topological space E, let $C_b(E;\mathbb{R})$ denote the space of bounded continuous functions on E, and turn $C_b(E;\mathbb{R})$ into a metric space by defining $\|g-f\|_{u,E}$ to be the distance between f and g. We will say that $\Lambda : C_b(E;\mathbb{R}) \longrightarrow \mathbb{R}$ is a **non-negative linear functional** if Λ is linear and $\Lambda(f) \geq 0$ for all $f \in C_b(E;[0,\infty))$. Furthermore, if Λ is a non-negative linear functional on $C_b(E;\mathbb{R})$, we will say that Λ is **tight** if it has the property that for every $\delta > 0$ there is a compact $K_\delta \subseteq E$ and an $A_\delta \in (0,\infty)$ for which

$$|\Lambda(f)| \leq A_\delta \|f\|_{u,K_\delta} + \delta\|f\|_{u,E} \quad \text{for all } f \in C_b(E;\mathbb{R}).$$

Notice that when E is itself compact, then every non-negative linear functional on $C_b(E;\mathbb{R})$ is tight.

8.1.24 Theorem (Riesz Representation). *Let E be a topological space, set $\mathcal{B} = \sigma\big(C_b(E;\mathbb{R})\big)$, and suppose that $\Lambda : C_b(E;\mathbb{R}) \longrightarrow \mathbb{R}$ is a non-negative linear functional which is tight. Then there is a unique finite measure μ on $(E,\mathcal{B})$ with the property that $\Lambda(f) = \int_E f\,d\mu$, $f \in C_b(E;\mathbb{R})$.*

PROOF: Clearly, all that we need to do is show that $\Lambda(f_n) \searrow 0$ whenever $\{f_n\}_1^\infty \subseteq C_b(E;\mathbb{R})$ is a non-increasing sequence which tends (pointwise) to 0. To this end, let $\epsilon > 0$ be given, set $\delta = \frac{\epsilon}{1+2\|f_1\|_{u,E}}$, and use Dini's Lemma to choose an $N(\delta) \in \mathbb{Z}^+$ so that $\|f_n\|_{u,K_\delta} \leq \frac{\epsilon}{2A_\delta}$ for all $n \geq N(\delta)$, where K_δ and A_δ are the quantities appearing in the tightness condition for Λ. Then, for $n \geq N(\delta)$, $|\Lambda(f_n)| \leq \epsilon$. □

The importance of Theorem 8.1.24 is hard to miss. Indeed, it seems to say that it is essentially impossible to avoid Lebesgue's theory of integration. On the other hand, it may not be evident how one might use Corollary 8.1.22. For this reason we close this section with a typical and important application of Carathéodory's Theorem. Namely, for each i from a non-empty index set $\mathfrak{I}$, let $(E_i, \mathcal{B}_i, \mu_i)$ be a probability space. Given $\emptyset \neq S \subseteq \mathfrak{I}$, set $\mathbf{E}_S = \prod_{i\in S} E_i$, and use Π_S to denote the natural projection map from $\mathbf{E} \equiv \mathbf{E}_\mathfrak{I}$ onto $\mathbf{E}_S$. Finally, let $\mathfrak{F}$ stand for the set of all non-empty, finite subsets F of $\mathfrak{I}$, and denote by $\mathcal{B}_\mathfrak{I}$ the σ algebra over $\mathbf{E}$ generated by sets of the form

$$\mathbf{\Gamma}_F \equiv \Pi_F^{-1}\left(\prod_{i\in F}\Gamma_i\right), \quad F \in \mathfrak{F} \text{ and } \Gamma_i \in \mathcal{B}_i,\ i \in F. \tag{8.1.25}$$

Our goal is to show there is a unique probability measure $\boldsymbol{\mu} \equiv \prod_{i\in\mathfrak{I}} \mu_i$ on $(\mathbf{E}, \mathcal{B}_\mathfrak{I})$ with the property that

$$\boldsymbol{\mu}(\mathbf{\Gamma}_F) = \prod_{i\in F} \mu_i(\Gamma_i) \tag{8.1.26}$$

for all choices of $F \in \mathfrak{F}$ and $\Gamma_i \in \mathcal{B}_i$, $i \in F$.

We begin by pointing out that uniqueness is clear. Indeed, the generating class described in (8.1.25) is obviously a π-system, and therefore (cf. Exercise

3.1.8) the condition in (8.1.26) can be satisfied by at most one measure. Secondly, observe that there is no problem when $\mathfrak{I}$ is finite. In fact, when $\mathfrak{I}$ has only one element there is nothing to do at all. Moreover, if we know how to handle $\mathfrak{I}$'s containing $n \in \mathbb{Z}^+$ elements and $\mathfrak{I} = \{i_1, \dots, i_{n+1}\}$, then we can take $\prod_1^{n+1} \mu_{i_m}$ to be the image (cf. Lemma 5.0.1) of (cf. Lemma 4.1.3)

$$\left(\prod_1^n \mu_{i_m}\right) \times \mu_{i_{n+1}}$$

under the mapping

$$\big((x_{i_1}, \dots, x_{i_n}), x_{i_{n+1}}\big) \in \left(\prod_1^n E_{i_m}\right) \times E_{i_{n+1}} \longmapsto (x_{i_1}, \dots, x_{i_{n+1}}) \in \prod_1^{n+1} E_{i_m}.$$

That is, we know how to construct $\boldsymbol{\mu}$ when $\mathfrak{I}$.

Thus, assume that $\mathfrak{I}$ is infinite. Given $F \in \mathfrak{F}$, use $\boldsymbol{\mu}_F$ to denote $\prod_{i \in F} \mu_i$. In order to construct $\boldsymbol{\mu}$, we first introduce the algebra

$$\mathcal{A} \equiv \bigcup_{F \in \mathfrak{F}} \Pi_F^{-1}(\mathcal{B}_F) \quad \text{where } \mathcal{B}_F \equiv \prod_{i \in F} \mathcal{B}_i,$$

and note that $\mathcal{B}_{\mathfrak{I}}$ is generated by $\mathcal{A}$. Next, observe that the map $\boldsymbol{\mu} : \mathcal{A} \longrightarrow [0,1]$ given by

$$\boldsymbol{\mu}(A) = \boldsymbol{\mu}_F(\boldsymbol{\Gamma}) \quad \text{if } A = \Pi_F^{-1}(\boldsymbol{\Gamma}) \text{ for some } F \in \mathfrak{F} \text{ and } \boldsymbol{\Gamma} \in \mathcal{B}_F$$

is well-defined and finitely additive. To see the first of these, suppose that, for some $\boldsymbol{\Gamma} \in \mathcal{B}_F$ and $\boldsymbol{\Gamma}' \in \mathcal{B}_{F'}$, $\Pi_F^{-1}(\boldsymbol{\Gamma}) = \Pi_{F'}^{-1}(\boldsymbol{\Gamma}')$. If $F = F'$, then it is clear that $\boldsymbol{\Gamma} = \boldsymbol{\Gamma}'$ and that there is no problem. On the other hand, if $F \subset F'$, then one has that $\boldsymbol{\Gamma}' = \boldsymbol{\Gamma} \times \mathbf{E}_{F' \setminus F}$ and therefore (since the μ_i's are probability measures) that

$$\boldsymbol{\mu}_{F'}(\boldsymbol{\Gamma}') = \boldsymbol{\mu}_F(\boldsymbol{\Gamma}) \prod_{i \in F' \setminus F} \mu_i(E_i) = \boldsymbol{\mu}_F(\boldsymbol{\Gamma}).$$

That is, $\boldsymbol{\mu}$ is well-defined on $\mathcal{A}$. Moreover, given disjoint A and A' from $\mathcal{A}$, choose $F \in \mathfrak{I}$ so that $A, A' \in \Pi_F^{-1}(\mathcal{B}_F)$, note that $\boldsymbol{\Gamma} = \Pi_F(A)$ is disjoint from $\boldsymbol{\Gamma}' = \Pi_{F'}(A')$, and conclude that

$$\boldsymbol{\mu}(A \cup A') = \boldsymbol{\mu}_F(\boldsymbol{\Gamma} \cup \boldsymbol{\Gamma}') = \boldsymbol{\mu}_F(\boldsymbol{\Gamma}) + \boldsymbol{\mu}_F(\boldsymbol{\Gamma}') = \boldsymbol{\mu}(A) + \boldsymbol{\mu}(A').$$

In view of the preceding paragraph and Corollary 8.1.22, all that remains is to show that if $\{A_n\}_1^\infty$ is a non-decreasing sequence from $\mathcal{A}$ and $\bigcap_1^\infty A_n = \emptyset$, then $\boldsymbol{\mu}(A_n) \searrow 0$. Equivalently, what we need to know is that if $\{A_n\}_1^\infty \subseteq \mathcal{A}$ is non-decreasing and, for all $n \in \mathbb{Z}^+$ and some $\epsilon > 0$, $\boldsymbol{\mu}(A_n) \geq \epsilon$, then $\bigcap_1^\infty A_n \neq \emptyset$. Thus, suppose that such a sequence is given, choose $\{F_n\}_1^\infty$ so that $A_n \in \Pi_{F_n}^{-1}(\mathcal{B}_{F_n})$ for each $n \in \mathbb{Z}^+$, and set $S = \mathfrak{I} \setminus \bigcup_1^\infty F_n$. Without loss of generality, we assume that $F_n \subset F_{n+1}$ for all n. Under the condition that $\mu(A_n) \geq \epsilon > 0$ for all $n \in \mathbb{Z}^+$, we must produce a sequence $\{\mathbf{a}_m\}_1^\infty$ such that $\mathbf{a}_1 \in \mathbf{E}_{F_1}$,

$$\mathbf{a}_m \in \mathbf{E}_{F_m \setminus F_{m-1}} \text{ for } m \geq 2, \text{ and } (\mathbf{a}_1, \dots, \mathbf{a}_n) \in \Pi_{F_n}(A_n) \text{ for all } n \in \mathbb{Z}^+.$$

In fact, given such a sequence $\{\mathbf{a}_m\}_1^\infty$, observe that by taking $\mathbf{a} \in \mathbf{E}$ so that $\Pi_{F_1}(\mathbf{a}) = \mathbf{a}_1$, $\Pi_{F_m \setminus F_{m-1}}(\mathbf{a}) = \mathbf{a}_m$ for $m \geq 2$, and (when $S \neq \emptyset$) $\Pi_S(\mathbf{a})$ is an arbitrary element of $\mathbf{E}_S$, one gets an element of $\bigcap_1^\infty A_n$.

To produce the $\mathbf{a}_m$'s, first choose and fix, for each $i \in \mathfrak{I}$, a reference point $e_i \in E_i$. Next, for each $n \in \mathbb{Z}^+$, define $\Phi_n : \mathbf{E}_{F_n} \longrightarrow \mathbf{E}$ so that

$$\Phi_n(\mathbf{x}_{F_n})_i = \begin{cases} x_i & \text{if } i \in F_n \\ e_i & \text{if } i \in \mathfrak{I} \setminus F_n, \end{cases}$$

and set $f_n = \mathbf{1}_{A_n} \circ \Phi_n$. Obviously,

$$\epsilon \leq \mu(A_n) = \int_{\mathbf{E}_{F_n}} f_n(\mathbf{x}_{F_n})\, \mu_{F_n}(d\mathbf{x}_{F_n}).$$

Furthermore, if, for each $m \in \mathbb{Z}^+$, we define the sequence $\{g_{m,n} : n \geq m\}$ of functions on $\mathbf{E}_{F_m}$ so that $g_{m,m}(\mathbf{x}_{F_m}) = f_m(\mathbf{x}_{F_m})$ and

$$g_{m,n}(\mathbf{x}_{F_m}) = \int_{\mathbf{E}_{F_n \setminus F_m}} f_n(\mathbf{x}_{F_m}, \mathbf{y}_{F_n \setminus F_m})\, \mu_{F_n \setminus F_m}(\mathbf{y}_{F_n \setminus F_m})$$

when $n > m$, then $g_{m,n+1} \leq g_{m,n}$ and

$$g_{m,n}(\mathbf{x}_{F_m}) = \int_{\mathbf{E}_{F_{m+1} \setminus F_m}} g_{m+1,n}(\mathbf{x}_{F_m}, \mathbf{y}_{F_{m+1} \setminus F_m})\, \mu_{F_{m+1} \setminus F_m}(d\mathbf{y}_{F_{m+1} \setminus F_m})$$

for all $1 \leq m < n$. Hence, $g_m \equiv \lim_{n \to \infty} g_{m,n}$ exists and, by the Monotone Convergence Theorem,

$$g_m(\mathbf{x}_{F_m}) = \int_{\mathbf{E}_{F_{m+1} \setminus F_m}} g_{m+1}(\mathbf{x}_{F_m}, \mathbf{y}_{F_{m+1} \setminus F_m})\, \mu_{F_{m+1} \setminus F_m}(d\mathbf{y}_{F_{m+1} \setminus F_m}). \tag{8.1.27}$$

Finally, since

$$\int_{\mathbf{E}_{F_1}} g_1(\mathbf{x}_{F_1})\, \mu_{F_1}(d\mathbf{x}_{F_1}) = \lim_{n \to \infty} \int_{\mathbf{E}_{F_n}} f_n(\mathbf{x}_{F_n})\, \mu_{F_n}(d\mathbf{x}_{F_n}) = \lim_{n \to \infty} \mu(A_n) \geq \epsilon,$$

there exists an $\mathbf{a}_1 \in \mathbf{E}_{F_1}$ such that $g_1(\mathbf{a}_1) \geq \epsilon$. In particular (since $g_1 \leq f_1$), this means that $\mathbf{a}_1 \in \Pi_{F_1}(A_1)$. In addition, from (8.1.27) with $m = 1$ and $\mathbf{x}_{F_1} = \mathbf{a}_1$, it means that there exists an $\mathbf{a}_2 \in \mathbf{E}_{F_2 \setminus F_1}$ for which $g_2(\mathbf{a}_1, \mathbf{a}_2) \geq \epsilon$ and therefore (since $g_2 \leq f_2$) $(\mathbf{a}_1, \mathbf{a}_2) \in \Pi_{F_2}(A_2)$. More generally, if $(\mathbf{a}_1, \dots, \mathbf{a}_m) \in \mathbf{E}_{F_m}$ and $g_m(\mathbf{a}_1, \dots, \mathbf{a}_m) \geq \epsilon$, then (since $g_m \leq f_m$) $(\mathbf{a}_1, \dots, \mathbf{a}_m) \in \Pi_{F_m}(A_m)$ and, by (8.1.27), there exists an $\mathbf{a}_{m+1} \in \mathbf{E}_{F_{m+1} \setminus F_m}$ such that $g_{m+1}(\mathbf{a}_1, \dots, \mathbf{a}_{m+1}) \geq \epsilon$. Hence, by induction, we are done and we have proved the following theorem.

8.1.28 Theorem. *Let $\mathfrak{I}$ be an arbitrary index set and, for each $i \in \mathfrak{I}$, let $(E_i, \mathcal{B}_i, \mu_i)$ be a probability space. If $\mathbf{E} = \prod_{i \in \mathfrak{I}} E_i$ and $\mathcal{B}_{\mathfrak{I}}$ is the σ-algebra over $\mathbf{E}$ generated by the sets $\mathbf{\Gamma}_F$ in* (8.1.25), *then there is a unique probability measure μ on $(\mathbf{E}, \mathcal{B}_{\mathfrak{I}})$ satisfying* (8.1.26).

The existence result proved in Theorem 8.1.28 plays an important role in probability theory, where it becomes a ubiquitous tool with which to construct infinite families of *mutually independent random variables.*

8.1.29 Remark: As the preceding discussion demonstrates, the version of Carathéodory's theorem which we have proved here is sufficient to handle the construction of non-trivial measures which are finite. Moreover, by an obvious localization procedure, one can reduce many problems involving σ-finite situations to finite ones. On the other hand, when confronting problems leading to measures which are not σ-finite, the results proved in this section are insufficient, and one must work harder. For an excellent account of both the methodology (as well as reason) for handling such problems, see L.C. Evans and R. Gariepy's *Measure Theory and Fine Properties of Functions*, published by the Studies in Advanced Math. Series of CRC Press (1991).

Exercises

8.1.30 Exercise: Assume that E is a metric space. Show that the Borel field $\mathcal{B}_E$ coincides with $\sigma\big(C_{\mathrm{b}}(E;\mathbb{R})\big)$. Next, assume, in addition, that E is separable and locally compact (i.e., every point $x \in E$ has a neighborhood whose closure is compact) and show that $\mathcal{B}_E = \sigma\big(C_{\mathrm{c}}(E;\mathbb{R})\big)$ (the space of $f \in C(E;\mathbb{R})$ which vanish off some compact subset of E).

8.1.31 Exercise: Here is another approach to the topic in Exercises 5.0.4 and 5.1.6. Let ψ be a bounded, right-continuous, non-decreasing function on $\mathbb{R}$. Set $\psi(\pm\infty) = \lim_{x\to\pm\infty} \psi(x)$.

(i) For each $\varphi \in C_{\mathrm{b}}(\mathbb{R};\mathbb{R})$, show that

$$I(\varphi) \equiv \lim_{R\to\infty} (\mathrm{R}) \int_{[-R,R]} \varphi\, d\psi \in \mathbb{R}$$

exists.

(ii) Show that $\varphi \in C_{\mathrm{b}}(\mathbb{R};\mathbb{R}) \longmapsto I(\varphi) \in \mathbb{R}$ is a non-negative, linear functional. Further, check that, for each $R > 0$,

$$\big|I(\varphi)\big| \le A\|\varphi\|_{\mathrm{u},[-R,R]} + \epsilon(R)\|\varphi\|_{\mathrm{u}}, \quad \varphi \in C_{\mathrm{b}}(\mathbb{R};\mathbb{R}),$$

where

$$A = \psi(\infty) - \psi(-\infty) \quad \text{and} \quad \epsilon(R) = A - \big(\psi(R) - \psi(-R)\big).$$

(iii) As a consequence of **(ii)** and Theorem 8.1.24, show that there is a unique measure μ_ψ on $(\mathbb{R},\mathcal{B}_{\mathbb{R}})$ with the property that $\mu\big((-\infty,x]\big) = \psi(x) - \psi(-\infty)$ for all $x \in \mathbb{R}$. Notice that the mapping $\psi \longmapsto \mu_\psi$ inverts the map discussed in Theorem 5.1.2.

8.2 The Radon–Nikodym Theorem

The goal of this section is to prove a converse to Exercise 3.3.16. That is, under suitable conditions, we want to show that if a measure μ is absolutely continuous with respect to a second measure ν, then there exists a non-negative f such that $d\mu = f\,d\nu$. In fact, we will show (cf. Exercise 8.2.9 for more information) that it suffices to know that $\nu(\Gamma) = 0 \implies \mu(\Gamma) = 0$.

Following J. von Neumann, we will base our reasoning on the following representation theorem which, like the one in Theorem 7.1.24, is due to F. Riesz. The result can be viewed as a sharpening of (6.2.8).

8.2.1 Theorem (F. Riesz). *Let* $\mathbf{H}$ *be a real Hilbert space and* $\Lambda : \mathbf{H} \longrightarrow \mathbb{R}$ *a linear mapping. Then*

$$\|\Lambda\| \equiv \sup\Bigl\{\Lambda(\mathbf{y}) : \mathbf{y} \in \mathbf{H} \text{ with } \|\mathbf{y}\|_{\mathbf{H}} \le 1\Bigr\} < \infty \tag{8.2.2}$$

if and only if there is an $\mathbf{x} \in \mathbf{H}$ *such that*

$$\Lambda(\mathbf{y}) = (\mathbf{y}, \mathbf{x})_{\mathbf{H}}, \quad \mathbf{y} \in \mathbf{H}, \tag{8.2.3}$$

in which case $\mathbf{x}$ *is uniquely determined by* (8.2.3) *and* $\|\mathbf{x}\|_{\mathbf{H}} = \|\Lambda\|$.

PROOF: Everything except the existence of $\mathbf{x}$ when (8.2.2) holds is clear. To prove this existence, first note that (8.2.2) implies that

$$\bigl|\Lambda(\mathbf{y}) - \Lambda(\mathbf{y}')\bigr| = \bigl|\Lambda(\mathbf{y} - \mathbf{y}')\bigr| \le \|\Lambda\|\|\mathbf{y} - \mathbf{y}'\|_{\mathbf{H}}, \quad \mathbf{y}, \mathbf{y}' \in \mathbf{H},$$

and therefore that Λ is continuous on $\mathbf{H}$. Hence, $\mathbf{F} \equiv \bigl\{\mathbf{y} \in \mathbf{H} : \Lambda(\mathbf{y}) = 0\bigr\}$ is a closed linear subspace of $\mathbf{H}$. Moreover, either $\mathbf{F} = \mathbf{H}$, in which case we may take $\mathbf{x} = \mathbf{0}$, or there is a $\mathbf{z} \in \mathbf{H} \setminus \mathbf{F}$. In the latter case, we take (cf. Theorem 7.1.3) $\tilde{\mathbf{z}} = \mathbf{z} - \Pi_{\mathbf{F}}\mathbf{z}$, and note that $\Lambda(\tilde{\mathbf{z}}) = \Lambda(\mathbf{z}) \ne 0$. In addition, observe that, for any $\mathbf{y} \in \mathbf{H}$,

$$\Lambda\left(\mathbf{y} - \frac{\Lambda(\mathbf{y})}{\Lambda(\tilde{\mathbf{z}})}\tilde{\mathbf{z}}\right) = 0.$$

Hence, $\mathbf{y} - \frac{\Lambda(\mathbf{y})}{\Lambda(\tilde{\mathbf{z}})}\tilde{\mathbf{z}} \in \mathbf{F}$ for any $\mathbf{y} \in \mathbf{H}$. But, since $\tilde{\mathbf{z}} \perp \mathbf{F}$, this means that

$$\bigl(\tilde{\mathbf{z}}, \mathbf{y}\bigr)_{\mathbf{H}} - \frac{\Lambda(\mathbf{y})}{\Lambda(\tilde{\mathbf{z}})}\|\tilde{\mathbf{z}}\|_{\mathbf{H}}^2 = \left(\tilde{\mathbf{z}}, \mathbf{y} - \frac{\Lambda(\mathbf{y})}{\Lambda(\tilde{\mathbf{z}})}\tilde{\mathbf{z}}\right)_{\mathbf{H}} = 0, \quad \mathbf{y} \in \mathbf{H};$$

and so we can take

$$\mathbf{x} = \frac{\Lambda(\tilde{\mathbf{z}})}{\|\tilde{\mathbf{z}}\|_{\mathbf{H}}^2}\tilde{\mathbf{z}}. \quad \square$$

Following J. von Neumann, we will now use Theorem 8.2.1 to derive an important property about the relationship between measures. Namely, given two measures μ and ν on the same measurable space $(E, \mathcal{B})$:

(a) we say ν **dominates** μ and write $\mu \le \nu$ if $\mu(\Gamma) \le \nu(\Gamma)$, $\Gamma \in \mathcal{B}$;
(b) we say μ is **absolutely continuous with respect to** ν and write $\mu \ll \nu$ if $\mu(\Gamma) = 0$ for all $\Gamma \in \mathcal{B}$ with $\nu(\Gamma) = 0$;
(c) we say μ and ν are **singular** and write $\mu \perp \nu$ if there is a $\Sigma \in \mathcal{B}$ with the property that $\mu(\Sigma) = \nu(\Sigma\complement) = 0$.

Obviously, both domination and absolute continuity express a relationship between μ and ν, the former being a much stronger statement than the latter. In fact, by starting with simple functions and passing to limits, it is easy to see that

$$(*)\qquad \mu \le \nu \implies \int f\,d\mu \le \int f\,d\nu \quad \text{for all } \mathcal{B}\text{-measurable } f : E \longrightarrow [0,\infty].$$

By contrast, singularity is a statement that the measures have nothing to do with one another and, in fact, *live on different portions of* E.

The result alluded to above comes in two parts and applies to μ's which are finite and ν's which are σ-finite. The first part says that μ can be written (in a unique way) as the sum of a measure μ_{a} which is absolutely continuous with respect to ν and a measure μ_σ which is singular to ν. The second part tells us that there is a unique non-negative $f \in L^1(\nu)$ with the property that

$$(8.2.4)\qquad \mu_{\mathrm{a}}(\Gamma) = \int_\Gamma f\,d\nu, \quad \Gamma \in \mathcal{B}.$$

In particular, if μ itself is absolutely continuous with respect to ν, then $\mu = \mu_{\mathrm{a}}$ and so (8.2.4) holds with μ in place of μ_{a}.

The key to von Neumann's proof of these results is the observation that everything can be reduced to consideration of μ's which are dominated by ν, in which case the existence of f becomes a simple application of Theorem 8.2.1.

8.2.5 Lemma. *Suppose that $(E,\mathcal{B},\nu)$ is a σ-finite measure space and that μ is a finite measure on $(E,\mathcal{B})$ which is dominated by ν. Then there is a unique $[0,1]$-valued $h \in L^1(\nu)$ with the property that*

$$(8.2.6)\qquad \int_E \varphi\,d\mu = \int_E \varphi h\,d\nu$$

for every $\mathcal{B}$-measurable $\varphi : E \longrightarrow [0,\infty]$.

PROOF: Since we can write E as the union of countably many, mutually disjoint, $\mathcal{B}$-measurable sets of finite ν-measure, we assume, without loss of generality, that ν is finite on $(E,\mathcal{B})$. But (cf. (*) above), in that case, $L^2(\nu) \subseteq L^1(\nu) \subseteq L^1(\mu)$, and the linear map

$$\varphi \in L^2(\nu) \longrightarrow \Lambda(\varphi) \equiv \int_E \varphi\,d\mu \in \mathbb{R}$$

satisfies $\left|\Lambda(\varphi)\right| \le \nu(E)^{\frac{1}{2}}\|\varphi\|_{L^2(\nu)}$. Hence, by Theorem 8.2.1 with $\mathbf{H} = L^2(\nu)$, there is an $h \in L^2(\nu)$ such that (8.2.6) holds for every $\varphi \in L^2(\nu)$ and, therefore, for every bounded $\mathcal{B}$-measurable function. We now want to show that h (which is determined only up to a set of ν-measure 0) can be chosen to take its values in $[0,1]$. To this end, set $A_n = \{h \le -\frac{1}{n}\}$ and $B_n = \{h \ge 1 + \frac{1}{n}\}$ for $n \in \mathbb{Z}^+$. Then, by (8.2.6),

$$-\frac{1}{n}\nu(A_n) \ge \mu(A_n) \ge 0 \quad \text{and} \quad \left(1 + \frac{1}{n}\right)\nu(B_n) \le \mu(B_n) \le \nu(B_n),$$

from which we conclude that $\nu(A_n) = \nu(B_n) = 0$, $n \in \mathbb{Z}^+$, and therefore that $\nu(h < 0) = \nu(h > 1) = 0$. In other words, we may assume that h takes its values in $[0,1]$, and, clearly, once we know this, (8.2.6) for all non-negative, $\mathcal{B}$-measurable φ's is an easy consequence of the Monotone Convergence Theorem.

Finally, the uniqueness assertion is trivial, since (8.2.6) determines the ν-integral of h over every $\Gamma \in \mathcal{B}$. $\square$

The first part of the next theorem is called **Lebesgue's Decomposition Theorem**, and the second part is the **Radon–Nikodym Theorem**.

8.2.7 Theorem. *Suppose that $(E, \mathcal{B}, \nu)$ is a σ-finite measure space, and let μ be a finite measure on $(E, \mathcal{B})$. Then there is a unique measure $\mu_a \le \mu$ on $(E, \mathcal{B})$ with the properties that $\mu_a \ll \nu$ and $\mu_\sigma \equiv (\mu - \mu_a) \perp \nu$. In addition, there is a unique non-negative $f \in L^1(\nu)$ for which (8.2.4) holds. In particular, $\mu \ll \nu$ if and only $\mu(\Gamma) = \int_\Gamma f\, d\nu$, $\Gamma \in \mathcal{B}$, for some non-negative $f \in L^1(\nu)$.*

PROOF: We first note that if $\mu(\Gamma) = \int_\Gamma f\, d\nu$, $\Gamma \in \mathcal{B}$, for some $f \in L^1(\nu)$, then (cf. Exercise 3.3.16) $\mu \ll \nu$ and (cf. Exercise 3.2.17) f is necessarily unique and (cf. the preceding) non-negative as an element of $L^1(\nu)$. Next, we prove that there is at most one choice of μ_a. To this end, suppose that $\mu = \mu_a + \mu_\sigma = \mu'_a + \mu'_\sigma$, where μ_a and μ'_a are both absolutely continuous with respect to ν and both μ_σ and μ'_σ are singular to ν. Choose Σ, $\Sigma' \in \mathcal{B}$ so that

$$\nu(\Sigma\complement) = \nu(\Sigma'\complement) = 0 \quad \text{and} \quad \mu_\sigma(\Sigma) = \mu'_\sigma(\Sigma') = 0,$$

and set $A = (\Sigma \cap \Sigma')\complement = \Sigma\complement \cup \Sigma'\complement$. Then $\nu(A) = \mu_\sigma(A\complement) = \mu'_\sigma(A\complement) = 0$; and therefore, for any $\Gamma \in \mathcal{B}$,

$$\mu_a(\Gamma) = \mu_a(A \cap \Gamma) = \mu(A \cap \Gamma) = \mu'_a(A \cap \Gamma) = \mu'_a(\Gamma).$$

In other words, $\mu_a = \mu'_a$.

To prove the existence statements, we first use Lemma 8.2.5, applied to μ and $\mu + \nu$, to find a $\mathcal{B}$-measurable $h : E \longrightarrow [0,1]$ with the property that

$$\int_E \varphi\, d\mu = \int_E \varphi h\, d\mu + \int_E \varphi h\, d\nu$$

for all non-negative, $\mathcal{B}$-measurable φ's. It is then clear that

$$\int_E \varphi(1-h)\, d\mu = \int_E \varphi h\, d\nu, \tag{8.2.8}$$

first for all $\varphi \in L^1(\mu) \cap L^1(\nu)$ and then for all non-negative, $\mathcal{B}$-measurable φ's. Now set $\Sigma = \{h < 1\}$ and $\mu_a(\Gamma) = \mu(\Sigma \cap \Gamma)$, $\Gamma \in \mathcal{B}$. Since

$$\nu(\Sigma\complement) = \nu(h = 1) = \int_{\{h=1\}} h\,d\nu = \int_{\{h=1\}} (1 - h)\,d\mu = 0,$$

it is clear that $\mu - \mu_a$ is singular to ν. At the same time, if $f \equiv (1-h)^{-1}\,h\mathbf{1}_\Sigma$, then, by (8.2.8) with $\varphi = (1-h)^{-1}\mathbf{1}_{\Gamma\cap\Sigma}$,

$$\mu_a(\Gamma) = \mu_a(\Gamma \cap \Sigma) = \int_E \varphi(1-h)\,d\mu = \int_\Gamma f\,d\nu$$

for each $\Gamma \in \mathcal{B}$. □

Given a finite measure μ and a σ-finite measure ν, the corresponding measures μ_a and μ_σ are called the **absolutely continuous** and **singular parts of μ with respect to ν**. Also, if μ is absolutely continuous with respect to ν, then the corresponding non-negative $f \in L^1(\nu)$ is called the **Radon–Nikodym derivative** of μ with respect to ν and is often denoted by $\frac{d\mu}{d\nu}$. The choice of this notation is explained by part **(ii)** of the Exercise 8.2.9 which follows.

Exercises

8.2.9 Exercise: Suppose that $\mathcal{C}$ is a countable partition of the non-empty set E, and use $\mathcal{B}$ to denote $\sigma(\mathcal{C})$.

(i) Show that $f : E \longrightarrow \overline{R}$ is $\mathcal{B}$-measurable if and only if f is constant on each $A \in \mathcal{C}$. Also, show that a measure ν is σ-finite on $(E, \mathcal{B})$ if and only if $\nu(A) < \infty$ for every $A \in \mathcal{C}$. Finally, if μ is a second measure on $(E, \mathcal{B})$, show that $\mu \ll \nu$ if and only if $\mu(A) = 0$ for all $A \in \mathcal{C}$ satisfying $\nu(A) = 0$.

(ii) Given any measures μ and ν on $(E, \mathcal{B})$ and a $\mathcal{B}$-measurable, ν-integrable $f : E \longrightarrow [0, \infty)$, show that $\mu(A) = \int_A f\,d\nu$ for all $A \in \mathcal{B}$ implies that, for every $A \in \mathcal{C}$, $\nu(A) \in (0, \infty) \implies f \restriction A = \frac{\mu(A)}{\mu(A)}$ and $\nu(A) = \infty \implies f \restriction A = 0$.

(iii) Using the preceding, show that, in general, one cannot dispense with the assumption in Theorem 8.1.13 that ν is σ-finite.

8.2.9 Exercise: Readers with good memories may be disturbed by the apparent difference between the notions of absolute continuity used here and that used earlier in Exercise 3.3.16. To allay such concerns, check that, as long as μ is finite, $\mu \ll \nu$ implies that for every $\epsilon > 0$ there is a $\delta > 0$ with the property that $\mu(\Gamma) < \epsilon$ whenever $\Gamma \in \mathcal{B}$ and $\nu(\Gamma) < \delta$. (Because ν is not assumed to be σ-finite, you should not use the Radon–Nikodym Theorem here.)

8.2.10 Exercise: The purpose of this exercise is to take a closer look at the relationship $\psi \longmapsto \mu_\psi$ (established in Exercise 8.1.31) taking a bounded, right-continuous, non-decreasing ψ on $\mathbb{R}$ into a finite measure μ_ψ on $(\mathbb{R}, \mathcal{B}_{\mathbb{R}})$.

(i) Given a measure space $(E, \mathcal{B}, \mu)$ and an $x \in E$, one says that x is an **atom** of $(E, \mathcal{B}, \mu)$ if $\mu(\Gamma) > 0$ for every $\Gamma \in \mathcal{B}$ which contains x. Further, one says that $(E, \mathcal{B}, \mu)$ is **non-atomic** if it has no atoms and that $(E, \mathcal{B}, \mu)$ is **purely atomic** if $\mu(\Gamma) = 0$ for every $\Gamma \in \mathcal{B}$ which does not contain any atoms of $(E, \mathcal{B}, \mu)$. Assuming that $\{x\} \in \mathcal{B}$ for every $x \in E$, show that x is an atom of $(E, \mathcal{B}, \mu)$ if and only if $\mu(\{x\}) > 0$, and, when μ is σ-finite, conclude that the set $A = \{x \in E : \mu(\{x\}) > 0\}$ of atoms is countable. In particular, if $\{x\} \in \mathcal{B}$ for every $x \in E$ and μ is σ-finite, show that $(E, \mathcal{B}, \mu)$ is purely atomic if an only if there is a countable set S such that

$$\mu(\Gamma) = \sum_{x \in \Gamma \cap S} \mu(\{x\}), \quad \Gamma \in \mathcal{B}.$$

(ii) Given a bounded, right-continuous, non-decreasing ψ on $\mathbb{R}$, show that ψ is continuous if and only if μ_ψ is non-atomic. Next, using either Lemma 1.2.20 or the preceding, show that the set $D(\psi)$ of $x \in \mathbb{R}$ for which $\psi(x) - \psi(x-) > 0$ is countable, and define the **discontinuous part** of ψ to be the function $\psi_{\mathrm{d}} : \mathbb{R} \longrightarrow \mathbb{R}$ given by

$$\psi_{\mathrm{d}}(x) = \sum_{t \in (-\infty, x] \cap D(\psi)} \big(\psi(t) - \psi(t-)\big), \quad x \in \mathbb{R}.$$

Show that ψ_{d} is a non-negative, bounded, right-continuous, non-decreasing function and that the **continuous part** $\psi_{\mathrm{c}} \equiv \psi - \psi_{\mathrm{d}}$ of ψ is a bounded, continuous, non-decreasing function. Finally, say that ψ is **purely discontinuous** if $\psi = \psi(-\infty) + \psi_{\mathrm{d}}$, and check that ψ is purely discontinuous if and only if μ_ψ is purely atomic.

(iii) We now want to see how to characterize $\mu_\psi \ll \lambda_{\mathbb{R}}$ in terms of ψ. For this purpose, we say that the bounded, right-continuous, non-decreasing function ψ is **absolutely continuous** if, for every $\epsilon > 0$, there is a $\delta > 0$ such that

$$\sum_{m=1}^{\infty} \big(\psi(b_m) - \psi(a_m)\big) \le \epsilon$$

whenever $\{(a_m, b_m)\}_1^\infty$ is a sequence of mutually disjoint, open intervals which satisfy $\sum_{m=1}^{\infty}(b_m - a_m) < \delta$. Obviously, every absolutely continuous ψ is uniformly continuous. Show that, in fact, ψ is absolutely continuous if and only if $\mu_\psi \ll \lambda_{\mathbb{R}}$. In this connection, we say that ψ is **singular** if $\mu_\psi \perp \lambda_{\mathbb{R}}$.

(iv) The preceding considerations lead to the following decomposition of a bounded, right-continuous, non-decreasing function ψ. Namely, let $\mu_{\psi,\mathrm{a}}$ and $\mu_{\psi,\sigma}$ be the absolutely continuous and singular parts of μ_ψ with respect to $\lambda_{\mathbb{R}}$, and define the **absolutely continuous part** ψ_{a} and **singular part** ψ_σ of ψ by

$$\psi_{\mathrm{a}}(x) = \mu_{\mathrm{a}}\big((-\infty, x]\big) \quad \text{and} \quad \psi_\sigma(x) = \mu_\sigma\big((-\infty, x]\big)$$

for $x \in \mathbb{R}$. Further, let $\psi_{\sigma,\mathrm{c}}$ and $\psi_{\sigma,\mathrm{d}}$ be the continuous and discontinuous parts of ψ_σ. Obviously,

$$\psi = \psi(-\infty) + \psi_{\mathrm{a}} + \psi_{\sigma,\mathrm{c}} + \psi_{\sigma,\mathrm{d}}. \tag{8.2.11}$$

Show that the decomposition in (8.2.11) is canonical in the sense that if $\psi = \psi(-\infty) + \psi_1 + \psi_2 + \psi_3$, where, for each $1 \le i \le 3$, ψ_i is right-continuous, non-decreasing, and tends to 0 at $-\infty$, ψ_1 is absolutely continuous, ψ_2 is continuous but singular, and ψ_3 is purely discontinuous, then $\psi_1 = \psi_{\mathrm{a}}$, $\psi_2 = \psi_{\sigma,\mathrm{c}}$, and $\psi_3 = \psi_{\sigma,\mathrm{d}}$.

8.2.12 Exercise: It is clear that absolutely continuous non-decreasing and purely discontinuous non-decreasing functions exist. But are there any continuous, singular, non-decreasing functions? In order to show that there are, we will now describe the Cantor–Lebesgue function. Referring to Exercise 2.1.20, recall that the closed set C_k is the union of 2^k disjoint closed intervals and that $[0,1] \setminus C_k$ is the union of $2^k - 1$ disjoint open intervals $(a_{k,j}, b_{k,j})$, $1 \le j < 2^k$, where we have ordered these so that $b_{k,j} < a_{k,j+1}$ for $1 \le j < 2^k - 1$. Next, set $a_{k,0} = -\infty$, $b_{k,0} = 0$, $a_{k,2^k} = 1$, $b_{k,2^k} = \infty$, and define $\psi_k : \mathbb{R} \longrightarrow [0,1]$ so that:

- **(a)** ψ_k is constant on each of the intervals $[a_{k,j}, b_{k,j}]$, $0 \le j \le 2^k$;
- **(b)** $\psi_k(b_{k,j}) = 2^{-k} j$ for $0 \le j \le 2^k$ and ψ_k is linear on each of the intervals $[b_{k,j}, a_{k,j+1}]$, $0 \le j < 2^k$.

Notice that each ψ_k is continuous and non-decreasing from $\mathbb{R}$ onto $[0,1]$. In addition, check that $\|\psi_{k+1} - \psi_k\|_{\mathrm{u},\mathbb{R}} = \frac{2^{-k}}{6}$; and conclude from this that ψ_k converges uniformly on $\mathbb{R}$ to a continuous, non-decreasing $\psi : \mathbb{R} \longrightarrow [0,1]$ with the property that $\mu_\psi(C) = 1$. At the same time, by Exercise 2.1.20, $\lambda_{\mathbb{R}}(C) = 0$, and therefore ψ is singular.

Solution Manual

§ 1.1

1.1.9: To prove Theorem 1.1.8, simply note that Lemma 1.1.7 says that

$$\varliminf_{\|\mathcal{C}\|\to 0} \mathcal{U}(f;\mathcal{C}) = \inf_{\mathcal{C}} \mathcal{U}(f;\mathcal{C})$$

and that

$$\varlimsup_{\|\mathcal{C}\|\to 0} \mathcal{L}(f;\mathcal{C}) = \sup_{\mathcal{C}} \mathcal{L}(f;\mathcal{C}).$$

To prove that $f \vee g$ is Riemann integrable if f and g are, observe that

$a' \vee b' - a \vee b \le (a'-a) \vee (b'-b) \le |a'-a| + |b'-b|$ for any a, a', b, and $b' \in \mathbb{R}$.

Thus, for any $\mathcal{C}$,

$$\mathcal{U}(f \vee g;\mathcal{C}) - \mathcal{L}(f \vee g;\mathcal{C}) \le \Big[\mathcal{U}(f;\mathcal{C}) - \mathcal{L}(f;\mathcal{C})\Big] + \Big[\mathcal{U}(g;\mathcal{C}) - \mathcal{L}(g;\mathcal{C})\Big],$$

from which it is clear that $f \vee g$ is Riemann integrable if f and g are. In addition, since

$$\mathcal{R}(f \vee g;\mathcal{C},\xi) \ge \mathcal{R}(f;\mathcal{C},\xi) \vee \mathcal{R}(g;\mathcal{C},\xi)$$

for every $\mathcal{C}$ and $\xi \in \Xi(\mathcal{C})$, the corresponding inequality is obvious.

One can handle $f \wedge g$ by either applying an analogous line of reasoning or simply noting that $f \wedge g = -(-f) \vee (-g)$. Finally, the linearity assertion follows immediately from the linearity of the approximating Riemann sums (cf. **(iii)** in Exercise 1.2.25 below).

1.1.10: Given $\mathcal{C}$ and $\epsilon > 0$, set

$$\mathcal{C}(\epsilon) = \left\{ I \in \mathcal{C} : \sup_I f - \inf_I f \ge \epsilon \right\}.$$

Assume that f is Riemann integrable, and let $\epsilon > 0$ be given. Then there exists a $\delta > 0$ such that

$$\mathcal{U}(f;\mathcal{C}) - \mathcal{L}(f;\mathcal{C}) < \epsilon^2$$

whenever $\|\mathcal{C}\| < \delta$. Thus, if $\|\mathcal{C}\| < \delta$, then

$$\begin{aligned} \epsilon \sum_{I\in\mathcal{C}(\epsilon)} \mathrm{vol}(I) &\le \sum_{I\in\mathcal{C}(\epsilon)} \left(\sup_I f - \inf_I f\right) \mathrm{vol}(I) \\ &\le \sum_{I\in\mathcal{C}} \left(\sup_I f - \inf_I f\right) \mathrm{vol}(I) = \mathcal{U}(f;\mathcal{C}) - \mathcal{L}(f;\mathcal{C}) < \epsilon^2 \end{aligned}$$

as long as $\|\mathcal{C}\| < \delta$.

Conversely, suppose that for each $\epsilon > 0$ there is a $\mathcal{C}_\epsilon$ for which

$$\sum_{I \in \mathcal{C}_\epsilon(\epsilon)} \mathrm{vol}(I) < \epsilon.$$

Then

$$\begin{aligned}
&\mathcal{U}(f;\mathcal{C}_\epsilon) - \mathcal{L}(f;\mathcal{C}_\epsilon)\\
&\quad = \sum_{I \in \mathcal{C}_\epsilon(\epsilon)} \left(\sup_I f - \inf_I f\right)\mathrm{vol}(I) + \sum_{I \in \mathcal{C}_\epsilon \setminus \mathcal{C}_\epsilon(\epsilon)} \left(\sup_I f - \inf_I f\right)\mathrm{vol}(I)\\
&\quad \le 2\|f\|_{\mathrm{u}} \sum_{I \in \mathcal{C}_\epsilon(\epsilon)} \mathrm{vol}(I) + \epsilon \sum_{I \in \mathcal{C}_\epsilon \setminus \mathcal{C}_\epsilon(\epsilon)} \mathrm{vol}(I)\\
&\quad \le 2\|f\|_{\mathrm{u}}\epsilon + \epsilon|J| = \epsilon\big(2\|f\|_{\mathrm{u}} + |J|\big) \longrightarrow 0 \quad \text{as } \epsilon \searrow 0.
\end{aligned}$$

Hence, $\inf_{\mathcal{C}} \mathcal{U}(f;\mathcal{C}) = \sup_{\mathcal{C}} \mathcal{L}(f;\mathcal{C})$, and so f is Riemann integrable.

1.1.12: Suppose that f is a bounded function on J which is continuous at all but a finite number of points, $\mathbf{a}_1, \dots, \mathbf{a}_m$. Given $\epsilon > 0$, choose non-overlapping cubes $Q_1, \dots, Q_m$ so that $\mathbf{a}_\ell$ is the center of Q_ℓ and $\mathrm{vol}(Q_\ell) < \frac{\epsilon}{m}$ for each $1 \le \ell \le m$. Next, observe that f is continuous on $\hat{J} \equiv \overline{J \setminus (\bigcup_1^m Q_\ell)}$ and therefore that

$$\sup\big\{f(y) - f(x) : x,\, y \in \hat{J} \text{ and } |y - x| < \delta\big\} < \epsilon$$

for some $\delta > 0$. Moreover, $\hat{J}$ admits a finite exact cover $\hat{\mathcal{C}}$ by non-overlapping cubes of diameter less than δ, and clearly

$$\mathcal{C} \equiv \hat{\mathcal{C}} \cup \{Q_1 \cap J, \dots, Q_m \cap J\}$$

is a finite, non-overlapping cover of J with the property that

$$\sum_{\{Q \in \mathcal{C} : \sup_Q f - \inf_Q f \ge \epsilon\}} \mathrm{vol}(Q) \le \sum_{\ell=1}^m \mathrm{vol}(Q_\ell) < \epsilon.$$

Now apply Exercise 1.1.10.

§ 1.2

1.2.25: **(i)** Given $\epsilon > 0$, choose $\delta > 0$ so that $|\psi'(y) - \psi'(x)| < \epsilon$ whenever $|y - x| < \delta$. Next, given $\mathcal{C}$ with $\|\mathcal{C}\| < \delta$, use the Mean Value Theorem to choose $\eta(I) \in \mathring{I}$ for each $I \in \mathcal{C}$ so that $\Delta_I \psi = \psi'(\eta(I))\mathrm{vol}(I)$. Then, for any $\xi \in \Xi(\mathcal{C})$,

$$\begin{aligned}
&\big|\mathcal{R}(\varphi \mid \psi; \mathcal{C}, \xi) - \mathcal{R}(\varphi\psi'; \mathcal{C}, \xi)\big|\\
&\qquad\qquad \le \sum_{I \in \mathcal{C}} |\varphi(\xi(I))|\big|\psi'(\eta(I)) - \psi'(\xi(I))\big| \le \epsilon\|\varphi\|_{\mathrm{u}}\mathrm{vol}(J).
\end{aligned}$$

Hence, since $\varphi\psi'$ is continuous and therefore Riemann integrable, we have now proved that

$$\lim_{\|\mathcal{C}\|\to 0} \sup_{\xi\in\Xi(\mathcal{C})} \left|\mathcal{R}(\varphi|\psi;\mathcal{C},\xi) - (\mathrm{R})\int_J \varphi(x)\psi'(x)\,dx\right| = 0.$$

(ii) Given $\epsilon > 0$, choose $0 < \delta < \min_{1\le m\le n}(a_m - a_{m-1})$ so that $|\varphi(y)-\varphi(x)| < \epsilon$ whenever $|y-x| < \delta$. Next, let $\mathcal{C}$ with $\|\mathcal{C}\| < \delta$ be given. Clearly, no $I \in \mathcal{C}$ contains more than one of the a_m's and each a_m is in at most two I's. In particular, for any $I \in \mathcal{C}$,

$$\Delta_I\psi = \begin{cases} 0 & \text{if } I \cap \{a_0,\dots,a_n\} = \emptyset \\ d_m & \text{if } a_m \in \mathring{I} \\ d_m^+ \equiv \psi(a_m+) - \psi(a_m) & \text{if } I^- = a_m \\ d_m^- \equiv \psi(a_m) - \psi(a_m-) & \text{if } I^+ = a_m. \end{cases}$$

Thus, if

$$M_0 = \{m : \, a_m \in \mathring{I} \text{ for some } I \in \mathcal{C}\}$$

and

$$M_\pm = \{m : \, a_m = I^\pm \text{ for some } I \in \mathcal{C}\},$$

and if $\xi_0 \in \mathcal{C}$ is chosen so that $a_m \in I \implies \xi_0(I) = a_m$, then

$$\begin{aligned}\mathcal{R}(\varphi \mid \psi;\mathcal{C},\xi_0) &= \sum_{m\in M_0} \varphi(a_m)d_m + \sum_{m\in M_+} \varphi(a_m)d_m^- + \sum_{m\in M_-} \varphi(a_m)d_m^+ \\ &= \sum_0^n \varphi(a_m)d_m,\end{aligned}$$

since (when $d_0^- = d_n^+ \equiv 0$) $d_m = d_m^- + d_m^+$, $0 \le m \le n$. At the same time, for any $\xi \in \Xi(\mathcal{C})$,

$$\big|\mathcal{R}(\varphi \mid \psi;\mathcal{C},\xi) - \mathcal{R}(\varphi \mid \psi;\mathcal{C},\xi_0)\big| \le \epsilon \sum_0^n \big(|d_m^-| + |d_m^+|\big).$$

Hence, we have now shown that

$$\lim_{\|\mathcal{C}\|\to 0} \sup_{\xi\in\Xi(\mathcal{C})} \left|\mathcal{R}(\varphi \mid \psi;\mathcal{C},\xi) - \sum_0^n \varphi(a_m)d_m\right| = 0.$$

(iii) Without loss of generality, assume that $\alpha = \beta = 1$. Given $\epsilon > 0$, choose $\delta > 0$ so that

$$\|\mathcal{C}\| < \delta \implies \max_{i\in\{1,2\}} \sup_{\xi\in\Xi(\mathcal{C})} \left|\mathcal{R}(\varphi_i \mid \psi;\mathcal{C},\xi) - (\mathrm{R})\int_J \varphi_i(x)\,d\psi(x)\right| < \epsilon.$$

Because $\mathcal{R}(\varphi_1 + \varphi_2 \mid \psi;\mathcal{C},\xi) = \mathcal{R}(\varphi_1 \mid \psi;\mathcal{C},\xi) + \mathcal{R}(\varphi_2|\psi;\mathcal{C},\xi)$, it is then clear that

$$\sup_{\xi\in\Xi(\mathcal{C})} \left|\mathcal{R}(\varphi_1 + \varphi_2 \mid \psi;\mathcal{C},\xi) - (\mathrm{R})\int_J \varphi_1(x)\,d\psi(x) - (\mathrm{R})\int_J \varphi_2(x)\,d\psi(x)\right| < 2\epsilon$$

whenever $\|\mathcal{C}\| < \delta$.

(iv) We begin by showing that φ is ψ-Riemann integrable on both J_1 and J_2. To this end, let $\epsilon > 0$ be given, and choose $\delta > 0$ so that

$$(*) \qquad \|\mathcal{C}\| < \delta \implies \sup_{\xi \in \Xi(\mathcal{C})} \left| \mathcal{R}(\varphi \mid \psi; \mathcal{C}, \xi) - (\mathrm{R}) \int_J \varphi(x)\, d\psi(x) \right| < \epsilon.$$

Now, suppose that $\mathcal{C}_1$ and $\mathcal{C}_1'$ are finite, non-overlapping covers of J_1 with mesh size less than δ. Let $\mathcal{C}_2$ be any finite, non-overlapping cover of J_2 with mesh size less than δ, and set $\mathcal{C} = \mathcal{C}_1 \cup \mathcal{C}_2$ and $\mathcal{C}' = \mathcal{C}_1' \cup \mathcal{C}_2$. Next, let $\xi_1 \in \Xi(\mathcal{C}_1)$ and $\xi_1' \in \Xi(\mathcal{C}_1')$ be given, and define $\xi \in \Xi(\mathcal{C})$ and $\xi' \in \Xi(\mathcal{C}')$ so that

$$\xi(I) = \begin{cases} \xi_1(I) & \text{if } I \in \mathcal{C}_1 \\ \xi_2(I) & \text{if } I \in \mathcal{C}_2 \end{cases} \quad \text{and} \quad \xi'(I) = \begin{cases} \xi_1'(I) & \text{if } I \in \mathcal{C}_1' \\ \xi_2(I) & \text{if } I \in \mathcal{C}_2. \end{cases}$$

Then, by (*),

$$\left| \mathcal{R}(\varphi|\psi; \mathcal{C}, \xi) - \mathcal{R}(\varphi|\psi; \mathcal{C}', \xi') \right| < 2\epsilon.$$

At the same time,

$$\mathcal{R}(\varphi|\psi; \mathcal{C}_1, \xi_1) - \mathcal{R}(\varphi|\psi; \mathcal{C}_1', \xi_1') = \mathcal{R}(\varphi|\psi; \mathcal{C}, \xi) - \mathcal{R}(\varphi|\psi; \mathcal{C}', \xi').$$

Thus, by Cauchy's convergence criterion, we have now shown that φ is ψ-Riemann integrable on J_1. The same argument shows that φ is ψ-Riemann integrable on J_2.

To prove (1.2.6), let $\epsilon > 0$ be given, and choose $\delta > 0$ so that (*) holds. Next, choose $\mathcal{C}_i$ for J_i and $\xi_i \in \Xi(\mathcal{C}_i)$ so that $\|\mathcal{C}_i\| < \delta$ and

$$(**) \qquad \max_{i \in \{1,2\}} \left| \mathcal{R}(\varphi|\psi; \mathcal{C}_i, \xi_i) - (\mathrm{R}) \int_{J_i} \varphi(x)\, d\psi(x) \right| < \epsilon.$$

Finally, set $\mathcal{C} = \mathcal{C}_1 \cup \mathcal{C}_2$, and define $\xi \in \Xi(\mathcal{C})$ by

$$\xi(I) = \begin{cases} \xi_1(I) & \text{if } I \in \mathcal{C}_1 \\ \xi_2(I) & \text{if } I \in \mathcal{C}_2. \end{cases}$$

Then,

$$\mathcal{R}(\varphi|\psi, \mathcal{C}, \xi) = \mathcal{R}(\varphi|\psi; \mathcal{C}_1, \xi_1) + \mathcal{R}(\varphi|\psi; \mathcal{C}_2, \xi_2),$$

and so (*) and (**) imply that

$$\left| (\mathrm{R}) \int_J \varphi(x)\, d\psi(x) - (\mathrm{R}) \int_{J_1} \varphi(x)\, d\psi(x) - (\mathrm{R}) \int_{J_2} \varphi(x)\, d\psi(x) \right| < 3\epsilon.$$

1.2.26: Suppose that φ is ψ-Riemann integrable on J. Given $\epsilon > 0$, choose $\delta > 0$ so that

$$\|\mathcal{C}\| < \delta \implies \left| \sum_{I \in \mathcal{C}} \left(\sup_I \varphi - \inf_I \varphi \right) \Delta_I \psi \right| < \epsilon^2.$$

Then, since

$$\epsilon \sum_{\{I\in\mathcal{C}:\, \sup_I \varphi - \inf_I \varphi \geq \epsilon\}} \Delta_I\psi \leq \sum_{I\in\mathcal{C}} \left(\sup_I \varphi - \inf_I \varphi\right) \Delta_I\psi,$$

it is clear that (1.2.27) holds whenever $\|\mathcal{C}\| < \delta$.

Conversely, suppose that, for every $\epsilon > 0$, there is a $\delta > 0$ for which (1.2.27) holds. Since

$$\sum_{I\in\mathcal{C}} \left(\sup_I \varphi - \inf_I \varphi\right) \Delta_I\psi \leq \epsilon\Delta_J\psi + 2\|\varphi\|_{\mathrm{u}} \sum_{\{I\in\mathcal{C}:\, \sup_I \varphi - \inf_I \varphi \geq \epsilon\}} \Delta_I\psi,$$

it follows that φ is ψ-Riemann integrable.

Finally, the argument just given shows that the existence for every $\epsilon > 0$ of a $\mathcal{C}$ for which (1.2.27) holds is equivalent to the existence for every $\epsilon > 0$ of a $\mathcal{C}$ for which

$$\sum_{I\in\mathcal{C}} \left(\sup_I \varphi - \inf_I \varphi\right) \Delta_I\psi < \epsilon$$

holds. Thus, what we must show when $\psi \in C(J)$ is that, for every $\mathcal{C}$,

$$\overline{\lim_{\|\mathcal{C}'\|\to 0}} \sum_{\{I\in\mathcal{C}':\, \sup_I \varphi - \inf_I \varphi \geq \epsilon\}} \Delta_I\psi \leq \sum_{\{I\in\mathcal{C}:\, \sup_I \varphi - \inf_I \varphi \geq \epsilon\}} \Delta_I\psi$$

and

$$\overline{\lim_{\|\mathcal{C}'\|\to 0}} \sum_{I\in\mathcal{C}'} \left(\sup_I \varphi - \inf_I \varphi\right) \Delta_I\psi \leq \sum_{I\in\mathcal{C}} \left(\sup_I \varphi - \inf_I \varphi\right) \Delta_I\psi.$$

To this end, let $\mathcal{C}$ be given and set $S = \{I^+ : I \in \mathcal{C} \text{ and } I^+ \neq J^+\}$. Given a $\mathcal{C}'$, set

$$\mathcal{C}'(\epsilon) = \left\{I' \in \mathcal{C}' : \sup_{I'} \varphi - \inf_{I'} \varphi \geq \epsilon\right\}.$$

Then

$$\begin{aligned}\sum_{I'\in\mathcal{C}'(\epsilon)} \Delta_{I'}\psi &\leq \sum_{I\in\mathcal{C}} \sum_{\{I'\in\mathcal{C}'(\epsilon):\, I'\subseteq I\}} \Delta_{I'}\psi + \sum_{t\in S} \sum_{\{I'\in\mathcal{C}'(\epsilon):\, t\in \mathring{I}'\}} \Delta_{I'}\psi \\ &\leq \sum_{I\in\mathcal{C}(\epsilon)} \Delta_I\psi + \operatorname{card}(S) \sup_{I'\in\mathcal{C}'} \Delta_{I'}\psi.\end{aligned}$$

But, by continuity, the last term tends to 0 as $\|\mathcal{C}'\| \longrightarrow 0$, and so we have proved the first of the above. The proof of the second is essentially the same, only this time one uses

$$\sum_{I\in\mathcal{C}} \sum_{\{I'\in\mathcal{C}':\, I'\subseteq I\}} \left(\sup_{I'} \varphi - \inf_{I'} \varphi\right) \Delta_{I'}\psi \leq \sum_{I\in\mathcal{C}} \left(\sup_I \varphi - \inf_I \varphi\right) \Delta_I\psi.$$

1.2.28: Let $\psi \in C(J)$. What we must show is that, for each $\mathcal{C}$ and $\epsilon > 0$, there is a $\delta > 0$ such that

$$\mathcal{S}(\psi;\mathcal{C}) - \mathcal{S}(\psi;\mathcal{C}') < \epsilon$$

whenever $\|\mathcal{C}'\| < \delta$. To this end, suppose that

$$\mathcal{C} = \Big\{[a_0, a_1], \dots, [a_{n-1}, a_n]\Big\}$$

where $J^- = a_0 < \cdots < a_n = J^+$. Next, given $\epsilon > 0$, choose $0 < \delta < \min_{1\le m\le n}(a_m - a_{m-1})$ so that

$$\omega_\psi(\delta) \equiv \sup\Big\{|\psi(t) - \psi(s)| : s, t \in J \text{ and } |t-s| < \delta\Big\} < \frac{\epsilon}{2n}.$$

If $\|\mathcal{C}'\| < \delta$ and $\mathcal{A}$ is the set of those $I' \in \mathcal{C}'$ for which there is an $I \in \mathcal{C}$ with $I' \subseteq I$, then for each $I' \in \mathcal{B} \equiv \mathcal{C}' \setminus \mathcal{A}$, there is precisely one $m \in \{1, \dots, n-1\}$ for which $a_m \in \mathring{I}'$. In particular, because $\mathcal{C}'$ is non-overlapping, $\mathcal{B}$ has at most n elements. Moreover, if $I' \in \mathcal{B}$, $a_m \in \mathring{I}'$, and we use $L(I')$ and $R(I')$ to denote $I' \cap [a_{m-1}, a_m]$ and $I' \cap [a_m, a_{m+1}]$, respectively, then

$$\mathcal{C} \vee \mathcal{C}' = \mathcal{A} \cup \bigcup_{I' \in \mathcal{B}} \{L(I'), R(I')\}.$$

Hence,

$$\begin{aligned}\mathcal{S}(\psi;\mathcal{C}) - \mathcal{S}(\psi;\mathcal{C}') &\le \mathcal{S}(\psi;\mathcal{C}'\vee\mathcal{C}) - \mathcal{S}(\psi;\mathcal{C})\\ &\le \sum_{I'\in\mathcal{B}} \big(|\Delta_{L(I')}\psi| + |\Delta_{R(I')}\psi|\big) \le 2n\omega_\psi(\delta) < \epsilon.\end{aligned}$$

Finally, by (1.2.14), (1.2.16), and (1.2.17), the analogous result for var_- and var_+ follow immediately.

Now, suppose that $\psi \in C^1(J)$. For $n \in \mathbb{Z}^+$ and $1 \le m < n$, set $a_{m,n} = J^- + \frac{m}{n}\Delta J$, and use the Mean Value Theorem to choose $\xi_{m,n} \in (a_{m,n}, a_{m+1,n})$ so that $\varphi(a_{m+1,n}) - \varphi(a_{m,n}) = \varphi'(\xi_{m,n})\frac{\Delta J}{n}$. Then

$$\mathrm{Var}_+(\psi) = \lim_{n\to\infty} \frac{\Delta J}{n} \sum_{m=0}^{n-1} \psi'(\xi_{m,n})^+ = (\mathrm{R})\int_J \varphi'(t)^+\,dt,$$

and similarly for $\mathrm{Var}_-(\psi)$ and $\mathrm{Var}(\psi)$.

1.2.29: Let ψ_1 and ψ_2 be given. Then, for any $c \le s < t \le d$,

$$\psi_2(t) - \psi_2(s) = \psi(t) - \psi(s) + \big(\psi_1(t) - \psi_1(s)\big) \ge \big(\psi(t) - \psi(s)\big)^+$$

since both $\psi_1(t) - \psi_1(s)$ and $\psi_2(t) - \psi_2(s)$ are non-negative. Hence, for any $c \leq s < t \leq d$ and any finite, non-overlapping, exact cover $\mathcal{C}$ of $[s,t]$,

$$\psi_2(t) - \psi_2(s) = \sum_{I\in\mathcal{C}} \Delta_I\psi_2 \geq \sum_{I\in\mathcal{C}} \bigl(\Delta_I\psi\bigr)^+;$$

and so $\psi_2(t) - \psi_2(s) \geq \psi_+(t) - \psi_+(s)$. Since $\Delta_I\psi_1 - \Delta_I\psi_- = \Delta_I\psi_2 - \Delta_I\psi_+$ for any interval $I \subseteq J$, we now see that $\psi_2 - \psi_+$ and $\psi_1 - \psi_-$ are non-decreasing.

Turning to the statements about the relationship between the jumps of ψ, ψ_+, and ψ_-, begin by observing that

$$\begin{aligned}\bigl|\psi_+(x\pm) - \psi_+(x)\bigr| &= \bigl(\psi(x\pm) - \psi(x)\bigr)^\pm \\ &\implies \bigl|\psi_-(x\pm) - \psi_-(x)\bigr| = \bigl(\psi(x\pm) - \psi(x)\bigr)^\mp.\end{aligned}$$

Further, if $\check{\psi}(x) \equiv \psi(c+d-x)$, $x \in [c,d]$, then $\check{\psi}$ again has bounded variation,

$$\begin{aligned}\psi_\pm(x) = \check{\psi}_\mp(c+d-x), \quad \psi(x-) = \check{\psi}\bigl((c+d-x)+\bigr),\\ \text{and } \psi_\pm(x-) = \check{\psi}_\mp\bigl((c+d-x)+\bigr).\end{aligned}$$

Thus, we need only show that $\psi_+(x+) - \psi_+(x) = \bigl(\psi(x+) - \psi(x)\bigr)^+$, and, because we can always restrict everything to the interval $[x,d]$, it suffices to handle $x = c$. To this end, note that $\psi_+(c+) - \psi_+(c) = \bigl(\psi(c+) - \psi(c)\bigr)^+$ is equivalent to $\beta \equiv \psi_+(c+) \wedge \psi_-(c+) = 0$. But, if

$$\psi_1(x) = \begin{cases} 0 & \text{if } x = c \\ \psi_-(x) - \beta & \text{if } x \in (c,d] \end{cases}$$

and

$$\psi_2(x) = \begin{cases} \psi(c) & \text{if } x = c \\ \psi(c) + \psi_+(x) - \beta & \text{if } x \in (c,d], \end{cases}$$

then ψ_1 and ψ_2 are both non-decreasing and $\psi = \psi_2 - \psi_1$. Hence, by the first part of this exercise, $\psi_- \leq \psi_1$, and so $\psi_-(d) \leq \psi_-(d) - \beta$. That is, $\beta = 0$.

1.2.30: Define $\psi(0) = 0$ and $\psi(t) = t\cos(\frac{\pi}{t})$, $t \in (0,1]$. Clearly ψ is continuous on $[0,1]$. On the other hand, if $t_n = \frac{1}{n}$ for $n \in \mathbb{N}$ and $\mathcal{C}_n$ consists of the intervals $[0,t_n]$, $[t_n,t_{n-1}], \ldots$, and $[t_1,t_0]$, it is easy to check that $\mathcal{S}(\psi;\mathcal{C}_n)$ diverges as $n \longrightarrow \infty$ at least as fast as the harmonic series does. Thus, ψ has unbounded variation on $[0,1]$.

To handle the second part of the exercise, consider the function

$$\psi(t) = \begin{cases} 0 & \text{if } t \in [0,1] \setminus \{\frac{1}{2}\} \\ 1 & \text{if } t = \frac{1}{2}. \end{cases}$$

Clearly $\operatorname{var}\bigl(\psi;[0,1]\bigr) = 2$. On the other hand, for any $\varphi \in C\bigl([0,1]\bigr)$, one has that $\mathcal{R}(\varphi|\psi;\mathcal{C},\xi) = 0$ as long as $\frac{1}{2}$ is not an endpoint of any $I \in \mathcal{C}$. Thus, (R) $\int_{[0,1]} \varphi(x)\,d\psi(x) = 0$ for every $\varphi \in C\bigl([0,1]\bigr)$.

§ 2.1

2.1.18: To prove the first part, let $\Gamma_1 \subseteq \Gamma_2$ be given, set $\Delta = \Gamma_2 \setminus \Gamma_1$, note that $\Gamma_2 = \Gamma_1 \cup \Delta$ and $\Delta \cap \Gamma_1 = \emptyset$, and conclude that

$$|\Gamma_2| = |\Gamma_1| + |\Delta|.$$

Thus, if $|\Gamma_1| < \infty$, then it can be subtracted from both sides to get the desired conclusion. To handle the second part, write

$$\Gamma_1 \cup \Gamma_2 = \Gamma_1 \cup (\Gamma_2 \setminus (\Gamma_1 \cap \Gamma_2)),$$

apply additivity to get

$$|\Gamma_1 \cup \Gamma_2| = |\Gamma_1| + |\Gamma_2 \setminus (\Gamma_1 \cap \Gamma_2)|,$$

and use the first part to complete the derivation.

2.1.19: Set $\Gamma_0 = \emptyset$, and define $A_m = \Gamma_m \cap \bigcup_0^{m-1} \Gamma_\ell$ and $\Gamma'_m = \Gamma_m \setminus A_m$ for $m \in \mathbb{Z}^+$. By assumption,

$$|A_m| \le \sum_{\ell=0}^{m-1} |\Gamma_\ell \cap \Gamma_m| = 0,$$

and therefore $|\Gamma_m| = |\Gamma'_m|$. Hence, since the $\Gamma'_m \cap \Gamma'_n = \emptyset$ for $m \neq n$ while $\bigcup_1^\infty \Gamma_m = \bigcup_1^\infty \Gamma'_m$, one has that

$$\left|\bigcup_1^\infty \Gamma_m\right| = \sum_1^\infty |\Gamma'_m| = \sum_1^\infty |\Gamma_m|.$$

2.1.20: Part (**i**) is obvious, since C_k is the union of 2^k mutually disjoint intervals of length 3^{-k}. Furthermore, the first assertion in part (**ii**) is clear, and the second one follows by induction on ℓ. To prove the characterization of $\mathring{C}_\ell$, note that a is the left hand endpoint of one of the intervals I making up C_ℓ if and only if $\alpha_k(a) \in \{0,2\}$ for $0 \le k \le \ell$, $\alpha_k(a) = 0$ for $k > \ell$, in which case the associated right hand endpoint is $a + 3^{-\ell}$. Thus, $x \in \mathring{C}_\ell$ if and only if $\alpha_k(x) = \alpha_k(a)$, $0 \le k \le \ell$, for such an a and there is a $k > \ell$ such that $\alpha_k(x) \neq 0$. Once one has these facts, the remainder of (**ii**) is easy. Moreover, as a consequence of (**ii**), we know that the subset $\bigcap_{\ell=1}^\infty \mathring{C}_\ell$ of C is isomorphic to $\mathring{A}$ which, in turn, differs from $\{0,2\}^{\mathbb{N}}$ by a countable set. Hence, the uncountability of C follows from that of $\{0,2\}^{\mathbb{N}}$.

§ 2.2

2.2.3: To handle part (**i**), first note that, by translation invariance, we may assume that $H = \{y \in \mathbb{R}^N : (y, \mathbf{e}_1)_{\mathbb{R}^N} = 0\}$ for some $\mathbf{e}_1 \in \mathbb{R}^N$ with unit length. Next, choose $\mathbf{e}_2, \dots, \mathbf{e}_N \in \mathbb{R}^N$ so that $\{\mathbf{e}_1, \dots, \mathbf{e}_N\}$ forms an orthonormal basis in $\mathbb{R}^N$, and consider the matrix $\mathcal{O}$ whose ith row is $\mathbf{e}_i$. Then $\mathcal{O}$ is orthogonal and $T_{\mathcal{O}}H = H_0 \equiv \{y \in \mathbb{R}^N : y_1 = 0\}$. In particular, $|H| = |H_0|$; and so, the problem reduces to proving that $|H_0| = 0$. But clearly $H_0 = \bigcup_1^\infty R_n$, where $R_n = \{y \in \mathbb{R}^N : y_1 = 0 \text{ and } |y_i| \le n \text{ for } 2 \le i \le N\}$. Hence, since each R_n has measure 0, H_0 does also.

The second part is proved as follows. By translation invariance, it is enough to treat the case when $c = \mathbf{0}$. Thus, set $B(r) = B(\mathbf{0}, r)$. Then $B(r) = T_r B(1)$, where $T_r x = rx$ for $x \in \mathbb{R}^N$, and so $|B(r)| = \Omega_N r^N$. Finally, note that

$$|\overline{B}(r)| \le \lim_{R \searrow r} |B(R)| = \Omega_N r^N = |B(r)| \le |\overline{B}(r)|.$$

2.2.4: Thinking of the $\mathbf{v}_k$'s as column vectors, let $[\mathbf{v}_1 \dots \mathbf{v}_N]$ denote the $N \times N$-matrix whose ith column is $\mathbf{v}_i$. It is then clear that $\mathcal{P}(\mathbf{v}_1 \dots \mathbf{v}_N)$ is the image of $[0,1]^N$ under $[\mathbf{v}_1 \dots \mathbf{v}_N]$; and, therefore, Lebesgue would assign $\mathcal{P}(\mathbf{v}_1, \dots, \mathbf{v}_N)$ volume equal to the absolute value of the determinant of $[\mathbf{v}_1 \dots \mathbf{v}_N]$.

In order to connect this with the classical theory, it is helpful to observe first that

$$\begin{aligned}(\det[\mathbf{v}_1 \dots \mathbf{v}_N])^2 &= \det\left(\begin{bmatrix}\mathbf{v}_1^{\mathrm{T}} \\ \vdots \\ \mathbf{v}_N^{\mathrm{T}}\end{bmatrix}\begin{bmatrix}\mathbf{v}_1 \dots \mathbf{v}_N\end{bmatrix}\right) \\ &= \det\left[\left(\left((\mathbf{v}_i, \mathbf{v}_j)_{\mathbb{R}^N}\right)\right)_{1 \le i,j \le N}\right],\end{aligned}$$

where $\mathbf{v}_i^{\mathrm{T}}$ denotes the row vector representation of $\mathbf{v}_i$ and $(\mathbf{v}_i, \mathbf{v}_j)_{\mathbb{R}^N}$ is the inner (i.e., "dot") product of $\mathbf{v}_i$ and $\mathbf{v}_j$. Next, we work by induction on N. Clearly there is no problem when $N = 1$. Thus, assume that the two coincide for some $N \in \mathbb{Z}^+$, and let $\mathbf{v}_1, \dots, \mathbf{v}_{N+1} \in \mathbb{R}^{N+1}$ be given. Since there is no problem in the linearly dependent case, we will assume that $\mathbf{v}_1, \dots, \mathbf{v}_{N+1}$ are linearly independent. Choose an orthogonal matrix $\mathcal{O}$ so that $T_{\mathcal{O}}\mathbf{v}_1, \dots, T_{\mathcal{O}}\mathbf{v}_N \in H(\mathbf{e}_1, \dots, \mathbf{e}_N)$, where $\{\mathbf{e}_1, \dots, \mathbf{e}_{N+1}\}$ is the standard, orthonormal basis in $\mathbb{R}^{N+1}$. Then, the Lebesgue volume of $\mathcal{P}(\mathbf{v}_1, \dots, \mathbf{v}_N)$, considered as a subset of $H(\mathbf{v}_1, \dots, \mathbf{v}_N)$, is the same as that of $\mathcal{P}(T_{\mathcal{O}}\mathbf{v}_1, \dots, T_{\mathcal{O}}\mathbf{v}_N)$, considered as a subset of $H(\mathbf{e}_1, \dots, \mathbf{e}_N)$, and is therefore equal to the square root of

$$\det\left[\left(\left((T_{\mathcal{O}}\mathbf{v}_i, T_{\mathcal{O}}\mathbf{v}_j)_{\mathbb{R}^N}\right)\right)_{1 \le i,j \le N}\right] = \det\left[\left(\left((\mathbf{v}_i, \mathbf{v}_j)_{\mathbb{R}^N}\right)\right)_{1 \le i,j \le N}\right].$$

Hence, all that remains to check is that

$$\det\left[\left(\left((\mathbf{v}_i, \mathbf{v}_j)_{\mathbb{R}^N}\right)\right)_{1 \le i,j \le N+1}\right] = |\mathbf{u}|^2 \det\left[\left(\left((\mathbf{v}_i, \mathbf{v}_j)_{\mathbb{R}^N}\right)\right)_{1 \le i,j \le N}\right]$$

where $\mathbf{u} = \mathbf{v}_{N+1} - \mathbf{w}$ and $\mathbf{w}$ is the perpendicular projection of $\mathbf{v}_{N+1}$ onto $H(\mathbf{v}_1, \dots, \mathbf{v}_N)$. But, since the determinant is a linear function of any one of its columns and because $\mathbf{w} \in H(\mathbf{v}_1, \dots, \mathbf{v}_N)$, we see that

$$\det\left[\Big(\Big((\mathbf{v}_i, \mathbf{v}_j)_{\mathbb{R}^{N+1}}\Big)\Big)_{1 \le i,j \le N+1}\right] = \big(\det[\mathbf{v}_1 \dots \mathbf{v}_{N+1}]\big)^2 = \big(\det[\mathbf{v}_1 \dots \mathbf{v}_N\, \mathbf{u}]\big)^2$$

$$= \det \begin{bmatrix} (\mathbf{v}_1, \mathbf{v}_1)_{\mathbb{R}^{N+1}} & \dots & (\mathbf{v}_1, \mathbf{v}_N)_{\mathbb{R}^{N+1}} & 0 \\ (\mathbf{v}_2, \mathbf{v}_1)_{\mathbb{R}^{N+1}} & \dots & (\mathbf{v}_2, \mathbf{v}_N)_{\mathbb{R}^{N+1}} & 0 \\ \vdots & \ddots & \vdots & \vdots \\ (\mathbf{v}_N, \mathbf{v}_1)_{\mathbb{R}^{N+1}} & \dots & (\mathbf{v}_N, \mathbf{v}_N)_{\mathbb{R}^{N+1}} & 0 \\ 0 & \dots & 0 & |\mathbf{u}|^2 \end{bmatrix}$$

$$= |\mathbf{u}|^2 \det\left[\Big(\Big((\mathbf{v}_i, \mathbf{v}_j)_{\mathbb{R}^{N+1}}\Big)\Big)_{1 \le i,j \le N}\right].$$

§ 3.1

3.1.7: If $\mathcal{B}$ is a σ-algebra, then it is obviously both an algebra and a monotone class. Conversely, suppose that $\mathcal{B}$ is an algebra and a monotone class. To see that it is a σ-algebra, let $\{\Gamma_n\}_1^\infty \subseteq \mathcal{B}$ be given, and set $A_n = \bigcup_1^n \Gamma_m$. Because $\mathcal{B}$ is an algebra, $A_n \in \mathcal{B}$ for every n. Moreover, because $\mathcal{B}$ is a monotone class and $A_n \nearrow \bigcup_1^\infty \Gamma_m$, $\bigcup_1^\infty \Gamma_m \in \mathcal{B}$. Thus, $\mathcal{B}$ is a σ-algebra.

Now suppose that $\mathcal{A}$ is an algebra and that $\mathcal{B}$ is the intersection of all the monotone classes containing $\mathcal{A}$. Clearly, $\mathcal{B}$ is itself a monotone class which contains $\mathcal{A}$. Thus, all that remains is to prove that $\mathcal{B}$ is an algebra. To this end, let $\mathcal{M}_1$ be the class of all $B \subseteq E$ with the properties that $B\complement \in \mathcal{B}$ and both $A \cup B \in \mathcal{B}$ and $A \cap B \in \mathcal{B}$ for all $A \in \mathcal{A}$. It is then a trivial matter to check that $\mathcal{M}_1$ is a monotone class. Moreover, because $\mathcal{A}$ is an algebra, $\mathcal{A} \subseteq \mathcal{M}_1$. Hence, $\mathcal{B} \subseteq \mathcal{M}_1$. Finally, let $\mathcal{M}_2$ be the class of $B \subseteq E$ with the properties that $B\complement \in \mathcal{B}$ and both $A \cup B \in \mathcal{B}$ and $A \cap B \in \mathcal{B}$ for all $A \in \mathcal{B}$. Again, it is clear that $\mathcal{M}_2$ is a monotone class. In addition, because $\mathcal{B} \subseteq \mathcal{M}_1$, we know that $\mathcal{A} \subseteq \mathcal{M}_2$. Thus, $\mathcal{B} \subseteq \mathcal{M}_2$, from which it is follows that $\mathcal{B}$ is an algebra.

3.1.8: Let $\mathcal{H}$ denote the class of $B \in \mathcal{B}$ for which $\mu(B) = \nu(B)$. By assumption, $\mathcal{H} \supseteq \{E\} \cup \mathcal{C}$. In addition, using the properties proved in Theorem 3.1.6, one sees that $\mathcal{H}$ satisfies each of the properties (**b**), (**c**), and (**d**) in the definition of a λ-system. Hence, $\mathcal{H}$ is a λ-system, and therefore, by Lemma 3.1.3, $\mathcal{H} = \mathcal{B}$.

3.1.9: Denote by $\mathcal{A}$ the set of all $\Gamma \subseteq E$ with the property that, for every $\epsilon > 0$, there exist a closed $F \subseteq \Gamma$ and an open $G \supseteq \Gamma$ such that $\mu(G \setminus F) < \epsilon$. It is clear that $\mathcal{A} \subseteq \overline{\mathcal{B}_E}^\mu$. To prove the opposite inclusion, we first check that $\mathcal{A}$ is a σ-algebra. Obviously, $\emptyset \in \mathcal{A}$, and $\Gamma \in \mathcal{A} \implies \Gamma\complement \in \mathcal{A}$. Moreover, if $\{\Gamma_n\}_1^\infty \subseteq \mathcal{A}$, $\Gamma = \bigcup_1^\infty \Gamma_n$, and $\epsilon > 0$ is given, choose $\{F_n\}_1^\infty \subseteq \mathfrak{F}$ and $\{G_n\}_1^\infty \subseteq \mathfrak{G}$ so that $F_n \subseteq \Gamma_n \subseteq G_n$ and

$$\mu(G_n \setminus F_n) < \frac{\epsilon}{2^{n+1}}, \qquad n \in \mathbb{Z}^+.$$

Next, set $G = \bigcup_1^\infty G_n$, $A = \bigcup_1^\infty F_n$, and $A_n = \bigcup_{m=1}^n F_m$, $n \in \mathbb{Z}^+$. Then, $G \in \mathfrak{G}$, $\{A_n\}_1^\infty \subseteq \mathfrak{F}$, $A \subseteq \Gamma \subseteq G$, $A_n \nearrow A$, and $\mu(G \setminus A) < \frac{\epsilon}{2}$. Hence, after choosing $n \in \mathbb{Z}^+$ so that $\mu(A \setminus A_n) < \frac{\epsilon}{2}$, we see that $\Gamma \in \mathcal{A}$ and therefore that $\mathcal{A}$ is a σ-algebra.

We next check that $\mathcal{B}_E \subseteq \mathcal{A}$, and, in view of the preceding paragraph, this comes down to checking that $\mathfrak{G} \subseteq \mathcal{A}$. But if $\Gamma \in \mathfrak{G}$ and we take F_n to be the set of $x \in E$ such that $\operatorname{dist}(x, \Gamma\complement) \equiv \inf\{\rho(x,y) : y \in \Gamma\complement\} \geq \frac{1}{n}$, then $F_n \nearrow \Gamma$ and so $\mu(F_n) \nearrow \mu(\Gamma)$. Hence, for given $\epsilon > 0$, we can take find an $n \in \mathbb{Z}^+$ such that $\mu(\Gamma \setminus F_n) = \mu(\Gamma) - \mu(F_n) < \epsilon$; and therefore we can take $G = \Gamma$ and $F = F_n$.

Finally, suppose that $\Gamma_1 \subseteq \Gamma \subseteq \Gamma_2$, where $\Gamma_1, \Gamma_2 \in \mathcal{B}_E$ and $\mu(\Gamma_2 \setminus \Gamma_1) = 0$. To see that $\Gamma \in \mathcal{A}$, let $\epsilon > 0$ be given, and use the preceding to find $F \in \mathfrak{F}$ and $G \in \mathfrak{G}$ so that $F \subseteq \Gamma_1$, $\Gamma_2 \subseteq G$,

$$\mu(\Gamma_1 \setminus F) < \frac{\epsilon}{2}, \text{ and } \mu(G \setminus \Gamma_2) < \frac{\epsilon}{2}.$$

Clearly, $F \subseteq \Gamma \subseteq G$ and $\mu(G \setminus F) < \epsilon$.

3.1.10: This is another easy application of Lemma 3.1.3. Namely, let $\mathcal{H}$ denote the class of $B \in \mathcal{B}_2$ for which $\Phi^{-1}(B) \in \mathcal{B}_1$. By elementary set theory, one sees that $\mathcal{H}$ is a λ-system. Hence, since $\mathcal{H} \supseteq \mathcal{C}$ and $\mathcal{C}$ generates $\mathcal{B}_2$, $\mathcal{B}_2 \subseteq \mathcal{H}$ and so Φ is measurable. By taking $\mathcal{C}$ to be the open subsets of E_2, one gets the second part as an application of the first. Finally, let $x \in \mathbb{R}^N$ be given, and note that $x + \Gamma = \Phi^{-1}(\Gamma)$ when $\Phi(y) \equiv x - y$. Hence, the third part is an application of the second.

3.1.11: To see that $\mu(\partial Q) = 0$, it suffices to check that $\mu(\{x \in \mathbb{R}^N : x_i = \alpha\}) = 0$ for every $1 \leq i \leq N$ and $\alpha \in \mathbb{R}$; and, by translation invariance, this comes down to checking that $\mu(\{x \in Q_0 : x_i = 0\}) = 0$ for $1 \leq i \leq N$. But, if $\rho = \mu(\{x \in Q_0 : x_i = 0\})$, then, by translation invariance,

$$\mu(\{x \in Q_0 : x_i = \beta\}) = \mu(\{\beta\mathbf{e}_i + x : x \in Q_0 \text{ and } x_i = 0\}) = \rho$$

for every $\beta \in [0,1]$. Thus,

$$1 = \mu(Q_0) \geq \sum_{m=0}^{n} \mu\left(\left\{x \in Q_0 : x_i = \frac{m}{n}\right\}\right) \geq (n+1)\rho$$

for every $n \in \mathbb{Z}^+$; and therefore $\rho = 0$.

Next, to see that $\mu([0,m\lambda]^N) = m^N \mu([0,\lambda]^N)$ for $m \in \mathbb{Z}^+$ and $\lambda \in (0,\infty)$, note that

$$[0,m\lambda]^N = \bigcup\left\{\lambda\mathbf{k} + [0,\lambda]^N : \mathbf{k} \in \mathbb{Z}^N \cap [0,m-1]^N\right\};$$

and so, by the preceding, Exercise 2.1.19, and translation invariance,

$$\mu([0,m\lambda]^N) = \sum_{\mathbf{k}\in\mathbb{Z}^N\cap[0,m-1]^N} \mu(\lambda\mathbf{k} + [0,\lambda]^N) = m^N\mu([0,\lambda]^N).$$

From the preceding, we now know that $\mu([0,q]^N) = q^N$ for any $q \in \mathbb{Q}\cap(0,\infty)$ ($\mathbb{Q}$ is used to denote the rational numbers). Thus, if r is any element of $(0,\infty)$, then, by choosing $\{q_n\}_1^\infty \subseteq \mathbb{Q}$ so that $q_n \searrow r$, we see that

$$\mu([0,r]^N) = \lim_{n\to\infty} \mu([0,q_n]^N) = \lim_{n\to\infty} q_n^N = r^N.$$

Combining this with translation invariance, we conclude that $\mu(Q) = \lambda_{\mathbb{R}^N}(Q)$ for every cube Q in $\mathbb{R}^N$; and, since every open set G in $\mathbb{R}^N$ is the countable union of non-overlapping cubes, it follows immediately from what we already know that $\mu(G) = \lambda_{\mathbb{R}^N}(G)$ for all open G's in $\mathbb{R}^N$. Thus, since, for every $R \in (0,\infty)$, the set of all open subsets of $(-R,R)^N$ is π-system which generates $\mathcal{B}_{\mathbb{R}^N}[(-R,R)^N]$, we now know that μ coincides with λ_{R^N} on $\mathcal{B}_{\mathbb{R}^N}[(-R,R)^N]$ for every $R \in (0,\infty)$ and therefore on the whole of $\mathcal{B}_{\mathbb{R}^N}$.

3.1.12: **(i)** For $m \in \mathbb{Z}^+$, set $A_m = \bigcap_{n\geq m}\Gamma_n$. Then $A_m \nearrow \underline{\lim}_{n\to\infty}\Gamma_n$ and $A_n \subseteq \Gamma_n$ for every $n \in \mathbb{Z}^+$. Hence

$$\mu\left(\underline{\lim}_{n\to\infty}\Gamma_n\right) = \lim_{n\to\infty}\mu(A_n) \leq \underline{\lim}_{n\to\infty}\mu(\Gamma_n).$$

(ii) Set $B_m = \bigcup_{n\geq m}\Gamma_n$. Then $B_m \searrow \overline{\lim}_{n\to\infty}\Gamma_n$ and $\Gamma_n \subseteq B_n$ for every $n \in \mathbb{Z}^+$. Hence, if $\mu(B_1) < \infty$, then

$$\mu\left(\overline{\lim}_{n\to\infty}\Gamma_n\right) = \lim_{n\to\infty}\mu(B_n) \geq \overline{\lim}_{n\to\infty}\mu_n(\Gamma_n).$$

(iii) If $\sum_1^\infty \mu(\Gamma_n) < \infty$, then $\mu\left(\bigcup_1^\infty \Gamma_n\right) < \infty$ and so (cf. part **(ii)**):

$$\mu\left(\overline{\lim}_{n\to\infty}\Gamma_n\right) = \lim_{n\to\infty}\mu(B_n) \leq \lim_{m\to\infty}\sum_{n\geq m}\mu(\Gamma_n) = 0.$$

3.1.14: Part **(i)** is clear. To prove **(ii)**, we work by induction on $n \geq 2$. Thus, set $S_n = \{1,\ldots,n\}$, assume the result for some $n \geq 2$, and, after replacing Γ_n

by $\Gamma_n \cup \Gamma_{n+1}$, conclude that $-\mu(\Gamma_1 \cup \cdots \cup \Gamma_{n+1})$ is equal

$$\begin{aligned}
&\sum_{\emptyset\neq F\subseteq S_{n-1}} (-1)^{\mathrm{card}(F)}\mu(\Gamma_F) - \sum_{\emptyset\neq F\subseteq S_{n-1}} (-1)^{\mathrm{card}(F)}\mu\Big(\Gamma_F \cap (\Gamma_n \cup \Gamma_{n+1})\Big) \\
&= \sum_{\emptyset\neq F\subseteq S_{n-1}} (-1)^{\mathrm{card}(F)}\mu(\Gamma_F) - \sum_{\emptyset\neq F\subseteq S_{n-1}} (-1)^{\mathrm{card}(F)}\mu(\Gamma_F \cap \Gamma_n) \\
&\qquad - \sum_{\emptyset\neq F\subseteq S_{n-1}} (-1)^{\mathrm{card}(F)}\mu(\Gamma_F \cap \Gamma_{n+1}) + \sum_{\substack{F\subseteq S_{n+1}\\ \{n,n+1\}\subseteq F}} (-1)^{\mathrm{card}(F)}\mu(\Gamma_F) \\
&= \sum_{\emptyset\neq F\subseteq S_{n+1}} (-1)^{\mathrm{card}(F)}\mu(\Gamma_F),
\end{aligned}$$

where we have used

$$\Gamma_F \cap (\Gamma_n \cup \Gamma_{n+1}) = (\Gamma_F \cap \Gamma_n) \cup (\Gamma_F \cap \Gamma_{n+1})$$

and part (**i**) in order to get the second line.

(**iii**) In this case,

$$\mu(\Gamma) = \frac{\mathrm{card}(\Gamma)}{n!} \quad \text{for all } \Gamma \subseteq E;$$

and, for each $1 \le m \le n$ and $F \subseteq \{1, \dots, n\}$ with m elements, $\mathrm{card}(\Gamma_F) = (n-m)!$. Since there are $\binom{n}{m} = \frac{n!}{m!(n-m)!}$ sets $F \subseteq \{1, \dots, n\}$ with exactly m elements, we now see that

$$\mu(\Gamma_1 \cup \cdots \cup \Gamma_n) = -\sum_{m=1}^{n} \frac{(-1)^m}{m!}.$$

§ 3.2

3.2.16: There is nothing to do when f is measurable. Now suppose that, for every a from a dense subset D of $\mathbb{R}$, $\{f > a\} \in \mathcal{B}$, and note that this assumption is equivalent to saying that $f^{-1}(\Gamma) \in \mathcal{B}$ for every $\Gamma \in \mathcal{C} \equiv \{(a, \infty] : a \in D\}$. Hence, by Exercise 3.1.10, one will know that f is measurable once one shows that $\mathcal{C}$ generates the Borel sets over $\overline{\mathbb{R}}$; and this comes down to checking that $(x, y) \in \sigma(\mathbb{R}; \mathcal{C})$ for every $-\infty < x < y < \infty$. But if $-\infty < x < y < \infty$ are given, then one can find $\{a_n\} \subseteq (x, \infty) \cap D$ and $\{b_n\} \subseteq (-\infty, y) \cap D$ so that $a_n \searrow x$ and $b_n \nearrow y$, which means that

$$(x, y) = \left(\bigcup_1^\infty (a_n, \infty]\right) \cap \left(\bigcap_1^\infty (b_n, \infty]\complement\right) \in \mathcal{B}.$$

The cases when $a = -\infty$ or $b = \infty$ can be handled in the same way.

If $\{f \geq a\} \in \mathcal{B}$ for each $a \in D$, let $b \in \mathbb{R}$ be given, choose $\{a_n\}_1^\infty \subseteq D \cap (b, \infty)$ so that $a_n \searrow b$, and conclude that $\{f > b\} = \bigcup_1^\infty \{f \geq a_n\} \in \mathcal{B}$. Hence, this case reduces to the preceding one. The cases when the inequalities are reversed follow when one replaces f by $-f$.

To handle the second part of the problem, first note that it suffices to show that $\{f < g\} \in \mathcal{B}$. Indeed, $\{f \leq g\} = \{g < f\}\complement$, $\{f = g\} = \{f \leq g\} \cap \{g \leq f\}$, and $\{f \neq g\} = \{f = g\}\complement$. But, if $\mathbb{Q}$ denotes the set of rational numbers in $\mathbb{R}$, then

$$\{f < g\} = \bigcup_{a \in \mathbb{Q}} \{f < a\} \cap \{g > a\} \in \mathcal{B}.$$

3.2.17: Following the hint, set $\varphi = g - f$ and note that it suffices to check that $\mu(\varphi < 0) = 0$ if $\int_\Gamma \varphi \, d\mu \geq 0$ for every $\Gamma \in \mathcal{B}$. To this end, take $\Gamma_n = \{\varphi \leq -\frac{1}{n}\}$ and observe that

$$0 \leq \int_{\Gamma_n} \varphi \, d\mu \leq -\tfrac{1}{n}\mu(\Gamma_n).$$

Hence, $0 = \mu(\Gamma_n) \nearrow \mu(\varphi < 0)$.

3.2.18 If $S(f)$ is countable, then it is obvious that f is μ-integrable if and only if $\sum_{x \in S(f)} |f(x)| < \infty$. Thus, the only problem is to show that f is μ-integrable only if $S(F)$ is countable; and, in checking this, we may and will assume that f is both μ-integrable and non-negative. Since $S(f) = \bigcup_{n=1}^\infty A_n$, where $A_n = \{f \geq n^{-1}\}$; and, by (3.2.9), $\operatorname{card}(A_n) = \mu(A_n) < \infty$ for each n. Hence, $S(f)$ is at most countable.

§ 3.3

3.3.16: To see that μ is a finite measure on $(E, \mathcal{B})$, it suffices to check that it is countably additive. Thus, let $\{\Gamma_n\}_1^\infty \subseteq \mathcal{B}$ be a sequence of mutually disjoint sets, and define

$$F_n = f \sum_{m=1}^{n} \mathbf{1}_{\Gamma_m} \text{ for } n \in \mathbb{Z}^+.$$

Then $F_n \nearrow f\mathbf{1}_\Gamma$ where $\Gamma = \bigcup_1^\infty \Gamma_n$; and so, by the Monotone Convergence Theorem,

$$\mu(\Gamma) = \int_\Gamma f \, d\nu = \lim_{n \to \infty} \int F_n \, d\nu = \lim_{n \to \infty} \sum_{m=1}^{n} \int_{\Gamma_m} f \, d\nu = \sum_1^\infty \mu(\Gamma_n).$$

To check the absolute continuity assertion, suppose that there were an $\epsilon > 0$ such that, for every $n \in \mathbb{Z}^+$, there exists a $\Gamma_n \in \mathcal{B}$ for which $\nu(\Gamma_n) < 2^{-n}$ and $\mu(\Gamma_n) \geq \epsilon$. Set $\Gamma = \overline{\lim}_{n \to \infty} \Gamma_n$. Since $\mu(E) < \infty$, one would know (cf. **(ii)** of Exercise 3.1.12) that $\mu(\Gamma) \geq \epsilon$. On the other hand, by the Borel–Cantelli Lemma (cf. **(iii)** of Exercise 3.1.12), one would also have that $\nu(\Gamma)$ and therefore $\int_\Gamma f \, d\nu$ both vanish. Since $\mu(\Gamma) = \int_\Gamma f \, d\nu$, this is clearly a contradiction.

3.3.17: Clearly, by Markov's Inequality, $\mu(|f| \geq \lambda) \longrightarrow 0$ as $\lambda \longrightarrow \infty$; and therefore, by the Lebesgue Dominated Convergence Theorem,

$$\int_{\{|f|\geq\lambda\}} |f|\, d\mu \longrightarrow 0 \text{ as } \lambda \longrightarrow \infty.$$

But, by a second application of Markov's inequality, this proves (3.3.18).

To produce the required example, consider the function

$$f(x) = \frac{1}{x(1-\log x)}, \quad x \in [0,1] \ (\equiv \infty \text{ when } x = 0).$$

Clearly, f is decreasing and $f(1) = 1$. Hence, for each $\lambda \in [1,\infty)$, $\{x : f(x) \geq \lambda\} = [0, x_\ell]$, where x_λ is the element of $(0,1]$ for which $x_\ell(1 + \log x_\ell) = \frac{1}{\lambda}$. In particular, since $x_\lambda \searrow 0$ as $\lambda \nearrow \infty$,

$$\lambda\big|\{x : f(x) \geq \lambda\}\big| = \lambda x_\lambda = \big(1 + \log x_\lambda\big)^{-1} \longrightarrow 0 \quad \text{as } \lambda \to \infty.$$

On the other hand,

$$\int_{[0,1]} f\, d\lambda_{\mathbb{R}} = \sum_1^\infty \int_{\{n\leq f<n+1\}} f\, d\lambda_{\mathbb{R}} \geq \sum_1^\infty n(x_n - x_{n+1}) = \sum_1^\infty x_n;$$

and, because $x_n \geq \big(n(1+\log n)\big)^{-1}$, this shows that f is not integrable.

Finally, if f is non-negative and measurable and one defines $\Gamma_n = \big\{f \in (n-1, n]\big\}$ for $n \in \mathbb{Z}^+$, then

$$\begin{aligned}\int f\, d\mu &= \sum_1^\infty \int_{\Gamma_n} f\, d\mu \leq \sum_1^\infty n\mu(\Gamma_n)\\ &= \sum_1^\infty n\mu(f > n-1) - \sum_1^\infty n\mu(f > n) = \sum_0^\infty \mu(f > n)\end{aligned}$$

when $\sum_1^\infty \mu(f > n) < \infty$; and, when $\mu(E) < \infty$,

$$\begin{aligned}\int_{\{f\leq N\}} f\, d\mu &= \sum_1^N \int_{\Gamma_n} f\, d\mu \geq \sum_1^N (n-1)\mu(\Gamma_n)\\ &\geq \sum_1^N n\mu(\Gamma_n) - \mu(E) \geq \sum_1^N \mu(f > n) - 2\mu(E)\end{aligned}$$

for every $N \in \mathbb{Z}^+$, from which the required result follows after $N \nearrow \infty$.

3.3.19: For $n \in \mathbb{N}$, set $\mathcal{K}_n = \{\mathbf{k} \in \mathbb{N}^N : \max_i k_i < 2^n\}$. Given $J = \prod_1^N [a_i, b_i]$, set $\ell_i = b_i - a_i$ and, for $n \in \mathbb{N}$ and $\mathbf{k} \in \mathcal{K}_n$, set

$$I_{\mathbf{k},n} = \prod_1^N \Big[a_i + 2^{-n}k_i\ell_i, a_i + 2^{-n}(k_i+1)\ell_i\Big).$$

Next, given $f \in C(J)$, set

$$f_n = \sum_{\mathbf{k}\in\mathcal{K}_n} f\big((a_1 + 2^{-n}k_1\ell_1, \dots, a_N + 2^{-n}k_N\ell_N)\big)\mathbf{1}_{I_{\mathbf{k},n}}.$$

Then, by Lebesgue's Dominated Convergence Theorem,

$$\sum_{\mathbf{k}\in\mathcal{K}_n} f\big((a_1 + 2^{-n}k_1\ell_1, \dots, a_N + 2^{-n}k_N\ell_N)\big)\mathrm{vol}(I_{\mathbf{k},n}) = \int_J f_n\, d\lambda_{\mathbb{R}^N} \longrightarrow \int_J f\, d\lambda_{\mathbb{R}^N}.$$

At the same time,

$$\sum_{\mathbf{k}\in\mathcal{K}_n} f\big((a_1 + 2^{-n}k_1\ell_1, \dots, a_N + 2^{-n}k_N\ell_N)\big)\mathrm{vol}(I_{\mathbf{k},n}) = \mathcal{R}(f;\mathcal{C}_n,\xi_n) \longrightarrow (\mathrm{R})\int_J f(x)\, dx,$$

where $\mathcal{C}_n = \{\overline{I}_{\mathbf{k},n} : \mathbf{k} \in \mathcal{K}_n\}$ and $\xi_n(I_{\mathbf{k},n}) = (a_1+2^{-n}k_1\ell_1, \dots, a_N+2^{-n}k_N\ell_N)$.

To prove the second part, let $\epsilon > 0$ be given and simply choose J_ϵ so that

$$\int_{J_\epsilon \complement} |f|\, d\lambda_{\mathbb{R}^N} < \epsilon.$$

3.3.20: Without loss of generality, we assume that all the g_n's are $\mathbb{R}$-valued and that, in part (**i**), $f_n \le g_n$, and, in part (**ii**), $|f_n| \le g_n$, everywhere.

(**i**) First choose a subsequence $\{f_{n'}\}$ of $\{f_n\}$ so that

$$\lim_{n'\to\infty} \int f_{n'}\, d\mu = \varlimsup_{n\to\infty} \int f_n\, d\mu,$$

and second choose a subsequence $\{g_{n''}\}$ of $\{g_{n'}\}$ so that $g_{n''} \longrightarrow g$ (a.e., μ). Then, by Fatou's Lemma applied to $h_n \equiv g_n - f_n \ge 0$,

$$\begin{aligned}\int g\, d\mu - \varlimsup_{n\to\infty} \int f_n\, d\mu &= \lim_{n''\to\infty} \int h_{n''}\, d\mu \ge \int \varliminf_{n''\to\infty} h_{n''}\, d\mu \\ &= \int \Big[g - \varlimsup_{n''\to\infty} f_{n''}\Big]\, d\mu \ge \int g\, d\mu - \int \varlimsup_{n\to\infty} f_n\, d\mu.\end{aligned}$$

(**ii**) Given the preceding, the proof of part (**ii**) in the case of almost everywhere convergence is exactly the same as the derivation of Lebesgue's Dominated Convergence Theorem from Fatou's Lemma. Namely, one sets $h_n = |f_n - f|$ and observes that $h_n \le g_n + g$. The case of convergence in measure is then handled in precisely the same way as it was in the proof of Theorem 3.3.11.

3.3.21: **(i)** If $\mathcal{K}$ is uniformly μ-absolutely continuous and $M \equiv \sup_{f\in\mathcal{K}} \|f\|_{L^1(\mu)} < \infty$, choose, for a given $\epsilon > 0$, $\delta > 0$ so that $\sup_{f\in\mathcal{K}} \int_\Gamma |f|\,d\mu < \epsilon$ whenever $\Gamma \in \mathcal{B}$ satisfies $\mu(\Gamma) < \delta$ and choose $R \in (0,\infty)$ so that $\frac{M}{R} < \delta$. Then, by Markov's inequality, $\sup_{f\in\mathcal{K}} \mu(|f| \geq R) \leq \delta$ and so $\sup_{f\in\mathcal{K}} \int_{|f|\geq R} |f|\,d\mu \leq \epsilon$.

Next, suppose that $\mathcal{K}$ is uniformly μ-integrable, and define

$$A(R) \equiv \sup_{f\in\mathcal{K}} \int_{\{|f|\geq R\}} |f|\,d\mu \quad \text{for } R \in (0,\infty).$$

Clearly

$$\sup_{f\in\mathcal{K}} \int_\Gamma |f|\,d\mu \leq R\mu(\Gamma) + A(R)$$

for any $\Gamma \in \mathcal{B}$ and $R \in (0,\infty)$. Hence, if, for given $\epsilon > 0$, we choose $R \in (0,\infty)$ so that $A(R) < \frac{\epsilon}{2}$, then $\int_\Gamma |f|\,d\mu < \epsilon$ for all $f \in \mathcal{K}$ and $\Gamma \in \mathcal{B}$ with $\mu(\Gamma) < \frac{\epsilon}{2R}$; and so $\mathcal{K}$ is uniformly μ-absolutely continuous. In addition, when $\mu(E) < \infty$, by choosing $R \in (0,\infty)$ so that $A(R) \leq 1$, we see that $\|f\|_{L^1(\mu)} \leq R\mu(E) + 1 < \infty$ for all $f \in \mathcal{K}$.

(ii) Note that, for any $R \in (0,\infty)$,

$$\int_{\{|f|\geq R\}} |f|\,d\mu \leq R^{-\delta} \int_{|f|\geq R} |f|^{1+\delta}\,d\mu \leq R^{-\delta} \int |f|^{1+\delta}\,d\mu.$$

(iii) Suppose that $f_n \longrightarrow f$ in $L^1(\mu)$. Clearly $f \in L^1(\mu)$ and $\sup_{n\in\mathbb{Z}^+} \|f_n\|_{L^1(\mu)} < \infty$. In addition, for given $\epsilon > 0$, choose $m \in \mathbb{Z}^+$ so that $\|f_n - f\|_{L^1(\mu)} < \frac{\epsilon}{4}$ for $n \geq m$; and, using Exercise 3.3.16, choose $\delta > 0$ so that

$$\max_{1\leq n\leq m} \int_\Gamma |f_n|\,d\mu < \tfrac{\epsilon}{4}$$

for all $\Gamma \in \mathcal{B}$ with $\mu(\Gamma) < \delta$, and note that $\int_\Gamma |f|\,d\mu < \frac{\epsilon}{2}$ whenever $\mu(\Gamma) < \delta$. Hence, for this choice of $\delta > 0$,

$$\sup_{n\geq m} \int_\Gamma |f_n|\,d\mu \leq \|f_n - f\|_{L^1(\mu)} + \int_\Gamma |f|\,d\mu < \epsilon$$

for all $\Gamma \in \mathcal{B}$ with $\mu(\Gamma) < \delta$.

Conversely, if $\mu(E) < \infty$, $f_n \longrightarrow f$ in μ-measure, and $\{f_n\}_1^\infty$ is uniformly μ-integrable, note that f is μ-integrable, and (cf. **(i)** above and Exercise 3.3.16) choose, for a given $\epsilon > 0$, a $\delta > 0$ so that

$$\mu(\Gamma) < \delta \implies \int_\Gamma |f_n - f|\,d\mu \leq \int_\Gamma |f_n|\,d\mu + \int_\Gamma |f|\,d\mu < \epsilon$$

for all $n \in \mathbb{Z}^+$. Hence, if $\Gamma_n = \left\{|f_n - f| \geq \frac{\epsilon}{\mu(E)}\right\}$ and m is chosen so that $\mu(\Gamma_n) < \delta$ for all $n \geq m$, then

$$\|f_n - f\|_{L^1(\mu)} = \int_{\Gamma_n\complement} |f_n - f|\,d\mu + \int_{\Gamma_n} |f_n - f|\,d\mu \leq \epsilon \quad \text{for } n \geq m.$$

(iv) Tightness enables one to reduce each of these assertions to the finite measure situation, in which case they have already been established.

3.3.22: If $f_n \longrightarrow f$ in μ-measure, then, because μ is finite, $\int |f_n - f| \wedge 1\, d\mu \longrightarrow 0$ by Lebesgue's Dominated Convergence Theorem. On the other hand, for each $\epsilon \in (0,1]$,

$$\mu(|f_n - f| \geq \epsilon) = \mu(|f_n - f| \wedge 1 \geq \epsilon) \leq \epsilon^{-1} \int |f_n - f| \wedge 1\, d\mu$$

follows from Markov's inequality. Thus, even if μ is not finite,

$$\int |f_n - f| \wedge 1\, d\mu \longrightarrow 0 \implies f_n \longrightarrow f \quad \text{in } \mu\text{-measure.}$$

3.3.23: Clearly the general case follows immediately from the case in which $\mu(E) \vee \nu(E) < \infty$; and therefore we will assume that μ and ν are both finite. Thus, by Exercise 3.1.8, all that we have to do is check that $\mu(G) = \nu(G)$ for every open subset of E.

Let G be an open set in E, and define

$$\varphi(x) = \left(\frac{\operatorname{dist}(x, G\complement)}{1 + \operatorname{dist}(x, G\complement)} \right)^{\frac{1}{n}}, \qquad x \in E \text{ and } n \in \mathbb{Z}^+,$$

where $\operatorname{dist}(x, \Gamma) \equiv \inf\{\rho(x,y) : y \in \Gamma\}$ is the ρ-distance from x to Γ. One then has that φ_n is uniformly ρ-continuous for each $n \in \mathbb{Z}^+$ and that $0 \leq \varphi_n(x) \nearrow \mathbf{1}_G(x)$ as $n \longrightarrow \infty$ for each $x \in E$. Thus, by the Monotone Convergence Theorem,

$$\mu(G) = \lim_{n\to\infty} \int \varphi_n\, d\mu = \lim_{n\to\infty} \int \varphi_n\, d\nu = \nu(G).$$

3.3.24: Simply observe that

$$\mu\left(\sup_{n \geq m} |f_n - f| \geq \epsilon \right) \leq \sum_{n=m}^{\infty} \mu(|f_n - f| \geq \epsilon) \longrightarrow 0 \quad \text{as } m \to \infty,$$

if $\sum_1^\infty \mu(|f_n - f| \geq \epsilon) < \infty$. In particular, since, by Markov's inequality,

$$\sum_m^\infty \mu(|f_n - f| \geq \epsilon) \leq \epsilon^{-1} \sum_m^\infty \|f_n - f\|_{L^1(\mu)},$$

this will be the case for every $\epsilon > 0$ when $\sum_1^\infty \|f_n - f\|_{L^1(\mu)} < \infty$.

§ 3.4

3.4.12: The first calculation is just Exercise 3.3.19 plus the Fundamental Theorem of Calculus. As for the second, note that, from the first calculation,

$$Mf(x) \geq \frac{1}{x} \int_{(0,x)} f(t)\,dt = (-x \log x)^{-1} \quad \text{for } x \in (0, e^{-1}),$$

and therefore, after a second application of the Fundamental Theorem of Calculus, that

$$\int_{(0,r)} Mf(x)\,dx = \lim_{\delta \searrow 0} \int_{(\delta,r)} (-x \log x)^{-1}\,dx = \lim_{\delta \searrow 0} \log\left(\frac{\log r}{\log \delta}\right) = \infty.$$

3.4.13: **(i)** Since

$$\begin{aligned}\left| \frac{1}{\delta} \int_{[x,x+\delta)} f(t)\,dt - f(x) \right| &\leq \frac{1}{\delta} \int_{[x,x+\delta)} \big|f(t) - f(x)\big|\,dt \\ &\leq 2\frac{1}{2\delta} \int_{(x-\delta,x+\delta)} \big|f(t) - f(x)\big|\,dt,\end{aligned}$$

it is clear that

$$x \in \text{Leb}(f) \implies \frac{1}{\delta} \int_{[x,x+\delta)} f(x)\,dx \longrightarrow f(x) \quad \text{as } \delta \searrow 0.$$

Similarly,

$$x \in \text{Leb}(f) \implies \frac{1}{\delta} \int_{(x-\delta,x]} f(x)\,dx \longrightarrow f(x) \quad \text{as } \delta \searrow 0.$$

(ii) Assume that f is continuous and has compact support. Obviously (cf. Exercise 5.0.3 below),

$$\begin{aligned}f_\epsilon(x) &= \int \rho_\epsilon(t) f(x-t)\,dt = \int \rho(t) f(x - \epsilon t)\,dt \\ &= \int_{[0,\infty)} \rho(-t) f(x + \epsilon t)\,dt + \int_{[0,\infty)} \rho(t) f(x - \epsilon t)\,dt.\end{aligned}$$

Now, set $\psi(t) = \frac{1}{\epsilon} \int_{[x,x+\epsilon t)} f(s)\,ds$, and use Exercise 3.3.19 and Theorem 1.2.7 to get

$$\begin{aligned}\int_{[0,\infty)} \rho(-t) f(x+\epsilon t)\,dt &= \lim_{R\to\infty} (\text{R}) \int_{[0,R]} \rho(-t)\,d\psi(t) \\ &= -\lim_{R\to\infty} (\text{R}) \int_{[0,R]} \psi(t)\,d\rho(-t) = \int_{[0,\infty)} t\rho'(-t) \frac{1}{t} \psi(t)\,dt,\end{aligned}$$

which is precisely (3.4.16). The derivation of (3.4.16') is essentially the same.

To extend the result to $f \in L^1(\mathbb{R})$ and $x \in \mathrm{Leb}(f)$, use Corollary 3.3.15 to find $\{f_n\}_1^\infty \subseteq C_c(\mathbb{R})$ so that $f_n \longrightarrow f$ in $L^1(\mathbb{R})$ and observe that

$$\begin{aligned}\int_{[0,\infty)} \rho(\mp t) f_n(x \mp \epsilon t)\,dt &= \int_{[x,\infty)} \rho_\epsilon(-x \mp t) f_n(\mp t)\,dt \\ &\longrightarrow \int_{[x,\infty)} \rho_\epsilon(-x \mp t) f(\mp t)\,dt = \int_{[0,\infty)} \rho(\mp t) f(x \mp \epsilon t)\,dt,\end{aligned}$$

while, for $x \in \mathrm{Leb}(f)$,

$$\begin{aligned}\frac{1}{t}\int_{[x,x+\epsilon t)} f_n(s)\,ds &\longrightarrow \frac{1}{t}\int_{[x,x+\epsilon t)} f(s)\,ds \\ \frac{1}{t}\int_{(x-\epsilon t,x]} f_n(s)\,ds &\longrightarrow \frac{1}{t}\int_{(x-\epsilon t,x]} f(s)\,ds\end{aligned}$$

boundedly and pointwise in $t \in (0,\infty)$. Thus, the required extension follows from the integrability of $t\rho'(t)$ plus Lebesgue's Dominated Convergence Theorem.

(iii) By **(i)**, **(ii)**, and Lebesgue's Dominated Convergence Theorem, it is clear that

$$\lim_{\epsilon \searrow 0} f_\epsilon(x) = -f(x) \int t\rho'(t)\,dt \quad \text{for } x \in \mathrm{Leb}(f).$$

Finally, again by Exercise 3.3.19 and Theorem 1.2.7,

$$-\int t\rho'(t)\,dt = -\lim_{R\to\infty} (\mathrm{R}) \int_{[-R,R]} t\,d\rho(t) \int \rho(t)\,dt = 1.$$

§ 4.1

4.1.8: Set $F(x,t) = f(x) - t$ for $(x,t) \in E \times [0,\infty)$, note that F is $\mathcal{B} \times \mathcal{B}_{[0,\infty)}$-measurable, and conclude that $\Gamma(f) = \{(x,t) : F(x,t) \geq 0\}$ and $\widehat{\Gamma}(f) = \{(x,t) : F(x,t) > 0\}$ are both $\mathcal{B} \times \mathcal{B}_{[0,\infty)}$-measurable. Finally, note that, for each $x \in E$,

$$\lambda_{\mathbb{R}}\big(\{t : (x,t) \in \Gamma(f)\}\big) = f(x) = \lambda_{\mathbb{R}}\big(\{t : (x,t) \in \widehat{\Gamma}(f)\}\big),$$

and get the desired result as an application of Tonelli's Theorem to $\mathbf{1}_{\Gamma(f)}$ and $\mathbf{1}_{\widehat{\Gamma}(f)}$, respectively.

4.1.10: In proving the first part, we begin by assuming that $\mu_i(E_i) < \infty$ for $i \in \{1,2\}$ and that $\nu(E_1 \times E_2) = \mu_1(E_1)\mu_2(E_2)$. Set $\mathcal{C} = \{C_1 \times C_2 : C_i \in \mathcal{C}_i \text{ for } i \in \{1,2\}\}$. Clearly, $\mathcal{C}$ is a π-system which generates $\mathcal{B}_1 \times \mathcal{B}_2$ and contains $E_1 \times E_2$. Thus, by Exercise 3.1.8, $\nu = \mu_1 \times \mu_2$ under the present assumptions. To remove the additional assumptions, note that, by the preceding,

$$\nu \restriction \mathcal{B}_1[E_{1,m}] \times \mathcal{B}_2[E_{2,n}] = (\mu_1 \times \mu_2) \restriction \mathcal{B}_1[E_{1,m}] \times \mathcal{B}_2[E_{2,n}] \quad \text{for all } m,\, n \in \mathbb{Z}^+.$$

Next, set

$$\tilde{E}_{i,1} = E_{i,1} \quad \text{and} \quad \tilde{E}_{i,n+1} = E_{i,n+1} \setminus \bigcup_1^n E_{i,m},$$

and use the above to see that

$$\nu \restriction \mathcal{B}_1[\tilde{E}_{1,m}] \times \mathcal{B}_2[\tilde{E}_{2,n}] = (\mu_1 \times \mu_2) \restriction \mathcal{B}_1[\tilde{E}_{1,m}] \times \mathcal{B}_2[\tilde{E}_{2,n}] \quad \text{for all } m,\, n \in \mathbb{Z}^+.$$

Finally, for $\Gamma \in \mathcal{B}_1 \times \mathcal{B}_2$, set $\Gamma_{(m,n)} = \Gamma \cap (\tilde{E}_{1,m} \times \tilde{E}_{2,n})$, and note that, since $\Gamma_{(m,n)} \in \mathcal{B}_1[\tilde{E}_{1,m}] \times \mathcal{B}_1[\tilde{E}_{2,n}]$,

$$\nu(\Gamma) = \sum_{(m,n)} \nu(\Gamma_{(m,n)}) = \sum_{(m,n)} \mu_1 \times \mu_2(\Gamma_{(m,n)}) = \mu_1 \times \mu_2(\Gamma).$$

To show that $\lambda_{\mathbb{R}^{M+N}} = \lambda_{\mathbb{R}^M} \times \lambda_{\mathbb{R}^N}$ on $\mathcal{B}_{\mathbb{R}^M} \times \mathcal{B}_{\mathbb{R}^N}$, take $\mathcal{C}_1$ and $\mathcal{C}_2$ in the preceding to be the rectangles in $\mathbb{R}^M$ and $\mathbb{R}^N$, respectively.

4.1.11: This is nothing more or less than Tonelli's Theorem applied to $f = \mathbf{1}_\Gamma$. The only point here is that, by passing through $\mu_1 \times \mu_2$, one is able to show that μ_1-almost every vertical slice has μ_2-measure 0 if and only if μ_2-almost every horizontal slice has μ_1-measure 0.

4.1.12: To see that μ_1 is a measure, let $\{\Gamma_n\} \subseteq \mathcal{B}_{(0,1)}$ be a sequence of mutually disjoint sets. We must show that

$$\mu_1\left(\bigcup_1^\infty \Gamma_n\right) = \sum_1^\infty \mu_1(\Gamma_n).$$

If any one of the Γ_n's has infinitely many elements, then both sides of the preceding are infinite. On the other hand, if each Γ_n is finite, then both sides of the above are expressions for the total number (infinite or not) of elements in $\bigcup_1^\infty \Gamma_n$. The hint given makes the rest of the exercise trivial.

4.1.13: To prove the first part, we may and will assume that $a = -\infty$ and $b = \infty$, since we can always extend f to $\mathbb{R} \times E$ by taking $f(t,x) = f(a,x)$ when $-\infty < t \le a$ and $f(t,x) = f(b,x)$ when $t \ge b < \infty$. Next, for $n \in \mathbb{Z}^+$, set

$$f_n(t,x) = f\left(\tfrac{k}{n}, x\right) \quad \text{for } k \in \mathbb{Z} \text{ and } k \le nt < k+1.$$

Obviously, each f_n is $\mathcal{B}_{\mathbb{R}} \times \mathcal{B}$-measurable. Moreover, as $n \to \infty$, $f_n \longrightarrow f$ pointwise. Thus, f is also $\mathcal{B}_{\mathbb{R}} \times \mathcal{B}$-measurable.

Now add the assumption that $f(\cdot\,, x) \in C^1\big((a,b)\big)$ for each $x \in E$. Given $n \in \mathbb{Z}^+$, set

$$g_n(t,x) = \begin{cases} n\Big(f\big(t+\frac{1}{n}, x\big) - f(t,x)\Big) & \text{if } a < t \le b - \frac{1}{n} \\ 0 & \text{if } a \vee \big(b - \frac{1}{n}\big) < t < b. \end{cases}$$

Then, by the preceding, each g_n is $\mathcal{B}_{(a,b)} \times \mathcal{B}$-measurable. In addition, $g_n \longrightarrow f'$ pointwise. Thus, f' is $\mathcal{B}_{(a,b)} \times \mathcal{B}$-measurable.

To complete the exercise when $|f(t,x)| \vee |f'(t,x)| \le g(x)$ for some $g \in L^1(\mu)$, let g_n be the function given in the preceding, and note that, by the Mean Value Theorem, $|g_n(t,x)| \le g(x)$. Hence, by Lebesgue's Dominated Convergence Theorem,

$$\frac{d}{dt}\int f(t,x)\,\mu(dx) = \lim_{n\to\infty}\int g_n(t,x)\,\mu(dx) = \int f'(t,x)\,\mu(dx).$$

Finally, the continuity of $t \in (a,b) \longmapsto \int f'(t,x)\,\mu(dx) \in \mathbb{R}$ is just another application of Lebesgue's Dominated Convergence Theorem.

§ 4.2

4.2.9: Because (cf. the paragraph preceding Lemma 4.2.6) $\mathbf{H}^N$ is countably sub-additive, it suffices to prove that $\mathbf{H}^N(\Gamma) = 0$ for every bounded subset Γ of a hyperplane in $\mathbb{R}^N$. In addition, since $\mathbf{H}^N$ is trivially translation and rotation invariant, we may and will assume that $\Gamma = [-R,R]^{N-1} \times \{0\}$ for some $R \in (0,\infty)$. But in that case, for each $n \in \mathbb{Z}^+$, Γ can be covered by the sets

$$I(\mathbf{k},n) \equiv \left(\prod_{i=1}^{N-1}\left[\tfrac{k_i}{n}, \tfrac{k_i+1}{n}\right]\right) \times \{0\}, \quad -nR \le k_i \le nR \text{ for } 1 \le i \le N-1.$$

But there are no more than $(2nR+1)^{N-1}$ such $I(\mathbf{k},n)$'s, and each one has radius $\frac{\sqrt{N}}{n}$. Hence, there is a constant $C < \infty$, which is independent of n, such that

$$\sum_{\mathbf{k}} \Omega_N \mathrm{rad}\big(I(\mathbf{k},n)\big)^N \le \frac{C}{n}.$$

§ 5.0

5.0.3: Note that, for any $\Gamma \in \overline{\mathcal{B}}_{\mathbb{R}^N}$, $T_A^{-1}(\Gamma) = T_{A^{-1}}(\Gamma)$, and therefore, by Theorem 2.2.2,

$$(T_A)_*\lambda_{\mathbb{R}}^N(\Gamma) = \big|\det(A^{-1})\big|\,|\Gamma| = \big|\det(A)\big|^{-1}\lambda_{\mathbb{R}^N}(\Gamma).$$

5.0.4: The first assertion is equivalent to $\psi^{-1}(t) \le s \iff \psi(s) \ge t$ for all $s \in [a,b]$ and $t \in [0,L]$. But, by right-continuity, $\psi\big(\psi^{-1}(t)\big) \ge t$, and therefore $s \ge \psi^{-1}(t) \implies \psi(s) \ge t$. At the same time, it is obvious that $\psi(s) \ge t \implies s \ge \psi^{-1}(t)$. Hence, we now know that the first assertion is true. Given the first assertion, it is clear that ψ^{-1} must be $\mathcal{B}_{[0,L]}$-measurable. Finally, since

$$\mu_\psi\big([a,s]\big) = \lambda_{[0,L]}\big(\{t \in [0,L] : \psi^{-1}(t) \le s\}\big) = \lambda_{\mathbb{R}}\big([0,\psi(s)]\big) = \psi(s),$$

and therefore

$$\mu_\psi(\{a\}) = \lim_{\epsilon \searrow 0} \mu_\psi\big([a,a+\epsilon]\big) = \lim_{\epsilon \searrow 0} \psi(\epsilon) = 0,$$

it follows that $\mu_\psi\big((a,s]\big) = \mu_\psi\big([a,s]\big) = \psi(s)$.

§ 5.1

5.1.6: **(i)** To prove (5.1.7), first note that $f \circ \psi$ is ψ-Riemann integrable on $[a,b]$ and that f is Riemann integrable on $\big[\psi(a),\psi(b)\big]$. Thus, all that we have to do is take $t_{m,n} = \big(1 - \frac{m}{n}\big)a + \frac{m}{n}b$ for $n \in \mathbb{Z}^+$ and $0 \le m \le n$, and note that

$$\mathcal{R}\big(f \circ \psi \,\big|\, \psi; \mathcal{C}_n, \xi_n\big) = \mathcal{R}\big(f; \tilde{\mathcal{C}}_n, \tilde{\xi}_n\big),$$

where

$$\mathcal{C}_n = \big\{[t_{m,n}, t_{m+1,n}] : 0 \le m < n\big\} \quad \text{and} \quad \xi_n\big([t_{m,n}, t_{m+1,n}]\big) = t_{m,n},$$

while

$$\tilde{\mathcal{C}}_n = \big\{\big[\psi(t_{m,n}), \psi(t_{m+1,n})\big] : 0 \le m < n\big\}$$
$$\text{and} \quad \tilde{\xi}_n\big(\big[\psi(t_{m,n}), \psi(t_{m+1,n})\big]\big) = \psi(t_{m,n}).$$

(ii) Φ is obviously non-decreasing. Moreover, $\lim_{b \searrow a} \Phi(b) - \Phi(a) = \mu(\{a\}) = 0$ and $\lim_{a \nearrow b} \Phi(b) - \Phi(a) = \mu(\{b\}) = 0$. Hence, Φ is continuous. To see that $\Phi_*\mu = \lambda_{\mathbb{R}^N} \upharpoonright \mathcal{B}_{\mathbb{R}^N}\big[\Phi(\mathbb{R})\big]$, let $-\infty < a < b < \infty$ be given, and apply **(i)** with $\psi = \Phi \upharpoonright [a,b]$ to see that

$$(\mathrm{R}) \int_{[a,b]} f \circ \Phi(t)\, d\Phi(t) = (\mathrm{R}) \int_{[\Phi(a),\Phi(b)]} f(t)\, dt \quad \text{for } f \in C\big([a,b];\mathbb{R}\big).$$

Finally, by Theorem 5.1.2 and Lemma 5.0.1,

$$(\mathrm{R}) \int_{[a,b]} f \circ \Phi(t)\, d\Phi(t) = \int_{[a,b]} f \circ \Phi \, d\mu = \int_{\big[\Phi(a),\Phi(b)\big]} f \, d\Phi_*\mu$$

for $f \in C\big([a,b];\mathbb{R}\big)$. From this it is immediate that

$$\int_{\Phi(\mathbb{R})} f \, d\Phi_*\mu = \int_{\Phi(\mathbb{R})} f \, d\lambda_{\mathbb{R}},$$

first for $f \in C_c\big(\Phi(\mathbb{R});\mathbb{R}\big)$ and then for general, non-negative, $\mathcal{B}_{\mathbb{R}}\big[\Phi(R)\big]$-measurable functions.

5.1.8: Without loss of generality, assume that f is non-negative. Then, by (5.1.9), $|f|^p$ is μ-integrable if and only if

$$\int_{(0,\infty)} t^{p-1}\mu(f>t)\,dt<\infty.$$

Now set $I_n=\left(\frac{1}{n+1},\frac{1}{n}\right]$ and $J_n=(n,n+1]$ for $n\in\mathbb{Z}^+$. Then,

$$\begin{aligned}\int_{(0,1]} t^{p-1}\mu(f>t)\,dt &= \sum_1^\infty \int_{I_n} t^{p-1}\mu(f>t)\,dt\\ &\le \sum_1^\infty n^{-p+1}\mu\left(f>\tfrac{1}{n+1}\right)\frac{1}{n(n+1)}\\ &\le 2^p\sum_1^\infty \frac{1}{(n+1)^{p+1}}\mu\left(f>\tfrac{1}{n+1}\right)\le 2^p\sum_1^\infty \frac{1}{n^{p+1}}\mu\left(f>\tfrac{1}{n}\right)\end{aligned}$$

and

$$\begin{aligned}\int_{(1,\infty)} t^{p-1}\mu(f>t)\,dt &= \sum_1^\infty \int_{J_n} t^{p-1}\mu(f>t)\,dt\\ &\le \sum_1^\infty (n+1)^{p-1}\mu(f>n)\le 2^{p-1}\sum_1^\infty n^{p-1}\mu(f>n).\end{aligned}$$

Thus,

$$\int_{(0,\infty)} t^{p-1}\mu(f>t)\,dt\le 2^p\sum_1^\infty \frac{1}{n^{p+1}}\mu\left(f>\tfrac{1}{n}\right)+2^{p-1}\sum_1^\infty n^{p-1}\mu(f>n).$$

Conversely,

$$\begin{aligned}\int_{(0,1]} t^{p-1}\mu(f>t)\,dt &= \sum_1^\infty \int_{I_n} t^{p-1}\mu(f>t)\,dt\\ &\ge \sum_1^\infty \frac{1}{(n+1)^{p-1}}\mu\left(f>\tfrac{1}{n}\right)\frac{1}{n(n+1)}\\ &\ge 2^{-p}\sum_1^\infty \frac{1}{n^{p+1}}\mu\left(f>\tfrac{1}{n}\right),\end{aligned}$$

and

$$\begin{aligned}\int_{(0,\infty)} t^{p-1}\mu(f>t)\,dt &\ge \int_{\left(\frac{1}{2},1\right]} t^{p-1}\mu(f>t)\,dt+\sum_1^\infty \int_{J_n} t^{p-1}\mu(f>t)\,dt\\ &\ge 2^{-p}\mu(f>1)+\sum_1^\infty n^{p-1}\mu(f>n+1)\ge 2^{-p}\sum_1^\infty n^{p-1}\mu(f>n).\end{aligned}$$

Thus,

$$2^{p+1}\int_{(0,\infty)} t^{p-1}\mu(f>t)\,dt \ge \sum_{1}^{\infty}\frac{1}{n^{p+1}}\mu\left(f>\tfrac{1}{n}\right)+\sum_{1}^{\infty}n^{p-1}\mu(f>n).$$

§ 5.2

5.2.4: **(i)** First, let Γ be a non-empty open subset of $\mathbf{S}^{N-1}$, and set $G=\{x\in B(0,1)\setminus\{0\}:\Phi(x)\in\Gamma\}$. Then, because Φ maps $B(0,1)\setminus\{0\}$ continuously onto $\mathbf{S}^{N-1}$, G is a non-empty open subset of $\mathbb{R}^N$, which means that $\lambda_{\mathbf{S}^{N-1}}(\Gamma)=N\lambda_{\mathbb{R}^N}(G)>0$.

Next, let $\mathcal{O}$ be an orthogonal matrix, and denote by $\mathbf{R}$ the corresponding rotation $T_{\mathcal{O}}$ on $\mathbb{R}^N$. Then $\Phi\circ\mathbf{R}=\mathbf{R}\circ\Phi$ on $\mathbb{R}^N\setminus\{0\}$, $\mathbf{1}_{B(0,1)}\circ\mathbf{R}=\mathbf{1}_{B(0,1)}$, and $\mathbf{R}_*\lambda_{\mathbb{R}^N}=\lambda_{\mathbb{R}^N}$; therefore the asserted invariance of $\lambda_{\mathbf{S}^{N-1}}$ follows directly from its definition. To prove the asserted orthogonality relations, define for each $\boldsymbol{\xi}\in\mathbb{R}^N\setminus\{0\}$ the reflection $\mathbf{R}_{\boldsymbol{\xi}}:\mathbb{R}^N\longrightarrow\mathbb{R}^N$ by

$$\mathbf{R}_{\boldsymbol{\xi}}x=x-\frac{2(\boldsymbol{\xi},x)_{\mathbb{R}^N}\boldsymbol{\xi}}{|\boldsymbol{\xi}|^2},\quad x\in\mathbb{R}^N.$$

Then, by rotation invariance,

$$\begin{aligned}\int_{\mathbf{S}^{N-1}}(\boldsymbol{\xi},\boldsymbol{\omega})_{\mathbb{R}^N}\,\lambda_{\mathbf{S}^{N-1}}(d\boldsymbol{\omega})&=\int_{\mathbf{S}^{N-1}}(\boldsymbol{\xi},\mathbf{R}_{\boldsymbol{\xi}}\boldsymbol{\omega})_{\mathbb{R}^N}\,\lambda_{\mathbf{S}^{N-1}}(d\boldsymbol{\omega})\\&=-\int_{\mathbf{S}^{N-1}}(\boldsymbol{\xi},\boldsymbol{\omega})_{\mathbb{R}^N}\,\lambda_{\mathbf{S}^{N-1}}(d\boldsymbol{\omega}).\end{aligned}$$

Similarly, for $\boldsymbol{\eta}\perp\boldsymbol{\xi}$,

$$\begin{aligned}\int_{\mathbf{S}^{N-1}}(\boldsymbol{\xi},\boldsymbol{\omega})_{\mathbb{R}^N}(\boldsymbol{\eta},\boldsymbol{\omega})_{\mathbb{R}^N}\,\lambda_{\mathbf{S}^{N-1}}(d\boldsymbol{\omega})&=\int_{\mathbf{S}^{N-1}}(\boldsymbol{\xi},\mathbf{R}_{\boldsymbol{\xi}}\boldsymbol{\omega})_{\mathbb{R}^N}(\boldsymbol{\eta},\mathbf{R}_{\boldsymbol{\xi}}\boldsymbol{\omega})_{\mathbb{R}^N}\,\lambda_{\mathbf{S}^{N-1}}(d\boldsymbol{\omega})\\&=-\int_{\mathbf{S}^{N-1}}(\boldsymbol{\xi},\boldsymbol{\omega})_{\mathbb{R}^N}(\boldsymbol{\eta},\boldsymbol{\omega})_{\mathbb{R}^N}\,\lambda_{\mathbf{S}^{N-1}}(d\boldsymbol{\omega});\end{aligned}$$

and therefore, for any $\boldsymbol{\eta}\in\mathbb{R}^N$,

$$\int_{\mathbf{S}^{N-1}}(\boldsymbol{\xi},\boldsymbol{\omega})_{\mathbb{R}^N}(\boldsymbol{\eta},\boldsymbol{\omega})_{\mathbb{R}^N}\,\lambda_{\mathbf{S}^{N-1}}(d\boldsymbol{\omega})=(\boldsymbol{\xi},\boldsymbol{\eta})_{\mathbb{R}^N}\int_{\mathbf{S}^{N-1}}\left(\frac{\boldsymbol{\xi}}{|\boldsymbol{\xi}|},\boldsymbol{\omega}\right)^2_{\mathbb{R}^N}\lambda_{\mathbf{S}^{N-1}}(d\boldsymbol{\omega}).$$

Finally, if $\{\mathbf{e}_1,\dots,\mathbf{e}_N\}$ is the standard orthonormal basis for $\mathbb{R}^N$, then, again by rotation invariance,

$$\int_{\mathbf{S}^{N-1}}\left(\frac{\boldsymbol{\xi}}{|\boldsymbol{\xi}|},\boldsymbol{\omega}\right)^2_{\mathbb{R}^N}\lambda_{\mathbf{S}^{N-1}}(d\boldsymbol{\omega})=\int_{\mathbf{S}^{N-1}}(\mathbf{e}_i,\boldsymbol{\omega})^2_{\mathbb{R}^N}\,\lambda_{\mathbf{S}^{N-1}}(d\boldsymbol{\omega})\text{ for }1\le i\le N;$$

and so

$$\int_{\mathbf{S}^{N-1}} \left(\frac{\boldsymbol{\xi}}{|\boldsymbol{\xi}|}, \boldsymbol{\omega}\right)^2_{\mathbb{R}^N} \lambda_{\mathbf{S}^{N-1}}(d\boldsymbol{\omega}) = \frac{1}{N}\sum_{i=1}^N \int_{\mathbf{S}^{N-1}} (\mathbf{e}_i, \boldsymbol{\omega})^2_{\mathbb{R}^N} \lambda_{\mathbf{S}^{N-1}}(d\boldsymbol{\omega})$$
$$= \frac{\lambda_{\mathbf{S}^{N-1}}(\mathbf{S}^{N-1})}{N} = \Omega_N.$$

(ii) Using $\mathbf{R}_\theta$ to denote the rotation determined by the matrix $\mathcal{O}_\theta$ described in the hint, we see, by rotation invariance and Tonelli's Theorem, that

$$\int_{\mathbf{S}^1} f\,d\nu = \frac{1}{2\pi}\int_{\mathbf{S}^1}\left(\int_{[0,2\pi]} f\circ\mathbf{R}_\theta(\boldsymbol{\omega})\,d\theta\right)\nu(d\boldsymbol{\omega})$$

for any non-negative, Borel measurable f on $\mathbf{S}^1$. Next, for fixed $\boldsymbol{\omega}\in\mathbf{S}^1$, choose $\eta_{\boldsymbol{\omega}}\in[0,2\pi)$ so that

$$\boldsymbol{\omega} = \begin{bmatrix}\cos\eta_{\boldsymbol{\omega}}\\ \sin\eta_{\boldsymbol{\omega}}\end{bmatrix}$$

and observe that

$$\int_{[0,2\pi]} f\circ\mathbf{R}_\theta(\boldsymbol{\omega})\,d\theta = \int_{[0,2\pi]} f\big(\cos(\eta_{\boldsymbol{\omega}}+\theta), \sin(\eta_{\boldsymbol{\omega}}+\theta)\big)\,d\theta$$
$$= \int_{[\eta_{\boldsymbol{\omega}},2\pi]} f\big(\cos\theta,\sin\theta\big)\,d\theta + \int_{[0,\eta_{\boldsymbol{\omega}}]} f\big(\cos(2\pi+\theta),\sin(2\pi+\theta)\big)\,d\theta$$
$$= \int_{[0,2\pi]} f\big(\cos\theta,\sin\theta\big)\,d\theta = \int_{\mathbf{S}^1} f\,d\mu.$$

Combined with the preceding, this now shows that

$$\int_{\mathbf{S}^1} f\,d\nu = \frac{\nu(\mathbf{S}^{N-1})}{2\pi}\int_{\mathbf{S}^1} f\,d\mu.$$

(iii) Let f be a non-negative, continuous function on $\mathbb{R}^N$ with compact support. Then, by Tonelli's Theorem and Theorem 5.2.2,

$$\int_{(0,\infty)} r^N\left(\int_{[-1,1]\times\mathbf{S}^{N-1}} (1-\rho^2)^{\frac{N}{2}-1} f\big(r\Xi(\rho,\boldsymbol{\omega})\big)\,d\rho\,\lambda_{\mathbf{S}^{N-1}}(d\boldsymbol{\omega})\right)dr$$
$$= \int_{[-1,1]} (1-\rho^2)^{\frac{N}{2}-1} F(\rho)\,d\rho$$

where

$$F(\rho) \equiv \int_{(0,\infty)\times\mathbf{S}^{N-1}} r^{N-1} r f\Big(r(1-\rho^2)^{\frac{1}{2}}\boldsymbol{\omega}, \rho r\Big)\,dr\times\lambda_{\mathbf{S}^{N-1}}(d\boldsymbol{\omega})$$
$$= \int_{\mathbb{R}^N\setminus\{0\}} |x| f\Big((1-\rho^2)^{\frac{1}{2}}x, \rho|x|\Big)\,dx$$
$$= (1-\rho^2)^{-\frac{N}{2}}\int_{\mathbb{R}^N\setminus\{\mathbf{0}\}} \frac{|y|}{(1-\rho^2)^{\frac{1}{2}}} f\left(y, \frac{\rho|y|}{(1-\rho^2)^{\frac{1}{2}}}\right)dy$$

Hence,

$$\int_{(0,\infty)} r^N \left(\int_{[-1,1]\times \mathbf{S}^{N-1}} (1-\rho^2)^{\frac{N}{2}-1} f\big(r\Xi(\rho,\omega)\big)\, d\rho\, \lambda_{\mathbf{S}^{N-1}}(d\omega) \right) dr$$

$$= \int_{\mathbb{R}^N\setminus\{0\}} \left((\mathrm{R}) \int_{[-\delta(y),\delta(y)]} f\big(y,\psi_y(\rho)\big)\, d\psi_y(\rho) \right) dy$$

where, for each $y \in \mathbb{R}^N \setminus \{0\}$ and $\rho \in [-\delta(y),\delta(y)]$, $\psi_y(\rho) = \frac{\rho|y|}{(1-\rho^2)^{\frac{1}{2}}}$, and $\delta(y) \in (0,1)$ is chosen so that

$$f\left(y, \frac{\rho|y|}{(1-\rho^2)^{\frac{1}{2}}} \right) = 0 \text{ for } \delta(y) \le \rho < 1.$$

Since, by part (**i**) of Exercise 5.1.6,

$$\begin{aligned}(\mathrm{R}) \int_{[-\delta(y),\delta(y)]} f\big(y,\psi_y(\rho)\big)\, d\psi_y(\rho) &= (\mathrm{R}) \int_{[\psi_y(-\delta(y)),\psi_y(\delta(y))]} f(y,\sigma)\, d\sigma \\ &= \int_{\mathbb{R}^1} f(y,\sigma)\, d\sigma\end{aligned}$$

for each $y \in \mathbb{R}^N \setminus \{0\}$, we now see that

$$\begin{aligned}&\int_{(0,\infty)} r^N \left(\int_{[-1,1]\times \mathbf{S}^{N-1}} (1-\rho^2)^{\frac{N}{2}-1} f\big(r\Xi(\rho,\omega)\big)\, d\rho \times \lambda_{\mathbf{S}^{N-1}}(d\omega) \right) dr \\ &\quad = \int_{\mathbb{R}^{N+1}} f(z)\, dz = \int_{(0,\infty)} r^N \left(\int_{\mathbf{S}^N} f(r\omega)\, \lambda_{\mathbf{S}^N}(d\omega) \right) dr;\end{aligned}$$

from which the desired result follows easily by taking f to be of the form $\eta(|z|)g\left(\frac{z}{|z|}\right)$.

5.2.6: (**i**) By Tonelli's Theorem, Theorem 5.2.2, and Theorem 1.2.7,

$$\begin{aligned}\left(\int_{\mathbb{R}} e^{-\frac{x^2}{2}}\, dx \right)^2 &= \int_{\mathbb{R}} e^{-\frac{x_1^2}{2}} \left(\int_{\mathbb{R}} e^{-\frac{x_2^2}{2}}\, dx_2 \right) dx_1 \\ &= \int_{\mathbb{R}^2} e^{-\frac{|x|^2}{2}}\, dx = \int_{\mathbf{S}^1} \left(\int_{(0,\infty)} r e^{-\frac{r^2}{2}}\, dr \right) d\omega \\ &= -2\pi \lim_{R\to\infty} (\mathrm{R}) \int_{[0,R]} d\left(e^{-\frac{r^2}{2}}\right) = 2\pi.\end{aligned}$$

Next, given A, set $\Sigma = A^{-\frac{1}{2}}$, the symmetric, positive definite square root of A^{-1}. Then, by Exercise 5.0.3 and Tonelli's Theorem,

$$\int_{\mathbb{R}^N} e^{-\frac{1}{2}(x,A^{-1}x)}\,dx = \int_{\mathbb{R}^N} e^{-\frac{1}{2}|T_\Sigma x|^2}\,dx$$
$$= \left|\det(\Sigma)\right|^{-1}\int_{\mathbb{R}^N} e^{-\frac{|x|^2}{2}}\,dx = \left(\det(A)\right)^{\frac{1}{2}}\left(\int_{\mathbb{R}} e^{-\frac{x^2}{2}}\,dx\right)^N,$$

which, together with the preceding, gives the required result.

(ii) To check the functional equation, use Theorem 1.2.7 to justify,

$$\Gamma(\gamma+1) = -\lim_{R\to\infty}(\mathrm{R})\int_{[R^{-1},R]} t^\gamma\,d\left(e^{-t}\right)$$
$$= \gamma\lim_{R\to\infty}(\mathrm{R})\int_{[R^{-1},R]} t^{\gamma-1}e^{-t}\,dt = \gamma\Gamma(\gamma).$$

Similarly, by (5.1.7) and (5.2.7),

$$\Gamma\left(\tfrac{1}{2}\right) = 2\lim_{R\to\infty}(\mathrm{R})\int_{[R^{-1},R]} e^{-t}\,d\left(t^{\frac{1}{2}}\right)$$
$$= 2\lim_{R\to\infty}(\mathrm{R})\int_{[R^{-2},R^2]} e^{-x^2}\,dx = \int_{\mathbb{R}} e^{-x^2}\,dx = \pi^{\frac{1}{2}}.$$

(iii) First observe that

$$(2\pi)^{\frac{N}{2}} = \int_{\mathbb{R}^N} e^{-\frac{x^2}{2}}\,dx = \omega_{N-1}\int_{(0,\infty)} r^{N-1}e^{-\frac{r^2}{2}}\,dr.$$

Second, use (5.1.7) to justify

$$\int_{(0,\infty)} r^{N-1}e^{-\frac{r^2}{2}}\,dr = \lim_{R\to\infty}(\mathrm{R})\int_{[R^{-1},R]} r^{N-2}e^{-\frac{r^2}{2}}\,d\left(\frac{r^2}{2}\right)$$
$$= 2^{\frac{N}{2}-1}\lim_{R\to\infty}(\mathrm{R})\int_{[2R^{-2},2^{-1}R^2]} t^{\frac{N}{2}-1}e^{-t}\,dt = 2^{\frac{N}{2}-1}\Gamma\left(\tfrac{N}{2}\right).$$

Hence, $(2\pi)^{\frac{N}{2}} = 2^{\frac{N}{2}-1}\omega_{N-1}\Gamma\left(\frac{N}{2}\right)$, and so

$$\omega_{N-1} = \frac{2\pi^{\frac{N}{2}}}{\Gamma\left(\frac{N}{2}\right)} \quad\text{and}\quad \Omega_N = \frac{\omega_{N-1}}{N} = \frac{\pi^{\frac{N}{2}}}{\frac{N}{2}\Gamma\left(\frac{N}{2}\right)} = \frac{\pi^{\frac{N}{2}}}{\Gamma\left(\frac{N}{2}+1\right)}.$$

(iv) Define ψ as in the hint, and observe that

$$\psi(s)^{\frac{1}{2}} = \frac{s+\left(s^2+4\alpha\beta\right)^{\frac{1}{2}}}{2\alpha}, \quad s\in\mathbb{R},$$

and therefore that

$$\frac{d}{ds}\psi(s)^{\frac{1}{2}} = \frac{1}{2\alpha}\left(1 + \frac{s}{\left(s^2 + 4\alpha\beta\right)^{\frac{1}{2}}}\right).$$

Thus, by Exercise 5.1.6, for any $R \in (0, \infty)$,

$$\begin{aligned}
e^{2\alpha\beta} & \int_{[\psi(-R)^{\frac{1}{2}}, \psi(R)^{\frac{1}{2}}]} t^{-\frac{1}{2}} \exp\left[-\alpha^2 t - \frac{\beta^2}{t}\right] dt \\
&= \int_{[\psi(-R)^{\frac{1}{2}}, \psi(R)^{\frac{1}{2}}]} t^{-\frac{1}{2}} \exp\left[-\left(\alpha t^{\frac{1}{2}} - \frac{\beta}{t^{\frac{1}{2}}}\right)^2\right] dt \\
&= (\mathrm{R}) \int_{[-R,R]} \exp\left[-\left(\alpha\psi(s)^{\frac{1}{2}} - \frac{\beta}{\psi(s)^{\frac{1}{2}}}\right)^2\right] d\psi(s)^{\frac{1}{2}} \\
&= \frac{1}{2\alpha}(\mathrm{R}) \int_{[-R,R]} e^{-s^2}\left(1 + \frac{s}{\left(s^2 + 4\alpha\beta\right)^{\frac{1}{2}}}\right) ds = \frac{1}{2\alpha}\int_{[-R,R]} e^{-s^2}\, ds.
\end{aligned}$$

Equivalently,

$$\int_{[\psi(-R)^{\frac{1}{2}}, \psi(R)^{\frac{1}{2}}]} t^{-\frac{1}{2}} \exp\left[-\alpha^2 t - \frac{\beta^2}{t}\right] dt = \frac{e^{-2\alpha\beta}}{2\alpha}\int_{[-R,R]} e^{-s^2}\, ds,$$

and, after differentiating with respect to β and applying Exercise 4.1.13, one also has that

$$\int_{[\psi(-R)^{\frac{1}{2}}, \psi(R)^{\frac{1}{2}}]} t^{-\frac{3}{2}} \exp\left[-\alpha^2 t - \frac{\beta^2}{t}\right] dt = \frac{e^{-2\alpha\beta}}{2\beta}\int_{[-R,R]} e^{-s^2}\, ds.$$

Clearly the desired results follow when one lets $R \nearrow \infty$.

§ 5.3

5.3.20: Since $\mathbf{\Phi}(G) \subseteq B(\mathbf{0}, 1)$ and $B(\mathbf{0}, 1) \setminus \mathbf{\Phi}(G) \subseteq \{(x_1, x_2) : x_2 = 0\}$, we know that $\Omega_N = |\mathbf{\Phi}(G)|$. In addition,

$$J\mathbf{\Phi}(r, \theta) = \begin{bmatrix} \cos\theta & r\sin\theta \\ \sin\theta & r\cos\theta \end{bmatrix},$$

and so

$$|\mathbf{\Phi}(G)| = \int_G \delta\mathbf{\Phi}(r, \theta)\, dr\, d\theta = \int_{(0,2\pi)}\left(\int_{(0,1)} r\, dr\right) d\theta = \pi.$$

5.3.21: Choose $r > 0$ and $F : B_{\mathbb{R}^N}(p,r) \longrightarrow \mathbb{R}$ so that (5.3.7) holds. Without loss of generality, we will assume that r is taken so that F has bounded second order derivatives on $B_{\mathbb{R}^N}(p,r)$. In particular, since F vanishes on $B_{\mathbb{R}^N}(p,r) \cap M$, there is a $C < \infty$ with the property that $|F(y)| \le C\,\mathrm{dist}(y,M)$ for all $y \in B_{\mathbb{R}^N}\left(p, \frac{r}{3}\right)$. Hence,

$$\varlimsup_{\xi\to 0} \frac{\mathrm{dist}(p+\xi\mathbf{v}, M)}{\xi^2} < \infty \implies \varlimsup_{\xi\to 0} \frac{F(p+\xi\mathbf{v}) - F(p)}{\xi^2} < \infty$$
$$\implies \big(\mathbf{v}, \nabla F(p)\big)_{\mathbb{R}^N} = 0 \implies \mathbf{v} \in \mathbf{T}_p(M).$$

Conversely, if $\mathbf{v} \in \mathbf{T}_p(M)$ and $\gamma : (-\epsilon,\epsilon) \longrightarrow M$ is chosen accordingly, then

$$\mathrm{dist}(p+\xi\mathbf{v}, M) \le |p+\xi\mathbf{v} - \gamma(\xi)| = \mathcal{O}(\xi^2) \quad \text{as } \xi \to 0.$$

5.3.22: If $f \in C(\mathbb{R}^N)$, then, by Theorem 5.2.2 and (5.3.10),

$$\begin{aligned}
\int_{\mathbf{S}^{N-1}(r)} f(\omega) r^{N-1} (\boldsymbol{\Phi}_r)_* \lambda_{\mathbf{S}^{N-1}}(d\omega) &= \int_{\mathbf{S}^{N-1}} f(r\omega) r^{N-1}\, \lambda_{\mathbf{S}^{N-1}}(d\omega) \\
&= \lim_{\delta\searrow 0} \frac{1}{2\delta} \int_{[r-\delta,r+\delta]} \rho^{N-1} \left(\int_{\mathbf{S}^{N-1}} f(\rho\omega)\, \lambda_{\mathbf{S}^{N-1}}(d\omega) \right) d\rho \\
&= \int_{\left[\mathbf{S}^{N-1}(r)\right](\delta)} f(x)\, dx = \int_{\mathbf{S}^{N-1}(r)} f(\omega)\, \lambda_{\mathbf{S}^{N-1}(r)}(d\omega);
\end{aligned}$$

and, given the preceding for all continuous f's, it is an easy matter to check it for all Borel measurable f's.

5.3.23: Without loss of generality, assume that $\Gamma = \boldsymbol{\Psi}(U)$, where $(\boldsymbol{\Psi}, U)$ is a coordinate chart for M. But then, since $\delta\boldsymbol{\Psi} > 0$ on U, (5.3.18) applies and says that

$$\lambda_M(\Gamma) = \int_U \delta\boldsymbol{\Psi}(u)\, du > 0.$$

5.3.24: (**i**) By Tonelli's Theorem,

$$\Gamma(\alpha)\Gamma(\beta) = \iint_{(0,\infty)^2} s^{\alpha-1} t^{\beta-1} e^{-s-t}\, ds\, dt.$$

Now let Φ be the mapping suggested in the hint, and observe that Φ maps $(0,\infty)\times(0,1)$ diffeomorphically onto $(0,\infty)^2$ and that $\delta\Phi(u,v) = u$. Hence,

$$\begin{aligned}
\Gamma(\alpha)\Gamma(\beta) &= \iint_{(0,\infty)^2} s^{\alpha-1} t^{\beta-1} e^{-(s+t)}\, dsdt \\
&= \iint_{(0,\infty)\times(0,1)} (uv)^{\alpha-1} \big(u(1-v)\big)^{\beta-1} e^{-u} u\, dudv \\
&= \int_{(0,\infty)} u^{\alpha+\beta-1} e^{-u}\, du \int_{(0,1)} v^{\alpha-1}(1-v)^{\beta-1}\, dv = \Gamma(\alpha+\beta) B(\alpha,\beta).
\end{aligned}$$

(**ii**) By Theorem 5.2.2,

$$\int_{\mathbb{R}^N} \left(1+|x|^2\right)^{-\lambda} dx = \omega_{N-1} \int_{(0,\infty)} \frac{r^{N-1}}{(1+r^2)^\lambda}\, dr.$$

Moreover, if Φ is the map in the hint, then, by Exercise 4.1.6 or Jacobi's Transformation Formula, the integral on the right is equal to

$$\frac{1}{2}\int_{(0,\infty)} \Phi(r)^{\frac{N-2}{2}} \left(1-\Phi(r)\right)^{\lambda-\frac{N}{2}-1} \Phi'(r)\, dr$$

$$= \frac{1}{2}\int_{(0,1)} u^{\frac{N}{2}-1}(1-u)^{\lambda-\frac{N}{2}-1}\, du = \frac{\Gamma\left(\frac{N}{2}\right)\Gamma\left(\lambda-\frac{N}{2}\right)}{2\Gamma(\lambda)}.$$

Hence, since (cf. (**iii**) in Exercise 4.2.16) $\omega_{N-1} = \frac{2\pi^{\frac{N}{2}}}{\Gamma\left(\frac{N}{2}\right)}$, the desired result follows.

A second approach to the same computation is to first observe that

$$\frac{\Gamma(\lambda)}{(1+|x|^2)^\lambda} = \int_{(0,\infty)} t^{\lambda-1} \exp\left[-t(1+|x|^2)\right] dt$$

and therefore that

$$\Gamma(\lambda)\int_{\mathbb{R}^N} \left(1+|x|^2\right)^{-\lambda} dx = \int_{(0,\infty)} t^{\lambda-1} e^{-t} \left(\int_{\mathbb{R}^N} e^{-t|x|^2}\, dx\right) dt$$

$$= \int_{(0,\infty)} t^{\lambda-1} \left(\tfrac{\pi}{t}\right)^{\frac{N}{2}} e^{-t}\, dt = \pi^{\frac{N}{2}}\Gamma\left(\lambda-\frac{N}{2}\right).$$

(**iii**) Set $\Phi(\xi)=\xi^2$ for $\xi\in(0,1)$. Then, by (5.3.4),

$$\int_{(-1,1)} (1-\xi^2)^{\lambda-1}\, d\xi = 2\int_{(0,1)} (1-\xi^2)^{\lambda-1}\, d\xi = \int_{(0,1)} (1-\eta)^{\lambda-1}\eta^{-\frac{1}{2}}\, d\eta$$

$$= \frac{\Gamma(\lambda)\Gamma\left(\frac{1}{2}\right)}{\Gamma\left(\lambda+\frac{1}{2}\right)} = \frac{\pi^{\frac{1}{2}}\Gamma(\lambda)}{\Gamma\left(\lambda+\frac{1}{2}\right)}.$$

In particular (cf. (**iii**) of Exercise 5.2.6),

$$\int_{(-1,1)} (1-\xi^2)^{\frac{N}{2}-1}\, d\xi = \frac{\pi^{\frac{1}{2}}\Gamma\left(\frac{N}{2}\right)}{\Gamma\left(\frac{N+1}{2}\right)} = \frac{\omega_N}{\omega_{N-1}}.$$

5.3.25: (**i**) To see that M is a hypersurface, simply take $F(x) = x_N - f(x_1,\dots,x_{N-1})$ for $x\in G\equiv U\times\mathbb{R}$. It is clear that $M=\{x\in G:\ F(x)=0\}$ and that $F_{,N}=1$ on G. Thus, (5.3.7) is trivially satisfied. Next, since $\mathbf{\Psi}$ is obviously 1–1 on U, we will know that $(\mathbf{\Psi},U)$ is a coordinate chart for M as soon as we check that $\mathbf{\Psi}_{,1}(u),\dots,\mathbf{\Psi}_{,N-1}(u)$ are linearly independent for each $u\in U$. But $\mathbf{\Psi}_{,j} = \mathbf{e}_j + f_{,j}(u)\mathbf{e}_N$, and so this linear independence is inherited from that of $\mathbf{e}_1,\dots,\mathbf{e}_N$.

(**ii**) Using the last part of (**i**), one sees that

$$\left(\mathbf{\Psi}_{,i},\mathbf{\Psi}_{,j}\right)_{\mathbb{R}^N} = \delta_{i,j} + f_{,i}f_{,j}, \quad 1\le i,j\le N-1,$$

which is equivalent to $\left(\!\left(\,(\mathbf{\Psi}_{,i},\mathbf{\Psi}_{,j})_{\mathbb{R}^N}\,\right)\!\right) = \mathbf{I}+\nabla f^{\mathrm{T}}\nabla f$. From here, the expression for $\delta\mathbf{\Psi}$ follows immediately from the hint given.

(**iii**) Given (**i**) and (**ii**), (**iii**) is just a re-statement of (5.3.18).

5.3.26: (**i**) After restricting to a connected component of M, assume $F_{,N} > 0$ on M, set $H = \{x \in G : F_{,N} > 0\}$, and define

$$\mathbf{\Phi}(x) = \begin{bmatrix} x_1 \\ \vdots \\ x_{N-1} \\ F(x) \end{bmatrix} \quad \text{for } x \in H.$$

If $x, y \in H$ and $\mathbf{\Phi}(x) = \mathbf{\Phi}(y)$, then $x_j = y_j$ for $1 \le j \le N-1$ and $F(x_1, \dots, x_{N-1}, x_N) = F(x_1, \dots, x_{N-1}, y_N)$, which is possible only if $x_N = y_N$. Thus, $\mathbf{\Phi}$ is 1–1 on H. In addition,

$$J\mathbf{\Phi}(x) = \begin{bmatrix} \mathbf{I}_{\mathbb{R}^{N-1}} & \mathbf{0} \\ \mathbf{v}(x) & F_{,N}(x) \end{bmatrix}, \quad \text{where } \mathbf{v}(x) = \big[F_{,1}(x), \dots, F_{,N-1}(x)\big].$$

In particular, $\delta\mathbf{\Phi}(x) = F_{,N}(x) > 0$ for all $x \in H$, and so $\mathbf{\Phi}$ is a diffeomorphism there. Finally, $M \subseteq H$, $F \restriction M = 0$, and so $\mathbf{\Phi}(M) = U \times \{0\}$. Hence, if $f(u) = (\mathbf{\Phi}^{-1})_N(u, 0)$, then $f \in C^3(U;\mathbb{R})$ and $M = \mathbf{\Phi}^{-1}(U \times \{0\}) = \big\{\big(u, f(u)\big) : u \in U\big\}$. Equivalently, $F\big(u, f(u)\big) = 0$ for $u \in U$.

(**ii**) In view of (5.3.25) and (**i**), all that remains is to check that

$$1 + \big|\nabla f(u)\big|^2 = \frac{\big|\nabla F\big(u, f(u)\big)\big|^2}{F_{,N}\big(u, f(u)\big)^2}.$$

But, from $F\big(u, f(u)\big) = 0$, $u \in U$, and the chain rule,

$$F_{,j}\big(u, f(u)\big) + F_{,N}\big(u, f(u)\big) f_{,j}(u) = 0,$$

and the desired relation is a matter of simple algebra from here.

5.3.27: (**i**) The argument given to prove (**i**) of Exercise 5.3.26 shows that $\mathbf{\Phi}$ is a diffeomorphism on G. Moreover, by (**ii**) of that exercise applied for fixed $t \in T$,

$$\int_{M_t} \varphi \, d\lambda_{M_t} = \int_{\mathbb{R}^{N-1}} \mathbf{1}_{\mathbf{\Phi}(G)}(u,t) \frac{\varphi|\nabla F|}{|F_{,N}|} \circ \mathbf{\Phi}^{-1}(u,t)\, du$$

for all $\varphi \in C_c(G;\mathbb{R})$. In particular,

$$t \in T \longmapsto \int_{M_t} \varphi \, d\lambda_{M_t} \in \mathbb{R}$$

is measurable.

(**ii**) Since (cf. (**i**) in the preceding solution) $\delta\mathbf{\Phi} = |F_{,N}|$, the equality

$$\int_G \varphi(x)\,dx = \int_{\mathbf{\Phi}(G)} \frac{\varphi}{|F_{,N}|} \circ \mathbf{\Phi}^{-1}(y)\,dy$$

is just Jacobi's Transformation Formula applied to $\mathbf{\Phi}^{-1}$. Hence, by Fubini's Theorem and (**i**) here,

$$\begin{aligned}\int_G \varphi(x)\,dx &= \int_T \left(\int_{\mathbb{R}^{N-1}} \mathbf{1}_{\mathbf{\Phi}(G)}(u,t)\frac{\varphi}{|F_{,N}|} \circ \mathbf{\Phi}^{-1}(u,t)\,du\right)\,dt \\ &= \int_T \left(\int_{M_t} \frac{\varphi}{|\nabla F|}\,d\lambda_{M_t}\right)\,dt.\end{aligned}$$

(**iii**) When $G = \{x \in \mathbb{R}^N : x_N \neq 0\}$ and $F(x) = |x|$, $T = (0,\infty)$ and $M_t = \mathbf{S}^{N-1}(t) \cap G$. Hence, the preceding combined with (5.3.22) yield

$$\begin{aligned}\int_G \varphi(x)\,dx &= \int_{(0,\infty)} \left(\int_{M_t} \varphi\,d\lambda_{\mathbf{S}^{N-1}(t)}\right)\,dt \\ &= \int_{(0,\infty)} t^{N-1}\left(\int_{M_1} \varphi(t\boldsymbol{\omega})\,\lambda_{\mathbf{S}^{N-1}}(d\boldsymbol{\omega})\right)\,dt.\end{aligned}$$

Finally, to pass from this to (5.2.3), note that $|\mathbb{R}^N \setminus G| = 0$ and that if $\Gamma = \{x \in \mathbf{S}^{N-1} : x_N = 0\}$ then (cf. (5.3.9)) $\Gamma(\delta) \subseteq \{x \in \mathbb{R}^N : x_N = 0\}$ and therefore

$$\lambda_{\mathbf{S}^{N-1}}(\Gamma) = \lim_{\delta \searrow 0} \frac{1}{2\delta}|\Gamma(\delta)| = 0.$$

§ 5.4

5.4.19: To prove the asserted inequality, we proceed in exactly the same way as we did in the proof of Theorem 5.4.17. Thus, assume that $x = \mathbf{0}$, choose $R > 0$ so that $\overline{B(\mathbf{0},R)} \subseteq G$, define the regions G_r, $r \in (0,R)$, as in the proof of Theorem 5.4.16, and note that this time $g_R\Delta u$ is less than or equal (not necessarily equal) to 0 on each G_r. In particular, this leads to the inequality

$$\begin{aligned}0 &\geq \int_{G_r} \big(g_R(y)\,\Delta u(y) - u(y)\,\Delta g_R(y)\big)\,dy \\ &= -R^{N-1}\int_{\mathbf{S}^{N-1}} u(R\boldsymbol{\omega})\frac{\partial g_R}{\partial \rho}(R\boldsymbol{\omega})\lambda_{\mathbf{S}^{N-1}}(d\boldsymbol{\omega}) \\ &\quad - r^{N-1}\int_{\mathbf{S}^{N-1}} g_R(r\boldsymbol{\omega})\frac{\partial u}{\partial \rho}(r\boldsymbol{\omega})\lambda_{\mathbf{S}^{N-1}}(d\boldsymbol{\omega}) \\ &\quad + r^{N-1}\int_{\mathbf{S}^{N-1}} u(r\boldsymbol{\omega})\frac{\partial g_R}{\partial \rho}(r\boldsymbol{\omega})\lambda_{\mathbf{S}^{N-1}}(d\boldsymbol{\omega}),\end{aligned}$$

which, after $r \searrow 0$, yields the desired result. Thus, we now know that if $\overline{B(x,r)} \subseteq G$, then

$$\Delta u \le 0 \text{ in } G \implies u(x) \ge \frac{1}{\omega_{N-1}} \int_{\mathbf{S}^{N-1}} u(x + r\omega)\, d\omega,$$

and equality holds when u is harmonic.

Starting from the preceding, multiplying through by Nr^{N-1}, integrating both sides with respect to $r \in (0, R)$ for some $R > 0$ satisfying $\overline{B(x,R)} \subseteq G$, and recalling that $\Omega_N = \frac{\omega_{N-1}}{N}$, one now gets

$$\Delta u \le 0 \text{ in } G \implies u(x) \ge \frac{1}{\Omega_N R^N} \int_{B(x,R)} u(y)\, dy,$$

with equality when $\Delta u = 0$.

Finally, to prove the strong minimum principle, assume that G is connected, that $u \in C^2(G;\mathbb{R})$ satisfies $\Delta u \le 0$ on G, and that $u(a) \le u(x)$ for some $a \in G$ and all $x \in G$. To see that $u \equiv u(a)$, set $A = \{x \in G : u(x) = u(a)\}$. By continuity, A is closed in G. Moreover, $a \in A$. Thus, by connectivity, we will know that $G = A$ as soon as we show that A is open. To this end, suppose that $x \in A$, and choose $R > 0$ so that $\overline{B(x,R)} \subseteq G$. By the preceding

$$u(a) \ge \frac{1}{|B(x,R)|} \int_{B(x,R)} u(y)\, dy.$$

In particular, if $u(y) > u(a) + \epsilon$ for some $y \in B(x,R)$ and $\epsilon > 0$, then, by continuity, $\Gamma \equiv \{y \in B(x,R) : u(y) > u(a) + \epsilon\}$ has positive measure, and so the preceding would lead to the contradiction

$$u(a) \ge u(a) + \epsilon \frac{|\Gamma|}{\Omega_N R^N} > u(a).$$

5.4.22: (**i**) First note that $(\gamma, (0,1))$ is a coordinate chart for ∂G and therefore that

$$\int_{\partial G \setminus \{\gamma(0)\}} \varphi\, d\lambda_{\partial G} = \int_{(0,1)} \varphi \circ \gamma(t) |\dot{\gamma}(t)|\, dt$$

for every non-negative, Borel measurable $\varphi : \partial G \longrightarrow \mathbb{R}$. At the same time, if $\mathbf{\Psi}(t) = \gamma(1+t)$ when $t \in \left(-\frac{1}{2}, 0\right]$ and $\mathbf{\Psi}(t) = \gamma(t)$ when $t \in \left(0, \frac{1}{2}\right)$, then $\left(\mathbf{\Psi}, \left(-\frac{1}{2}, \frac{1}{2}\right)\right)$ is also a coordinate system for ∂G, and so

$$\int_{\partial G \setminus \{\gamma(\frac{1}{2})\}} \varphi\, d\lambda_{\partial G} = \int_{(-\frac{1}{2}, \frac{1}{2})} \varphi \circ \mathbf{\Psi}(t) |\dot{\mathbf{\Psi}}(t)|\, dt.$$

for all non-negative, Borel measurable φ's. In particular, this latter shows that $\{\gamma(0)\}$ has $\lambda_{\partial G}$-measure 0 and therefore that

$$\int_{\partial G} \varphi\, d\lambda_{\partial G} = \int_{[0,1]} \varphi\big(\gamma(t)\big) |\dot{\gamma}(t)|\, dt$$

for all non-negative, Borel measurable $\varphi : \partial G \longrightarrow \mathbb{R}$.

(ii) Set

$$\mathbf{w}(t) = |\dot\gamma(t)|^{-1}\big(\dot\gamma_2(t), -\dot\gamma_1(t)\big).$$

Since $\dot\gamma(t) \in \mathbf{T}_{\gamma(t)}\partial G \setminus \{\mathbf{0}\}$ and $\mathbf{w}(t) \perp \dot\gamma(t)$ for every $t \in [0,1)$, this proves that $\mathbf{n}\big(\gamma(t)\big) = \pm\dot\gamma(t)$ for each t. But $t \in [0,1) \longmapsto \big(\mathbf{n}(t), \mathbf{w}(t)\big)_{\mathbb{R}^2} \in \{-1,1\}$ is continuous, and so the same sign holds for all $t \in [0,1)$.

Now assume that $\mathbf{n}\big(\gamma(t)\big) = \mathbf{w}(t)$ for all $t \in [0,1)$, and set

$$\mathbf{F}(x) = \begin{bmatrix} u(x) \\ -v(x) \end{bmatrix}.$$

Then, by the Divergence Theorem,

$$\begin{aligned}\int_G \Delta h(x)\,dx &= \int_G \operatorname{div}\mathbf{F}(x)\,dx = \int_{\partial G} \big(\mathbf{F}(x), \mathbf{n}(x)\big)_{\mathbb{R}^2}\,d\lambda_{\partial G}(dx) \\ &= \int_{[0,1]} \Big(u\big(\gamma(t)\big)\dot\gamma_2(t) + v\big(\gamma(t)\big)\dot\gamma_1(t)\Big)\,dt = \int_{[0,1]} \mathfrak{Im}\Big[f\big(\gamma(t)\big)\dot\Gamma(t)\Big]\,dt.\end{aligned}$$

At the same time,

$$\begin{aligned}\int_{[0,1]} \mathfrak{Re}\Big[f\big(\gamma(t)\big)\dot\Gamma(t)\Big]\,dt &= \int_{[0,1]} \Big(u\big(\gamma(t)\big)\dot\gamma_1(t) - v\big(\gamma(t)\big)\dot\gamma_2(t)\Big)\,dt \\ &= (\mathrm{R})\int_{[0,1]} \frac{d}{dt} h\big(\gamma(t)\big)\,dt = h\big(\gamma(1)\big) - h\big(\gamma(0)\big) = 0.\end{aligned}$$

Hence, after combining these two, we get the desired result.

5.4.24: To see that each M_t is a hypersurface, let $p \in M_t$ be given, choose $r > 0$ so that $B(p,r) \subseteq G$, and consider $y \in B(p,r) \longmapsto F(y) - t \in \mathbb{R}$ in (5.3.7).

Given a compact subset K of G, use the Heine–Borel property to choose $p_1, \dots, p_n \in K$, $r_1, \dots, r_n \in (0,\infty)$, and $j_1, \dots, j_n \in \{1, \dots, N\}$ so that

$$K \subseteq \bigcup_1^n B(p_m, r_m) \subseteq \bigcup_1^n \overline{B(p_m, 3r_m)} \subseteq G$$

and

$$|F_{,j_m}| > 0 \quad \text{on } B(p_m, 3r_m) \text{ for } 1 \le m \le n.$$

Finally, set

$$\psi_m(x) = \begin{cases} \operatorname{dist}\big(x, B(p_m, 2r_m)\complement\big) & \text{if } 1 \le m \le n \\ \operatorname{dist}(x, K) & \text{if } m = 0 \end{cases} \quad \text{and} \quad s(x) = \sum_0^n \psi_m.$$

If $\varphi : G \longrightarrow \mathbb{R}$ is a continuous function which vanishes off K, define $\varphi_m = \frac{\psi_m \varphi}{s}$ for $1 \le m \le n$, note that $\psi_m \in C_c\big(B(p_m, 3r_m)\big)$ and $\varphi = \sum_1^m \varphi_m$, and apply Exercise 5.3.27 (with j_m replacing N) to see that

$$t \longmapsto \int \varphi\,d\lambda_{M_t} = \sum_1^n \mathbf{1}_{T_m}(t) \int \varphi_m\,d\lambda_{M_{t,m}}$$

is measurable and that

$$\begin{aligned}\int \varphi\, d\lambda_{\mathbb{R}^N} &= \sum_1^n \int_{B(p_m,3r_m)} \varphi_m\, d\lambda_{\mathbb{R}^N} \\ &= \sum_1^n \int_{T_m} \left(\int_{M_{t,m}} \frac{\varphi_m(t,u)}{|\nabla F(t,u)|}\, \lambda_{M_{t,m}}(du) \right) dt \\ &= \int_T \left(\int_{M_t} \frac{\varphi(t,u)}{|\nabla F(t,u)|}\, \lambda_{M_t}(du) \right) dt,\end{aligned}$$

where $T_m = \{F(x) : x \in B(p_m, 3r_m)\}$ and $M_{t,m} = M_t \cap B(p_m, 3r_m)$.

§ 6.1

6.1.6: **(i)** Let $\epsilon > 0$ be given. Since log is obviously smooth on $[\epsilon, \infty)$ and has negative second derivative there, it is clear that $x \in [\epsilon, \infty) \longmapsto \log_\epsilon x \equiv \log \frac{x}{\epsilon} \in [0, \infty)$ is continuous and concave.

Next, suppose that $a_1, \dots, a_n \in [\epsilon, \infty)$ and $\mu_1, \dots, \mu_n \in [0, 1]$ are given, and define

$$\mu(\Gamma) = \sum_{\{m:\, a_m \in \Gamma\}} \mu_m$$

for $\Gamma \subseteq \mathbb{R}$. Clearly, μ is a probability measure on $(\mathbb{R}, \mathcal{P}(\mathbb{R}))$ and, by Jensen's inequality,

$$\begin{aligned}\log\left(\prod_1^n a_m^{\mu_m}\right) &= \log\epsilon + \sum_1^n \mu_m \log_\epsilon a_m = \log\epsilon + \int \log_\epsilon x\, \mu(dx) \\ &\leq \log\epsilon + \log_\epsilon\left(\int x\, \mu(dx)\right) = \log\left(\sum_1^n \mu_m a_m\right).\end{aligned}$$

Thus, the desired inequality results after one exponentiates both sides.

(ii) One can do this problem by simply reproducing the argument given to prove Hölder's inequality. Alternatively, one can work by the induction on $n \geq 2$. Indeed, when $n = 2$, the result is just Hölder's inequality. Next, assume the result for some $n \geq 2$, and define $q \in [1, \infty)$ by $\frac{1}{q} = \frac{1}{p_n} + \frac{1}{p_{n+1}}$. Then $\sum_1^{n-1} \frac{1}{p_m} + \frac{1}{q} = 1$, and so, by induction hypothesis,

$$\int f_1 \cdots f_{n-1}\, g\, d\mu \leq \prod_1^{n-1} \left(\int f_m^{p_m}\, d\mu \right)^{\frac{1}{p_m}} \left(\int g^q\, d\mu \right)^{\frac{1}{q}},$$

where $g \equiv f_n f_{n+1}$. Finally, set $p = \frac{p_n}{q} = 1 + \frac{p_n}{p_{n+1}}$, note that the Hölder conjugate $p' = \frac{p_{n+1}}{q}$, and apply Hölder's inequality to obtain

$$\begin{aligned}\left(\int g^q\, d\mu \right)^{\frac{1}{q}} &\leq \left(\int f_n^{pq}\, d\mu \right)^{\frac{1}{pq}} \left(\int f_{n+1}^{p'q}\, d\mu \right)^{\frac{1}{p'q}} \\ &= \left(\int f_n^{p_n}\, d\mu \right)^{\frac{1}{p_n}} \left(\int f_{n+1}^{p_{n+1}}\, d\mu \right)^{\frac{1}{p_{n+1}}}.\end{aligned}$$

6.1.7: Starting from the observation made, the inequality

$$2\int f_1 f_2\,d\mu \le t\int f_1^2\,d\mu + \frac{1}{t}\int f_2^2\,d\mu, \quad t\in(0,\infty),$$

is obvious when one takes $t=\alpha^2$. Next, note that if either f_1 or f_2 vanishes μ-a.e., then $f_1 f_2 = 0$ μ-a.e. and there is nothing to do. On the other hand, if neither f_1 nor f_2 vanishes μ-a.e., then the preceding leads to

$$(*)\qquad \int f_1 f_2\,d\mu \le \left(\int f_1^2\,d\mu\right)^{\frac{1}{2}}\left(\int f_2^2\,d\mu\right)^{\frac{1}{2}}$$

when one takes

$$t^2 = \frac{\int f_2^2\,d\mu}{\int f_1^2\,d\mu}.$$

To remove the assumption that the f_i's are bounded, replace f_i by $f_i\wedge R$, use (*) to see that

$$\int (f_1\wedge R)(f_2\wedge R)\,d\mu \le \left(\int f_1^2\,d\mu\right)^{\frac{1}{2}}\left(\int f_2^2\,d\mu\right)^{\frac{1}{2}},$$

and use the Monotone Convergence Theorem to pass to the limit as $R\nearrow\infty$. Finally, to handle μ's which are not finite, note that there is nothing to do when either $\int f_1^2\,d\mu$ or $\int f_2^2\,d\mu$ is infinite or zero. Thus, we assume that both lie in $(0,\infty)$, in which case $\mu(f_1\ge\epsilon)<\infty$ for every $\epsilon>0$, and the result already established yields

$$\int_{\{f_1\ge\epsilon\}} f_1 f_2\,d\mu \le \left(\int_{\{f_1\ge\epsilon\}} f_1^2\,d\mu\right)^{\frac{1}{2}}\left(\int f_2^2\,d\mu\right)^{\frac{1}{2}}.$$

Thus, the general case follows from the Monotone Convergence Theorem after one lets $\epsilon\searrow 0$.

To complete the proof, suppose that $\left(\int f_1 f_2\,d\mu\right)^2 = \int f_1^2\,d\mu\int f_2^2\,d\mu$. If either factor on the right vanishes, then it is clear that either $f_1+0f_2=0$ or $0f_1+f_2=0$ μ-a.e. Thus, assume that $\int f_i^2\,d\mu\in(0,\infty)$, $i\in\{1,2\}$, take $\alpha>0$ so that

$$\alpha^2 = \frac{\int f_2^2\,d\mu}{\int f_1^2\,d\mu},$$

and conclude that $\int(\alpha f_1 - f_2)^2\,d\mu = 0$.

Given Schwarz's inequality, one gets Minkowski's inequality for $p=2$ by the simple observation that

$$\begin{aligned}\int(\alpha f_1+\beta f_2)^2\,d\mu &\le \alpha^2\int f_1^2\,d\mu + 2\alpha\beta\int |f_1|\,|f_2|\,d\mu + \beta^2\int f_2^2\,d\mu\\ &\le \left[\left(\alpha^2\int f_1^2\,d\mu\right)^{\frac{1}{2}} + \left(\beta^2\int f_2^2\,d\mu\right)^{\frac{1}{2}}\right]^2.\end{aligned}$$

6.1.9: Let C be a closed, convex subset of $\mathbb{R}^N$. Given $q \notin C$, let $p \in C$ be chosen so that $|q-p| \le |q-x|$ for all $x \in C$, and set $\mathbf{w}_q = \frac{q-p}{|q-p|}$. To see that

$$\left(\mathbf{w}_q, q-x\right)_{\mathbb{R}^N} > 0 \quad \text{for all } x \in C,$$

first note that

$$\left(\mathbf{w}_q, q-x\right)_{\mathbb{R}^N} = |q-p| + \left(\mathbf{w}_q, p-x\right)_{\mathbb{R}^N}.$$

Thus, it suffices to check that $\left(\mathbf{w}_q, p-x\right)_{\mathbb{R}^N} \ge 0$ for all $x \in C$. But, for $x \in C$, the smooth function $t \in [0,1] \longmapsto \varphi(t) \equiv |q-(1-t)p-tx|^2$ achieves its minimum at $t=0$. Hence,

$$2\left(\mathbf{w}_q, p-x\right)_{\mathbb{R}^N} = \frac{\varphi'(0)}{|q-p|} \ge 0.$$

Now let $(E, \mathcal{B}, \mu)$ be a probability space and $\mathbf{F} : E \longrightarrow C$ a μ-integrable map. To see that $\int \mathbf{F}\, d\mu \in C$, suppose not, set $q = \int \mathbf{F}\, d\mu$, and apply the preceding to derive the contradiction

$$0 < \int \left(\mathbf{w}_q, q-\mathbf{F}\right)_{\mathbb{R}^N} d\mu = \left(\mathbf{w}_q, q-q\right)_{\mathbb{R}^N} = 0.$$

To complete the proof of Jensen's inequality, let $g : C \longrightarrow [0,\infty)$ be a continuous, concave function and set $\widehat{C} = \left\{(x,t) \in \mathbb{R}^{N+1} : t \le g(x)\right\}$. $\widehat{C}$ is then a closed, convex subset of $\mathbb{R}^{N+1}$. Next, set $p = \int \mathbf{F}\, d\mu$, and, for each $\epsilon > 0$, use the first assertion applied to $\widehat{C}$ and $\left(p, g(p)+\epsilon\right)$ to find $\widehat{\mathbf{w}}_\epsilon = (\mathbf{w}_\epsilon, \rho_\epsilon) \in \mathbb{R}^N \times \mathbb{R}$ so that

$$\left(\mathbf{w}_\epsilon, p-x\right)_{\mathbb{R}^N} + \rho_\epsilon\left(g(p)+\epsilon-t\right) > 0 \quad \text{for all } (x,t) \in \widehat{C}.$$

Note that this is possible only if $\rho_\epsilon > 0$, set $\mathbf{v}_\epsilon = \frac{\mathbf{w}_\epsilon}{\rho_\epsilon}$, and conclude that

$$g(x) \le g(p) + \epsilon + \left(\mathbf{v}_\epsilon, p-x\right)_{\mathbb{R}^N}, \quad x \in C.$$

Finally, replace x is the preceding by $\mathbf{F}$, integrate with respect to μ, and come to

$$\int g \circ \mathbf{F}\, d\mu \le g(p) + \epsilon + (\mathbf{v}_\epsilon, p-p)_{\mathbb{R}^N} = g(p) + \epsilon.$$

Now, let $\epsilon \searrow 0$.

§ 6.2

6.2.20: Without loss of generality, assume that $q_1 < \infty$. Set $f_1 = |f|^{tp_t}$, $f_2 = |f|^{(1-t)p_t}$, and apply Hölder's inequality with $p = \frac{q_1}{tp_t}$ to get

$$\|f\|_{L^{p_t}(\mu)}^{p_t} = \int |f_1|\,|f_2|\,d\mu \le \|f_1\|_{L^p(\mu)}\|f_2\|_{L^{p'}(\mu)} = \|f\|_{L^{q_1}(\mu)}^{tp_t}\|f\|_{L^{q_2}(\mu)}^{(1-t)p_t}.$$

6.2.22: **(i)** Let $1 \le p \le q < \infty$ and set $\alpha = \frac{p}{q}$. Then, since $\mu(E) = 1$, Jensen's inequality applies and yields

$$\|f\|_{L^p(\mu)}^p = \int_E \left(|f|^q\right)^\alpha d\mu \le \left(\int_E |f|^q\,d\mu\right)^\alpha = \|f\|_{L^q(\mu)}^{\frac{p}{q}}.$$

(ii) First note that, by Hölder's inequality or direct computation,

$$\|f\|_{L^q(\mu)} \le \|f\|_{L^p(\mu)}^{\frac{p}{q}}\|f\|_{L^\infty(\mu)}^{1-\frac{p}{q}}$$

for any measure μ and any $1 \le p < q \le \infty$. Second, note that, for the particular measure under consideration, $\|f\|_{L^\infty(\mu)} \le \|f\|_{L^p(\mu)}$ for any $p \in [1,\infty]$.

(iii) Since there is nothing to do when $\mu(E) = 0$, we will assume that $\mu(E) \in (0,\infty]$. To prove the first assertion, suppose that $0 \le M < \|f\|_{L^\infty(\mu)}$ and set $\Gamma = \{|f| > M\}$. Then $\mu(\Gamma) > 0$, and so

$$\underline{\lim}_{p\to\infty} \|f\|_{L^p(\mu)} \ge M \underline{\lim}_{p\to\infty} \left(\mu(\Gamma)\right)^{\frac{1}{p}} \ge M.$$

In other words, $\underline{\lim}_{p\to\infty}\|f\|_{L^p(\mu)} \ge M$ for every $M \in [0, \|f\|_{L^\infty(\mu)})$, and so $\underline{\lim}_{p\to\infty}\|f\|_{L^p(\mu)} \ge \|f\|_{L^\infty(\mu)}$. To prove the second assertion, assume that $\mu(E) < \infty$ or $\|f\|_{L^1(\mu)} < \infty$, and note that, for any $p \in [1,\infty)$,

$$\|f\|_{L^p(\mu)} \le \left(\|f\|_{L^\infty(\mu)}\mu(E)^{\frac{1}{p}}\right) \wedge \left(\|f\|_{L^\infty(\mu)}^{1-\frac{1}{p}}\|f\|_{L^1(\mu)}^{\frac{1}{p}}\right),$$

and let $p \nearrow \infty$.

(iv) Suppose that $(E_1, \mathcal{B}_1)$ is a measurable space and that $(E_2, \mathcal{B}_2, \mu_2)$ is a σ-finite measure space. Choose $\{E_{2,n}\}_1^\infty \subseteq \mathcal{B}_2$ so that $E_{2,n} \nearrow E_2$ and $\mu_2(E_{2,n}) < \infty$, $n \in \mathbb{Z}^+$. Given a measurable f on $(E_1 \times E_2, \mathcal{B}_1 \times \mathcal{B}_2)$, observe that

$$\begin{aligned}\left\|f(x_1,\cdot_2)\right\|_{L^\infty(\mu)} &= \lim_{n\to\infty}\left\|f(x_1,\cdot_2)\mathbf{1}_{E_{2,n}}(\cdot_2)\right\|_{L^\infty(\mu)}\\ &= \lim_{n\to\infty}\lim_{p\to\infty}\left\|f(x_1,\cdot_2)\mathbf{1}_{E_{2,n}}(\cdot_2)\right\|_{L^p(\mu)}.\end{aligned}$$

Thus, since, by Lemma 4.1.2,

$$x_1 \in E_1 \longmapsto \int_{E_{2,n}} \left|f(x_1,x_2)\right|^p \mu_2(dx_2)$$

is $\mathcal{B}_1$-measurable for each $n \in \mathbb{Z}^+$ and $p \in [1, \infty)$, so is

$$x_1 \in E_1 \longmapsto \left\|f(x_1, \cdot_2)\right\|_{L^\infty(\mu_2)}.$$

6.2.23: (**i**) The steps leading to

$$\|f\|^p_{L^p(\mu)} \leq \frac{Cp}{p-1}\|f\|^{p-1}_{L^{p-1}(\nu)} \leq \frac{Cp}{p-1}\|f\|^{p-1}_{L^p(\mu)}\|g\|_{L^p(\mu)}$$

are easy applications of Exercise 5.1.8 and Hölder's inequality.

(**ii**) When either $\|f\|_{L^p(\mu)} = 0$ or $\|g\|_{L^p(\mu)} = \infty$, there is nothing to do. Thus, assume that $\|f\|_{L^p(\mu)} > 0$ and that $\|g\|_{L^p(\mu)} < \infty$. When, in addition, $\|f\|_{L^p(\mu)} < \infty$, (6.2.24) follows from the preceding after division by $\|f\|^{p-1}_{L^p(\mu)}$. More generally, set $f_R = f \wedge R$, note that

$$\{f_R \geq t\} = \begin{cases} \emptyset & \text{if } t > R \\ \{f_R \geq t\} & \text{if } t \in (0, R], \end{cases}$$

and conclude that $\mu(f_R \geq t) \leq \frac{C}{t}\int_{\{f_R \geq t\}} g\,d\mu$. Hence, when μ is finite, and therefore $\|f_R\|_{L^p(\mu)} \leq R\mu(E)^{\frac{1}{p}} < \infty$, we have first that $\|f_R\|_{L^p(\mu)} \leq \frac{Cp}{p-1}\|g\|_{L^p(\mu)}$ for all $R \in (0, \infty)$ and then, after $R \nearrow \infty$, $\|f\|_{L^p(\mu)} \leq \frac{Cp}{p-1}\|g\|_{L^p(\mu)}$.

(**iii**) Finally, assume that $\mu(f \geq \epsilon) < \infty$ for each $\epsilon > 0$, set $\mu_\epsilon(\Gamma) = \mu\big(\Gamma \cap \{f \geq \epsilon\}\big)$ for $\Gamma \in \mathcal{B}$, note that

$$\begin{aligned} \mu_\epsilon(f \geq t) &= \begin{cases} \mu(f \geq t) & \text{if } t \geq \epsilon \\ \mu(f \geq \epsilon) & \text{if } t \in [0, \epsilon) \end{cases} \\ &\leq C \begin{cases} \frac{1}{t}\int_{\{f \geq t\}} g\,d\mu & \text{if } t \geq \epsilon \\ \frac{1}{\epsilon}\int_{\{f \geq \epsilon\}} g\,d\mu & \text{if } t \in [0, \epsilon) \end{cases} \leq \frac{C}{t}\int_{\{f \geq t\}} g\,d\mu_\epsilon, \end{aligned}$$

and conclude from (**ii**) that

$$\|f\|_{L^p(\mu_\epsilon)} \leq \frac{Cp}{p-1}\|g\|_{L^p(\mu_\epsilon)} \leq \frac{Cp}{p-1}\|g\|_{L^p(\mu)}$$

for all $\epsilon > 0$. Finally, let $\epsilon \searrow 0$.

6.2.27: By the first inequality in (3.4.7),

$$\big|\{Mf \geq t\}\big| \leq \frac{2}{t}\int_{\{Mf \geq t\}} |f|\,d\lambda_{\mathbb{R}}, \quad f \in L^1(\mathbb{R}).$$

Thus, by (**iii**) in Exercise 5.2.23, for any $f \in L^1(\mathbb{R})$,

$$\|Mf\|_{L^p(\mathbb{R})} \leq \frac{2p}{p-1}\|f\|_{L^p(\mathbb{R})}, \quad p \in (1, \infty).$$

More generally, if $p \in (1,\infty)$ and $f \in L^p(\mathbb{R})$, use (**v**) in Theorem 6.2.2 to find $\{f_n\}_1^\infty \subseteq C_c(\mathbb{R};\mathbb{R})$ so that $f_n \longrightarrow f$ in $L^p(\mathbb{R})$. Then, for any $x \in \mathbb{R}$ and any ball $B \ni x$,

$$\frac{1}{|B|}\int_B |f|\,d\lambda_{\mathbb{R}} = \lim_{n\to\infty}\frac{1}{|B|}\int_B |f_n|\,d\lambda_{\mathbb{R}} \le \varliminf_{n\to\infty} Mf_n(x),$$

and so $Mf \le \varliminf_{n\to\infty} Mf_n$. Thus, by the preceding combined with Fatou's Lemma,

$$\|Mf\|_{L^p(\mathbb{R})} \le \varliminf_{n\to\infty}\|Mf_n\|_{L^p(\mathbb{R})} \le \lim_{n\to\infty}\frac{2p}{p-1}\|f_n\|_{L^p(\mathbb{R})} = \frac{2p}{p-1}\|f\|_{L^p(\mathbb{R})}.$$

§ 6.3

6.3.17: By Tonelli's Theorem and the translation invariance of Lebesgue's measure, for any non-negative, measurable functions f and g,

$$\int f * g\,d\lambda_{\mathbb{R}^N} = \int g(y)\left(\int f(x-y)\,dx\right)dy = \int f\,d\lambda_{\mathbb{R}^N}\int g\,d\lambda_{\mathbb{R}^N}.$$

To handle $f,\ g \in L^1(\mathbb{R}^N)$, note that $f*g = f^+*g^+ + f^-*g^- - f^+*g^+ - f^-*g^+$, apply the preceding to each term, and combine.

6.3.18: (**i**) Note that

$$\gamma_{\sqrt{s}} * \gamma_{\sqrt{t}}(x) = \frac{1}{2\pi\sqrt{st}}\int_{\mathbb{R}^N}\exp\left[-\frac{|x-y|^2}{2s} - \frac{|y|^2}{2t}\right]dy.$$

Next, by elementary algebra,

$$\frac{|x-y|^2}{s} + \frac{|y|^2}{t} = \frac{s+t}{st}\left|y - \frac{t}{s+t}x\right|^2 + \frac{|x|^2}{s+t}.$$

Hence, by the translation invariance of Lebesgue's measure,

$$\begin{aligned}\gamma_{\sqrt{s}} * \gamma_{\sqrt{t}}(x) &= \gamma_{\sqrt{s+t}}(x)\sqrt{\frac{s+t}{2\pi st}}\int \exp\left[-\frac{(s+t)\left|y - \frac{t}{s+t}x\right|^2}{st}\right]dy \\ &= \gamma_{\sqrt{s+t}}(x)\sqrt{\frac{s+t}{2\pi st}}\int \exp\left[-\frac{(s+t)|y|^2}{st}\right]dy = \gamma_{\sqrt{s+t}}(x).\end{aligned}$$

(**ii**) By definition, $\nu_{s^2} * \nu_{t^2}(\eta) = 0 = \nu_{(s+t)^2}(\eta)$ if $\eta \le 0$. Thus, assume that $\eta > 0$. Then

$$\nu_{s^2} * \nu_{t^2}(\eta) = \frac{st}{\pi}\int_{(0,\eta)}[\xi(\eta-\xi)]^{-\frac{3}{2}}\exp\left[-\frac{s^2}{\eta-\xi} - \frac{t^2}{\xi}\right]d\xi.$$

Now define $\Phi(\xi) = \frac{\xi}{\eta-\xi}$ for $\xi \in (0,\eta)$, note that

$$[\xi(\eta-\xi)]^{-\frac{3}{2}} = \eta^{-2}\frac{\big(1+\Phi(\xi)\big)\Phi'(\xi)}{\Phi(\xi)^{\frac{3}{2}}}$$

while

$$\frac{s^2}{\eta-\xi} + \frac{t^2}{\xi} = \frac{s^2+t^2}{\eta} + \frac{s^2\Phi(\xi)}{\eta} + \frac{t^2}{\eta\Phi(\xi)},$$

and conclude, from Jacobi's Transformation Formula and **(iv)** in Exercise 5.2.6, that

$$\begin{aligned}\nu_{s^2} * \nu_{t^2}(\eta) &= \frac{st}{\pi\eta^2}\exp\left[-\frac{s^2+t^2}{\eta}\right]\int_{(0,\infty)}\big(u^{-\frac{1}{2}}+u^{-\frac{3}{2}}\big)\exp\left[-\frac{s^2u}{\eta}-\frac{t^2}{\eta u}\right]d\xi \\ &= \frac{st}{\pi\eta^2}\exp\left[-\frac{(s+t)^2}{\eta}\right]\left(\frac{\sqrt{\pi\eta}}{s}+\frac{\sqrt{\pi\eta}}{t}\right) = \nu_{(s+t)^2}(\eta).\end{aligned}$$

(iii) The first part is covered by **(ii)** in Exercise 5.3.24 and follows, anyhow, as an immediate consequence of the second part. To prove the second part, make a couple of simple changes of variable to justify,

$$\begin{aligned}\int_{(0,\infty)}\gamma_{\sqrt{\frac{\xi}{2}}}(x)\nu_{t^2}(\xi)\,d\xi &= \frac{t}{\pi^{\frac{N+1}{2}}}\int_{(0,\infty)}\xi^{-\frac{N+3}{2}}e^{-\frac{|x|^2+t^2}{\xi}}\,d\xi \\ &= \frac{t}{\big(\pi(|x|^2+t^2)\big)^{\frac{N+1}{2}}}\int_{(0,\infty)}\xi^{-\frac{N+3}{2}}e^{-\frac{1}{\xi}}\,d\xi \\ &= \frac{t}{\big(\pi(|x|^2+t^2)\big)^{\frac{N+1}{2}}}\int_{(0,\infty)}\eta^{\frac{N+1}{2}-1}e^{-\eta}\,d\eta \\ &= \frac{t\Gamma\left(\frac{N+1}{2}\right)}{\pi^{\frac{N+1}{2}}}\big(|x|^2+t^2\big)^{-\frac{N+1}{2}} = \frac{2t}{\omega_N}\big(|x|^2+t^2\big)^{-\frac{N+1}{2}} = P_t(x),\end{aligned}$$

where, in the second to last equality, we used **(iii)** of Exercise 5.2.6 .

Next, given the preceding, applying Tonelli's Theorem, and using parts **(i)** and **(ii)** above, one has

$$\begin{aligned}P_s * P_t(x) &= \iint_{(0,\infty)^2}\gamma_{\sqrt{\frac{\xi}{2}}} * \gamma_{\sqrt{\frac{\eta}{2}}}(x)\nu_{s^2}(\xi)\nu_{t^2}(\eta)\,d\xi d\eta \\ &= \iint_{(0,\infty)^2}\gamma_{\sqrt{\frac{\xi+\eta}{2}}}(x)\nu_{s^2}(\xi)\nu_{t^2}(\eta)\,d\xi d\eta \\ &= \int_{(0,\infty)}\gamma_{\sqrt{\frac{u}{2}}}(x)\left(\int_{(0,\infty)}\nu_{s^2}(u-\xi)\nu_{t^2}(\xi)\,d\xi\right)du \\ &= \int_{(0,\infty)}\gamma_{\sqrt{\frac{u}{2}}}(x)\nu_{s^2+t^2}(u)\,du = P_{s+t}(x).\end{aligned}$$

(iv) Assume that $x > 0$. Then

$$\begin{aligned} g_\alpha * g_\beta(x) &= \frac{e^{-x}}{\Gamma(\alpha)\Gamma(\beta)} \int_{(0,x)} (x-t)^{\alpha-1} t^{\beta-1}\, dt \\ &= \frac{x^{\alpha+\beta-1}e^{-x}}{\Gamma(\alpha)\Gamma(\beta)} \int_{(0,1)} (1-t)^{\alpha-1} t^{\beta-1}\, dt = \frac{\Gamma(\alpha+\beta)B(\alpha,\beta)}{\Gamma(\alpha)\Gamma(\beta)} g_{\alpha+\beta}(x). \end{aligned}$$

Finally, integrate both sides in x over $(0,\infty)$ and obtain

$$1 = \frac{\Gamma(\alpha+\beta)}{\Gamma(\alpha)\Gamma(\beta)}$$

as a consequence of Exercise 6.3.17.

6.3.22: When $f \in C_c(\mathbb{R}^N;\mathbb{R})$, the continuity of $f * \mu$ is an immediate consequence of Lebesgue's Dominated Convergence Theorem. Furthermore, it is obvious that $|f * \mu(x)| \le \mu(\mathbb{R}^N)\|f\|_u = \mu(\mathbb{R}^N)\|f\|_{L^\infty(\mathbb{R}^N)}$; and, when $p \in [1,\infty)$, one can use Tonelli's Theorem followed by Theorem 6.2.14 applied to $F(x,y) = f(x-y)$ to get

$$\|f * \mu\|_{L^p(\mu)} \le \|F\|_{L^{1,p}(\mu,\lambda_{\mathbb{R}^N})} \le \|F\|_{L^{p,1}(\lambda_{\mathbb{R}^N},\mu)} = \mu(\mathbb{R}^N)\|f\|_{L^p(\mathbb{R}^N)},$$

since $\|F(\,\cdot\,,y)\|_{L^p(\mathbb{R}^N)} = \|f\|_{L^p(\mathbb{R}^N)}$ for all $y \in \mathbb{R}^N$.

Next, let $p \in [1,\infty)$ be fixed, and define $\mathcal{K}_\mu f = f * \mu$ for $f \in C_c(\mathbb{R}^N;\mathbb{R})$. Clearly, $\mathcal{K}_\mu$ is linear, and, by the preceding, $\|\mathcal{K}_\mu f\|_{L^p(\mathbb{R}^N)} \le \mu(\mathbb{R}^N)\|f\|_{L^p(\mathbb{R}^N)}$. In particular, $\mathcal{K}_\mu$ is uniformly Lipschitz continuous from $C_c(\mathbb{R}^N;\mathbb{R})$ to $L^p(\mathbb{R}^N)$ with respect to the metric induced by $\|\cdot\|_{L^p(\mathbb{R}^N)}$. Hence, because (cf. **(v)** in Theorem 6.2.2) $C_c(\mathbb{R}^N;\mathbb{R})$ is dense in $L^p(\mathbb{R}^N)$, elementary point-set topology says that $\mathcal{K}_\mu$ admits a unique extension as a continuous map from $L^p(\mathbb{R}^N)$ into itself.

Finally, to see that $f * \mu$ need not be continuous in general, even though $f \in L^1(\mathbb{R}^N) \cap L^\infty(\mathbb{R}^N)$, take μ to be the measure given by $\mu(\Gamma) = \mathbf{1}_\Gamma(\mathbf{0})$, and observe that $f = f * \mu$. Thus, for this choice of μ, $f * \mu$ is continuous precisely when f itself is.

6.3.24: **(i)** Set $\varphi(t) = \exp\left[-\frac{1}{t}\right]$ for $t > 0$ and $\varphi(t) = 0$ for $t \le 0$. Proving that $\rho \in C_c^\infty(\mathbb{R}^N;\mathbb{R})$ comes down to checking that

$$\lim_{t \searrow 0} \frac{\varphi^{(n)}(t)}{t} = 0 \quad \text{for all } n \in \mathbb{N},$$

where $\varphi^{(n)} = \frac{d^n\varphi}{dt^n}$. But this is obvious when $n = 0$. Moreover, by induction on $n \in \mathbb{Z}^+$,

$$\varphi^{(n)}(t) = \varphi(t) \sum_{m=1}^{n} \frac{m!}{t^{m+1}}, \quad t \in (0,\infty).$$

Thus, since $\lim_{t\searrow 0} t^{-m}\varphi(t) = 0$ for every $m \in \mathbb{N}$, we are done.

(ii) Set $\epsilon = \frac{1}{3}\mathrm{dist}(F, G\complement)$ and $\eta = \rho_\epsilon * \mathbf{1}_{F^{(\epsilon)}}$. By (6.3.13), $\eta \in C^\infty(\mathbb{R}^N;\mathbb{R})$. In fact, $\eta : \mathbb{R}^N \longrightarrow [0,1]$,

$$\eta(x) = \int_{F^{(\epsilon)}} \rho_\epsilon(x-y)\,dy \geq \int_{B(\mathbf{0},1)} \rho(y)\,dy = 1 \quad \text{for } x \in F,$$

and

$$\eta(x) = \int_{F^{(\epsilon)}} \rho_\epsilon(x-y)\,dy \leq \int_{B(\mathbf{0},1)\complement} \rho(y)\,dy = 0 \quad \text{for } x \notin G.$$

Thus, $\mathbf{1}_F \leq \eta \leq \mathbf{1}_G$.

(iii) Given $f \in L^p(\mathbb{R}^N)$ and $n \in \mathbb{Z}^+$, set $f_n = \mathbf{1}_{B(\mathbf{0},n)} f$ and $\psi_n = \rho_{\frac{1}{n}} * f_n$. By (6.3.13), $\psi_n \in C^\infty_{\mathrm{c}}(\mathbb{R}^N;\mathbb{R})$. In addition, by (6.3.15),

$$\begin{aligned}\|f - \psi_n\|_{L^p(\mathbb{R}^N)} &\leq \|f - \rho_{\frac{1}{n}} * f\|_{L^p(\mathbb{R}^N)} + \|\rho_{\frac{1}{n}} * (f - f_n)\|_{L^p(\mathbb{R}^N)} \\ &\leq \|f - \rho_{\frac{1}{n}} * f\|_{L^p(\mathbb{R}^N)} + \|f - f_n\|_{L^p(\mathbb{R}^N)} \longrightarrow 0 \quad \text{as } n \to \infty.\end{aligned}$$

Notice that if, in addition, $f \in L^q(\mathbb{R}^N)$ for some other $q \in [1,\infty)$, then $\psi_n \longrightarrow f$ in $L^q(\mathbb{R}^N)$ as well.

(iv) We first show that

$$(*) \qquad \overline{B(x,2t)} \subseteq G \implies \big[(u\mathbf{1}_{B(x,2t)}) * \rho_t\big](\xi) = u(\xi) \quad \text{for } \xi \in B(x,t).$$

To this end, let $\xi \in B(x,t)$ be given, and note that

$$\begin{aligned}\big[(u\mathbf{1}_{B(x,2t)}) * \rho_t\big](\xi) &= \int_{B(\mathbf{0},1)} u(\xi - ty)\rho(y)\,dy \\ &= c_N \int_{(0,1)} s^{N-1} e^{-(1-s^2)^{-1}} \left(\int_{\mathbf{S}^{N-1}} u(\xi - st\boldsymbol{\omega})\,\lambda_{\mathbf{S}^{N-1}}(d\boldsymbol{\omega})\right) ds \\ &= \omega_{N-1} u(\xi) c_N \int_{(0,1)} s^{N-1} e^{-(1-s^2)^{-1}}\,ds = u(\xi)\int_{B(\mathbf{0},1)} \rho(y)\,dy = u(\xi),\end{aligned}$$

where, in the passage to the last line, we have used the Mean Value Property for u.

Given (*), it is obvious (from (6.3.13)) that $u \in C^\infty(G;\mathbb{R})$. In addition, by the Mean Value Property,

$$\int_{\mathbf{S}^{N-1}} \big(u(x + t\boldsymbol{\omega}) - u(x)\big)\,\lambda_{\mathbf{S}^{N_1}}(d\boldsymbol{\omega}) = 0 \quad \text{whenever } \overline{B(x,t)} \subseteq G.$$

Thus, everything reduces to proving (6.3.25). But, by Taylor's Theorem,

$$f(x + t\boldsymbol{\omega}) - f(x) = t\sum_1^N \omega_j f_{,j}(x) + \frac{1}{2}\sum_1^N \omega_i\omega_j f_{,ij}(x) + o(t^2) \quad \text{as } t \searrow 0.$$

Hence, by (5.2.5),

$$\int_{\mathbf{S}^{N-1}} \big(f(x + t\boldsymbol{\omega}) - f(x)\big)\,\lambda_{\mathbf{S}^{N-1}}(d\boldsymbol{\omega}) = \frac{t^2\Omega_N}{2}\Delta f(x) + o(t^2),$$

and so (6.3.25) follows after one divides through by t^2 and lets $t \searrow 0$.

6.3.26: Given $\Gamma \in \overline{\mathcal{B}}_{\mathbb{R}^N}$ with $0 < |\Gamma| < \infty$, note that both $\mathbf{1}_{-\Gamma}$ and $\mathbf{1}_\Gamma$ are square integrable, and apply Theorem 6.3.7 to conclude that $u = \mathbf{1}_{-\Gamma} * \mathbf{1}_\Gamma \in C\big(\mathbb{R}; [0,\infty)\big)$. Next, observe that

$$u(z) = \big|\{\xi \in \Gamma : \xi - z \in \Gamma\}\big| \le |\Gamma|.$$

In particular, $z \notin \Delta \implies u(z) = 0$, and so $u \le |\Gamma|\mathbf{1}_\Delta$. At the same time, $u(0) = |\Gamma| > 0$, and so there exists a $\delta > 0$ such that $0 < u(z) \le |\Gamma|\mathbf{1}_\Delta(z)$ for $|z| < \delta$. That is, $(-\delta, \delta) \subseteq \Delta$.

6.3.27: To begin, let $f : (0,\infty) \longrightarrow (0,\infty)$ be a Borel measurable function. Then, for $\alpha \in (0,\infty)$,

$$\begin{aligned}\int_{(0,\infty)} f(\alpha x)\,\mu(dx) &= \int_{(0,\infty)} \frac{f(\alpha x)}{x}\,dx \\ &= \int_{(0,\infty)} \frac{f(\alpha x)}{\alpha x}\alpha\,dx = \int_{(0,\infty)} \frac{f(x)}{x}\,dx = \int_{(0,\infty)} f\,d\mu;\end{aligned}$$

and, when $\Phi(x) = \frac{1}{x}$,

$$\int_{(0,\infty)} f\left(\tfrac{1}{x}\right)\,\mu(dx) = \int_{(0,\infty)} \frac{f\circ\Phi(x)}{\Phi(x)}|\Phi'(x)|\,dx = \int_{(0,\infty)} \frac{f(y)}{y}\,dy = \int_{(0,\infty)} f\,d\mu.$$

Thus, the required invariance properties have been established.

Turning to the product $f \bullet g$, we use the preceding invariance properties to justify

$$\begin{aligned}\int_{(0,\infty)} |f(y)|\left|g\left(\tfrac{x}{y}\right)\right|\,\mu(dy) &= \int_{(0,\infty)} \left|f\left(\tfrac{1}{y}\right)\right|\,|g(xy)|\,dy \\ &= \int_{(0,\infty)} \left|f\left(\tfrac{x}{y}\right)\right|\,|g(y)|\,dy\end{aligned}$$

for all $x \in (0,\infty)$. Thus, $\Lambda_\mu(f,g) = \Lambda_\mu(g,f)$; and a similar argument shows that $f \bullet g(x) = g \bullet f(x)$ for $x \in \Lambda_\mu(f,g)$.

Next, let $q \in [1,\infty)$ and $f \in L^q(\mu)$ be given. Set $K(x,y) = f\left(\frac{x}{y}\right)$, and note that

$$\|K(\,\cdot\,,y)\|_{L^q(\mu)} = \|f\|_{L^q(\mu)} = \|K(x,\,\cdot\,)\|_{L^q(\mu)} \quad \text{for all } x,\, y \in (0,\infty).$$

Thus, Corollary 6.2.18 applies and says that

$$\mu\big(\Lambda_\mu(f,g)\complement\big) = 0 \quad \text{and} \quad \|f \bullet g\|_{L^r(\mu)} \le \|f\|_{L^q(\mu)}\|g\|_{L^p(\mu)}$$

when $p \in [1,\infty]$ satisfies $\frac{1}{r} \equiv \frac{1}{p} + \frac{1}{q} - 1 \ge 0$ and $g \in L^p(\mu)$. When $q = \infty$, the same conclusion can be drawn by using $f \bullet g = g \bullet f$ and reversing the roles of f and g.

Finally, to prove Hardy's inequality, take f and g as suggested in the hint, and observe that

$$\begin{aligned}
&\left[\int_{(0,\infty)} \frac{1}{x^{1+\alpha}}\left(\int_{(0,x)} \varphi(y)\,dy\right)^p dx\right]^{\frac{1}{p}} \\
&\quad = \left[\int_{(0,\infty)}\left(\int_{(0,\infty)} \left(\tfrac{x}{y}\right)^{-\frac{\alpha}{p}} \mathbf{1}_{[1,\infty)}\left(\tfrac{x}{y}\right) y^{1-\frac{\alpha}{p}}\varphi(y)\,\mu(dy)\right)^p \mu(dx)\right]^{\frac{1}{p}} \\
&\quad = \|f \bullet g\|_{L^p(\mu)} \le \|f\|_{L^1(\mu)}\|g\|_{L^p(\mu)}.
\end{aligned}$$

Since

$$\|f\|_{L^1(\mu)} = \int_{[1,\infty)} x^{-(1+\frac{\alpha}{p})}\,dx = \frac{p}{\alpha},$$

while

$$\|g\|_{L^p(\mu)} = \left[\int_{(0,\infty)} y^{p-\alpha-1}\varphi(y)^p\,dy\right]^{\frac{1}{p}} = \left[\int_{(0,\infty)} \frac{\left(y\varphi(y)\right)^p}{y^{1+\alpha}}\,dy\right]^{\frac{1}{p}},$$

the derivation is complete.

§ 7.1

7.1.15: If $\bar{\mathbf{F}} \neq \mathbf{H}$, then there exists a $\mathbf{y} \in \mathbf{H}\setminus\bar{\mathbf{F}}$, and so $\mathbf{x} = \mathbf{y} - \Pi_{\bar{\mathbf{F}}}\mathbf{y} \neq \mathbf{0}$ and yet $\mathbf{x} \perp \mathbf{F}$. Conversely, if $\mathbf{x} \in \mathbf{H}\setminus\{\mathbf{0}\}$ and $\mathbf{x} \perp \mathbf{F}$, then

$$\|\mathbf{y}-\mathbf{x}\|_{\mathbf{H}}^2 = \|\mathbf{x}\|_{\mathbf{H}}^2 - 2\mathfrak{Re}(\mathbf{x},\mathbf{y})_{\mathbf{H}} + \|\mathbf{y}\|_{\mathbf{H}}^2 = \|\mathbf{x}\|_{\mathbf{H}}^2 + \|\mathbf{y}\|_{\mathbf{H}}^2 \ge \|\mathbf{x}\|_{\eta}^2 \quad \text{for all } \mathbf{y} \in \mathbf{F}.$$

Hence, $\mathbf{x} \notin \bar{\mathbf{F}}$.

Next suppose that $\{\mathbf{e}_m : m \in \mathbb{Z}^+\}$ is an orthonormal sequence in $\mathbf{H}$, and let $\mathbf{F} = \operatorname{span}\left(\{\mathbf{e}_m : m \in \mathbb{Z}^+\}\right)$. If $\{\mathbf{e}_m : m \in \mathbb{Z}^+\}$ is complete, the $\mathbf{H} = \bar{\mathbf{F}}$, and so

$$\|\mathbf{x}\|_{\mathbf{H}}^2 = \left\|\Pi_{\bar{\mathbf{F}}}\mathbf{x}\right\|_{\mathbf{H}}^2 = \sum_{m=1}^{\infty}\left|(\mathbf{x},\mathbf{e}_m)_{\mathbf{H}}\right|^2$$

follows from (7.1.8). Conversely, if, for all $\mathbf{x} \in \mathbf{H}$, equality holds in Bessel's inequality, then there cannot be any non-zero $\mathbf{x} \perp \mathbf{F}$. Hence, by the above, $\bar{\mathbf{F}} = \mathbf{H}$, and so $\mathbf{x} = \Pi_{\bar{\mathbf{F}}}\mathbf{x} = \sum_{m=1}^{\infty}(\mathbf{x},\mathbf{e}_m)_{\mathbf{H}}\mathbf{e}_m$, where the final equality is (7.1.11).

7.1.16: The first assertion is simply the trivial remark that closed subsets of a complete metric space are complete with respect to the metric obtained by restriction. Turning to the second assertion, first observe that it suffices to handle real Hilbert spaces, since complex ones can be written as the sum of their real and imaginary parts. Now suppose that $\mathbf{F}$ is a finite dimensional subspace of the real Hilbert space $\mathbf{H}$. To see that $\mathbf{F}$ is closed, set $N = \dim(\mathbf{F})$, and choose an orthonormal basis $\{\mathbf{e}_1, \dots, \mathbf{e}_N\}$ for $\mathbf{F}$. Then

$$\mathbf{x} \in \mathbf{F} \longmapsto \Phi(\mathbf{x}) = \big((\mathbf{x}, \mathbf{e}_1)_{\mathbf{H}}, \dots, (\mathbf{x}, \mathbf{e}_N)_{\mathbf{H}}\big) \in \mathbb{R}^N$$

is an isometric isomorphism. In particular, if $\{\mathbf{x}_n\}_1^\infty \subseteq \mathbf{F}$ and $\mathbf{x}_n \longrightarrow \mathbf{x}$ in $\mathbf{H}$, then there exists a $\boldsymbol{\xi} \in \mathbb{R}^N$ such that $\Phi(\mathbf{x}_n) \longrightarrow \mathbf{x}$ in $\mathbb{R}^N$. But this means that $\mathbf{x} = \Phi^{-1}(\boldsymbol{\xi}) \in \mathbf{F}$.

Finally, to see that $C([0,1];\mathbb{R})$ is not closed in $L^2([0,1];\mathbb{R})$, remember (cf. part (**v**) of Theorem 6.2.2) that $C([0,1];\mathbb{R})$ is dense in $L^2([0,1];\mathbb{R})$. Hence, if $C([0,1];\mathbb{R})$ were closed, then every element of $L^2([0,1];\mathbb{R})$ would have to be continuous.

7.1.17: Suppose that $\Pi = \Pi_{\mathbf{F}}$ for some closed subspace $\mathbf{F}$. Clearly, $\text{Range}(\Pi_{\mathbf{F}}) \subseteq \mathbf{F}$ and $\Pi_{\mathbf{F}}\mathbf{x} = \mathbf{x}$ for $\mathbf{x} \in \mathbf{F}$. Hence, $\text{Range}(\Pi_{\mathbf{F}}) = \mathbf{F}$ and $\Pi_{\mathbf{F}}^2 = \Pi_{\mathbf{F}}$. In addition, for any $\mathbf{x}, \mathbf{x}' \in \mathbf{H}$,

$$\big(\Pi_{\mathbf{F}}\mathbf{x}, \mathbf{x}'\big)_{\mathbf{H}} = \big(\Pi_{\mathbf{F}}\mathbf{x}, \mathbf{x}'\big)_{\mathbf{H}} + \big(\Pi_{\mathbf{F}}\mathbf{x}, \mathbf{x}' - \Pi_{\mathbf{F}}\mathbf{x}'\big)_{\mathbf{H}} = \big(\Pi_{\mathbf{F}}\mathbf{x}, \Pi_{\mathbf{F}}\mathbf{x}'\big)_{\mathbf{H}},$$

which means, by symmetry, that $\big(\Pi_{\mathbf{F}}\mathbf{x}, \mathbf{x}'\big)_{\mathbf{H}} = \big(\mathbf{x}, \Pi_{\mathbf{F}}\mathbf{x}'\big)_{\mathbf{H}}$.

Conversely, suppose that $\Pi^2 = \Pi$ and $\big(\Pi\mathbf{x}, \mathbf{x}'\big)_{\mathbf{H}} = \big(\mathbf{x}, \Pi\mathbf{x}'\big)_{\mathbf{H}}$, and set $\mathbf{F} = \text{Range}(\Pi)$. Clearly $\mathbf{F}$ is a subspace. Furthermore,

$$\big(\mathbf{y} - \Pi\mathbf{y}, \mathbf{x}'\big)_{\mathbf{H}} = \big(\mathbf{y}, \mathbf{x}'\big)_{\mathbf{H}} - \big(\mathbf{y}, \Pi\mathbf{x}'\big)_{\mathbf{H}} = 0 \quad \text{for all } \mathbf{x}' \in \mathbf{F}.$$

Hence, $\Pi = \Pi_{\bar{\mathbf{F}}}$, which means first that $\mathbf{F} = \bar{\mathbf{F}}$ and second that $\Pi = \Pi_{\mathbf{F}}$.

Finally, let $\mathbf{F}$ be a subspace, and define $\mathbf{F}^\perp$ accordingly. Because, for each $\mathbf{x} \in \mathbf{H}$, $\mathbf{y} \rightsquigarrow (\mathbf{y}, \mathbf{x})_{\mathbf{H}}$ is linear and continuous, it is clear that $\mathbf{F}^\perp$ is a closed subspace. To prove that $\Pi_{\mathbf{F}^\perp} = I - \Pi_{\bar{\mathbf{F}}}$, set $\Pi = I - \Pi_{\bar{\mathbf{F}}}$. Clearly, $\text{Range}(\Pi) \subseteq \mathbf{F}^\perp$. Hence, because $\Pi_{\bar{\mathbf{F}}}\mathbf{x} = \mathbf{0}$ for every $\mathbf{x} \in \mathbf{F}^\perp$, we see that $\text{Range}(\Pi) = \mathbf{F}^\perp$ and $\Pi^2 = \Pi$. Finally, for any $\mathbf{x}, \mathbf{x}' \in \mathbf{H}$,

$$\begin{aligned}\big(\Pi\mathbf{x}, \mathbf{x}'\big)_{\mathbf{H}} &= \big(\mathbf{x}, \mathbf{x}'\big)_{\mathbf{H}} - \big(\Pi_{\bar{\mathbf{F}}}\mathbf{x}, \mathbf{x}'\big)_{\mathbf{H}} \\ &= \big(\mathbf{x}, \mathbf{x}'\big)_{\mathbf{H}} - \big(\mathbf{x}, \Pi_{\bar{\mathbf{F}}}\mathbf{x}'\big)_{\mathbf{H}} = \big(\mathbf{x}, \Pi\mathbf{x}'\big)_{\mathbf{H}}.\end{aligned}$$

Hence, by the first part, $\Pi = \Pi_{\text{Range}(\Pi)}$.

7.1.18: Let $\{\mathbf{x}_n\}_1^\infty$ be a dense sequence in $\mathbf{H}$. Given $E \subseteq \mathbf{H}$ with the stated properties, set

$$E_{m,n} = \big\{\mathbf{e} \in E : \big|(\mathbf{x}_n, \mathbf{e})_{\mathbf{H}}\big| \geq \tfrac{1}{m}\big\} \quad \text{for } (m,n) \in (\mathbb{Z}^+)^2.$$

Then, for each (m,n), $E_{m,n}$ is finite. Indeed, if $\text{card}(E_{m,n}) > m^2\|\mathbf{x}_n\|_{\mathbf{H}}^2$, then we could find $L > m^2\|\mathbf{x}_n\|_{\mathbf{H}}^2$ and distinct $\mathbf{e}_1, \dots, \mathbf{e}_L \in E_{m,n}$. But, by Bessel's inequality, this would lead to the contradiction

$$\|\mathbf{x}_n\|_{\mathbf{H}}^2 \geq \sum_{\ell=1}^{L} \big|(\mathbf{x}_n, \mathbf{e}_\ell)_{\mathbf{H}}\big|^2 \geq \frac{L}{m^2} > \|\mathbf{x}_n\|_{\mathbf{H}}^2.$$

Knowing that each $E_{m,n}$ is finite, we know that

$$E' = \bigcup_{(m,n)\in(\mathbb{Z}^+)^2} E_{m,n} \quad \text{is countable}$$

and $\mathbf{x}_n \perp E \setminus E'$ for all $n \in \mathbb{Z}^+$. But, since $\{\mathbf{x}_n\}_1^\infty$ is dense, this latter fact means that $\mathbf{x} \perp E \setminus E'$ for all $\mathbf{x} \in \mathbf{H}$. That is, $E \setminus E' \subseteq \{\mathbf{0}\}$, and so $E = E'$. In other words, we have now proved that E is a countable basis for $\mathbf{H}$. Finally, because we have assumed that $\mathbf{H}$ is not finite dimensional, we can conclude that E must be countable and infinite, and therefore isomorphic to $\mathbb{Z}^+$.

7.1.19: Given an orthonormal basis E in $\mathbf{F}$ and an orthonormal basis E' in $\mathbf{F}^\perp$, it is clear that $\tilde{E} = E \cup E'$ is again orthonormal. To see that it is complete, simply observe that (remember that, by Exercise 7.1.18, both E and E' are countable)

$$\mathbf{x} = \Pi_{\mathbf{F}}\mathbf{x} + \Pi_{\mathbf{F}^\perp}\mathbf{x} = \sum_{\mathbf{e}\in E} (\mathbf{x}, \mathbf{e})_{\mathbf{H}}\mathbf{e} + \sum_{\mathbf{e}'\in E'} (\mathbf{x}, \mathbf{e}')_{\mathbf{H}}\mathbf{e}'$$

for every $\mathbf{x} \in \mathbf{H}$.

Conversely, if E is an orthonormal basis in $\mathbf{F}$ and $\tilde{E} \supseteq E$ is an orthonormal basis in $\mathbf{H}$, set $E' = \tilde{E} \setminus E$. Clearly $\operatorname{span}(E') \subseteq \mathbf{F}^\perp$. At the same time, if $\mathbf{x} \in \mathbf{F}^\perp$, then $\mathbf{x} \perp E$ and therefore

$$\mathbf{x} = \sum_{\mathbf{e}'\in E'} (\mathbf{x}, \mathbf{e}')_{\mathbf{H}}\mathbf{e}'.$$

Hence E' is an orthonormal basis in $\mathbf{F}^\perp$.

§ 7.2

7.2.6: Because

$$(2L)^{-\frac{1}{2}} e^{\frac{\sqrt{-1}\,\pi m x}{L}} = e^{\frac{\sqrt{-1}\,\pi}{2L}} \left(\mathfrak{e}_{-L,2L}\right)_m(x), \quad m \in \mathbb{Z},$$

it is clear that

$$\left\{(2L)^{-\frac{1}{2}} e^{\frac{\sqrt{-1}\,\pi m x}{L}} : m \in \mathbb{Z}\right\}$$

is an orthonormal basis in $L^2\big([-L, L]; \mathbb{C}\big)$. Hence, by the same reasoning as was used to prove Corollary 7.2.5,

$$\left\{(2L)^{-\frac{1}{2}} \cos\frac{\pi m x}{L} : m \in \mathbb{N}\right\} \cup \left\{(2L)^{-\frac{1}{2}} \sin\frac{\pi m x}{L} : m \in \mathbb{Z}^+\right\}$$

is an orthonormal basis in both $L^2(\mathbb{R}^N)$ and $L^2(\mathbb{R}^N; \mathbb{C})$. Moreover, by symmetry,

$$f \in \mathbf{E} \implies \int_{[-L,L]} f(x) \sin\frac{\pi m x}{L}\, dx = 0, \quad m \in \mathbb{Z}^+$$

$$f \in \mathbf{O} \implies \int_{[-L,L]} f(x) \cos\frac{\pi m x}{L}\, dx = 0, \quad m \in \mathbb{N}.$$

Hence,

$$\left\{(2L)^{-\frac{1}{2}}\cos\frac{\pi m x}{L} : m\in\mathbb{N}\right\} \quad\text{and}\quad \left\{(2L)^{-\frac{1}{2}}\sin\frac{\pi m x}{L} : m\in\mathbb{Z}^+\right\}$$

are bases for $\mathbf{E}$ and $\mathbf{O}$, respectively. Finally, define $\Phi_{\mathbf{O}} : \mathbf{O} \longrightarrow L^2([0,1];\mathbb{C})$ and $\Phi_{\mathbf{E}} : \mathbf{E} \longrightarrow L^2([0,1];\mathbb{C})$ so that

$$\Phi_{\mathbf{O}}(f) = \Phi_{\mathbf{E}}(f) = 2^{\frac{1}{2}} f \restriction [0,L],$$

and observe that each of these is an isometric isomorphism.

7.2.7: Using integration by parts, one finds that

$$\begin{aligned}
&-\frac{\sqrt{-1}\,\pi m}{L}\big(f,(\mathfrak{e}_{-L,2L})_m\big)_{L^2([-L,L];\mathbb{C})} = \big(f,(\mathfrak{e}_{-L,2L})'_m\big)_{L^2([-L,L];\mathbb{C})}\\
&= f(L)-f(-L)-\big(f',(\mathfrak{e}_{-L,2L})_m\big)_{L^2([-L,L];\mathbb{C})} = -\big(f,(\mathfrak{e}_{-L,2L})_m\big)_{L^2([-L,L];\mathbb{C})}.
\end{aligned}$$

Hence, by induction on $n\in\mathbb{N}$,

$$\left(\frac{\sqrt{-1}\,\pi m}{L}\right)^n \big(f,(\mathfrak{e}_{-L,2L})_m\big)_{L^2([-L,L];\mathbb{C})} = \big(f^{(n)},(\mathfrak{e}_{-L,2L})_m\big)_{L^2([-L,L];\mathbb{C})}.$$

In particular,

$$\left|\big(f,(\mathfrak{e}_{-L,2L})_m\big)_{L^2([-L,L];\mathbb{C})}\right| \le \left(\frac{L}{\pi m}\right)^2 \|f''\|_{L^2([-L,L];\mathbb{C})}, \quad m\in\mathbb{Z}\setminus\{0\};$$

and so

$$\sum_{m\in\mathbb{Z}} \big(f,(\mathfrak{e}_{-L,2L})_m\big)_{L^2([-L,L];\mathbb{C})}(\mathfrak{e}_{[-L,2L]})_m$$

is absolutely and uniformly convergent.

§ 7.3

7.3.12: Because $(\mathbf{x},\mathbf{y})\in\mathbb{R}^N\times\mathbb{R}^N \longmapsto f(\mathbf{x}-\mathbf{y})g(\mathbf{y})\in\mathbb{C}$ is $\lambda_{\mathbb{R}^N}\times\lambda_{\mathbb{R}^N}$-integrable, Fubini's Theorem justifies the computation:

$$\begin{aligned}
\widehat{f*g}(\boldsymbol{\xi}) &= \iint\limits_{\mathbb{R}^N\times\mathbb{R}^N} f(\mathbf{x}-\mathbf{y})g(\mathbf{y})\mathfrak{e}\big((\boldsymbol{\xi},\mathbf{x})_{\mathbb{R}^N}\big)\,d\mathbf{x}d\mathbf{y}\\
&= \int_{\mathbb{R}^N} g(\mathbf{y})\mathfrak{e}\big((\boldsymbol{\xi},\mathbf{y})_{\mathbb{R}^N}\big)\left(\int_{\mathbb{R}^N} f(\mathbf{x}-\mathbf{y})\mathfrak{e}\big((\boldsymbol{\xi},\mathbf{x}-\mathbf{y})_{\mathbb{R}^N}\big)\,d\mathbf{x}\right)d\mathbf{y} = \hat f(\boldsymbol{\xi})\hat g(\boldsymbol{\xi}).
\end{aligned}$$

In particular, if $f,\,g\in L^1(\mathbb{R}^N;\mathbb{C})$, then $\hat f\hat g = \widehat{f*g}$ is the Fourier transform of the function $f*g\in L^1(\mathbb{R}^N;\mathbb{C})$.

7.3.13: By the change of variable $\tau = \frac{1+|\mathbf{x}|^2}{2t}$, we have

$$\int_{(0,\infty)} t^{-\frac{N+3}{2}} e^{-\frac{1}{2t}} e^{-\frac{|\mathbf{x}|^2}{2t}}\, dt = \int_{(0,\infty)} t^{-\frac{N+1}{2}} e^{-\frac{1+|\mathbf{x}|^2}{2t}} \frac{dt}{t}$$

$$= \left(\frac{2}{1+|\mathbf{x}|^2}\right)^{\frac{N+1}{2}} \int_{(0,\infty)} \tau^{\frac{N+1}{2}} e^{-\tau} \frac{d\tau}{\tau} = 2^{\frac{N+1}{2}} \Gamma\left(\frac{N+1}{2}\right) (1+|\mathbf{x}|^2)^{-\frac{N+1}{2}}$$

Since, by part (**iii**) of Exercise 5.2.6,

$$\frac{2}{\omega_N} = \pi^{-\frac{N+1}{2}} \Gamma\left(\frac{N+1}{2}\right),$$

we now see that

$$P(\mathbf{x}) = (2\pi)^{-\frac{N+1}{2}} \int_{(0,\infty)} t^{-\frac{N+3}{2}} e^{-\frac{1}{2t}} e^{-\frac{|\mathbf{x}|^2}{2}}\, dt.$$

By combining this with (7.3.9) and applying Fubini's Theorem, we find that

$$\begin{aligned}\hat{P}(\boldsymbol{\xi}) &= (2\pi)^{-\frac{1}{2}} \int_{(0,\infty)} t^{-\frac{3}{2}} e^{-\frac{1}{2t}} \hat{\gamma}_{2\pi t}(\boldsymbol{\xi})\, dt \\ &= (2\pi)^{-\frac{1}{2}} \int_{(0,\infty)} t^{-\frac{3}{2}} e^{-\frac{1}{2t}} e^{-2|\pi\boldsymbol{\xi}|^2}\, dt = e^{-2\pi|\boldsymbol{\xi}|},\end{aligned}$$

where, in the last step, we have used part (**iv**) of Exercise 5.2.6 with $\alpha = 2^{-\frac{1}{2}}$ and $\beta = 2^{\frac{1}{2}}\pi|\boldsymbol{\xi}|$.

Finally, to see how this computation leads to (6.3.21), observe that, by an easy change of variables, $\hat{P}_t(\boldsymbol{\xi}) = \hat{P}(t\boldsymbol{\xi}) = e^{-2\pi|\boldsymbol{\xi}|}$. Hence, $\widehat{P_s * P_t}(\boldsymbol{\xi}) = \hat{P}_{s+t}(\boldsymbol{\xi})$, and therefore, by Exercise 7.3.12 and Theorem 7.3.10, $P_s * P_t = P_{s+t}$.

7.3.14: The equality $\widehat{g_t}(\boldsymbol{\xi}) = \hat{g}(t\boldsymbol{\xi})$ is nothing but an obvious change of variable. Moreover, given this identity, $\widehat{g_t} \longrightarrow 1$ uniformly on compacts because $\hat{g}$ is continuous and $\hat{g}(\mathbf{0}) = \int_{\mathbb{R}^N} g(\mathbf{x})\, d\mathbf{x} = 1$.

§ 7.4

7.4.22: When $k = \ell = 0$, there is nothing to do, and $c_0(m,0,0) = 1$. Now suppose that

$$(2\pi x)^k h_m^{(\ell)} = \sum_{n=0}^{k+\ell} c_n(m,k,\ell) h_{m-k-\ell+2n}.$$

Then, by the second line of (7.4.14),

$$\begin{aligned}(2\pi x)^{k+1} h_m^{(\ell)} &= \pi^{\frac{1}{2}} \sum_{n=0}^{k+\ell} c_n(m,k,\ell)\Big(\big((m-k-\ell+2n+1)^+\big)^{\frac{1}{2}} h_{m-k-\ell+2n+1} \\ &\qquad + \big((m-k-\ell+2n)^+\big)^{\frac{1}{2}} h_{m-k-\ell+2n-1}\Big) \\ &= \sum_{n=0}^{k+1+\ell} c_n(m,k+1,\ell) h_{m-(k+1)-\ell+2n},\end{aligned}$$

where $c_n(m, k+1, \ell)$ is given by

$$\begin{aligned}
&\pi^{\frac{1}{2}}(m+k+\ell+1)^{\frac{1}{2}} c_{m+k+\ell}(m,k,\ell) && \text{if } n = m+k+1+\ell \\
&\pi^{\frac{1}{2}}\left((m-k-\ell+2n-1)^+\right)^{\frac{1}{2}} c_{n-1}(m,k,\ell)) && \\
&\qquad + \pi^{\frac{1}{2}}\left((m-k-\ell+2n)^+\right)^{\frac{1}{2}} c_n(m,k,\ell) && \text{if } 1 \le n \le k+\ell \\
&\pi^{\frac{1}{2}}\left((m-k-\ell)^+\right)^{\frac{1}{2}} c_0(m,k,\ell) && \text{if } n = 0.
\end{aligned}$$

Hence,

$$\left|c_n(m, k+1, \ell)\right| \le 2\pi^{\frac{1}{2}}(m+k+\ell+1)^{\frac{1}{2}} \max_{0 \le n \le k+\ell} \left|c_n(m,k,\ell)\right|.$$

Essentially the same argument, only this time using the first line of (7.4.14), leads to

$$(2\pi x)^k h^{(\ell+1)} = \sum_{n=0}^{k+\ell+1} c_n(m,k,\ell+1) h_{m-k-(\ell+1)+2n},$$

where

$$\left|c_n(m,k,\ell+1)\right| \le 2\pi^{\frac{1}{2}}(m+k+\ell+1)^{\frac{1}{2}} \max_{0 \le n \le k+\ell} \left|c_n(m,k,\ell)\right|.$$

Thus, by induction, we have now proved the desired representation with

$$\left|c_n(m,k,\ell)\right| \le \left(2\pi^{\frac{1}{2}}\right)^{k+\ell} \prod_{\mu=0}^{k+\ell} (m+\mu)^{\frac{1}{2}},$$

from which the stated estimate is trivial.

7.4.23: **(i)** Obviously

$$f \in L^2(\Gamma;\mathbb{C}) \longmapsto \Phi(f) \equiv 2^{\frac{1}{4}} e^{-\pi x^2} \in L^2(\mathbb{R};\mathbb{C})$$

is an isometric, isomorphism. Thus, since $H_n = \Phi^{-1} h_n$, $n \in \mathbb{N}$, $\{H_n : n \in \mathbb{N}\}$ is an orthonormal sequence in $L^2(\Gamma;\mathbb{C})$. Moreover, if $\psi \in L^2(\Gamma;\mathbb{C})$ is orthogonal to $\{H_n : n \in \mathbb{N}\}$, then $f = \Phi(\psi) \perp \{h_n : n \in \mathbb{N}\}$, which means that $f = 0$ Lebesgue almost everywhere. Hence $\psi = \mathbf{0}$ as an element of $L^2(\Gamma;\mathbb{C})$, and so, by Exercise 7.1.15, we have proved that $\{H_n : n \in \mathbb{N}\}$ is an orthonormal basis in $L^2(\Gamma;\mathbb{C})$.

(ii) Because $L = -A \circ D$, (7.4.10) plus (7.4.7) and (7.4.12) combine to give

$$\begin{aligned}
\left(Lf, G_n\right)_{L^2(\Gamma;\mathbb{C})} &= -\left(Df, DG_n\right)_{L^2(\Gamma;\mathbb{C})} = -4\pi n\left(Df, G_{n-1}\right)_{L^2(\Gamma;\mathbb{C})} \\
&= -4\pi n\left(f, AG_{n-1}\right)_{L^2(\Gamma;\mathbb{C})} = -4\pi n\left(f, G_n\right)_{L^2(\Gamma;\mathbb{C})}.
\end{aligned}$$

Hence, after multiplying through by β_n^{-1}, we get

$$\big(Lf, H_n\big)_{L^2(\Gamma;\mathbb{C})} = -4\pi n\big(f, H_n\big)_{L^2(\Gamma;\mathbb{C})}.$$

(iii) Observe that another expression for u_f is

$$(*) \qquad u_f(t,x) = \int_{\mathbb{R}} f\left(e^{-4\pi t}x + \sqrt{\frac{1-e^{-8\pi t}}{2}}\,y\right) e^{-\pi y^2}\,dy.$$

Hence the facts that $u_f \in C^\infty\big([0,\infty)\times\mathbb{R};\mathbb{C}\big)$ and

$$\sup_{t\geq 0}\left|\frac{\partial^\ell u_f}{\partial x^\ell}(t,x)\right| \leq C_\ell\big(1+|x|^2\big)^{\nu_\ell}$$

for suitably chosen $C_\ell < \infty$ and $\nu_\ell \in \mathbb{N}$ can be read off directly from the fact that $f \in C^\infty_{\nearrow}(\mathbb{R};\mathbb{C})$.

(iv) First observe that, from (*), it is clear that $u_f(t,\,\cdot\,) \longrightarrow f$ uniformly on compacts as $t \searrow 0$. Hence, because, by the estimate derived in **(iii)** when $\ell = 1$, it is clear that $\sup_{t\geq 0}|u_f(t,\,\cdot\,)| \in L^4(\Gamma;\mathbb{C})$, it follows (cf. part **(ii)** of Exercise 3.3.21) that $u_f(t,\,\cdot\,) \longrightarrow f$ in $L^2(\Gamma;\mathbb{C})$. Next, suppose that we knew that $\frac{\partial u_f}{\partial t} = Lu_f$. Then we would have

$$\frac{u_f(t,\,\cdot\,) - u_f(s,\,\cdot\,)}{t-s} - Lf = \frac{1}{t-s}\int_{[s,t]}\big(Lu_f(\tau,\,\cdot\,) - u_f(s,\,\cdot\,)\big)\,d\tau,$$

and so

$$\left\|\frac{u_f(t,\,\cdot\,) - u_f(s,\,\cdot\,)}{t-s} - Lf\right\|_{L^2(\Gamma;\mathbb{C})} \longrightarrow 0$$

would again follow from the estimates in part **(iii)**. Thus, we turn our attention to checking that $\frac{\partial u_f}{\partial t} = Lu_f$. But this comes down to showing that, for each $y \in \mathbb{R}$,

$$\frac{\partial}{\partial t}\gamma_{2^{-1}(1-e^{-8\pi t})}\big(y - e^{-4\pi t}x\big) = \left(\frac{\partial^2}{\partial x^2} - 4\pi x\frac{\partial}{\partial x}\right)\gamma_{2^{-1}(1-e^{-8\pi t})}\big(y - e^{-4\pi t}x\big),$$

which is an easy application of elementary calculus once one verifies (also with elementary calculus) that $\frac{\partial \gamma_t}{\partial t} = (4\pi)^{-1}\frac{\partial^2\gamma_t}{\partial x^2}$.

With the preceding at hand, we now know that

$$\frac{d}{dt}\big(u_f(t,\,\cdot\,), H_n\big)_{L^2(\Gamma;\mathbb{C})} = \big(Lu_f(t,\,\cdot\,), H_n\big)_{L^2(\Gamma;\mathbb{C})} = -4\pi n\big(u_f(t,\,\cdot\,), H_n\big)_{L^2(\Gamma;\mathbb{C})}$$

and $\big(u_f(t,\,\cdot\,), H_n\big)_{L^2(\Gamma;\mathbb{C})} \longrightarrow (f, H_n)_{L^2(\Gamma;\mathbb{C})}$ as $t \searrow 0$. Thus, we have now proved that

$$\big(u_f(t,\,\cdot\,), H_n\big)_{L^2(\Gamma;\mathbb{C})} = e^{-4\pi n t}\big(f, H_n\big)_{L^2(\Gamma;\mathbb{C})}.$$

In particular, when $f = H_m$, this proves first that

$$\big(u_{H_m}(t,\,\cdot\,), H_n\big)_{L^2(\Gamma;\mathbb{C})} = \delta_{m,n}e^{-4\pi n t},$$

and thence that $u_{H_m}(t,\,\cdot\,) = e^{-4\pi m t}H_m$.

(**v**) Notice that

$$u_f(t,x) = \int_{\mathbb{R}} M(t,x,y)f(y)\,\Gamma(dy), \quad f \in C^\infty_{\nearrow}(\mathbb{R};\mathbb{C}).$$

Hence, by the final part of (**iv**),

$$\big(M(t,x,\,\cdot\,),H_m\big)_{L^2(\Gamma;\mathbb{C})} = e^{-4\pi m t}H_m(x), \quad (t,x)\in(0,\infty)\times\mathbb{R}.$$

Since $M(t,x,\,\cdot\,)\in L^2(\Gamma;\mathbb{C})$, it follows that

$$M(t,x,\,\cdot\,) = \sum_{m=0}^{\infty} e^{-4\pi m t}H_m(x)H_m \quad \text{in } L^2(\Gamma;\mathbb{C}).$$

Finally, by the estimate in (7.4.18), we know that, for each $R>0$, there is a $C_R<\infty$ such that

$$\sup_{|x|\vee|y|\le R}\big|H_m(x)H_m(y)\big| \le C_R(1+m), \quad m\in\mathbb{N}.$$

Hence the series converges uniformly on compacts.

§ 7.5

7.5.10: We begin by remarking that, after combining the estimates in Exercise 7.4.22 with the one in (7.4.18), we know that, for each $(\boldsymbol{\alpha},\boldsymbol{\beta})\in(\mathbb{N}^N)^2$, there exists a $C_{\boldsymbol{\alpha},\boldsymbol{\beta}}<\infty$ such that

$$\big\|\mathbf{x}^{\boldsymbol{\alpha}}\partial^{\boldsymbol{\beta}}h_{\mathbf{n}}\big\|_{\mathrm{u}} \le C_{\boldsymbol{\alpha},\boldsymbol{\beta}}\big(1+\|\mathbf{n}\|\big)^{\frac{\|\boldsymbol{\alpha}\|+\|\boldsymbol{\beta}\|+1}{2}}, \quad \mathbf{n}\in\mathbb{N}^N.$$

In particular, if (7.5.11) holds, then it is clear that, for each $(\boldsymbol{\alpha},\boldsymbol{\beta})\in(\mathbb{N}^N)^2$, the series

$$\sum_{\mathbf{n}\in\mathbb{N}^N}\big(f,h_{\mathbf{n}}\big)_{L^2(\mathbb{R}^N;\mathbb{C})}\mathbf{x}^{\boldsymbol{\alpha}}\partial^{\boldsymbol{\beta}}h_{\mathbf{n}}$$

converges uniformly on $\mathbb{R}^N$. Hence, since, when $\boldsymbol{\alpha}=\boldsymbol{\beta}=\mathbf{0}$, it converges to f, we now see that $f\in\mathcal{S}(\mathbb{R}^N;\mathbb{C})$. In fact, for arbitrary $(\boldsymbol{\alpha},\boldsymbol{\beta})\in(\mathbb{N}^N)^2$, the corresponding series converges uniformly to $\mathbf{x}^{\boldsymbol{\alpha}}\partial^{\boldsymbol{\beta}}f$. Conversely, if $f\in\mathcal{S}(\mathbb{R}^N;\mathbb{C})$, then, by (3) in Theorem 7.5.9, Green's Identity, and the last line of (7.4.14), it is easy to see that

$$(*)\qquad \Big(\big(-\Delta+|2\pi\mathbf{x}|^2\big)f,h_{\mathbf{n}}\Big)_{L^2(\mathbb{R}^N;\mathbb{C})} = 4\pi\big(\|\mathbf{n}\|+\tfrac{1}{2}\big)\big(f,h_{\mathbf{n}}\big)_{L^2(\mathbb{R}^N;\mathbb{C})}.$$

Hence, after iteration,

$$\Big(\big(-\Delta+|2\pi\mathbf{x}|^2\big)^k f,h_{\mathbf{n}}\Big)_{L^2(\mathbb{R}^N;\mathbb{C})} = \Big(4\pi\big(\|\mathbf{n}\|+\tfrac{1}{2}\big)\Big)^k\big(f,h_{\mathbf{n}}\big)_{L^2(\mathbb{R}^N;\mathbb{C})}.$$

Obviously, this proves that (7.5.11) for $f \in \mathcal{S}(\mathbb{R}^N;\mathbb{C})$.

Finally, suppose that $f \in C^\infty(\mathbb{R}^N;\mathbb{C})$ and that $(-\Delta+|2\pi\mathbf{x}|^2)^k f \in L^2(\mathbb{R}^N;\mathbb{C})$ for every $k \in \mathbb{N}$. To see that f satisfies (7.5.11), take $\{\eta_R : R > 0\}$ as in the proof of Theorem 7.5.6. Then, by Green's Identity,

$$\begin{aligned}\Big(\big(-\Delta+|2\pi\mathbf{x}|^2\big)f,\eta_R h_{\mathbf{n}}\Big)_{L^2(\mathbb{R}^N;\mathbb{C})} &= \Big(f,(-\Delta+|2\pi\mathbf{x}|^2)\eta_R h_{\mathbf{n}}\Big)_{L^2(\mathbb{R}^N;\mathbb{C})} \\ = 4\pi\big(\|\mathbf{n}\|+\tfrac{1}{2}\big)\big(f,\eta_R h_{\mathbf{n}}\big)_{L^2(\mathbb{R}^N;\mathbb{C})} \\ +2\Big(f,\big(\nabla\eta_R,\nabla h_{\mathbf{n}}\big)_{\mathbb{R}^N}\Big)_{L^2(\mathbb{R}^N;\mathbb{C})} + \big(f,h_{\mathbf{n}}\Delta\eta_R\big)_{L^2(\mathbb{R}^N;\mathbb{C})}.\end{aligned}$$

Since, from the first line of (7.4.14), it is clear that

$$\big(\nabla\eta_R,\nabla h_{\mathbf{n}}\big)_{\mathbb{R}^N} \longrightarrow 0 \quad \text{in } L^2(\mathbb{R}^N;\mathbb{C}),$$

while certainly $\eta_R h_{\mathbf{n}} \longrightarrow h_{\mathbf{n}}$ and $h_{\mathbf{n}}\Delta\eta_R \longrightarrow 0$ in $L^2(\mathbb{R}^N;\mathbb{C})$, it follows that (*) holds. Hence, just as before, iteration leads to (7.5.11).

§ 8.1

8.1.30: Since every continuous function is Borel measurable, it suffices to show that $\mathcal{B}_E \subseteq \sigma\big(C_b(E;\mathbb{R})\big)$. To this end, let G be a non-empty open set in E, and set

$$f_n(x) = \left(\frac{\operatorname{dist}(x,G\complement)}{1+\operatorname{dist}(x,G\complement)}\right)^{\frac{1}{n}}, \qquad n \in \mathbb{Z}^+.$$

Clearly, $f_n \in C_b(E;\mathbb{R})$ and $G = \bigcap_1^\infty\{f_n > 0\} \in \sigma\big(C_b(E;\mathbb{R})\big)$. Thus, every open set is an element of $\sigma\big(C_b(E;\mathbb{R})\big)$, and so $\mathcal{B}_E \subseteq \sigma\big(C_b(E;\mathbb{R})\big)$. Moreover, when G is relatively compact, each $f_n \in C_c(E;\mathbb{R})$. Hence, when, in addition, every open set in E can be written as the countable union of relatively compact open sets, the preceding argument shows that $\mathcal{B}_E = \sigma\big(C_c(E;\mathbb{R})\big)$. In particular, this will be the case when E is a separable, locally compact, metric space.

8.1.31: (i) Given $0 < r < R < \infty$, we have that

$$\begin{aligned}\left|(\mathrm{R})\int_{[-R,R]}\varphi\,d\psi - \int_{[-r,r]}\varphi\,d\psi\right| &= \left|(\mathrm{R})\int_{[-R,-r]}\varphi\,d\psi + (\mathrm{R})\int_{[r,R]}\varphi\,d\psi\right| \\ &\le \|\varphi\|_{\mathrm{u}}\big(\psi(\infty)-\psi(r)+\psi(-r)-\psi(-\infty)\big).\end{aligned}$$

Thus, by Cauchy's convergence criterion, we know that $I(f)$ exists.

(**ii**) Clearly $I(\varphi) \ge 0$ when $\varphi \ge 0$, and, by (**ii**) in Examples 1.2.2, $\varphi \in C_b(\mathbb{R};\mathbb{R}) \longmapsto I(\varphi) \in \mathbb{R}$ is linear. Next, just as in the preceding, for $r > 0$,

$$\begin{aligned}\left| I(\varphi) - (\mathrm{R})\int_{[-r,r]} \varphi\, d\psi \right| &= \lim_{R\to\infty} \left| (\mathrm{R})\int_{[-R,R]} \varphi\, d\psi - (\mathrm{R})\int_{[-r,r]} \varphi\, d\psi \right| \\ &\le \|\varphi\|_{\mathrm{u}}\big(\psi(\infty) - \psi(r) + \psi(-r) - \psi(-\infty)\big) = \epsilon(R)\|\varphi\|_{\mathrm{u}}.\end{aligned}$$

At the same time,

$$\left| (\mathrm{R})\int_{[-r,r]} \varphi\, d\psi \right| \le \|\varphi\|_{\mathrm{u},[-r,r]}\big(\psi(\infty) - \psi(-\infty)\big) = A\|\varphi\|_{\mathrm{u},[-r,r]}.$$

(**iii**) By Theorem 8.1.24, Exercise 8.1.30, and (**ii**), we know that there is a finite measure μ_ψ on $\mathcal{B}_{\mathbb{R}}$ with the property that $I(\varphi) = \int \varphi\, d\mu$ for all $\varphi \in C_b(\mathbb{R};\mathbb{R})$. To see that $\mu_\psi\big((-\infty,x]\big) = \psi(x) - \psi(-\infty)$ for all $x \in \mathbb{R}$, define, for each $n \in \mathbb{Z}^+$, $x_n = x + \frac{1}{n}$, and

$$\varphi_n(t) = \begin{cases} 1 & \text{if } t \in (-\infty, x] \\ 1 - n(t-x) & \text{if } t \in (x, x_n] \\ 0 & \text{if } x \in (x_n, \infty). \end{cases}$$

Then φ_n is continuous, $\mathbf{1}_{(-\infty,x]} \le \varphi_n \le \mathbf{1}_{(-\infty,x_n]}$, and $\varphi_n \searrow \mathbf{1}_{(-\infty,x]}$ as $n \to \infty$. In particular, $I(\varphi_n) \searrow \mu_\psi\big((-\infty,x]\big)$. At the same time,

$$\begin{aligned}\psi(x) - \psi(-\infty) &= \lim_{R\to\infty}\big(\psi(x) - \psi(-R)\big) \\ &\le \lim_{R\to\infty} (\mathrm{R})\int_{[-R,R]} \varphi_n\, d\psi = I(\varphi_n) \\ &= \lim_{R\to\infty} (\mathrm{R})\int_{[-R,R]} \varphi_n\, d\psi \le \psi(x_n) - \psi(-\infty).\end{aligned}$$

Hence, because ψ is right-continuous, we also know that $\lim_{n\to\infty} I(\varphi_n) = \psi(x) - \psi(-\infty)$, which, combined with the preceding, shows that $\mu\big((-\infty,x]\big) = \psi(x) - \psi(-\infty)$.

To see that there is only one finite measure μ on $(\mathbb{R},\mathcal{B}_{\mathbb{R}})$ satisfying

$$\mu\big((-\infty,x]\big) = \psi(x) - \psi(-\infty) \quad \text{for all } x \in \mathbb{R},$$

suppose that μ is a second. Then

$$\mu(\mathbb{R}) = \lim_{x\nearrow\infty} \mu\big((-\infty,x]\big) = \lim_{x\nearrow\infty} \mu_\psi\big((-\infty,x]\big) = \mu_\psi(\mathbb{R}).$$

Thus, if $\mathcal{C} = \big\{(-\infty,x] : x \in \mathbb{R}\big\}$, then μ and μ_ψ agree on $\{\mathbb{R}\} \cup \mathcal{C}$. But $\mathcal{C}$ is a π-system and (cf. Exercise 3.2.16) $\mathcal{B}_{\mathbb{R}} = \sigma(\mathcal{C})$. Hence, by Exercise 3.1.8, $\mu = \mu_\psi$.

§ 8.2

8.2.14: Assume, without loss of generality, that $\emptyset \in \mathcal{C}$. Next, for $\Gamma \subseteq E$, let $\mathcal{C}(\Gamma) = \{A \in \mathcal{C} : A \subseteq \Gamma\}$. The key to this exercise lies in the observation that

$$(*) \qquad \Gamma \in \mathcal{B} \iff \Gamma = \bigcup \mathcal{C}(\Gamma) \equiv \bigcup_{A \in \mathcal{C}(\Gamma)} A.$$

To see (*), first note that, since $\mathcal{C}(\Gamma)$ is always countable, $\bigcup \mathcal{C}(\Gamma) \in \mathcal{B}$ for every $\Gamma \subseteq E$. Thus, it suffices to show that $\Gamma = \bigcup \mathcal{C}(\Gamma)$ if $\Gamma \in \mathcal{B}$. To this end, let $\mathcal{A}$ denote the class of $\Gamma \subseteq E$ for which $\Gamma = \bigcup \mathcal{C}(\Gamma)$. Because $\mathcal{C}$ is a partition, $\mathcal{C} \subseteq \mathcal{A}$. Moreover, it is obvious that

$$\{\Gamma_n\}_1^\infty \subseteq \mathcal{A} \implies \bigcup_1^\infty \Gamma_n \in \mathcal{A}.$$

Finally, if $\Gamma \in \mathcal{A}$, then, again because $\mathcal{C}$ is a partition, $\mathcal{C}(\Gamma\complement) = \mathcal{C} \setminus \mathcal{C}(\Gamma)$ and so

$$\bigcup \mathcal{C}(\Gamma\complement) = E \setminus \Gamma = \Gamma\complement.$$

In other words, $\mathcal{A}$ is a σ-algebra containing $\mathcal{C}$, and therefore also $\mathcal{B} = \sigma(\mathcal{C})$.

(i) If $f : E \longrightarrow \overline{\mathbb{R}}$ is $\mathcal{B}$-measurable and $A \in \mathcal{C} \setminus \{\emptyset\}$, let $a \in A$ and set $\Gamma = \{x \in A : f(x) \neq f(a)\}$. Then $\Gamma \in \mathcal{B}$ and so (*) holds. But, because $\Gamma \subseteq A \in \mathcal{A}$, the only possible elements of $\mathcal{C}(\Gamma)$ are $\emptyset$ and A. Hence, since $\Gamma \neq A$, it must be that $\Gamma = \emptyset$. Equivalently, $f \restriction A$ is constant.

Next, suppose that $\nu(A) < \infty$ for every $A \in \mathcal{C}$. Since $\mathcal{C}$ is countable and $E = \bigcup \mathcal{C}$, ν is σ-finite. Conversely, suppose that ν is σ-finite, and choose $\{E_n\}_1^\infty \subseteq \mathcal{B}$ so that $E_n \nearrow E$ and $\nu(E_n) < \infty$ for each $n \in \mathbb{Z}^+$. Given $A \in \mathcal{C}$, we will know that $\nu(A) < \infty$ once we show that $A \subseteq E_n$ for some $n \in \mathbb{Z}^+$. But, because $E_n = \mathcal{C}(E_n)$, either $A \cap E_n = \emptyset$ or $A \subseteq E_n$. Hence, either $A \cap E_n = \emptyset$ for all $n \in \mathbb{Z}^+$, in which case $A = \bigcup_1^\infty (A \cap E_n) = \emptyset$ and $A \subseteq \bigcap_1^\infty E_n$, or $A \subseteq E_n$ for some $n \in \mathbb{Z}^+$.

Finally, suppose that μ and ν are a pair of measures on $(E, \mathcal{B})$. If $\mu \ll \nu$, then certainly $\mu(A) = 0$ whenever $A \in \mathcal{C}$ and $\nu(A) = 0$. Conversely, if $\mu(A) = 0$ whenever $A \in \mathcal{C}$ and $\nu(A) = 0$, then for $\Gamma \in \mathcal{B}$ with $\nu(\Gamma) = 0$, $\mu(A) = 0$ for all $A \in \mathcal{C}(\Gamma)$ and so, because of (*),

$$\mu(\Gamma) = \sum_{A \in \mathcal{C}(\Gamma)} \mu(A) = 0.$$

(ii) Let ν be a measure on $(E, \mathcal{B})$ and $f : E \longrightarrow [0, \infty)$ a $\mathcal{B}$-measurable, ν-integrable function. If $\mu(\Gamma) = \int_\Gamma f \, d\nu$ for every $\Gamma \in \mathcal{B}$, then, because $f \restriction A$ is constant whenever $A \in \mathcal{A}$, $\mu(A) = f(a)\nu(A)$ for every $a \in A \in \mathcal{C}$. Hence, if $A \in \mathcal{C}$, then $\nu(A) \in (0, \infty) \implies f \restriction A \equiv \frac{\mu(A)}{\nu(A)}$ and $\nu(A) = \infty \implies f \restriction A \equiv 0 \implies \mu(A) = 0$.

(iii) Let $\mathcal{C} = \{\emptyset, E\}$, and define the measures μ and ν on $(E, \mathcal{B})$ so that $\mu(E) = 1$ and $\nu(E) = \infty$. Clearly, $\mu \ll \nu$. On the other hand, if $\mu(\Gamma) = \int_\Gamma f \, d\nu$, $\Gamma \in \mathcal{B}$, for some non-negative, $\mathcal{B}$-measurable f, then, by the last part of **(ii)**, $\mu(E) = 0$. Thus the conclusion of the Radon–Nikodym Theorem fails in this situation.

8.2.15: Suppose not. Then there exists an $\epsilon > 0$ and a sequence $\{A_n\}_0^\infty \subseteq \mathcal{B}$ such that $\mu(A_n) \geq \epsilon$ but $\nu(A_n) \leq 2^{-n}$. Hence, if $B = \overline{\lim}_{n\to\infty} A_n$, then (cf. part (**ii**) of Exercise 3.1.12) $\mu(B) \geq \epsilon$ while, by the Borel–Cantelli Lemma (cf. part (**iv**) of Exercise 3.1.12), $\nu(B) = 0$.

8.2.16: (**i**) Assume that $\{x\} \in \mathcal{B}$ for every $x \in E$ and that $(E, \mathcal{B}, \mu)$ is a σ-finite. In proving that the set A of atoms is countable, it is easy to reduce to the case when μ is finite. Thus, we will assume that $\mu(E) < \infty$. Next, set $A_n = \{x \in E : \mu(\{x\}) \geq \frac{1}{n}\}$, and observe that $\operatorname{card}(A_n) \leq n\mu(E)$. Hence, since A is the union over $n \in \mathbb{Z}^+$ of the A_n's, it follows that A must be countable. Moreover, when μ is purely atomic, $\mu(A\complement) = 0$ and therefore

$$\mu(\Gamma) = \mu(\Gamma \cap A) = \sum_{x\in A} \mu(\Gamma \cap \{x\}) = \sum_{x\in\Gamma\cap A} \mu(\{x\}).$$

Conversely, if there is a countable S such that

$$\mu(\Gamma) = \sum_{x\in\Gamma\cap S} \mu(\{x\}), \quad \Gamma \in \mathcal{B}_{\mathbb{R}},$$

then $A = \{x \in S : \mu(\{x\}) > 0\}$ is the set of atoms, and it is clear that $\Gamma \cap A = \emptyset \implies \mu(\Gamma) = 0$.

(**ii**) Clearly,

$$\mu_\psi(\{x\}) = \lim_{\epsilon\searrow 0} \mu_\psi((x-\epsilon, x]) = \lim_{\epsilon\searrow 0} (\psi(x) - \psi(x-\epsilon)) = \psi(x) - \psi(x-).$$

Hence, x is an atom if and only if $\psi(x) - \psi(x-) > 0$, in which case $\mu_\psi(\{x\}) = \psi(x) - \psi(x-)$. In particular, μ_ψ is non-atomic if and only if ψ is continuous. On the other hand, if μ_ψ is purely atomic and A is its set of atoms, then $A = D(\psi)$ and

$$\begin{aligned}\psi(x) - \psi(-\infty) &= \sum_{t\in D(\psi)\cap(-\infty,x]} \mu_\psi(\{t\}) \\ &= \sum_{t\in D(\psi)\cap(-\infty,x]} (\psi(t) - \psi(t-)) = \psi_{\mathrm{d}}(x).\end{aligned}$$

Conversely, if ψ is purely discontinuous, then

$$\mu_\psi((-\infty, x]) = \sum_{t\in D(\psi)\cap(-\infty,x]} (\psi(t) - \psi(t-)) = \nu((-\infty, x]), \quad x \in \mathbb{R},$$

where ν is the purely atomic measure given by

$$\nu(\Gamma) = \sum_{x\in D(\psi)\cap\Gamma} (\psi(x) - \psi(x-)), \quad \Gamma \in \mathcal{B}_{\mathbb{R}}.$$

Hence, since $\mathcal{C} = \{(-\infty, x] : x \in \mathbb{R}\}$ is a π-system which generates $\mathcal{B}_{\mathbb{R}}$ and $\mu_\psi \restriction \mathcal{C} \cup \{\mathbb{R}\} = \nu \restriction \mathcal{C} \cup \{\mathbb{R}\}$, it follows, by Exercise 3.1.8, that $\mu_\psi = \nu$ on $\mathcal{B}_{\mathbb{R}}$.

(iii) First suppose that $\mu_\psi \ll \lambda_{\mathbb{R}}$. Clearly μ_ψ is then non-atomic, and therefore $\mu_\psi((a,b)) = \psi(b) - \psi(a)$ for all $a < b$. Next, set $f = \frac{d\mu_\psi}{d\lambda_{\mathbb{R}}}$. Given $\epsilon > 0$, choose $R > 0$ so that $\int_{\{f \ge R\}} f\,dx < \frac{\epsilon}{2}$ and take $\delta = \frac{\epsilon}{2R}$. If $\{(a_m, b_m)\}$ is a sequence of mutually disjoint intervals with $\sum(b_m - a_m) < \delta$, then

$$\begin{aligned}\sum\bigl(\psi(b_m) - \psi(a_m)\bigr) &= \sum \int_{(a_m,b_m)} f\,dx \\ &\le \sum\left(\int_{(a_m,b_m)\cap\{f\le R\}} f\,dx + \int_{(a_m,b_m)\cap\{f\ge R\}} f\,dx\right) \\ &\le R\sum(b_m - a_m) + \int_{\{f\ge R\}} f\,dx < \epsilon,\end{aligned}$$

where, in the passage to the last line, we have used the fact that the intervals are mutually disjoint. Conversely, suppose that ψ is absolutely continuous, and choose, for given $\epsilon > 0$, $\delta > 0$ so that

$$\sum(b_m - a_m) < \delta \implies \sum\bigl(\psi(b_m) - \psi(a_m)\bigr) < \epsilon$$

for sequences $\{(a_m, b_m)\}$ of mutually disjoint open intervals. Next, given $\Gamma \in \mathcal{B}_{\mathbb{R}}$ with $|\Gamma| = 0$, choose an open set $G \supseteq \Gamma$ so that $\lambda_{\mathbb{R}}(G) < \delta$, and find (cf. Lemma 2.1.10) a sequence $\{(a_m, \beta_m)\}$ of mutually disjoint open intervals so that $G = \bigcup(a_m, b_m)$. Then $\sum(b_m - a_m) = \lambda_{\mathbb{R}}(G) < \delta$ and so

$$\mu_\psi(\Gamma) \le \sum \mu_\psi\bigl((a_m, b_m)\bigr) = \sum\bigl(\psi(b_m) - \psi(a_m)\bigr) < \epsilon.$$

Hence $\mu_\psi \ll \lambda_{\mathbb{R}}$.

(iv) Set $\nu_1 = \mu_{\psi_1}$ and $\nu_2 = \mu_{\psi_2+\psi_3}$. Clearly $\mu_\psi = \nu_1 + \nu_2$, $\nu_1 \ll \lambda_{\mathbb{R}}$, $\nu_2 \perp \lambda_{\mathbb{R}}$, and therefore $\nu_1 = \mu_{\psi,\mathrm{a}}$ while $\nu_2 = \mu_{\psi,\sigma}$. In particular, this means that $\psi_1 = \psi_{\mathrm{a}}$ and that $\psi_2 + \psi_3 = \psi_\sigma$. Finally, because ψ_2 is continuous and ψ_3 is purely discontinuous, this second equality means that $\psi_{\sigma,\mathrm{d}} = \psi_{3,\mathrm{d}} = \psi_3$, from which $\psi_2 = \psi_{\sigma,\mathrm{c}}$ is trivial.

8.2.17: To prove that $\|\psi_{k+1} - \psi_k\|_{\mathrm{u},\mathbb{R}} = \frac{2^{-k}}{6}$, note first that $\psi_{k+1} \restriction \mathbb{R}\setminus C_k = \psi_k \restriction \mathbb{R}\setminus C_k$ and that

$$\|\psi_{k+1} - \psi_k\|_{\mathrm{u},[b_{k,j},a_{k,j+1}]} = \|\psi_{k+1} - \psi_k\|_{\mathrm{u},[0,3^{-k}]}, \quad 0 \le j < 2^k.$$

Second, check that $\psi_k(x) = \left(\frac{3}{2}\right)^k x$ for $x \in [0, 3^{-k}]$ and that

$$\psi_{k+1}(x) = \begin{cases} \left(\frac{3}{2}\right)^{k+1} x & \text{if } x \in [0, 3^{-k-1}] \\ 2^{-k-1} & \text{if } x \in (3^{-k-1}, 23^{-k-1}) \\ 2^{-k} - \left(\frac{3}{2}\right)^{k+1}(3^{-k} - x) & \text{if } x \in [23^{-k-1}, 3^{-k}]. \end{cases}$$

Finally, combine these two to complete the computation.

Given the preceding, it is obvious that $\psi_k \longrightarrow \psi$ uniformly on $\mathbb{R}$ and, therefore, that ψ is a continuous, non-decreasing function with $\psi(-\infty) = \psi(0) = 0$ and $\psi(1) = \psi(\infty) = 1$. Moreover, ψ is constant on each of the intervals $[a_{k,j}, b_{k,j}]$, $k \in \mathbb{N}$ and $0 \le j \le 2^k$. Hence, $\mu_\psi(\mathbb{R}\setminus C_k) = 0$ for every k, and so $\mu_\psi(\mathbb{R}\setminus C) = 0$ also.

Notation

Notation	Description	See†
(a.e., μ)	To be read *almost everywhere with respect to* μ	§3.3
α^+ & α^-	The positive and negative parts of $\alpha \in \mathbb{R}$	§1.2
$(\mathbf{v}, \mathbf{w})_{\mathbb{R}^N}$	The inner product of $\mathbf{v}$ and $\mathbf{w}$ in $\mathbb{R}^N$	
$\overline{\lim_{n\to\infty}} A_n$ & $\underline{\lim_{n\to\infty}} A_n$	The limit superior $\bigcap_{m=1}^{\infty} \bigcup_{n\geq m} A_n$ and limit inferior $\bigcup_{m=1}^{\infty} \bigcap_{n\geq m} A_n$ of the sets $\{A_n\}_1^{\infty}$	E 3.1.12
$B_E(a, r)$	The ball of radius r around a in the metric space E	
$\mathcal{B}_E$	The Borel σ-algebra over the topological space E	§3.1
$\mathcal{B}[E']$	The restriction of the σ-algebra $\mathcal{B}$ to the subset E'	§3.1
$\overline{\mathcal{B}}^{\mu}$	The completion of the σ-algebra $\mathcal{B}$ with respect to the measure μ	§3.1
$\overline{\mathcal{B}}_{\mathbb{R}^N}$	The σ-algebra of Lebesque measurable subsets in $\mathbb{R}^N$	§2.1
$\mathcal{B}_1 \times \mathcal{B}_2$	The product σ-algebra generated by sets of the form $\Gamma_1 \times \Gamma_2$ for $\Gamma_i \in \mathcal{B}_i$	§3.2
$C(E)$ & $C(E; F)$	The spaces of continuous $\mathbb{R}$-valued and F-valued functions on E	
$C_{\mathrm{b}}(E; \mathbb{R})$	The space of bounded, continuous, $\mathbb{R}$-valued functions on the topological space E.	
$C_{\mathrm{c}}(G)$	The space of $f \in C(G)$ with compact support in G.	
$C^n(G; \mathbb{R}^N)$	The space of $f : G \longrightarrow \mathbb{R}^N$ with n continuous derivatives	
$\|\mathcal{C}\|$	The mesh size of the collection $\mathcal{C}$	§1.1
$\mathcal{C}_1 \leq \mathcal{C}_2$	To be read $\mathcal{C}_2$ *is a refinement of* $\mathcal{C}_1$	§1.1
$\mathcal{C}_1 \vee \mathcal{C}_2$	The least common refinement of $\mathcal{C}_1$ and $\mathcal{C}_2$	§1.1
ΔI & $\Delta_I \psi$	The length of the interval I and the change $\psi(I^+) - \psi(I^-)$ of ψ over I	§§1.2 & 1.3
$\delta\Phi$	The Jacobian of Φ	§5.3
∂^{α}	Abreviation for $\frac{\partial^{\lvert\alpha\rvert}}{\partial x_1^{\alpha_1} \cdots \partial x_N^{\alpha_N}}$	§5.2
$\partial\Gamma$	The boundary $\overline{\Gamma} \setminus \mathring{\Gamma}$ of the set Γ	

$f \upharpoonright \Gamma$	The restriction of the function f to the set Γ	
$f(x+)$ & $f(x-)$	The right and left limits of f at $x \in \mathbb{R}$	§1.2
$\|f\|_{\mathrm{u}}$ & $\|f\|_{\mathrm{u},E}$	The uniform norms of the function f & $f \upharpoonright E$	
$f \wedge g$ & $f \vee g$	The minimum & maximum of f and g	
$\|f\|_{L^p(\mu)}$	The p-Lebesgue norm of f with respect to the measure μ	§3.2 & §5.2
$(f, g)_{L^2(\mu)}$	The inner product of f and g in $L^2(\mu)$	§7.2
$\|f\|_{L^{(p_1,p_2)}(\mu_1,\mu_2)}$	The mixed (p_1, p_2)-Lebesgue norm with respect to the pair of measures (μ_1, μ_2)	§5.2
$\mathfrak{F}$ & $\mathfrak{F}_\sigma$	The collections of all closed subsets and all countable unions of closed subsets	§2.1
$f * g$	The convolution of the functions f and g	(5.3.3)
$\Phi^*\mu = \mu \circ \Phi^{-1}$	The image of the measure μ under the map Φ	§5.1
$\|\Gamma\|_e$	The exterior or outer Lebesgue measure of the set Γ	§2.1
$\Gamma\complement$	The complement of the set Γ	
$\mathring{\Gamma}$	The interior of the set Γ	
$\overline{\Gamma}$	The closure of the set Γ	
$\Gamma^{(\delta)}$	The open δ-hull of the set Γ	§5.3
$\mathfrak{G}$ & $\mathfrak{G}_\delta$	The collections of all open sets and all countable intersections of open sets	§2.1
$\mathbf{1}_\Gamma$	The indicator or characteristic function of the set Γ	§3.2
I^- & I^+	The left and right end points of the interval I	
$\mathbf{I}_{\mathbb{R}^N}$	The identity matrix on $\mathbb{R}^N$	§2.2
$\int_\Gamma f\, dx$	The Lebesgue integral of f over $\Gamma \in \overline{\mathcal{B}}_{\mathbb{R}^N}$	
$J\Phi$	The Jacobian matrix of Φ	§5.3
$L^p(\mu)$	The Lebesgue space of f with $\|f\|_{L^p(\mu)} < \infty$	§3.2 & §5.2
$L^p(\mathbb{R}^N)$	The Lebesgue space of $L^p(\lambda_{\mathbb{R}^N})$	
$L^{(p_1,p_2)}(\mu_1, \mu_2)$	Mixed norm Lebesgue space	§6.2

λ_E	Lebesgue's measure restricted to $E \in \overline{\mathcal{B}}_{\mathbb{R}^N}$	
$\mu_1 \times \mu_2$	The product of the measures μ_1 and μ_2.	§3.4
$\mu \ll \nu$	To be read μ *is absolutely continuous with respect to* ν.	§7.2
$\frac{d\mu}{d\nu}$	The Radon–Nikodym derivative of μ with respect to ν	§7.2
$\mu \perp \nu$	To be read μ *is singular to* ν.	§7.2
$\overset{\mu}{\sim}$	Equivalence relation determined by the measure μ.	§3.2
$\overline{\mu}$	The completion of the measure μ.	§3.1
$\mathbb{N}$	The set of non-negative integers.	
∇F	The gradient of the function F.	§5.3
$\mathbf{n}(x)$	The outer normal at the point x.	§5.3
ω_{N-1} & Ω_N	The surface area of the sphere and the volume of the unit ball in $\mathbb{R}^N$	E. 5.2.6
$\mathcal{P}(E)$	The power set (set of all subsets) of E	§3.1
$\mathbb{Q}$	The set of all rational numbers in $\mathbb{R}$	
$\overline{\mathbb{R}}$	The extended real line	§3.2
$\widehat{\mathbb{R}^2}$	The extended plane $\overline{\mathbb{R}}^2$ with the two points $(\infty, -\infty)$ and $(-\infty, \infty)$ removed	§3.2
$(\mathrm{R})\int_J \varphi(x)\,d\psi(x)$	The Riemann–Stieljes integral of φ with respect to ψ	§1.2
$\mathrm{sgn}(x)$	The signum of $x \in \mathbb{R}$	§1.2
$\sigma(E;\mathcal{C})$	The σ-algebra over E generated by the collection $\mathcal{C}$	§3.1
$\mathbf{S}^{N-1}$	The unit sphere in $\mathbb{R}^N$	
T_A	The linear transformation determined by the matrix A	§2.2
$\mathbf{T}_x$	The tangent space at x	§5.3
$\Xi(\mathcal{C})$	The set of choice maps for the collection $\mathcal{C}$	§1.1
ζ^α	$\prod_{j=1}^N \zeta_j^{\alpha_j}$	§ 6.3
$\mathbb{Z}$ & $\mathbb{Z}^+$	The set of all integers and its subset of positive integers	

Index

F

G

H

I

J

L

My mother used to make the most delicious pickle out of green tomatoes. I can smell it now. All day long she would be cooking, wrapping her spices in little bits of cheesecloth. She never liked to have the spices loose in the pickle; she didn't like biting on those bits of spice. All the tomatoes, onions, and spices would be in this huge kettle. All the neighbors wanted some after she cooked a batch, and of course she gave.

Green Tomato Pickle

2 dozen green tomatoes, sliced
3 medium green peppers, sliced
3 medium onions, sliced
4 cloves garlic, smashed
1 cup sugar
2 tablespoons salt
1 quart vinegar
1 ounce pickling spices

Place the tomatoes, green peppers, onions, garlic, sugar, salt, and vinegar in a large pot and mix well. Wrap the pickling spices in cheesecloth to form a bouquet and add to the pot. Bring to a boil, lower the heat, and simmer for 3 hours, stirring every 20 minutes. Use the side of the spoon to break up any large pieces. Turn off the heat and allow to cool somewhat. Remove the bouquet and place the mixture in mason jars. Refrigerate. Serve cold.

1 can cream of celery soup
1 egg
2 tablespoons milk
1 cup bread crumbs
2 cups vegetable oil
½ cup crumbled cheddar cheese

In a bowl combine the ham, mustard, and half the soup. Blend well and chill for 30 minutes or so. Shape into 8 croquettes. Dip each croquette in a mixture of the egg and milk, then coat with the bread crumbs. Heat the oil in a large frying pan and fry the croquettes until they are golden brown. Drain the croquettes on paper towels. Heat the remaining soup in a saucepan and add the cheese. Thin with milk if necessary. Spoon over the croquettes to serve.

Serves 4

Grandpa's Famous Leftover Steak Hash

4 tablespoons bacon fat
3 large potatoes, cut into ½-inch cubes
1 medium onion, chopped
2 cloves garlic, minced
½ cup beef gravy or stock
3 cups cubed leftover steak
salt, to taste
freshly ground pepper, to taste
dash of Tabasco

Heat the bacon fat in a large frying pan. Add the potatoes, onion, and garlic. Cook over a medium flame, turning occasionally, until potatoes are just tender. Add the gravy or stock, leftover steak, salt, pepper, and dash of Tabasco. Turn up the flame slightly and cook until the potatoes have browned and the meat is hot.

Serves 4

Greens—they were one of my husband's specialties. He loved all kinds, even what I would call weeds. Down in Shadyside he'd go out and pick all kinds of things. He'd come back with a mess of dandelions or poke salad and he'd handle them so lovingly, picking through them so carefully, gently washing them. Wild greens always seemed bitter to me, but he'd parboil them first and let them sit in some hot water to take out that bitter taste. I guess it worked 'cause everyone just loved his greens. He liked his greens rich, you know. Seasoned well. No salt pork, mind you —he felt that was tasteless—but a leftover ham bone or slab of bacon was just what the doctor ordered.

Yes, we fried tomatoes, not green ones, mind you, but nice, firm, ripe ones. Laura (my grandmother's cousin) knows how much I like them. Every time I go there, she has fried tomatoes waiting; she just spoils me so!

Fried Tomatoes

4 large tomatoes, sliced into ½-inch slices
salt, to taste
freshly ground pepper, to taste
½ cup flour
4 tablespoons bacon fat
2 tablespoons sugar
2 tablespoons butter

Wash tomatoes and slice. Sprinkle with salt and pepper, then dredge in flour, trying not to handle the slices too much. Heat the bacon fat in a frying pan.

When the bacon fat is hot, add the tomatoes and brown on both sides. As the tomato slices are done, remove to a serving platter, sprinkle with sugar, and dot with butter. (If you desire, deglaze the frying pan with 2 tablespoons of water and pour this over the tomatoes.

Serves 6

Sliced Tomatoes

3 large, firm, ripe tomatoes
1 medium sweet onion, thinly sliced
salt, to taste
freshly ground white pepper, to taste
sugar, to taste
vegetable oil or mayonnaise

Rub the skin of the tomatoes with the dull edge of a knife until the skin wrinkles, then peel it off. Slice the tomatoes ½ inch thick and arrange with the sliced onion. Sprinkle with salt, pepper, and just a little sugar. Serve by themselves or with salad oil or mayonnaise.

Serves 4

I don't know why we peeled the tomatoes. That's the way we did it, that's all.

Gee's Watermelon Pickle

1 medium watermelon
  water
1 cup salt
1 quart plus 1 cup vinegar
5 pounds sugar
2 tablespoons whole allspice
2 tablespoons cloves
1 2-inch stick cinnamon
1 3-inch piece fresh ginger

Peel the watermelon, removing both the green outer skin and the red meat. Cut the rind into bite-size pieces. Place the rind in a large bowl and cover with water. Add the salt and let stand overnight. Drain and wash the rind thoroughly. Place in a large pot, cover with cold water, and add 1 cup of vinegar. Cook for 2 hours. Drain. Rinse out the pot and place in it the quart of vinegar, sugar, and 2 quarts of water. Bring to a boil and add the allspice, cloves, cinnamon, and ginger. Boil for 5 minutes, then add the rind. Cook until the rind is tender and easily pierced by a fork. Put up in mason jars. Allow the jars to cool, then refrigerate.

Gee's Cole Slaw

1 small head green cabbage
2 teaspoons salt
1 medium onion, finely chopped
1 cup vinegar
1 cup water
½ teaspoon freshly ground pepper
¼ teaspoon dry mustard
4 tablespoons butter
½ cup sugar
2 eggs

Shred the cabbage, sprinkle with 1 teaspoon of the salt, and add onions. Set aside. In a saucepan heat the vinegar and water. Add the other teaspoon of salt, the pepper, mustard, butter, and sugar. In a bowl beat the eggs and set aside. When the vinegar mixture is hot and all the ingredients are dissolved, remove from the heat. Gradually add 4–5 tablespoons of the vinegar mixture to the eggs, beating constantly, then add all the egg mixture to the vinegar mixture, again beating constantly. Place the mixture over a medium flame and heat until the mixture thickens, stirring constantly. Pour over the cabbage and onion, and refrigerate overnight. (Carrots, green peppers, and so forth, may also be added.)

Everyday Cucumber Salad

1 large cucumber, thinly sliced
1 small onion, thinly sliced
salt, to taste
freshly ground pepper, to taste
1 cup cider vinegar
1 tablespoon sugar

Place the sliced cucumber and onion in a bowl and sprinkle with the salt, pepper, and sugar. Allow to sit for a few minutes. Pour the vinegar over the cucumber and onion. Place several ice cubes on the cucumbers to crisp them. Refrigerate for 1 hour. Remove the ice cubes and serve.

Note: Add red pepper flakes if you desire.

Sweet Potato and Orange Casserole

2 pounds sweet potatoes (4 cups when mashed)
4 tablespoons butter
½ cup brown sugar
1 teaspoon salt
1 teaspoon grated orange rind
1 teaspoon grated lemon rind
½ cup fresh orange juice

Place the potatoes in water to cover. Bring to a boil, lower the heat, and cook until the potatoes are soft and easily pierced with a fork. Peel the potatoes while they are still hot and mash them in a mixing bowl. Add the butter, sugar, salt, orange rind, lemon rind, and orange juice. Beat until the mixture is fluffy and smooth. Fold into a shallow greased casserole and bake in a preheated oven at 350° for 30 minutes, until golden brown on top.

Serves 6

Leftover Yam Pudding

2 cups mashed leftover sweet potatoes or yams
2 tablespoons butter
¼ cup heavy cream
2 tablespoons orange juice
2 teaspoons grated orange or lemon rind
1 cup milk
½ cup raisins
¼ teaspoon nutmeg
¼ teaspoon salt
½ cup chopped nuts
7 tablespoons flour
4 tablespoons sugar

In a large bowl mix together the yams, butter, heavy cream, orange juice, orange or lemon rind, milk, raisins, nutmeg, salt, and nuts. Blend well. Sift in the flour and sugar. Fold into a greased casserole or baking pan and place in a preheated oven at 325°. Bake 30–40 minutes.

Serves 4

Corn Patties

3 eggs, separated
2 cups fresh corn, cut from the cob
¼ cup flour
1 teaspoon salt
¼ teaspoon freshly ground pepper
oil for frying

Beat the egg yolks well. Add the corn, flour, salt, and pepper. Beat the egg whites until stiff. Fold the egg whites into the corn mixture. Drop a spoonful at a time onto a lightly greased frying pan. Fry over a medium flame until brown; turn and brown the other side.

Serves 4

Spanish Rice

8 slices bacon
1 cup rice
1 medium onion, chopped
2 cups crushed tomatoes
1 small green pepper, chopped
1 clove garlic, minced
1 small jar pimentos, chopped (4 ounces)
1 cup water
2 teaspoons salt
freshly ground pepper, to taste

In a large frying pan cook the bacon until crisp. Remove to a paper towel and set aside. Wash the rice and place in a frying pan with the bacon fat. Add the onion and cook slowly until the onion starts to brown. Add the tomatoes, green pepper, garlic, pimentos, water, salt, and pepper. Stir well. Cover tightly and cook about 15 minutes over a very low flame until all the liquid is absorbed and the rice is tender. Serve with the bacon slices.

Serves 4

Oh, we used to fix all kinds of things, Grandpa and I. He was quite a cook and could do as much or more than I could. We'd plan two, three days ahead. Eggs, fried ham, soft crabs, if we could get them, peeled sliced tomatoes which Charlie had to have with most every meal. Creamed chipped beef, and spot, those little fish the kids used to catch, fried so crisp you'd just eat the whole thing, bones and all. Bacon, fried potatoes, fried tomatoes, corn, and always Grandpa's biscuits.

Charlie's Biscuits

2 cups sifted flour
2 teaspoons baking powder
½ teaspoon baking soda
½ teaspoon salt
4 tablespoons solid shortening
⅔ cup sour milk or buttermilk

In a bowl mix the flour, baking powder, baking soda, and salt. Cut in the solid shortening using 2 knives or a pastry blender. Do not overmix. Pour in the milk and mix until the dough is free from the sides of the bowl. Turn the dough onto a lightly floured board or table top. Roll the dough out to a thickness of about ½–¾ inch. Cut out biscuits with a cutter or use a small jar. Bake in a preheated oven at 350° until the biscuits are brown (approximately 10–15 minutes).

Serves 6

My husband loved to make ice cream. I remember once, before we moved to Washington, we went out for a drive in the country. It was strawberry time, and he bought a whole crate of berries. When we got home he said to me, "You can fix some strawberry preserves, you and Mama, if you want to. I'm going to make ice cream." He took all the overripe berries and made the most delicious ice cream out of those berries. Everyone thought they had died and gone to heaven.

Macaroon Ice Cream

1 box Almond Macaroon cookies
2 quarts heavy cream
2 cups sugar
4 eggs
almond extract, to taste

Place the cookies on a baking sheet and dry them in a warm oven. Roll over them with a rolling pin and crush them until they are fine crumbs. Soak the crumbs in some heavy cream. Mix all the ingredients together and place in a churn or ice cream maker, and add the ice and salt according to the directions for your machine. Churn until thick and frozen.

Serves 12

Charlie would make his ice cream as often as he could. Back in those days it was a real project. He'd give Papa this great big bowl of ice cream in the evenings, and of course my papa would say, "Charlie, I can't eat all this ice cream!" Charlie would laugh and say, "If you keep working on it you can. If you just keep working on it." And slowly but surely he would. Papa really loved Charlie's ice cream.

I guess you'd call the cooking I did homey; not Southern, but just plain homey.

Gram T

JULIA ELLEN RUFFIN THOMPSON

August 30, 1893 Warrenton, Virginia.

Gram T, my maternal grandmother, was the sweet queen. I remember so vividly the smells of pies, cakes, and cookies coming from her kitchen. And I learned early that she would give me, her "first," almost anything I wanted. When I was very young, I didn't spend too much time in her kitchen, but in later years I did come back for those lessons that passed on many old family traditions. I'm lucky to have had her too.

Gram T

Mama's Stuffed Shad
Sautéed Shad Roe
Corn Pudding
Rice Pudding
Gingerbread
Lemonade
Salmon Croquettes
Virginia-Style Crab Cakes
Baked Virginia Ham
Fruitcake
Mashed Potatoes
Potato Cakes
Refrigerator Rolls
Curried Hash
Chitlins
Brown Betty
Hard Sauce
Virginia Cake
Christmas Fruit Cup
Vanilla Cookies
Quick Blackberry Pie
Dr. Stanfield's Eggnog
Aunt Pam's Eggnog

I was born a long time ago. I was born at home, in my father's house. We didn't go to any hospital, didn't have any hospital. All the babies were born in their homes in those days. I had a doctor though. There were twelve of us, five boys and seven girls, and I was number ten so I had lots of brothers and sisters to look after me.

My mama did all the cooking, but we all had to help. The little ones had to set the table. Mama would put the tablecloth on, then Liza and I would put the knives and forks around. We always sat together at the table.

Mama's Stuffed Shad

1 4-pound whole shad, cleaned, split, and boned if possible
water
salt
8 tablespoons (1 stick) butter
1 small onion, minced
1 stalk celery, chopped
2 cups prepared stuffing
½ teaspoon thyme
freshly ground pepper, to taste
2 slices bacon
1 medium onion, sliced
1 small green pepper, sliced
1 cup white wine

Spread the fish open and place it in a deep baking pan. Cover with water. Add 1 teaspoon of salt. Allow the fish to sit for about 30 minutes. Meanwhile, heat 4 tablespoons of butter in a small skillet and sauté the minced onion and celery until the onion wilts. Add the stuffing and the thyme. Mix well.

Add a little water or wine if the mixture is too dry. Set aside. Remove the shad from the water and pat dry. Sprinkle the inside with salt and pepper. Fold the stuffing mixture into the cavity of the fish and cover as much of one side as possible. Place the bacon strips over the stuffing and add a layer of onion and green pepper slices. Close the fish and secure the edge. Place the fish in a greased baking pan and dot with butter. Pour the wine over the fish. Cover and bake in a preheated oven for 40–60 minutes, depending on the thickness of the fish. The fish is done when it flakes easily. Do not overcook.

Serves 6

Sautéed Shad Roe

2 pair shad roe
water
salt
4 slices bacon
2 tablespoons butter
½ lemon
freshly ground pepper, to taste

Place the roe in a bowl and cover with water. Add 1 teaspoon of salt and turn gently to distribute the salt but making sure not to break the skin of the roe. Soak for 30 minutes. Meanwhile, fry the bacon in a cast-iron skillet. Remove the bacon but retain the bacon fat. Allow the skillet to cool a little. Drain the roe and place in the skillet. Cover and sauté gently for about 5 minutes. Turn and sauté 5 minutes more. Add the butter and the juice of the half lemon. Sauté another few minutes. Turn off the heat and allow the roe to sit for a few minutes. Season with salt and pepper.

Serves 4

Note: This is especially good served with bacon and eggs on a weekend morning.

I can remember being sent to the store in the mornings for sausage. The sausage came in big tubs, and they'd take the paddle and pat it until you knew you had a pound or two. All homemade.

Yes, Mama did all the cooking, she cooked for everybody. For breakfast we always had sausage, hominy grits, homemade biscuits, and plenty of fresh milk. We grew up like that. For dinner we had everything from the garden. We ate cabbage a lot. Cabbage was always cooked with some sort of meat and potatoes, loads of potatoes. Mama would make a corn pudding every now and then, not every day but every now and then.

Corn Pudding

2 cups corn, freshly cut from the cob
¼ cup sugar
1 teaspoon salt
½ teaspoon freshly ground pepper
2 eggs, lightly beaten
1 cup milk
½ teaspoon baking powder
4 tablespoons butter

Place the corn in a large bowl and add the sugar, salt, and pepper. Mix well and then add the eggs and milk. Mix well. Add the baking powder and pour into a casserole that has been greased with butter. Dot the top with butter and place in a preheated oven at 350° for 30 minutes or until set.

Serves 6

Rice Pudding

2 cups leftover rice
3 eggs, lightly beaten
½ cup sugar
3 cups milk
¼ teaspoon cinnamon
1 teaspoon vanilla extract
½ teaspoon salt
½ cup raisins
1 tablespoon butter
¼ teaspoon nutmeg

Break up the leftover rice. In a large bowl combine the eggs, sugar, milk, cinnamon, vanilla extract, and salt. Add the raisins and rice. Grease a casserole with half of the butter and fold in the rice mixture. Sprinkle the top with nutmeg and dot with the remaining butter. Place in a preheated oven at 300° and bake until the pudding has set (about 2 hours).

Serves 6

Mama's favorite dessert was gingerbread. Boy, did we like for her to make that. She'd cut it into squares from that great big pan of hers, and she always had a sauce to go over it.

Gingerbread

2 cups flour
½ cup sugar
1 teaspoon baking soda
1 teaspoon cinnamon
2 teaspoons ground ginger
8 tablespoons (1 stick) butter, at room temperature
½ cup hot water
1 large egg, lightly beaten
1 cup molasses

Sift together the flour, sugar, baking soda, cinnamon, and ground ginger. Set aside. Melt the butter in the hot water. Stir the water and butter into the flour. Mix well. Add the beaten egg and then the molasses. Again, mix well. Fold the mixture into a 9-inch square baking pan that has been greased with butter. Bake in a preheated oven at 350° for 30–40 minutes until done.

Serves 6

Life was hard, but I would say I had a really nice childhood.

We went to church all the time because my father was the minister. When he first went to Warrenton, it was just a log cabin. Since my father was educated, that didn't suit him. So he went to work and built that big church, and the people thought he was terrible. They said he could never pay for it. He said he could and they could. They went to work, and they did pay for it too.

We used to have these beautiful picnics out in the groves, and they'd have lemonade in barrels. And all the barrels had been smoked out or something 'cause they were dark inside. They would have a million tin cups hanging around the barrels. You could go and get your lemonade, get as much as you wanted. And everyone would bring something for lunch. Different things would be spread all over. All over the tables. Mama would cook half a ham and potato salad. There'd be cakes, fried chicken, slaw, everything. You talk about an awful good time. The men in our church, oh, they'd work so. They would put up swings and croquet on the grass. That was one thing I used to love to play. We'd just play that for hours.

Lemonade

- 2½ cups freshly squeezed lemon juice
- 2 cups sugar
- 5 cups water
- 1 lemon, sliced

Combine the lemon juice, sugar, and water. Stir to mix well and dissolve all the sugar. Add the lemon slices and chill well. Serve in tall glasses over ice.

Serves 6

When I was a teenager, I worked for Mrs. Carter. She had a private house where people came for vacation. An old lady named Mrs. Brown, she cooked. She could really cook. Lucy and I, we served in the dining room. People sat all around, and we'd bring the dishes in to pass around to each person. We served. Not the way we do now, where you go up and get what you want. We'd start with the oatmeal. Lucy would go with the oatmeal, and I would go with the Cream of Wheat. They'd look at me so hard. My hair hung down to here and I didn't fix it up, and I'd go

right in with those ole crackers and serve 'em. But they were nice, just as nice and polite as they could be. Please and thank you, that was the order of the day.

Mrs. Brown made fried sausage and hot biscuits or rolls, whichever, and eggs, any kind of way you wanted, and liver pudding—lots of that. In the spring when the shad was runnin' we had that. They had it fried and stuffed. It was stuffed with bread crumbs and onions. A stuffin' like you put in your chicken. We had the roe too. And when we didn't have any of that, well, she'd make salmon from a can—like my mother would do it—and make croquettes. They were just delicious.

Salmon Croquettes

1 8-ounce can salmon
cooked potatoes (leftover mashed potatoes, seasoned with a little butter, milk, and salt)
1 small onion, minced
1 large egg (or 2 small), lightly beaten
½ teaspoon celery seed
salt, to taste
freshly ground pepper, to taste
bread crumbs
bacon fat

Pour the juice off the salmon and pick out the bones. Break up the salmon with a fork, then mash and mix it together with the mashed potatoes. Add the onion, egg, celery seed, salt, and pepper. Mix well. Shape into croquettes and roll them in the bread crumbs. Place in the refrigerator to stiffen them. Heat the bacon fat in a skillet and fry the croquettes slowly until golden brown.

Serves 4

Virginia-Style Crab Cakes

1 pound lump or backfin crab meat
1 small onion, minced
1 stalk celery, minced
¼ medium green pepper, minced
1 cup milk
1 tablespoon flour
1 teaspoon salt
1 teaspoon freshly ground pepper
dash of cayenne pepper
1 cup bread crumbs
bacon fat or vegetable oil for frying

Pick the crab meat, removing all bits of shell and hard pieces. Break the crab meat into pieces and add the onion, celery, and green pepper. Put this mixture through a meat grinder. Set aside. Heat the milk over a medium flame and add the flour, stirring constantly until the mixture thickens. Fold in the crab meat mixture and the salt, pepper, and cayenne pepper. Mix well. Form the mixture into cakes and roll in bread crumbs. Fry over a medium flame until golden brown all around.

Serves 4

We had hogs. My brothers would raise them, and then they'd take them to the place where they killed them. They had a group of men who would kill the hogs and strip them and put them up on a pole, and you had to pay for it all. Your name would be on your hogs, so when they were finished, you'd go and get them. We never had but two at a time. And we'd have chitlins and liver and everything. Mama would make liver pudding, and my brothers would help. Then later on we'd get our ham hocks and things. Real Southern, you know!

Baked Virginia Ham

1 10–15-pound Virginia ham
water
2 tablespoons plus 1 cup brown sugar
1 cup cider
freshly ground pepper, to taste
1 cup sherry or cider

Scrub the ham with a stiff brush under running water. Place the ham in a large deep pot and cover with water. Soak for 12–24 hours, changing the water at least once. The longer you soak the ham, the less salty it will be. After soaking, place the ham in the pot, skin side down, with fresh water to cover. Bring to a boil, add 2 tablespoons of brown sugar, lower the flame, and simmer 20 minutes to the pound. Make sure that the ham is always covered with water. When the ham is about half done, add 1 cup of cider. When done, allow the ham to cool in its cooking liquid. Before the ham is cold, remove it from the cooking liquid and cut away the skin. Cut away most of the fat, leaving about ¼ inch of fat. Sprinkle with pepper and cover with a thick layer of brown sugar. Bake in a preheated oven at 350° for 30–40 minutes, basting often with cider or sherry.

Serves many

Mama always cooked a pot of beans: navy beans and black-eyed peas. She always kept a great big potful. They were good in those days because they were cooked with Virginia meat, and you could get some of the meat if you dug down in there.

At Christmas we didn't have a tree and all like they do now. We had stockings. You're supposed to bring one stocking, but all of us would bring two stockings down. We didn't have a mantel to hang them on, we'd hang 'em over the chair and we'd put our name on them.

I remember one time I got roller skates and Liza didn't. Liza got a little hat, and she was fussin' and making Christmas terrible. So Mama called me into the kitchen and said, "Listen, darling, I know you'll do this for Mother. Please give Liza your skates and I'll get you another pair soon as I can." Lots of children would have fussed but I didn't. I went right in and gave Liza my roller skates. She took them with glee and I took her hat. About a month later Mama did get me another pair.

Fruitcake

1 pound (4 sticks) butter, at room temperature
1 pound light brown sugar
10 eggs, lightly beaten
1 cup B & O black molasses
1 teaspoon baking powder
1 teaspooon baking soda, dissolved in a little water
4 cups flour, sifted
½ cup plus 2 tablespoons brandy, rum, or whiskey
1½ pounds raisins
1 box currants
1 cup citron
½ cup grated orange peel
½ cup grated lemon peel
1 cup chopped almonds
½ cup chopped black walnuts
1 teaspoon ground cloves
1 teaspoon nutmeg
1 teaspoon mace
1 teaspoon allspice
2 teaspoons cinnamon

Cream together the butter and sugar. Add the beaten eggs and mix well. Add the molasses and baking powder and mix well. Add the baking soda and 3 cups of sifted flour and mix well. Add the brandy, rum, or whiskey. Mix and set aside. Mix together the raisins, currants, citron, orange peel, lemon peel, almonds, walnuts, ground cloves, nutmeg, mace, allspice, and cinnamon. Add the remaining cup of flour and mix thoroughly. Fold the fruit mixture into the cake mixture and mix well (it should be very stiff). Fold this mixture into 2 8-inch greased cake molds and bake in a preheated oven at 325° for 2 hours. Allow the cakes to cool. Pour a little more liquor over the cakes and store in cake tins until ready to use.

Makes 2 cakes

I got married at my mother's house, in the parlor. We took pictures in the backyard. The whole family was there. Of course I had such a big family that that was enough.

Know how most get married, then have a reception? After we got married, we changed clothes and went away on the train. There was no eating afterward. The eating was before. My mother and my brothers all had that dinner for us, you know. The dinner was at lunchtime, 1:30. I was married at four o'clock. We ate at 1:30 'cause my brothers kept saying the ice cream hadn't come. My brothers had ordered it for me from Washington. It was called the velvet kind and was awful good. They all worked, my brothers. They all had jobs and brought in the money. They gave us all they had.

I really learned to cook after I was married, that's the truth. My sister was already married and lived in Orange [New Jersey]. I'd call her up and ask how to cook this or that. But I cooked for many years afterward because I was married for many years.

Doc (my husband) and the kids used to love the peaches I did. I'd hate to see 'em coming. They'd have these great big baskets of peaches, and you know you have to put them up. You can't fool with them, 'cause they'll go bad on you. I'd peel 'em and cut 'em in half, then make a syrup of sugar—sugar and water—and I'd put a couple of sticks of cinnamon in 'em. I'd put the peaches in after the syrup was made. You have to top those jars off while they're still good and hot or they won't keep.

I made jelly. Deed I did. I made so much jelly, I turned to jelly.

Oh, I had big dinners when Doc was alive 'cause I was younger then and could do it. He's been gone a long time now. But I used to have the nicest dinners. I used to have a leg of lamb, mashed potatoes, string beans, and corn pudding. That would be for dinner; and iced tea, plenty of iced tea.

Mashed Potatoes

6 large Idaho potatoes
salt
16 tablespoons (2 sticks) butter (or more, to taste)
½ cup milk
freshly ground pepper, to taste

Peel and quarter the potatoes. Rinse, place in a large saucepan, and cover with cold water. Add 1 teaspoon of salt and bring to a boil. Lower the flame and cook gently for 15 minutes or until the potatoes can be just pierced with a fork. Do not overcook. Drain the potatoes and immediately return to the pot. Add the butter and mash the potatoes with a potato masher. Add the milk, salt, and pepper. Mix well.

Serves 6

Variation: For a different taste, mix in chopped parsley, chives, or scallions.

Potato Cakes

2 cups leftover potatoes
2 eggs
1 tablespoon chopped parsley

salt, to taste
freshly ground pepper, to taste
1 cup seasoned bread crumbs
4 tablespoons bacon fat

Mash the potatoes if not already mashed. Mix together the potatoes and eggs. Add the chopped parsley, salt, and pepper. Shape into cakes and roll in seasoned bread crumbs. Heat the bacon fat in an iron skillet and slowly brown the cakes on both sides.

Serves 4

Refrigerator Rolls

1 cup mashed potatoes
⅔ cup solid shortening
½ cup sugar
1 teaspoon salt
2 eggs, lightly beaten
1 yeast cake
½ cup warm water
1 cup hot milk
6 cups flour (or more as needed)

Allow the mashed potatoes to come to room temperature. Add the shortening, sugar, salt, and eggs. Mix well. Dissolve the yeast cake in the warm water and add to the hot milk. Add the milk-yeast mixture to the potatoes. Sift the flour and add to the potato mixture. Mix well to form a stiff dough. Toss on a floured counter and knead well. Place in a large tub, rub the top with butter, and place in the refrigerator until the dough has doubled in size. One hour before baking, pull off chunks of dough and shape into rolls as desired. Cover the rolls with a damp cloth and allow them to rise again. Place the rolls on a cookie sheet and bake in a preheated oven at 325° for 15–20 minutes.

Serves 8

Doc used to have me cooking all kinds of things. I remember one time somebody gave him some curry. I had never tasted it before. Never. Never tasted anything quite like it before. Oooh, I just thought it was delicious. And I put it in my ground meat to make hash. Doc loved that. He really loved that hash. I just put that curry in and added the egg and everything was beat all together, then I put bread crumbs and a little butter on top, put it in the oven, and brother, was it good.

Curried Hash

- 3 tablespoons bacon fat
- 2 tablespoons butter
- 1 medium onion, chopped
- 2 cloves garlic, minced
- 3 cups cubed leftover steak or roast
- 1 cup beef broth
- 3 large potatoes, cut into ½-inch cubes
- salt, to taste
- freshly ground pepper, to taste
- ½ teaspoon thyme
- 2 tablespoons curry powder

Heat the bacon fat and butter in a cast-iron skillet. Sauté the onion and garlic for 5 minutes or until the onion starts to wilt. Place the cubed meat in a greased casserole. Pour the onion and garlic over the meat. Pour in the beef broth and add the potatoes. Season with salt, pepper, thyme, and curry powder. Mix well. Bake in a preheated oven at 350° for 30–40 minutes.

Serves 6

Now Doc had peculiar tastes. He loved sauerkraut and corned beef. They didn't have that where I grew up, but I learned to cook it for him. And he loved chitlins. He'd bring those damn things home, and I'd have to clean them and fix them up for him. But he loved them, so I'd do 'em.

Chitlins

20 pounds chitlins
water
1 large onion, coarsely chopped
1 small green pepper, quartered
3 cloves garlic, smashed
1 stalk celery, whole with leaves
1 bay leaf
1 teaspoon red pepper flakes
4 teaspoons salt
2 teaspoons freshly ground pepper
1 cup cider vinegar

Place the chitlins in cold water. Pull them out a piece at a time; if too long, cut in half. Scraping with a sharp knife, clean the chitlins, removing food particles and excess fat. Make sure to leave a little fat. Wash them. Place them in a large heavy pot and cover with water. Add the onion, green pepper, garlic, celery, bay leaf, red pepper flakes, salt, and pepper. Bring to a boil, lower the flame, and simmer for 30 minutes. Add the vinegar and simmer for 2 hours or until tender.

Serves 6–8

No meal was complete without dessert, and Doc had a sweet tooth. He had to have his dessert right after dinner. I used to make bread pudding for dessert all the time, and apple pie, which he'd put butter on, rice pudding, and Brown Betty. He loved my Brown Betty.

Brown Betty

5 apples
1 tablespoon lemon juice
1 cup brown sugar
1 teaspoon nutmeg
½ cup raisins
3 cups fresh ½-inch bread crumbs
4 tablespoons butter
½ cup water or sweet wine

Peel, core, and coarsely chop the apples. Toss them with the lemon juice. Place the apples in a deep baking dish. Add the sugar, nutmeg, and raisins. Toss gently and top with the bread crumbs. Sprinkle a little extra sugar over the top of the bread crumbs and dot with butter. Pour the water or wine over the top. Cover and bake in a preheated oven at 350° for about 40 minutes. Uncover and brown 10–20 minutes more. Serve hot with a hard sauce (see below), heavy cream, or whipped cream.

Serves 4–6

Hard Sauce

8 tablespoons (1 stick) butter, at room temperature
1 cup sugar

½ teaspoon nutmeg
1 tablespoon cognac

Cream the butter and sugar together. Add the nutmeg and cognac. Mix thoroughly. Place in the refrigerator and cool before serving. (This sauce is not cooked.) Serve over hot Brown Betty.

Virginia Cake

16 tablespoons (2 sticks) butter, at room temperature
1½ cups flour
1 teaspoon baking powder
6 eggs, lightly beaten
1½ cups sugar
vanilla extract

Cream together the butter, flour, and baking powder. Add the eggs and sugar. Using an egg beater, mix thoroughly. Add the vanilla extract and mix well. Pour the mixture into a lightly greased cake pan and bake in a preheated oven at 275° for 1 hour.

Serves 6–8

Christmas was a great time. My sister lived with me, and on Christmas Eve she and I would cook and cook. The young people would be out, and she and I would be home cooking so that we'd be ready for the whole family on Christmas day. I would go to bed only after the fruit cup was all made, the last thing. We'd use apples, oranges, and everything—all in little squares. The kids used to love that.

Christmas Fruit Cup

12 juice oranges
3 large grapefruit
2 Granny Smith apples
2 Red Delicious apples
2 tablespoons lemon juice
3 bananas
1 pound seedless grapes
1 cup halved cherries
½ cup sugar
additional orange juice, to taste and as needed

Cut the oranges and grapefruit in half and scoop the sections into a large bowl. Make sure to scoop out only the meat. Squeeze the fruit halves into the bowl, getting out all the juice. Peel the apples, cut into eighths, and cut crosswise into small pieces. Toss the pieces of apple in the lemon juice, then add to the orange, grapefruit, and juice mixture. Peel and slice the bananas into the bowl. Cut the seedless grapes in half and add them. Add the cherries. Add the sugar and additional orange juice. Stir until the sugar is dissolved.

Serves 10–12

Vanilla Cookies

16 tablespoons (2 sticks) butter, at room temperature
½ cup light brown sugar
½ cup white sugar
2 eggs, lightly beaten
1 teaspoon vanilla extract
2½ cups flour, sifted
¼ teaspoon salt
½ teaspoon baking powder
½ teaspoon baking soda

Place the butter and sugars together in a bowl. Cream well with a fork. Add the eggs and vanilla extract. Mix the flour with the salt, baking powder, and baking soda. Sift flour mixture again. Add the flour to the butter mixture and cream well. Chill the cookie mixture. Roll the dough out on a flat surface and cut into cookies, or fill a cookie press and form them on an ungreased cookie sheet. Bake in a preheated oven at 400° for 8–10 minutes.

Serves 6

Quick Blackberry Pie

1 quart blackberries
1½ cups sugar
1 tablespoon plus 1 teaspoon cornstarch
1 teaspoon water
¼ teaspoon nutmeg
1 ready-made pastry shell

Wash the blackberries and select the best ones. Place these berries in a large deep pot along with the sugar. Bring to a boil. Mix the cornstarch and water and add to the berries. As soon as the berries begin to thicken, remove from the heat, add the nutmeg, and set aside to cool. When they are cool, fold into a pastry shell and bake in a preheated oven at 400° for 25 minutes.

Serves 6–8

Dr. Stanfield's Eggnog

6 eggs, separated
8 tablespoons sugar
16 tablespoons whiskey or rum
9 tablespoons heavy cream
nutmeg

Beat the egg yolks and whites separately. To the egg yolks add 4 tablespoons of sugar and the whiskey. To the beaten whites add the remaining sugar. Beat both mixtures well. Add the heavy cream to the egg yolk mixture and beat well. Add the egg white mixture to the egg yolk mixture and again beat well. Add a pinch of nutmeg.

Serves 6

Note: This recipe makes a rich concoction. Milk may be added to make a thinner drink.

Aunt Pam's Eggnog

5 eggs, separated
1 cup sugar
2 cups whiskey or rum
1 teaspoon nutmeg
2 cups heavy cream
1 cup milk

Beat the egg yolks and whites separately. Add ½ cup of sugar to the egg yolks and beat well. Add the whiskey and nutmeg. Beat well. Beat the heavy cream until it thickens. Add the cream and milk to the yolk mixture and mix well. Add the remaining sugar to the egg whites, then add this mixture to the egg yolk-whiskey mixture. Mix well.

Serves 10

Aunt C

CAROLYN MILDRED HARDY CASSIO

April 14, 1917 Kansas City, Missouri.

Aunt C had been my friend and mentor for as long as I can remember. It was she who got me started with many of my artistic pursuits, my first paintings, my first seashells, and so forth. I don't remember her cooking much when I was small, but as I grew older and began to spend time with her on Martha's Vineyard, her abundant skills in the kitchen became apparent. I spent lots of time in her kitchen, which was always filled with family, friends, love, good cheer, and wonderful food. She was always quick to give me a pointer and share in my cooking. Her words of encouragement on matters large and small won't be forgotten.

Aunt C

Fried Corn
Applesauce
New England Boiled Dinner
Citrus Fruit Jelly
Roast Pork with Orange Glaze
Pepper Steak
Curried Shrimp
Apple-Carrot Salad
Brandied Raisins
Short Ribs with Noodles
Striped Bass
Simple Baked Vineyard Bluefish
Fish Broth Soup
Bluefish Salad
Clam Chowder
Black Beans with Pork Chops
Cranberry Sauce
Aunt C's Punch
Sautéed New Potatoes

Grandma Jones raised us. We lived right next door. Grandma was very good. She would let us help, and she insisted that we learn how to do everything. Grandma did it because Mother was busy with her music and Mother never was a good cook anyway. She didn't like to cook, and she didn't like the kitchen. Grandma Jones was a house person. Both my sister Harriet and I started with her.

One of the best things, I remember, was our garden. We learned early how to make use of all the fruits and vegetables, the chickens and other animals too. We even had pigs—which was unheard of in Ohio. Usually, you think of people having pigs in Virginia or one of the Carolinas. But we had them.

Fried Corn

4 slices bacon
4 ears sweet corn (leftovers)
½ medium green pepper, chopped
1 medium onion, chopped
½ teaspoon salt, or to taste
½ teaspoon freshly ground pepper
2 tablespoons milk

Fry the bacon in a black cast-iron skillet. Set aside, leaving the bacon fat in the skillet. Meanwhile, slice the corn off the cob. Add the green pepper and onion to the skillet and sauté for several minutes. Add the corn, salt, and pepper. Cover and sauté for 5 minutes, stirring several times so that the corn does not stick. Add the milk and sauté 5 minutes more.

Serves 4 (or 2 if you really love it)

Variation: Other leftover vegetables may be added to this mixture as well.

As a child I think the first thing they started you doing was picking the ripe fruit off the trees and going to the garden for the vegetables. Then when I was about ten, I had the chore of picking the eggs. You had to go out and pick the eggs from under the hens, not let them set on them because, at a special time of the year, you put eggs under the hen to make a new crop of chickens. But every morning you had to go out there to get the eggs and take care of the chicken coop. And that meant also cleaning up and putting down fresh straw for them and stuff like that.

Applesauce

1 dozen apples
1 tablespoon lemon juice
1 cup water
½ cup sugar
2 tablespoons butter
½ teaspoon cinnamon
½ teaspoon nutmeg
1 tablespoon brandied raisins (see page 114)

Wash, peel, and core the apples. Slice them and place in a saucepan. Toss with the lemon juice to keep them from turning brown. Add the water and bring to a boil. Lower the heat and simmer until the apples are soft. Add the sugar, butter, cinnamon, nutmeg, and brandied raisins. Mix well and simmer 5–10 minutes more, until most of the water is absorbed.

We didn't have a bathroom, we had an outhouse. Cleaning it was one of the other chores that had to be done.

As you grow up you have to take on more and more responsibility. Everybody in our household had things to do. As you got older they increased the things you could and did do.

With everybody helping in that old kitchen at the same time, I don't think I ever really cooked a meal—a whole meal—until I was married.

New England Boiled Dinner

1 6-pound smoked ham (bone in)
6 cups water
14 new potatoes
8 carrots
1 medium head cabbage, quartered
1 medium onion, quartered
salt, to taste
1 teaspoon seasoned pepper

Place the ham in a deep covered cast-iron pot. Add the water. Cover and simmer for 2–3 hours or until done. Remove the ham and set aside. Meanwhile, peel the potatoes and carrots. Cut the carrots into 2-inch lengths. Add the potatoes, carrots, cabbage, onion, salt, and seasoned pepper to the pot.

Bring to a boil, lower the heat, cover, and simmer for 30 minutes. Reheat the ham in the stock. Slice the ham and serve with the vegetables.

Serves 8

Some things don't change much. The jelly I make, well Grandma Jones taught me how to do that. It's the same old recipe. You do it the same way.

Citrus Fruit Jelly

2 oranges
1 lime
1 lemon
1 grapefruit
zest of 1 lime, slivered and chopped
1 cup bottled lime juice
7 cups sugar
1 package Certo non-flavored gelatin
paraffin

Slice the fruit, remove the fruit sections, and scoop into a large saucepan. Squeeze the juice from the fruit. There should be about 4 cups of fruit. Mash the fruit and add the lime juice. Bring to a slow boil, lower the heat, and cook for 5 minutes. Add the sugar. Slowly bring back to a rolling boil, stirring constantly. Allow it to boil for about 2 minutes, then add the Certo. Stir well. Return to a rolling boil. Remove from the heat and pour into jelly containers. Top with the paraffin.

Oh, I always loved chicken and greens. That was a Sunday thing with us—every Sunday, almost.

Sunday dinner was always a big event because on Saturday we got to go to market and pick up things like fresh horseradish. If it was summer and watermelon was in, we would get a watermelon. Whatever we didn't grow, we could buy, and that was a big event. We always pigged out on Sunday—everything.

Now, we didn't have a lot of meat. Meat was very expensive, and if we had it once a week, that was fine—I mean like steak.

We ate a lot of fish, you could get that fresh. The fish in the Midwest is similar to the porgies we get here; the fish was not expensive, of course.

When it came time to slaughter, you had all the parts of the pig to use. That was a special time. Grandpa—Uncle Shad, we called him—would do all the smoking himself—bacon, hams, whole sides. And later, when that was all done of course, you got to eat all those things.

Roast Pork with Orange Glaze

1 3–4-pound pork roast (boned rib roast)
4 cloves garlic, cut into slivers
1 mild jalapeño pepper, sliced
1 teaspoon salt
½ teaspoon freshly ground pepper
½ teaspoon flour
1 cup chicken stock
1 orange
½ cup Cointreau or other orange liqueur
½ cup sugar

Punch holes in the roast with the tip of a knife. Stuff the holes with a slice of garlic and a small piece of jalapeño pepper. Score the top in two directions and rub into the scores a mixture of the salt, pepper, and flour. Place in a roasting pan and add ½ cup of the chicken stock to the bottom of the pan, adding more as necessary. Place the roasting pan in an oven preheated to 350°. Cook for 1 hour. Meanwhile, cut the orange in half and scoop out the meat. Place in a saucepan along with the Cointreau and sugar. Add a little more orange juice if necessary. Simmer slowly until reduced by half. After the roast has cooked 1 hour, brush with some of the orange glaze. Return to the oven and continue cooking, brushing with glaze each half hour until the roast is done. Deglaze the bottom of the pan with a little water and thicken slightly with a little flour. Season with salt and pepper and add the remaining orange glaze. Serve over the sliced pork.

Serves 6–8

My parents, Grandma and Grandpa Hardy, would put in a garden. They could handle that. But they didn't do anything toward puttin' down stuff for the winter. Like we would put down eggs in a big crock with water to keep over the winter when the hens didn't lay. We had a potato cellar too.

I started cooking after I got married. Cass was a senior at Meharry Medical College. We had a room and cooking privileges. So I fixed our meals. We were on an allowance from both parents. I had to cook and evidently it worked out okay.

Traveling with Cass while he was in the army, every place had cooking privileges, so I cooked, but it was not quite like settling in and doing a family thing. That didn't come until later. Then came the real stuff where we had planned meals and had to make sure you had everything, doing three meals a day, a full responsibility.

Pepper Steak

2 pounds round steak or other inexpensive steak
1 cup flour
2 tablespoons olive oil
2 tablespoons vegetable oil
2 medium green peppers, sliced
2 medium onions, sliced lengthwise
1 stalk celery, sliced
1 cup beef stock or broth
salt, to taste
freshly ground pepper, to taste

Pound the steak with a tenderizing mallet. Turn and pound the other side. Cut the steak into strips, dredge in flour, and fry in the olive oil and vegetable oil mixed together. Remove the steak and set aside. Add the vegetables and sauté for 5 minutes. Remove and set aside. Add ½ cup of flour and the beef stock. Brown slowly, scraping the bottom of the pan, until the gravy has begun to thicken. Return the steak to the pan and cover with the vegetables. Season with salt and pepper. Cover and simmer over a low flame for 30 minutes, watching that it does not thicken too much. Serve over rice, noodles, or mashed potatoes.

Serves 4

Cass had had beautiful experiences in Europe right after the war, and he loved to dine. Dining consisted of the table setting and all the good stuff that goes with it—wine and he liked to have an appetizer. It didn't have to be much—it could be fruit or soup, it could be anything. But that was a course. He liked that kind of service. He liked an entrée, he liked two vegetables, a starch, a salad, and a dessert.

Curried Shrimp

1 pound shrimp, in the shell
1 can flat beer
8 tablespoons (1 stick) butter
½ cup flour
2 teaspoons curry powder
salt, to taste
1 teaspoon freshly ground pepper
½ cup chopped black olives
1 tablespoon capers

Boil the shrimp in the beer for 10 minutes. Shell and devein the shrimp. Strain the stock and set aside. Make a roux: melt the butter in a black cast-iron skillet. When it begins to bubble, add the flour, stirring constantly for about 5 minutes, until the flour has browned some. Add the curry powder. Stir, then add enough stock until the mixture reaches a gravy consistency. Add the shrimp, salt, and pepper. Add the chopped olives and capers.

Serves 4

Note: Serve over rice.

Apple-Carrot Salad

1 Granny Smith apple
1 Red Delicious apple
1 McIntosh apple
2 tablespoons lemon juice
2 medium carrots, peeled and grated
1 stalk celery, chopped
1 teaspoon Dijon mustard
½ cup mayonnaise

Core, slice, and coarsely chop the apples. Sprinkle and toss with the lemon juice to keep from discoloring. Add the grated carrots, celery, mustard, and mayonnaise. Mix well.

Serves 6

Brandied Raisins

1 pound raisins
brandy to cover

Place the raisins in a jar just big enough to hold them. Pour brandy over the raisins, covering them. Place wax paper or plastic wrap over the top and cover tightly with the lid. Allow the raisins to sit for at least six months before use.

Note: May be added to applesauce, bread pudding, cranberry sauce, and so forth, or served with ham or roast chicken, or over ice cream.

Well, when you have to start thinking about doing that kind of cooking seven days a week, you really have to get your act together. I have a cookbook that I used to use a lot, called the Little Settlement Cook Book. *In the back, I wrote out menus so that I'd be able to refer to it if I didn't know what I was going to have. I found using a menu and setting it up for a week made my shopping easier.*

Shopping was something else again. Most everything was available nearby. There was a butcher and the supermarket. The only thing Aunt Lela and I had problems with was getting green vegetables. So we would take the car and go up on Grove Street to the farm for our tomatoes and cabbage and stuff. Then we'd drive down to Newark for our collard or turnip greens.

I used to love to do beef and noodles. It's a heavy dish. Always put wine in the gravy to give it a little kick. And we were so conscious of how the food looked on the plate. That particular dish is brown so you always serve green beans with that to make the plate look pretty; a salad, then lemon pie for dessert, and that makes a pretty table. And if you wanted you could take your broth and serve it as an appetizer. And that's part of being poor. You learn to use everything. You learn to have a pretty table no matter what. So much of the cooking you do depends on how much money you have to spend.

Short Ribs with Noodles

4 beef short ribs, cut into 3 pieces each
1 cup flour
3 cups water
1 large onion, coarsely chopped
1 small bay leaf
2 carrots, peeled and cut into 1-inch lengths
2 stalks celery with leaves, coarsely chopped
3 sprigs parsley
½ teaspoon oregano
½ teaspoon thyme
½ teaspoon marjoram
1 teaspoon salt, or to taste
1 teaspoon seasoned pepper
1 apple, cored and cut into eighths
½ pound thin flat noodles

Dredge the ribs in flour. Place them in a large cast-iron pot or casserole. Place, uncovered, in an oven preheated to 400°. Brown for 15 minutes. Add the water, onion, and bay leaf. Cover, lower the heat to 350°, and cook for 1 hour. Add the carrots, celery, parsley, oregano, thyme, marjoram, salt, and seasoned pepper. Cover and cook for another 30 minutes. Add the apple and cook 10 minutes more or until the meat is almost falling off the bone. Meanwhile, boil and drain the noodles. When done, place the noodles on a platter and put the meat and vegetables in the center. Pour the thickened liquid over everything.

Serves 4

When Cass and Uncle Frank started their medical practice, Aunt Lela and I had $20 a week food allowance, and you know we had you children. And we also wanted our flowers. So we went up to the Red Cross and had them help us work out a food

budget for four people for $20 a week, and we found ways to make that $20 go a little further. Also we managed to have a couple of bucks left over—after our marketing—for those flowers. Do you remember how beautiful Aunt Lela's garden was? She had the most beautiful flowers, and we laugh about it because $20 a day wouldn't handle it now.

Striped Bass

2 stalks celery
1 carrot
1 leek
4 tablespoons butter
2 cups white wine
4 striped bass fillets, approximately 2 pounds
2 cups fresh fish stock
½ teaspoon thyme
1 bay leaf
2 cups heavy cream
salt, to taste
½ teaspoon freshly ground white pepper

Julienne the celery, carrot, and leek. Heat the butter in a skillet and sauté the vegetables until soft. Add the wine and simmer until reduced by half. In another, larger skillet arrange the fillets in one layer. Pour in the fish stock. Add the thyme and bay leaf. Bring to a simmer, cover, and poach for 5 minutes or until the fish just flakes. Remove the fish to a heated platter and set aside. Reduce the poaching liquid by half. Add the heavy cream and the vegetable mixture. Cook slowly, not boiling, until the mixture reduces to about 3 cups total and has thickened. Add the salt and pepper. Pour over the fish and serve.

Serves 4

We raised you kids and we ate well. You had your fruit and everything that you needed—as healthy as you are. So we laugh about how times changed and how much you can do if you want to. We pinched pennies; it was incredible when you stop to think about it. But if you want to, you can do it.

You know, a lot of times you can take a recipe that you get in a magazine or book, or whatever, and then make your own. Which is what I do a lot. It's not exactly that recipe, you have to change it according to what you have or what you like. But it can be pretty consistent and good.

Nothing gets thrown away. Nothing goes to waste, nothing. And to me, that's the art of a real cook—a person who knows how to take something, take the first thing, cook it one way, and have leftovers. Then take the leftovers and turn them into something else again.

There's no reason to waste anything. When you have fish, after you clean it and you take that fillet, then you take what's left and boil that up with a little onion and stuff, and next thing you have a delicious fish stock that you can do anything with. It's really great in clam chowder.

Simple Baked Vineyard Bluefish

2 1-pound bluefish fillets
1 teaspoon salt
1 teaspoon freshly ground pepper
½ teaspoon garlic powder
1 lemon
2 tablespoons mayonnaise
1 teaspoon Worcestershire sauce

Rinse the fish well. Pat dry with a paper towel. Place skin side down in a baking pan greased with butter. Sprinkle all over with salt, pepper, and garlic powder. Squeeze lemon juice over the fillets and allow to sit for several minutes. Mix the mayonnaise and Worcestershire sauce together and spread over the fish. Bake for 10–15 minutes or 10 minutes for each inch of thickness.

Serves 4

Leftover fish makes a fantastic fish salad the next day. You can do that with cod or with any fish. When you bake it, it comes out just the right consistency for salad. And it all depends on how you spice it up and what your taste buds tell you to do.

Fish Broth Soup

1 2–3-pound weakfish (sea trout)
6 cups water
2 medium onions, chopped
2 cloves garlic, minced
2 ribs celery with leaves, chopped
4 carrots, peeled and chopped
½ cup coarsely chopped parsley
1 small bay leaf
1 teaspoon thyme
1 teaspoon salt
1 teaspoon freshly ground pepper
3 cups white wine
3 tablespoons butter
1 tomato, skinned, seeded, and chopped

Clean the fish. Fillet the fish and then skin the fillets. Place the fish head with the gills removed, the bones, and the skin in a large pot. Add the water, 1 medium onion, garlic, 1 rib of celery, 2 carrots, parsley, bay leaf, thyme, salt, and pepper. Bring to a boil, lower the flame, and simmer for 15 minutes uncovered. Add the wine. Simmer another 15 minutes. Remove from the heat and strain the cooking liquid. Meanwhile, melt the butter in a large pot. Add the remaining onion, celery, and carrots. Sauté for 10 minutes. Add the tomato. Sauté for 5 minutes. Add stock and bring to a slow boil. Add the fish, cut into ½-inch cubes. Simmer for 5 minutes. Check for seasonings and add salt if needed.

Serves 4

Bluefish Salad

2 bluefish fillets (about 2 pounds)
salt, to taste
½ teaspoon freshly ground pepper, or to taste
2 limes
1 Granny Smith apple, cored and finely chopped
2 stalks celery, finely chopped
½ cup sliced scallions, green parts only
1 cup white wine
½ cup mayonnaise
½ cup sliced olives
1 dozen cherry tomatoes, sliced in half
4 large lettuce leaves

Sprinkle the fillets with salt and pepper. Place them in a greased baking pan. Bake in a preheated oven at 325° for about 15 minutes or until the fish just flakes. Break up the fish, retaining the cooking juices. Over the fish pour a marinade of the juice of 2 limes, chopped apple, chopped celery, green scallions, and white wine. Marinate for 2–3 hours, turning twice. Pour off all liquid not absorbed and add the mayonnaise. Season with salt and pepper to taste. Serve on a bed of lettuce, garnished with the sliced olives and cherry tomatoes.

Serves 4

Variations: May be served in half an avocado. Serve with hot sauce or Tabasco to taste.

Now that clam chowder is really the thing, some people like to put their clams in a blender. I don't like the blender because then it's all mush. And another thing I always do is add a little wine, just at the end.

Clam Chowder

24 chowder clams
4 slices thick bacon, cut into ¼-inch dice
1 large onion, chopped
2 stalks celery, chopped
1 clove garlic, minced
6 large potatoes, cut into ½-inch cubes
¾ cup white wine
8 tablespoons (1 stick) butter
½ teaspoon freshly ground pepper
1 cup heavy cream
salt, to taste

Scrub the clams well to remove all grit and sand. Place them in a large pot and cover halfway with water. Cover the pot and bring to a boil. Steam until all the clams are open. If all the clams will not fit at one time, steam them in several batches. Remove the clams and set them aside in a large bowl, allowing them to cool. Strain the cooking liquid through several layers of cheesecloth. When the clams are cool, remove them from the shells and coarsely chop them. In the now empty and cleaned pot, fry the bacon until about half done. Add the onion, celery, and garlic. Sauté for 10 minutes. Add the strained cooking liquid along with the potatoes. Simmer for 10 minutes, until the potatoes are almost done. Add the wine, butter, and pepper. Simmer for 5 minutes, then add the heavy cream, stirring constantly. Do not allow the pot to return to a boil. Simmer until the liquid thickens. If needed, add a little flour mixed with milk to the liquid. Add the clams and check for seasonings. Add a litle salt if necessary.

Serves 8

Lamb? You can't do very much with leftover lamb except cold lamb sandwiches, with sliced onions. And you simply must have that onion on them. And, oh yes, lamb stew.

There are only two consistent things in my black bean soup—the black beans and the pork. Now sometimes you have it when I toss in leftover salad or rice. Sometimes I add chicken or pork chops. Almost anything can go into that pot.

Black Beans with Pork Chops

1 pound black beans
3 quarts water
2 32-ounce cans College Inn chicken broth
2 medium onions, chopped
1 green pepper, chopped
2 cloves garlic, chopped
2 teaspoons seasoned pepper
1 teaspoon oregano
½ teaspoon thyme
salt, to taste
8 pork chops (country-style spareribs)

Pick the beans. Place them in a large cast-iron pot and cover with the water. Soak overnight. Drain the beans and return them to the pot. Add the chicken broth, onions, green pepper, garlic, and seasonings. Bring to a slow boil, cover, and lower the flame. Simmer for 2 hours, until the beans are almost soft, adding more broth if necessary. Add the pork chops. Cover and simmer another 1–2 hours. Taste and add salt. Correct seasonings if necessary.

Serves 6

Note: Serve with rice and a salad. Better the second day.

Yes, I put leftover avocado from the salad in the soup. Oh, honey, I put everything in the soup.

I always keep cans of chicken broth on hand. If you cook with chicken or beef broth or stock, that gives you all the nourishment that you would get from having meat.

Another thing that is very funny. As you get older, you can't eat so much. So you'd better have had all those goodies while you were younger.

I use curry a lot. Now that I don't use salt, I use curry on a lot of things, especially on my plate. I use spices much more now. Garlic and ginger, cinnamon and mustard.

Would you believe I even put a little mustard in my cranberry sauce? It makes all the difference in the world. It just gives it another taste. It's really great. Of course cranberry sauce can be sweet and tart from the berries, and just a little bit of mustard makes it all go together. Tangs it up.

Cranberry Sauce

1 pound fresh cranberries
1 cup water
1 cup sugar
½ teaspoon dry mustard
½ teaspoon ground cardamom
1 teaspoon lemon juice

Wash and stem the berries. Place them in a saucepan and add the water. Bring to a boil, lower the heat, and cover. Simmer until the skin splits, about 10 minutes. Add the sugar, dry mustard, cardamom, and lemon juice. Simmer 5 minutes more, then remove from the heat. Pour into a container and allow to cool and set.

Serves 8

Marinate meats and chicken in lemon juice or lime juice. Fish, everything. It's a natural tenderizer, and it brings out the flavor of whatever you're having. You never get a sour taste.

Aunt C's Punch

2 quarts sun tea (see Note)
1 12-ounce can ginger ale
1 6-ounce can frozen orange juice concentrate
1 6-ounce can frozen Hawaiian Punch concentrate
1 quart lemon or lime sherbet

In a large punch bowl, mix together the sun tea, ginger ale, and orange juice and Hawaiian Punch concentrates. Add the sherbet in two pieces and allow to soften.

Serves 8–10

Note: For sun tea, place 9 Lipton tea bags in a 1-gallon jar of cold water. Place the jar in the sun for 3–4 hours. Remove the tea bags and add sugar and lemon to taste. Refrigerate.

I had a wonderful batch of potatoes last year. I was so tickled. I hadn't planted potatoes since I left Ohio. In Montclair (New Jersey) we just didn't do that. We mostly had flowers and that sort of stuff. So last year I got these seed potatoes, and over in front of the cottage I planted them. And they came out just beautifully. We had sweet potatoes and new potatoes—the whole thing—till the end of October. It was just wonderful.

Sautéed New Potatoes

8 tablespoons (1 stick) butter
2 pounds small new potatoes, unpeeled
½ cup chicken stock
1 clove garlic
1 teaspoon salt, or to taste
½ teaspoon freshly ground pepper
1 tablespoon minced parsley

Melt half the butter in a black cast-iron skillet. Add the potatoes in one layer. Cover and sauté for 5 minutes. Add the chicken stock. Cover and cook for 15 minutes, shaking the skillet often to keep the potatoes from sticking. Remove the cover and add the remaining butter and the garlic. Turn up the heat and brown the potatoes until all liquid is absorbed. Sprinkle with salt, pepper, and parsley. Turn to make sure all the potatoes are coated.

Serves 6

You have to doodle with a dish. When I tell you what I use in a recipe, I will come pretty close, but if you ask me a teaspoon of this and a tablespoon of that, I haven't got time for that. You'll have to doodle with it to get it the way you *like it.*

Basically, the best meals are the big old-fashioned sit-down kind.

Frenchy

OLIVER WENDELL LESESNE

February 22, 1918 South Island, South Carolina.

Frenchy is the father of one of my best friends. I had met him when I was a teenager, but it wasn't until I was in my twenties that I truly came to know the man who affectionately called me Number 2. There was a period of uncertainty in my life when things seemed pretty tough. Frenchy took me in, gave me a place to live, fed and cared for me, counseled me, let me borrow his car—in short—treated me like his own son. It was during this time that I learned his way with a pot, for there always seemed to be a pot on the stove no matter what. He was always there with a joke or warm words of encouragement.

Frenchy

Stuffed Pork Chops
Grits and Cheese
Smoked Pork Butt with Lima Beans
Simple Broiled Flounder
Sautéed Mushrooms
The Frenchman's Zucchini Pot
Boiled Lobster
Lobster and Spaghetti 1
Broiled Lobster
Chicken Breasts Stuffed with Prosciutto
Lobster and Spaghetti 2
Flounder Stuffed with Crab Meat
Lobster Omelet
Chicken with Mustard

My growing up was segmented, let's say. My formative years were spent in Charleston, South Carolina, where I lived with my aunt. She was a "bodice"—as they called them at that time—not a dressmaker, that was a different thing. She had a sitting room where she taught all the affluent blacks the art of sewing. She created all the gowns for what they called the Beaux Arts Ball, in Charleston. After my aunt died I went to New York to be with my mother, sisters, and brothers—which I had quite a few of: seven sisters and two brothers.

My real introduction to cooking was by my sisters, my mother was too busy trying to keep the whole family together. But my mother's influence was handed down to my sisters. You see, I was the baby. They raised me, fed me, and taught me to cook. And they kicked my butt if I didn't go to school.

My sisters in New York taught me to cook, but my basics came from being in South Carolina with my aunt's housekeeper, Ma Bowley. Being the inquisitive person I was, whenever she would be cooking something I'd go see what was going on. Whatever was for dinner was always good. There was no cooking by recipes, like we do today. No, Ma Bowley just cooked by instinct. She couldn't tell you what she put in—a pinch of this or a pinch of that; she could never tell you what was part of it. But when you sat down to eat, boy, was it good.

Stuffed Pork Chops

2 tablespoons butter
1 medium onion, chopped
2 stalks celery, chopped
1 cup raisins
1 cup diced apple
½ cup chopped walnuts
1 cup seasoned bread crumbs
4 tablespoons honey
8 pork chops, ¾ inch thick with a pocket cut into them
salt
freshly ground pepper, to taste
1 cup orange juice

Melt the butter in a skillet and add the onion and celery. Sauté for 5 minutes, until the onion just begins to lose its color. Remove from the heat and add the raisins, apple, walnuts, bread crumbs, and honey. Mix well and sprinkle with ½ teaspoon of salt. Sprinkle the inside of the pork chop pockets with just a little salt and pepper and stuff with the mixture. Place the chops in a greased baking pan and place in an oven preheated to 350°. Bake for 30 minutes. Lower the heat to 325° and pour the orange juice over and around the chops. Continue baking for about 1 hour, basting every 15 minutes. Add more orange juice if necessary.

Serves 4

I think the very first thing I can remember learning how to cook was hominy grits. "Don't cook it too fast. Cook it real slow, and keep stirring it so's it don't lump up," I can hear Ma Bowley saying. I can remember that.

♣

Grits and Cheese

3 cups water
1 cup grits
4 tablespoons butter
1 teaspoon salt
1 teaspoon freshly ground pepper
2 eggs
½ cup milk
1 cup grated cheddar cheese
1 teaspoon paprika

Bring the water to a boil and add the grits. Lower the flame and cook slowly until the mixture thickens. Add the butter, salt, and pepper. In a bowl beat the eggs and then add the milk. Add the grits and the cheese. Pour the mixture into a greased casserole and sprinkle the top with paprika. Bake in a preheated oven at 300° for about 30 minutes, until the grits are set.

Serves 4

Seems I can remember lots of things when I stop to think; discipline to the ninth degree. I had my chores to do. I knew exactly what I had to do. On Saturday I had to scrub down the porches: two in the front and two in the back. And we had twenty-three scatter rugs in the house, and we had to sweep them off with salt, common salt, and put them up on the line and beat them. Every Saturday I went to the market with my oldest sister to do the shopping. I had a paper route every day before school, the Charleston News and Courier.

I thought my aunt was a tyrant when I first went there, but when I left I really could not have loved her more. She taught me a sense of values. My aunt always told me, as far as the food and necessities of life were concerned, "Waste not, want not." She had lots of clichés, like "Don't let your eyes be bigger than your stomach." I don't care how many people she had in the house, when I went to bed, she would always come up to kiss me good night and listen to my prayers. These things had a great effect on me.

Oh, when I was little, I would watch and see the little things, like making biscuits or something like that. Ma Bowley would let me watch.

My mother died about a year after we went to New York, so that left me with my brothers and sisters. Cooking at that time was a necessity. My brothers and sisters were working so when I came home from school . . . well, I just had to learn how to cook things, simple things.

But I imagined from my family background that I would always be interested in cooking, just as it is a tradition in my family to have coffee. Wherever I go in my family today, there is always a pot of coffee on. I don't know whether it's inherited or not, but I wonder about it. My father was a fantastic cook. I like to cook and I like to experiment with different things. Know what I mean?

When I was coming up, it was a different era altogether. We would have to cook things that were sustenance and economical. We would have to cook things that would last two or three days, in a family with maybe nine or ten to feed. You had to cook ham hocks and beans and rice and things like that, greens and such.

Smoked Pork Butt with Lima Beans

1 2-pound pork shoulder butt (also called a daisy ham)
water
2 10-ounce packages frozen lima beans
1 medium onion, chopped
1 teaspoon seasoned salt
½ teaspoon freshly ground pepper
½ bay leaf
dash of hot sauce or Tabasco

Place the ham in a large cast-iron Dutch oven and cover with water. Bring to a boil, cover, and lower the heat. Simmer for 1 hour, turning occasionally, then remove the ham and add the remaining ingredients. Remove the covering from the ham. Bring to a boil again and return the ham to the pot. Lower the flame and simmer, uncovered, for 30 minutes, until the lima beans are done and the liquid has thickened, but before the limas get too soft.

After I came out of high school, I went to work at a club called the Union League Club at 37th Street and Park Avenue. I started down there as a bar boy, just cleaning up the bar and the glasses, things like that. Then I got involved with the head chef. He loved brandy and I would send brandy up on the elevator, and he would send me my meals down, and I got to be very friendly with him. He was very instrumental in my learning about a variety of foods. He was mostly into French, but he taught me a lot of things about different ways of cooking—how vegetables should be cooked. Much of what I learned was not indigenous to the black society as far as eating was concerned. That was my first step over the bridge into different kinds of cooking.

Now chicken is one of the things that can be cooked in so many different ways. Fish is such a large category as far as cooking is concerned. A chicken is a chicken, don't care where it comes from, it's still a chicken. You don't say what kind a chicken it is, you just say chicken. Now fish, how many damn fish do you know?

Now, I'll tell you another influence on my life. When I went up to the Cape, it was Arthur "Tutter" Lopes. He was a Portuguese boy out of Pawtucket, and he and I were inseparable friends. I learned a lot of different ways to cook fish from him. Tutter and I would go out to Menemsha and catch flounder, and Tutter would take that flounder and fillet it in about thirty seconds. We would take water-ground cornmeal and bacon fat, or something like that, with us, and we would build a fire out of driftwood, and twenty minutes after that damn fish came out of the ocean, we'd be eatin' him.

Simple Broiled Flounder

4 ½-pound flounder fillets
salt, to taste
4 tablespoons butter
1 teaspoon lemon juice
freshly ground pepper, to taste

Wash the fillets and pat dry. Place them in a baking pan greased with butter, and sprinkle with salt. Melt the butter in a small saucepan and add the lemon juice. Spoon the butter mixture over the fish, reserving a little to baste the fish with later. Sprinkle with pepper. Place in a preheated broiler 4 inches from the flame and broil for 5–8 minutes, depending on the thickness. Baste once during cooking.

Serves 4

When Loraine and I were married, Rose [Loraine's sister] and the whole family would get together. We used to have a woman called Miss Scott to cook Thanksgiving dinner for us. I never stopped and watched, but I know it was the best dinner I ever ate. She could never write down on a piece of paper what she did; hell, she just did it. She couldn't even say "a pinch of this or a pinch of that." She just knew how to cook greens, sweet potatoes, turkey. She knew how to cook onions; she knew how to cook string beans, she knew how to make an apple pie and biscuits. Everything was there. That was a God-given talent. I don't think it was something you read in a cookbook like Julia, whatever her name is. This woman could just cook!

Sautéed Mushrooms

1 pound small mushrooms
4 tablespoons butter
1 teaspoon minced shallots
1 clove garlic, minced
1 tablespoon cognac
½ teaspoon flour
salt, to taste
freshly ground pepper, to taste

Wash mushrooms and slice them into thick slices. Heat the butter in a skillet over a low flame. Add the shallots and garlic. Sauté for 3–4 minutes. Turn up the flame and add the mushrooms. Cook over a medium-high flame so the liquid from the mushrooms does not collect. Lower the flame and add the cognac. Stir well. Add the flour to thicken and season with salt and pepper.

Serves 4

The pressure cooker was something I used when I got married. It was just a fad, a quick thing. But pressure cookers are not my stick, never been my stick. I don't think I have used that pressure cooker in ten years 'cause I found out that slow cooking with the lid of the pot on would retain more nutrients. I meant to bring a pot back from the Cape this year. We had a cast-iron Dutch oven that was so heavy. It's like a steamless cooker. You put the lid on it, and believe me, nothing comes out. Nothing! It is the same principle as a pressure cooker, I guess, but on a different level.

The Frenchman's Zucchini Pot

2 tablespoons olive oil
¼ pound slab bacon, cubed
2 medium onions, chopped
4 Italian sweet peppers, sliced
2 ears corn, cut into 1-inch slices
1 2-pound can whole tomatoes, coarsely chopped
1 bay leaf
salt, to taste
freshly ground pepper, to taste
3 medium zucchinis, sliced
1 10-ounce box frozen okra
½ teaspoon red pepper flakes

Heat the olive oil in a large cast-iron pot. Add the bacon and sauté for 10 minutes. Add the onions, sweet peppers, and corn. Sauté for 15 minutes. Add the tomatoes, bay leaf, salt, and pepper. Simmer for 20–30 minutes. Add the zucchinis, cover, and simmer over a low flame for another 30 minutes. Add the okra and red pepper flakes. Simmer 15 minutes more or until the okra is done.

Serves 6

Note: Serve as soup or over rice.

Now cast iron is really the best to cook in, as far as real cooking is concerned. I would say that the old-fashioned type of cook, even in the deep South, had to have cast-iron pots and a cast-iron frying pan. And I know some great chefs who wouldn't cook in anything else. I don't think anything distributes the heat better than a cast-iron pan.

I spent my summers on the boat, a lobster boat, with my brother. It was then that I found out that lobster can be cooked so many ways. Everybody thinks that lobster has to be boiled or broiled. Baloney. There's lobster with sauce and served over spaghetti, lobster fra diablo. There's lobster omelets. There are so many ways to cook a lobster. My brother showed me how. My brother's still there, in Manchester with all the lobstercrats.

Boiled Lobster

4 1½-pound lobsters
5 quarts water
2 12-ounce cans Narragansett ale (or any other ale)
2 teaspoons salt

Make sure that the lobsters are alive and kicking. Place the water, ale, and salt in a large, deep stockpot. Bring the pot to a boil. One at a time but in rapid succession grab the lobsters behind the head and drop them in the pot. Allow the pot to return to a boil. Cover and boil for 10–15 minutes. Drain the lobsters and serve with drawn butter immediately.

Serves 4

Note: The ale should be flat. So, leave open overnight.

Lobster and Spaghetti 1

4 tablespoons butter
1 tablespoon plus 1 teaspoon olive oil
2 tablespoons minced onion
1 teaspoon minced green pepper
2 cloves garlic, minced
½ pound mushrooms, sliced
1 pound lobster meat, coarsely chopped
2 tablespoons sherry
1 cup heavy cream
salt
freshly ground pepper, to taste
3 quarts water
½ pound spaghetti (capellini)
1 teaspoon finely chopped parsley

Heat the butter and 1 tablespoon of olive oil in a skillet. Add the onion, green pepper, and garlic. Sauté over a low flame for 3–4 minutes. Add the mushrooms and turn up the flame to medium. Sauté for 5 minutes. Add the lobster and stir to coat. Sauté for 3 minutes. Add the sherry and heavy cream. Season with salt and pepper. Meanwhile, bring the water to a boil in a large pot. Add the teaspoon of olive oil and a dash of salt. Add the spaghetti and return to a boil. Lower the flame and cook for 10 minutes. Serve the lobster in sauce over the spaghetti, sprinkled with the parsley.

Serves 4

Broiled Lobster

8 tablespoons (1 stick) butter
1 tablespoon vermouth
pinch of salt
4 1½-pound lobsters, split (while still alive), cleaned, and with the claws cracked
1 teaspoon paprika

Melt the butter in a small saucepan. Add the vermouth and salt. Brush the inside of the lobsters with the butter mixture and sprinkle with paprika. Place the lobsters under a preheated broiler about 6 inches from the flame. Broil for 8–15 minutes, basting several times with the butter mixture, until the flesh has turned white and the top is just slightly charred.

Serves 4

I worked in Wells restaurant in Harlem for ten years, and I don't think anybody in the world could cook fried chicken better than Wells. You know Wells was famous for its chicken and waffles. In fact, I never learned the secret to that chicken, though I tried. It was in the spices and the mixture they put the chicken in. I learned that much. We wouldn't call it a batter. It was like flour and spices with a little sugar in it. Seems, when that chicken hit that grease, the sugar formed a crisp outside. You know what I mean? I would get off work sometimes and wouldn't feel like eating, take some chicken home, then get up the next morning and eat that chicken. The crispness outside would still be there, and the inside would still be juicy. I am not being prejudiced for the mere fact that I worked there, but I have never tasted chicken like that nowhere, nowhere!

We had a fellow in Harlem who never got his just deserts. His name was Creole Pete, and he had a restaurant on 131st Street and 7th Avenue. He was from New

Orleans, and he could make the best Creole food, like gumbo. People used to come from all over just to eat that gumbo of his. I used to go in there with a friend of mine named Red Allen, a trumpet player from New Orleans. Red was always telling me that the secret to the gumbo was the filé powder . . . you couldn't make that good gumbo without the filé powder.

Chicken Breasts Stuffed with Prosciutto

- 2 chicken breasts (4 halves)
- 4 slices prosciutto
- 8 tablespoons (1 stick) butter
- salt, to taste
- freshly ground pepper, to taste
- 1 teaspoon flour
- 1 teaspoon minced onion
- 1 pound small mushrooms
- 1 cup chicken stock
- ½ teaspoon thyme

Cut a pocket into each breast. Stuff each pocket with a slice of prosciutto and a thin slice of butter. Sprinkle the outside of the chicken with a little salt and pepper. Place a pat of butter on each breast and place under a preheated broiler. Broil for 10 minutes, then turn. Broil until done. Meanwhile, make a simple mushroom sauce: Melt 2 tablespoons of butter in a skillet. Add the flour and brown it, stirring constantly. Add the onion and mushrooms and cook for 5 minutes. Add the chicken stock and season with salt, pepper, and thyme. Stir well and simmer to thicken. Serve the sauce over the chicken breasts.

Serves 4

Lobster and Spaghetti 2

3 tablespoons olive oil
1 small onion, chopped
1 small green pepper, chopped
3 cloves garlic, minced
1 28-ounce can puréed tomatoes
1 cup white wine
salt, to taste
freshly ground pepper, to taste
½ teaspoon thyme
½ bay leaf
1 pound lobster meat, coarsely chopped
1 teaspoon finely chopped parsley
3 quarts water
½ pound spaghetti

Heat the olive oil in a skillet over a medium flame. Add the onion, green pepper, and garlic. Lower the flame and sauté for 5 minutes. Add the puréed tomatoes and wine. Stir well and simmer for 5 minutes. Season with salt, pepper, thyme, and bay leaf. Simmer for 5 minutes. Add the lobster and parsley and simmer 10–15 minutes more. Serve over spaghetti.

Serves 4

I like to use spices galore. See, when you're going to cook a fish, you have to decide what you're going to do to that fish, and then you have to know what you're going to have with that fish. You can't just start throwin' stuff on that fish. Are you going to broil it, or are you going to stuff it and with what? Bread? Shrimp? Crab meat?

Now snapper should be cooked with a spicy seasoning, something in the spicy vein because redfish itself has a kinda bland taste to it. It's not like cooking a fillet of sole. Fillet of sole you can cook with oregano or rosemary or tarragon. As far as I'm concerned redfish would taste much better with spices, whether you make it with what you call a Spanish sauce or bake it with tomatoes or something. I like it baked with crab meat or shrimp.

Flounder Stuffed with Crab Meat

½ pound crab meat
1 teaspoon minced onion
½ teaspoon Dijon mustard
2 eggs, beaten
1 cup bread crumbs
salt
freshly ground pepper
4 ½-pound flounder fillets
½ cup flour
4 tablespoons butter
½ cup white wine

Pick the crab meat, removing all shells. Mix the crab meat together with the onion, mustard, 2 tablespoons of the beaten egg mixture, 1 teaspoon of bread crumbs, and ½ teaspoon each of salt and pepper. Mix well. Wash the fillets and pat dry. Spread an equal amount of the crab meat mixture over each fillet. Roll up each fillet and secure with a toothpick. Dredge in flour, then dip in the remaining egg and in the bread crumbs. Grease a small baking dish with butter and place the rolled fillets in one layer. Top with pats of butter and place under a broiler 6 inches from the flame. Cook for 5 minutes. Pour the wine over the fillets and continue cooking 5–7 minutes more, until done.

Serves 4

I once worked for a woman called Estelle Thompson Lee. She had a catering business, and she catered for the Rockefellers. I was in charge of all the whiskey. I was in charge of all the bars. But I used to watch how she would do things. She taught me a lot. She taught me how to cook lamb and how to cook a crown roast. How to cook steaks on an outdoor grill. The grill was as long as a dining room table. On this end of the grill were all the charcoal beds. She would put four bricks on this end so that the grill came down at an angle; one end higher than the other. Anybody wanting a well-done steak got it from one end, rare was at the other end. She knew where everything was, the rare, the well done, and the medium. She didn't even have to look.

Lobster Omelet

6 tablespoons butter
1 tablespoon minced onion
1 teaspoon minced green pepper
½ pound lobster meat, coarsely chopped
½ teaspoon finely chopped parsley
4 eggs
salt, to taste
freshly ground pepper, to taste

In a small skillet heat 2 tablespoons of butter and add the onion and green pepper. Sauté for 3–4 minutes. Add the lobster and sauté 3–4 minutes more. Sprinkle with the parsley and set aside. Beat together 2 eggs at a time. Add a sprinkle of salt and pepper to the eggs and beat again. Heat 2 tablespoons of butter in an omelet pan over a medium flame. Lower the flame and pour in the eggs. Allow the eggs to begin to set. Place half the lobster mixture on one side, up to the middle of the omelet. Fold the omelet in half and continue cooking until the omelet is set and done on one side. Turn and finish cooking. Repeat to make the second omelet.

Serves 2

If Mrs. Lee taught me anything, she taught me, "Nothing goes out unless it's right." There are certain wines that had to be chilled to certain temperatures. Now you can go into a good restaurant today and order a bottle of wine, and it is expensive wine and they bring the bottle out and show it to you and let you taste it and smell the cork, and then they put it in a bucket of ice. The temperature is lowered so that the flavor of the wine is completely lost. That's all for show. Forty degrees is the temperature that any champagne should be chilled to, no less. A lot of people don't know that, they accept it in the bucket.

Chicken with Mustard

1 3-pound chicken, cut into serving pieces
1 lemon
1 cup chicken stock
1 cup white wine
1 medium onion, chopped
2 cloves garlic, minced
1 teaspoon thyme
salt, to taste
1 teaspoon freshly ground pepper
2 tablespoons honey
3 teaspoons Dijon mustard
1 pound small mushrooms
2 tablespoons flour

Wash the chicken well and squeeze the lemon over the chicken. Place it in a large skillet and add the chicken stock, wine, onion, garlic, thyme, salt, and pepper. Bring to a boil and cook for several minutes. Lower the flame and cover. Simmer for 15–20 minutes, until the chicken is done. Remove the chicken and set aside. Strain the cooking liquid and return to the skillet. Add the honey, mustard, and mushrooms. Cook over a high flame for 5 minutes. Lower the flame and add the flour, stirring constantly. Stir until the flour is absorbed and the sauce has thickened.

Serves 4

The same damn thing tastes different to different people, for the mere fact that people are different.

I have found out that some of the best cooks in the world are those who have that God-given talent to cook.

Damu

ERUNDINA PADRON JACOBSON

October 29, 1908 Tampa, Florida.

It seemed as if I already knew Damu by the time I actually met her. I had heard so much about her. She was the grandmother of two very close friends of mine, two beautiful women I had known since my early teens. At some point they took me to Damu's for dinner, probably before a night of partying. I was totally taken by her. I was fascinated by her Cuban background, the Spanish she spoke, and wonderful food that was new to me. By then I had eaten in local Cuban-Chinese restaurants, but this was food of a different character altogether. It was wonderful. I think I asked her for one of her recipes that very first night. I became a frequent visitor, with and without her granddaughters. We often talked on the phone as she counseled me on one or another of her dishes that I was attempting to make. More than the specific dishes, though, I learned a style from her, a new way of approaching things.

Damu

Escabeche
Beef Heart
Brain Fritters
Arroz con Pollo with Damu's Blessings
Arroz con Camarones à la Damu
Black Beans
White Beans and Collard Greens Soup
Barbecued Pig's Feet
Barbecue Sauce
Salad Dressing
Damu's Basic Pork Shoulder
Sweet Potato Biscuits
Bread Pudding with Honey Rum Sauce
Baked Porgies
Souse
Pig's Feet with Garbanzo Beans

Oh yes, I come from a big family. I had two sisters and four brothers, aunts, uncles, cousins. After Castro we lost track of many of my relatives who were still in Cuba. I had one uncle, I remember, who fought in the Spanish-American War. I don't know where he is now or if he is still alive. Before, when I was a child, we used to go to Cuba often, and relatives from Cuba would come to visit us in Florida.

In my household it was my aunt who did most of the cooking. She was very, very good, a great cook. In other words she was good at everything. She was a seamstress by trade. She made men's clothes and she cooked. That's where I started learning about cooking—my sister and I.

Escabeche

1 lime, cut in half
4 ¾-inch kingfish steaks
½ cup flour
1 cup olive oil
1 large onion, thinly sliced
1 large green pepper, thinly sliced
5 cloves garlic, minced
1 small hot pepper, cut in half
2 bay leaves
½ teaspoon oregano
½ teaspoon basil
½ teaspoon thyme
1 teaspoon crushed allspice
1 tablespoon chopped parsley
salt, to taste
freshly ground pepper, to taste
¾ cup vinegar
½ cup water

Squeeze the lime over the kingfish and wash well under running water. Drain and pat dry. Dredge in flour. Heat the oil in a large cast-iron skillet and fry the fish until done and well browned on both sides. Remove from the skillet to a deep ceramic or glass dish, large enough to hold the fish in one layer. Reheat the remaining oil and sauté the onion, green pepper, garlic, hot pepper, and bay leaves until the onion wilts. Remove the skillet from the flame and add the oregano, basil, thyme, allspice, parsley, salt, pepper, vinegar, and water. Return the skillet to the flame and bring almost to a boil. Pour the mixture over the fish and allow the fish and the mixture to cool. Place in the refrigerator overnight. May be eaten cold or allowed to come to room temperature.

Serves 4

Back in those days you could make a meal for a quarter. That's right, twenty-five cents. My aunt would buy two beef hearts, so there would be enough for all the children. She'd take them, boil them, chop them up, and simmer them in a sauce. That was served over rice or grits or whatever we happened to have. She fed us all.

Beef Heart

1 beef heart
water
salt
2 tablespoons olive oil
1 medium onion, chopped
1 small green pepper, chopped
2 cloves garlic, minced
1 stalk celery, chopped
1 1-pound can whole tomatoes
½ teaspoon oregano
1 bay leaf
freshly ground pepper, to taste

Place the heart in water to cover and add ½ teaspoon of salt. Bring the water to a boil. Lower the flame, cover loosely, and cook for 30 minutes. Remove from the water and allow to cool. Cut into ½-inch pieces, discarding all tough veins and gristle. Heat the oil in a small Dutch oven. Add the onion, green pepper, garlic, and celery. Sauté for 10 minutes. Chop the tomatoes coarsely and add them. Add the oregano, bay leaf, salt, and pepper. Simmer gently for 40 minutes. Add the heart pieces and simmer 30–40 minutes more. Serve over rice.

Serves 4

When I was little, one of my favorites was brain fritters. My aunt would take the brains and drop them in hot water, then take them right out and peel all the skin off. You know, let them cool, get cold. Then she would slice them, salt and pepper them, and dip them in egg and bread crumbs. Then she would fry them in oil. They would come out soooo good. Sometimes she would garnish them with parsley. She would make those, and we kids would go crazy over them.

Brain Fritters

1 set brains
water
salt
freshly ground pepper
2 eggs
2 cups cornflakes
4 tablespoons vegetable oil
4 tablespoons butter
1 tablespoon chopped parsley

Cover the brains with cold water to which 1 teaspoon of salt has been added. Allow the brains to soak for about 1 hour. Bring enough salted water to cover the brains to a boil. Place the brains in the boiling water for 5 minutes. Remove from the hot water and place under cold running water to cool. Carefully peel away the membrane covering the brains. Let the brains stand until they are cold. Slice into ½-inch slices and sprinkle each side with salt and pepper. Meanwhile, beat the eggs and crush the cornflakes with a rolling pin, making sure they are crushed fine. Heat the oil and butter in a frying pan. Dip the brain slices in the egg mixture and then in crushed cornflakes. Fry until golden brown. Sprinkle with chopped parsley. Serve.

Serves 4–6

There was a time when I was alone and didn't have anybody to cook for. I mean, there was no man. I had moved to New York in 1930, and I would cook, but just plain cooking, you know—rice and beans, steak, whatever was easy. It was just the two of us, my daughter and me.

As much as I watched my aunt, I don't think I picked up too much. I used to eat her food and I used to love to watch her cooking, but I really didn't learn much about cooking until I married Chris. He was a person who loved to cook and he was very good, and that's where I really started to learn to cook. I picked things up from him, and then that also brought back the things I had learned from my aunt, my Spanish background, you know.

My father made cigars. He was in the trade in those days. That was what he did in Florida. He also had a darkroom and took pictures. He was always into something.

My first husband was Lora's father. I was married at a very early age, fifteen, and Lora was born before I was sixteen. I was so young. I really wasn't interested in much, definitely not cooking. I did cook, but that was simply survival cooking. And anyway, I didn't stay with my first husband too long.

I married Chris in 1940. Being with someone who was interested in cooking really made me blossom. He gave me an incentive to cook. He never liked to go to restaurants, never. Maybe a couple of times if someone invited us. But as far as him saying, "Let's go out to have dinner," never. So we cooked at home all the time. Most times, together. He really made me into a good cook.

Arroz con Pollo with Damu's Blessings

½ pound bacon, cut into ¼-inch dice
2 whole chickens, cut into pieces
4 cloves garlic, smashed
salt, to taste
freshly ground pepper, to taste
2 large onions, chopped
1 large green pepper, chopped
2 stalks celery, chopped
2 bay leaves
1 2-pound can whole tomatoes, chopped
2 teaspoons bijol (see Note)
1 large jar Alcaparrado (see Note)
3 tablespoons soy sauce
½ bottle (12.7 ounces) dry white wine
⅔ 1-pound box Uncle Ben's rice
⅓ 10-ounce box frozen green peas
1 4-ounce jar pimentos

Fry the bacon in a large cast-iron skillet. Wash the chicken and pat dry. Rub with garlic and sprinkle with salt and pepper. When the bacon is almost done, add the chicken and sauté for about 10 minutes, until browned on all sides. Add the garlic, onions, green pepper, celery, and bay leaves. Sauté for 10 minutes. Add the tomatoes with their liquid and the bijol. Drain the Alcaparrado and add along with the tomatoes. Stir and simmer for 5 minutes. Add the soy sauce, salt, pepper, and wine. Cover and cook slowly for 15–20 minutes or until the chicken is almost done. Add the rice, cover, and cook until all the liquid is absorbed, about 30 minutes. Meanwhile, cook the frozen peas. Garnish with the green peas and pimentos.

Serves 10–12

Note: Bijol is a yellow-orange food coloring. Alcaparrado is a mixture of olives, capers, and pimentos. Both are available in Spanish food stores.

Arroz con Camarones à la Damu

2 pounds shrimp
¼ cup vegetable oil
1 large onion, chopped
1 medium green pepper, chopped
4 cloves garlic, minced
2 stalks celery, chopped
1 bay leaf
1 teaspoon oregano
1 teaspoon thyme
salt, to taste
freshly ground pepper, to taste
2 cups water
1 12-ounce can flat beer
bijol (see Note on page 161)
1 8-ounce jar pitted olives
2 cups rice
1 dozen clams

Shell and devein the shrimp. Wash, drain, and set aside. Heat the vegetable oil in a large Dutch oven. Add the onion, green pepper, garlic, celery, and bay leaf. Sauté for 10–15 minutes. Add the oregano, thyme, salt, pepper, water, and beer. Add bijol until the color looks right. Simmer for 30 minutes. Add the shrimp, olives, and rice. Mix well. Cover, lower the flame, and cook until the water is absorbed. Place the clams in a single layer over the top of the rice, cover, and bake in an oven preheated to 300° for 20–30 minutes.

Serves 6–10

Once I got interested in cooking, it all came back to me . . . everything that my aunt had taught me. It all came back, and my aunt was still alive so I would go to her to

talk about cooking. She would tell me to do it this or that way. She died at the age of ninety-two, but she was still lucid. She still knew how to make everything, so I learned a lot more from her and my sister too.

Black Beans

1 pound black beans
water
4 chorizo sausages, sliced
2 tablespoons olive oil
½ pound bacon, cut into ¼-inch dice
1 large onion, chopped
1 medium green pepper, chopped
5 cloves garlic, chopped
1 bay leaf
½ teaspoon oregano
freshly ground pepper, to taste
2 tablespoons vinegar
salt, to taste

Pick through the beans, removing all stones. Rinse the beans. Place them in a large Dutch oven and cover with water to about 2 inches above the beans. Bring the beans to a boil and simmer for 1 hour. Add the chorizo sausage slices and a little more water if necessary. Cover and simmer another hour, until the beans begin to get soft. Meanwhile, heat the olive oil in a large cast-iron skillet. Add the diced bacon and fry until the bacon is half done. Add the onion, green pepper, garlic, bay leaf, oregano, and pepper. Sauté until the onion begins to wilt. When the beans begin to soften, add the sauté mixture and continue cooking about 1 hour. When the beans are done, add the vinegar and salt. Allow the beans to sit for 20–30 minutes before serving.

Serves 10–12

I think the best thing my aunt taught me was how to use my imagination, because cooking is a lot about imagination, you know. Sometimes you don't have recipes to rely on. You have to make your own recipes. Even if you do have a recipe, you can still use your imagination. I never follow anything to the letter.

White Beans and Collard Greens Soup

1 pound small white beans
water
2 tablespoons olive oil
¼ pound bacon, cut into ¼-inch dice
1 large onion, chopped
1 medium green pepper, chopped
4 cloves garlic, minced
3 medium chorizo sausages, sliced
½ teaspoon thyme
salt, to taste
freshly ground pepper, to taste
2 cups chopped collard greens

Pick through the beans, removing all stones. Rinse the beans. Place them in a large Dutch oven and cover with water to about 2 inches above the beans. Soak the beans for 6 hours. Drain and cover again with water. Bring to a boil, lower the flame, and cover. Simmer gently for 1 hour. Meanwhile, heat the oil in a large cast-iron skillet. Fry the bacon until it is half done. Add the onion, green pepper, and garlic. Sauté until the onion wilts. When the beans begin to soften, add the sauté mixture along with the chorizo sausage slices, thyme, salt, and pepper. Wash the greens. Discard all thick stems and brown leaves. Roll the greens and slice them into thin strips. When the beans are almost done, add the greens and simmer 20–30 minutes more. Mash some of the beans against the side of the pot to thicken the soup. Adjust the seasonings before serving.

Serves 10–12

In my aunt's day they didn't have things like barbecue sauce. That was just something I picked up. I even put it on some Spanish dishes. I got a recipe for this sauce to barbecue spareribs once. Well, I changed it, and it was so delicious. Everybody just raved over them. So I thought, if the spareribs could be done that way, why not pig's feet too. So I took the plain feet, parboiled them, and barbecued them with my sauce. They were fabulous, just wonderful. They were so good that my family thought I should put the recipe in the paper.

Barbecued Pig's Feet

8 front pig's feet, split in half
water
1 teaspoon salt
2 tablespoons vinegar
1 bay leaf
1 small onion, quartered
barbecue sauce (recipe follows)

Wash the feet. Place the feet in a large pot and cover with cold water. Add salt, vinegar, bay leaf, and onion. Bring the pot to a boil, lower the flame and simmer for 40–45 minutes to just parboil them. Remove the feet from the water and set aside to cool. Place the feet in a large baking pan. Pour barbecue sauce over the feet and place in an oven preheated to 325°. Cook for 1½ hours or until the feet are done. Baste with more barbecue sauce every 20 minutes or so. Also pour in a little stock if the feet start to become too dry.

Serves 4–6

Barbecue Sauce

2 cans tomato soup
2 cans stock (water from pig's feet)
½ cup ketchup
⅓ cup vinegar
2 tablespoons Worcestershire sauce
1 cup minced onion
½ cup minced green pepper
¼ cup minced celery
1 bay leaf
2 tablespoons brown sugar
1 teaspoon chili powder
salt, to taste
freshly ground pepper, to taste

Place all ingredients in a saucepan. Bring to a boil, lower the flame, and simmer for 30 minutes, stirring occasionally so the sauce does not stick. Add more stock if necessary.

Makes 6 cups

Now, my salad dressing is something that I just duddled up. A pinch of this and a dash of that. I mean, it just comes out delicious. In fact, Tom [granddaughter's husband] said, "Oh, my goodness! This is voodoo dressing." He just goes crazy over it.

Salad Dressing

¾ cup olive oil
¼ cup wine vinegar
2 tablespoons water
3 tablespoons Dijon mustard
3 tablespoons ketchup
2 teaspoons horseradish
5 cloves garlic, minced
1 teaspoon oregano
salt, to taste
freshly ground pepper, to taste

Place all ingredients in a blender or food processor. Blend until thick and creamy. Refrigerate, tightly closed, until ready to use.

Makes 1½ cups

When I'm cooking, I stay right there in the kitchen. I get angry when anything comes out wrong or if anything burns on me. I just scream. I don't want anybody to call me or talk to me. I fall apart if anything goes wrong when I'm cooking. I always want everything to be perfect—right down to the last grain of rice.

Damu's Basic Pork Shoulder

1 4–6-pound pork shoulder (fresh picnic)
1 lemon, cut in half
4 cloves garlic, sliced
2 teaspoons oregano
salt, to taste
freshly ground pepper, to taste

Rinse the pork shoulder in cold water. Puncture the shoulder all over with a small knife, forming small pockets. Squeeze the lemon over the shoulder allowing some of the juice to get into the pockets. Place a slice of garlic in each pocket along with a little oregano, salt and pepper. Sprinkle the remaining oregano and additional salt and pepper over the shoulder and set aside. Allow the shoulder to marinate for several hours. Place in a baking pan, skin side up, and roast in a preheated oven at 350° for 2 hours (20 minutes per pound), or until a meat thermometer says well done for pork, and the top skin is nice and crisp.

Serves 4–6

Sweet Potato Biscuits

1 large sweet potato
water
2 cups flour
2 teaspoons baking powder
¼ teaspoon salt
1 tablespoon sugar
¾ cup solid shortening

Boil the sweet potato until tender, about 15 minutes. Remove from the water and mash. In a large bowl mix together the flour, baking powder, salt, and sugar. Mix well and then sift. Add 1 cup of the mashed sweet potato and the shortening, using a knife and fork to cut them into the flour mixture. Work until a dough is formed. Add a little water if the dough gets too stiff. Roll the dough out on a floured board to a thickness of about ½ inch. Cut out biscuits and place them on an ungreased baking sheet. Put in an oven preheated to 400° and bake until the biscuits are golden in color, about 15 minutes.

Serves 6–8

Note: Spread butter or honey on the biscuits. These biscuits are great with almost anything, especially greens.

Bread Pudding with Honey Rum Sauce

4 cups ¾-inch pieces stale bread
3 cups milk
1 cup heavy cream
2 tablespoons butter
3 eggs, lightly beaten
¾ cup sugar
½ teaspoon salt
1 teaspoon cinnamon
½ teaspoon nutmeg
1 teaspoon vanilla extract
½ cup golden raisins
3 tablespoons butter, melted
½ cup dark rum
½ cup water
½ cup honey

Mix the bread with the milk and heavy cream. Spread the 2 tablespoons of butter in a casserole. Pour in the bread mixture. Add the eggs, sugar, salt, cinnamon, nutmeg, vanilla extract, raisins, and melted butter. Mix well. Place the casserole in a baking pan with about 1 inch of water in it. Put in an oven preheated to 350°. Bake for about 1 hour, until the pudding is brown on top but still moist. Combine the rum, water, honey, and remaining tablespoon of butter in a saucepan. Bring to a boil, then simmer for 10 minutes, stirring constantly, until the sauce starts to thicken. Serve over the pudding.

Serves 6–8

Note: In this case you might say the proof of the pudding is in the sauce.

I always try to take my time—even when I'm doing something that is real quick and easy—like those porgies that Chris used to like so much. Fry the porgies fast. Let them brown but quickly. Then you just put them in a large pan, add onion, green

pepper, and garlic. Then make a sauce to go over them, with tomatoes and wine, whatever you have, then bake them. There you have it.

Baked Porgies

4 medium porgies, cleaned but with head intact
1 lime
salt, to taste
freshly ground pepper, to taste
garlic powder, to taste
1 cup flour
2 cups vegetable oil
2 tablespoons olive oil
1 large onion, sliced
1 medium green pepper, sliced
4 cloves garlic, minced
1 bay leaf
2 8-ounce cans tomato sauce
1 teaspoon oregano
½ teaspoon basil
½ cup white wine
½ teaspoon Worcestershire sauce (or slightly more to taste)

Rinse the fish in cold water. Squeeze the lime over them and rinse again. Pat dry. Sprinkle the fish inside and out with salt, pepper, and garlic powder. Let them sit for 30 minutes. Heat the vegetable oil in a large frying pan until almost smoking. Dredge the fish in flour and fry in the hot oil until golden brown. Set aside. Heat the olive oil in a large saucepan. Add the onion, green pepper, garlic, and bay leaf. Sauté for 10 minutes. Add the tomato sauce, oregano, basil, salt, and pepper. Simmer for 5 minutes. Add the wine and Worcestershire sauce. Simmer the sauce for 10–15 minutes. Place the fish in one layer in a large baking pan. Pour the sauce over the fish and bake in a preheated oven at 325° for 45 minutes.

Serves 4

A long time ago, someone taught me to put food away before it gets completely cold. The heat helps seal the food and keeps it fresher.

Do you know the secret of a good chicken? If you get a chicken that is too white, has no color, then that's no good. You need a chicken that has some color to it, and it has to smell fresh too.

When you go to cook a turkey, always buy a tom turkey. That's the best turkey, a tom turkey.

Chris used to make souse every New Year's. I feel sorry now that I took a lot of things for granted. I'd always say, "Oh, he's going to always make the souse," and I never learned to make it as good as he did. I guess I know what he did, but it will never be the same. I just never learned to make it the way he made it.

If I didn't like it, I didn't learn to cook it.

Souse

1 pig's head, chopped into pieces
water
salt
1 medium onion, quartered
1 stalk celery, with leaves
2 tablespoons vegetable oil or bacon fat
1 small onion, minced
½ small green pepper, minced
2 cloves garlic, minced
1 stalk celery, chopped fine
1 bay leaf
freshly ground pepper, to taste

Place the head pieces in a large pot. Cover with water and add 1 teaspoon of salt, quartered onion, and celery stalk. Bring to a boil. Lower the flame to medium and boil for 2 hours or until the meat is falling off the bones. Remove the head pieces and reserve the cooking liquid. Pick the meat off the bones and set aside. Heat the oil or bacon fat in a large sauce pot. Add the onion, green pepper, garlic, chopped celery, and bay leaf. Sauté for 20 minutes. Add the meat along with salt and pepper. Strain the cooking liquid and add it to the pot. Simmer for about 1 hour, until the liquid has been reduced by about one-third. Pour into a glass loaf pan and allow to cool. Refrigerate until the souse has jelled. Cut into slices to serve.

Serves 6–8

Variation: May also be served hot over peas and rice.

Pig's Feet with Garbanzo Beans

6 pig's feet, split in half
3 cups water
1 16-ounce can tomato sauce
1 large onion, chopped
1 medium green pepper, chopped
4 cloves garlic, minced
2 stalks celery, chopped
½ teaspoon oregano
½ teaspoon thyme
2 bay leaves
½ cup Al Caparrado (see Note on page 161)
dash of Worcestershire sauce
dash of hot sauce or Tabasco
salt, to taste
freshly ground pepper, to taste
1 6-ounce can garbanzo beans, drained

Place the pig's feet in a large Dutch oven. Add the remaining ingredients except the garbanzo beans. Stir and bring the Dutch oven to a boil. Lower the flame and cover. Simmer for 2 hours, until the pig's feet are tender. Add the garbanzo beans and simmer, uncovered, 20–30 minutes more. Adjust the seasonings, cover, and allow to sit for 15 minutes before serving.

Serves 6

I started taking care of Pam when she was nine months old, you know. And she used to walk behind me and she didn't know how to talk, and that's when she started saying, "Damu, Damu." And I've been called Damu ever since.

You have to like to cook, I mean really like it. Then you can do anything.

Verta

VERTA MAE SMART-GROSVENOR

April 4, 19—　Fairfax, South Carolina.

To tell the truth, I'm not sure exactly when I met Verta; it just seems like I've known her forever. I do know there was a time when we would be cooking together almost every day; definitely every weekend. I do know that I learned a ton of things looking over her shoulder, and I must have gotten 'em down fairly good because whenever she left my daughter Chandra with me, she would wolf down everything I cooked. Over the years Verta and I have continued to share and compare our culinary adventures, sometimes quietly, sometimes quite loudly, but always with love and passion.

Verta

Stewed Guinea Hen
My Father's Stuffed Fish
Short Ribs
Simple Fried Chicken
Fried Okra
Mixed Greens
Shrimp, Okra, and Eggplant Rice
South Carolina Okra Soup
Home Fries
Ground Nut Stew
Boiled Fish
Stuffed Eggplant (Guinea Squash)
Shepherd's Pie
Turnip Greens with Corn Bread Dumplings

I was born in South Carolina and I never give the year of my birth—the date is April 4—because a woman who tells her age will tell anything.

It seems like whenever people get together it's around food, good food. When someone dies or gets married, the food, the spread, has to be just right.

Shellin' peas on the porch is my earliest recollection of bein' involved with food. When I was very little, oh, we used to do lots of little things—go to the garden and help pick peas and okra or go get a tomato or even eggs from the hen house. But mostly the garden. I do remember we had a smoke house too. So I have very early recollections of bein' involved.

I've always been interested in food. My father was kind of a food nut, he just made it seem so exciting.

Cookin' the food? I would say I began by helping my grandmother Sula. I used to help my father cook because my mother worked during the day and my father worked at night after we moved to Philadelphia, so then he used to cook dinner and I would go with him to help shop and then I would help him cook. So I really learned to cook more from my father.

Stewed Guinea Hen

1 4-pound guinea hen,
1 small onion, chopped
2 cloves garlic, minced
1 rib celery, chopped
1 teaspoon salt
1 teaspoon freshly ground pepper
1 teaspoon garlic powder
½ teaspoon thyme
½ teaspoon crushed sage
½ cup flour
½ cup bacon fat
2 cups chicken stock
1 cup white wine

Wash the hen and cut it into quarters. Sprinkle the pieces all over with salt, pepper, garlic powder, and sage, and set aside for several hours. Dredge the guinea hen pieces in the flour. Heat the bacon fat in a cast-iron casserole and add the hen when the fat is nice and hot. Brown well on all sides. Cover and place in an oven preheated to 325°. If all the bacon fat has cooked off, add a little liquid. Cook for 1 hour. Add the onion, celery, and garlic. Cook 5 minutes more. Add the stock and wine, stirring to scrape up all the browned material off the bottom of the pot. Add the thyme, taste for salt and pepper, and add if necessary. Cover and cook 20 minutes or until the meat is beginning to fall off the bones.

Serves 4

As I got older my responsibility would be to start things—start and wait for my father to come and finish up. Then later I would be put in charge of one particular thing. It would be my job to cook that particular thing. And then I graduated into being in charge of the dinner, which I messed up sometimes . . . a lot . . . a whole lot.

I've just never had spaghetti and meatballs quite like my father's. I've been to Italian restaurants here and in Italy. I've been in Italian homes, but he made the best spaghetti and meatballs you can imagine.

We had all kinds of fish. We would have eel, we would have roe. And then sometimes with porgies, he would take and pan-smother them with onions and peppers. He would make this baked fish. Yeah, his baked fish was fabulous, just fabulous. He used to wrap the fish in peppers and onions, and cover it in one of those blue tin roasters and bake it, and it would just come out perfect. You would take the juices from that and put it over your rice. And we had rice every day, that was understood.

My Father's Stuffed Fish

1 3–4-pound whole fish (bluefish, red snapper, etc.), cleaned and split
salt, to taste
freshly ground pepper, to taste
2 teaspoons thyme
8 tablespoons (1 stick) butter
1 small onion, chopped
1 small green pepper, chopped
1 rib celery, chopped
1 clove garlic, minced
½ pound mushrooms, chopped
2 cups leftover rice
1 medium onion, sliced
1 medium green pepper, sliced
1 rib celery, including leaves, cut into 2-inch slices
1 large carrot, peeled and cut into 2-inch slices
1 medium zucchini, cut into 2-inch slices
2 cups fish stock

Clean and wash the fish, and pat dry. Make several cuts in the flesh of the fish and rub a mixture of salt, pepper, and a little thyme into them. Sprinkle the fish inside with salt and pepper. In a skillet melt 2 tablespoons of butter. Add the chopped onion, green pepper, celery, and garlic. Sauté for 5 minutes. Add the mushrooms and continue sautéeing for several minutes, until the mushrooms just begin to give up their liquid. Add the rice, remove from the heat, and mix well. Stuff the fish with the rice mixture. Grease a covered baking pan with a little butter and lay the fish in it. Scatter the sliced vegetables around the fish and sprinkle with salt, pepper, and thyme. Slice the remaining butter into pats and place over the fish and the vegetables, making sure some of the butter gets into the cuts in the fish. Add 1 cup of the fish stock, cover, and place in an oven preheated to 350°. Cook for 30–40 minutes depending on the size of the fish, until the flesh just flakes. Baste the fish and vegetables with the pan juices several times during cooking. Add more fish stock if necessary.

Serves 6

I was an only child—just me. That accounted for my bein' close to both my parents. See, when we were down South, my cousins were like siblings. But when we moved North, I only saw them in the summers. When school closed I would go back down South and always come back the day before school opened. I never spent summers in the city, never went to camp or any of that stuff.

Short Ribs

4 pounds lean short ribs, cut into several pieces
salt, to taste
freshly ground pepper, to taste
1 large onion, chopped
3 cloves garlic, chopped
3 tablespoons vegetable oil
½ cup flour
2 cups beef stock or water
3 large carrots, peeled and cut into large pieces
1 teaspoon thyme
½ teaspoon basil
1 teaspoon paprika

Place the ribs in a large bowl and sprinkle with salt and pepper. Add the onion and garlic. Mix well and set aside for several hours. Heat the vegetable oil in a cast-iron casserole. Shake the onion and garlic off the ribs and dredge the ribs in the flour. Place the ribs in the casserole and brown well on all sides. Lower the flame and add the onion and garlic. Cook for 5 minutes. Add the beef stock, cover, and place the casserole in an oven preheated to 350°. Cook for 1 hour. Add the carrots, thyme, basil, and paprika. Cover and cook

20 minutes more. Check for salt, pepper, and liquid. Add more salt, pepper, and liquid as needed. Cover and cook until the meat is falling off the bones, about 15 minutes.

Serves 6

The first thing I learned to really cook was rice. That was my first job. It was very important that you learned how to cook rice perfectly. We were from Carolina, and it would be a disgrace not to know how to cook rice—if it was mushy or it was whatever.

I was eighteen when I left home. I went to Paris. I began cooking there because it was cheaper. I was just so accustomed to cooking, I didn't really think about it. It was great to eat out, to find great cheap restaurants and all that, but I always found a room in Paris that had a space to cook. I couldn't conceive of havin' a place where you could not cook. I was excited about the markets and found it interesting to find new things and then go try and cook them. It wasn't enough to eat them in a restaurant, I had to try to cook them.

I never fried a chicken the whole time I was in Paris. Somehow it just didn't seem right. I don't know what it was. There's this restaurant in Paris called Haynes', a soul food restaurant, and I remember having fried chicken there once, but I never ordered it again. Just didn't seem right, I don't know. The French just grow chickens differently. They're perfect for doin' all those French things, but they just don't fry right.

Simple Fried Chicken

1 3½-pound frying chicken, cut into 8 serving pieces
1 lemon, cut in half
1 teaspoon salt
½ teaspoon freshly ground pepper
½ cup flour
2 cups vegetable oil

Wash the chicken pieces and pat dry. Rub the chicken pieces all over with the lemon halves. Sprinkle with the salt and pepper. Let the chicken sit at room temperature for at least 30 minutes. Place the chicken in a paper bag and add the flour. Shake well to coat each piece evenly. In a large cast-iron skillet heat the oil over a high flame until the oil is very hot but not smoking. Add the chicken, skin side down, cover, and fry for 10 minutes. Turn the pieces, cover, and fry 10 minutes more. Uncover the skillet, lower the flame, and turn the chicken again. Fry another 5 minutes or so, until golden brown. Drain the chicken on brown paper or paper towels.

Serves 4

I liked very much what the French did with vegetables. I used to like the idea that the vegetable was just not some side dish. You know, I could appreciate that, coming from the South. You can go to many American restaurants and homes where a vegetable is just something that you say to your child, "At least take a spoonful, Johnny."

You eat corn, okra, and greens, as you would a piece of meat. It is an important part of what you are eating for dinner; it is not just some accompaniment.

Fried Okra

2 pounds fresh okra
2 eggs, lightly beaten
2 cups cornmeal
salt, to taste
freshly ground pepper, to taste
bacon fat

Cut off the top stems and wash the okra, discarding any that are dry and woody. If the okras are large, cut on a diagonal into 1-inch lengths. If they are small, leave them whole. Dip in the beaten egg and roll or shake in cornmeal seasoned with salt and pepper. Fry in hot oil or bacon fat until golden brown.

Serves 6

Mixed Greens

- 1 pound bacon ends (smoked bacon) or 1 large ham bone (Virginia ham) with plenty of meat still on it
- 1 small onion, chopped
- 2 quarts water
- 2 bunches turnip greens or rape salad
- 1 bunch mustard greens
- 1 bunch collard greens
- 1 bunch kale
- 2 small red pepper pods
- 1 teaspoon salt
- 1 teaspoon freshly ground pepper

Place the bacon ends or ham bone in a large heavy pot and fry for 5 minutes. Add the onion and sauté for a few minutes. Add the water. Bring to a boil, lower the heat, cover loosely, and simmer for 40–60 minutes, until the liquid is reduced by about half. Meanwhile, clean the greens, making sure to discard all dead and yellow leaves, and remove all large stems. Wash the greens in several changes of water to remove all sand and grit. Drain and coarsely chop the greens. Add them to the pot in 4 or 5 batches, allowing each batch to cook down some before adding the next. When all the greens are in the pot, mix them well and add the red pepper pods, salt and pepper. Cover and simmer for 30–40 minutes or until the greens are just tender. Serve with chopped onion, hot sauce, or pepper sauce.

Serves 6 greens lovers; serves 10 as a side dish

Note: Any combination and/or proportion of collard greens, mustard greens, turnip greens, kale, rape salad, spinach, beet tops, or cabbage may be used according to your taste. Greens may be cooked without meat. Bouillon may be substituted.

I was very young when I was in France, very impressionable. I didn't have a lot of prejudices to get over, so it was all fine with me.

When I came back to this country and got married and had children, well, that was the first time as a family that I had, really had, my own permanent kitchen. So it became a mix of different styles. My base was always Southern, but it was kinda influenced by all the places and people I had known.

Shrimp, Okra, and Eggplant Rice

3 tablespoons bacon fat
1 medium onion, chopped
½ small green pepper, chopped
2 cloves garlic, minced
1 medium eggplant, peeled and cut into ½-inch cubes
1 teaspoon salt
½ teaspoon freshly ground pepper
½ teaspoon thyme
½ teaspoon red pepper flakes
½ pound okra, sliced into ½-inch slices
2 cups long-grain rice
1 pound small shrimp
4 cups liquid (chicken stock, seafood stock, or boiled shrimp shells)

Heat 2 tablespoons of the bacon fat in a large heavy pot. Add the onion, green pepper, and garlic. Sauté until the onion has wilted. Add the eggplant. Sauté for 5 minutes. Season with the salt, pepper, thyme, and red pepper flakes. Stir and add the okra. Sauté 5 minutes more. Add the remaining tablespoon of bacon fat, rice, and shrimp. Cook until the shrimp just turn pink. Add the liquid. Stir once. Allow the pot to come to a boil, then lower the heat to very low. Cover and cook for 20 minutes or until the rice is done and all the liquid is absorbed.

Serves 6–8

I like having a soup or something, then getting into the food. I don't particularly like serving crackers and cheeses and stuff at the table before dinner.

I'm a big soup person. I might start with a soup—so my kids got used to that—then we might have rice and greens and fried chicken.

South Carolina Okra Soup

2 pieces beef soup bones
2 quarts water
3 beef short ribs, cut into 9 pieces
½ pound stew beef, cubed
1 small onion, quartered
1 stalk celery with leaves
½ small green pepper, coarsely chopped
2 cloves garlic, smashed
2 teaspoons salt
1 teaspoon freshly ground pepper
½ pound bacon, cut into ¼-inch dice
1 medium onion, chopped
1 stalk celery, chopped
½ small green pepper, finely chopped
2 cloves garlic, minced
1 teaspoon thyme
½ teaspoon red pepper flakes
3 pounds plum tomatoes, peeled and chopped (2 cups)
1 cup canned crushed tomatoes
1 pound fresh okra, sliced
1 medium can white corn (1–1½ cups)
1 pound shrimp, shelled

Place the soup bones in a large heavy pot and add the water. Bring the pot to a boil, lower the heat, and simmer for 20 minutes. Add the short ribs, stew beef, quartered onion, celery stalk with leaves, coarsely chopped green pepper, smashed garlic, 1 teaspoon of salt, and ½ teaspoon of pepper. Loosely cover and cook over a low flame for 1½–2 hours, until all the meat is tender. Remove the meat and set aside. When cooled, pull the meat from the bones and cube. Strain the liquid and reserve. In the same but now empty pot, fry the bacon until half done. Add the chopped onion, chopped celery, finely chopped green pepper, and minced garlic. Sauté for 5 minutes. Add the thyme, red pepper flakes, and remaining salt and pepper. Continue to sauté until the onion has wilted. Add the plum tomatoes and crushed tomatoes, okra, and corn. Stir and cook for 5 minutes. Add the reserved liquid. Stir well and add the shrimp and cubed meat. Simmer, uncovered, for 10–15 minutes. Check the seasonings. Serve over rice or cornbread.

Serves 8–10

In the early days all the cooking was geared around school. I liked the idea of a heavy meal during the day, around lunchtime. Most of the time the kids came home for lunch, and that was good 'cause at night they could go to bed early. Even if I had company for dinner.

They always had breakfast, right through high school. Breakfast of grits. I used to keep boiled potatoes in the refrigerator so that I could make quick home fries. I was never a big cereal, that is, cornflakes person. I liked Familia—that Swiss line of cereal—with the nuts and raisins. Yeah, Familia. Well, I found out recently that they hated Familia and still hate it—just cannot stand to even see it, even in the store.

Home Fries

4 large potatoes
1 tablespoon vegetable oil
3 tablespoons olive oil
1 medium onion, thinly sliced
1 teaspoon salt
½ teaspoon freshly ground pepper
½ teaspoon garlic powder
½ teaspoon paprika

Wash the potatoes well; do not peel them. Cut the potatoes lengthwise into quarters and then cut into ¼-inch slices. Mix the oils together and heat in an iron skillet over a medium flame. Add the potatoes and cover. Fry gently over a medium-low flame. Cook until the potatoes begin to soften. Remove the cover and add the onion, salt, pepper, garlic powder, and paprika. Turn the potatoes and mix well. Cook, allowing the bottom layer to brown well. Turn and allow the next layer to brown. Cook until all the potatoes are brown and crisp.

Serves 4

Sometimes we had leftovers, which is left over from my childhood. There is no reason why breakfast has to be bacon and eggs.

I can always remember my grandmother doing a quick hash in the morning of leftover meat and potatoes—that's right! Some rice and a little meat gravy left from last night and a piece of meat, and you just have it.

When the children were about nine or ten, I guess that's when I went into my experimentation stage, a country a month. The rule was that that month I would try all kinds of stuff. I would just experiment, not just a dinner thing but all the meals. I had great success with China, that was my best shot.

There are but so many ways to cook, but so many things you can do with food. It's just about your style of doing it.

What I started doing then, in about 1968, was to study not only the cooking style but the history of the food. So no matter what country it just became a question of how they dealt with food and why—culinary anthropology.

So this culinary imperialism that this country—the Western world—has, well, that's French, you know. I mean, you can't own . . . eggs belong to the world. So just because a thing is a soufflé, I mean, that's just how they deal with it, but other people do other things with it. Things just as good.

You can always learn something from another culture. It doesn't make you right or them wrong. It's simply how you *do it. It's just a matter of taste, after all. If you're so locked into your taste that you can't appreciate the goodness in someone else's taste, that's too bad for you.*

Ground Nut Stew

1 4-pound chicken, cut into 10–12 pieces
1 lemon
1 teaspoon salt
1½ teaspoons white pepper
1 teaspoon ground ginger
½ cup vegetable oil
1 medium onion, chopped
1 teaspoon minced garlic
1 1-pound can puréed tomatoes
2 tablespoons tomato paste
1 teaspoon minced fresh ginger
½ teaspoon cayenne pepper
1 tablespoon dried shrimp, ground in a blender
4 cups hot water
1 cup peanut butter plus 1 cup water
1 medium eggplant, peeled and cubed
2 cups fresh or frozen okra
½ pound small shrimp, shelled

Wash the chicken and pat dry. Rub the chicken with the lemon and then with a mixture of salt, 1 teaspoon of white pepper, and ground ginger. Allow the chicken to sit for 30 minutes or so. Heat the oil in a large heavy casserole. Brown the chicken on both sides. Remove the chicken and set aside. Discard all but ¼ cup of the remaining oil. Add the onion and garlic. Cook for 5 minutes, until the onion begins to wilt. Add the puréed tomatoes, tomato paste, minced ginger, cayenne pepper, the remaining ½ teaspoon of white pepper, and ground dried shrimp. Stir well and simmer for 5 minutes. Add the hot water and stir well. Mix the peanut butter with 1 cup of water to thin it. Add the peanut butter, stirring constantly. Simmer over a low flame for 30 minutes. Add the eggplant and simmer 5 minutes more. Add the okra and simmer another 5 minutes. Add the chicken. Cover and simmer for 30 minutes, until the chicken is done. Add the shrimp and cook for 5 minutes. Cover the pot and allow it to sit for 10 minutes before serving.

Serves 4–6

I eventually began serious culinary investigations of foods. I guess you could say I became a militant about the whole thing. My job, I felt then, as an investigator was to demystify the culinary imperialism, and that's about the time that I decided to write my book, Vibration Cooking *[Ballantine, 1986].*

I didn't ever think of my book being published. I was just going to write this stuff up, then I was going to get it copied, and I would give it out to friends for birthdays or wedding presents. This would be my way of sounding off. It was to be something that I would exchange with my cooking friends. But it turned out otherwise, and that was all right too because it forced me to truly back up what I had concluded from my investigations.

I like pepper. I like it coarse ground, fine ground, red, white, and black. Dried flakes, green peppers, those long skinny ones, peppers.

I love paprika. I always get mine in the Hungarian section because it's really good and they like paprika a lot. Not just to throw on top of potato salad or deviled eggs. It has its own properties.

Onions. You always have to have onions. Yellow onions, red onions, green onions . . . onions.

Boiled Fish

2 limes
2 pounds firm fleshed fish, such as cod
½ teaspoon salt
½ teaspoon freshly ground pepper
1 medium onion, thinly sliced
2 tablespoons vegetable oil
1 cup water
½ teaspoon red pepper flakes

Squeeze one of the limes over the fish and wash gently. Drain and place in a bowl. Sprinkle with salt and pepper. Add the sliced onion and 1 tablespoon of vegetable oil. Turn several times and set aside to marinate for 1 hour. Pour the remaining tablespoon of vegetable oil in a frying pan and add the fish. Cook over a low flame, not browning but just cooking. When the fish begins to whiten, add 1 cup of water and the juice of 1 lime. Add the red pepper flakes and test for salt. Cover and simmer until the fish flakes. Serve over grits.

Serves 4

Speakin' of greens, when I was in Scandinavia, I was dyin' for some greens. I mean, all the stuff they had was great, but I just wanted some greens. Well, it occurred to me that with all the beets they had, they had to have some beet greens. So when I went to the market, I just asked and sure enough they had piles of beet greens just sittin' there waiting to be thrown out. Well, did I ever make up a pot that night.

I'll tell you what I do with beet greens. I mix 'em with some spinach or some dandelion greens or both. And the kids just loved that.

I'm always takin' the traditional way and adapting it. I think it's just an instinct you have to have.

Stuffed Eggplant (Guinea Squash)

2 large, firm eggplants (guinea squash)
salt
½ cup bread crumbs
3 tablespoons chopped parsley
½ teaspoon thyme
½ teaspoon oregano
freshly ground pepper, to taste
pinch of cayenne pepper
1 tablespoon vegetable oil
½ pound ground lamb
1 small onion, chopped
1 tablespoon chopped green pepper
1 stalk celery, chopped
½ cup grated mozzarella cheese

Slice the eggplants in half lengthwise. Scoop out the meat, being careful not to break the skins. Sprinkle the eggplant with salt and set aside for 30 minutes. Pat dry with a paper towel. Mash the eggplant meat in a large bowl. Add the bread crumbs, parsley, thyme, oregano, salt, pepper, and cayenne pepper. Heat the oil in a cast-iron skillet and sauté the lamb until lightly browned. Remove the lamb and set aside, Pour off all but 2 tablespoons of the grease. Add the onion, green pepper, and celery. Sauté until tender. Add the lamb and sautéed vegetables to the eggplant mixture. Mix well. Add a little water if the mixture is not moist enough. Fill the eggplant skins with the mixture and place in an oiled baking dish. Bake in an oven preheated to 350° for 10 minutes, remove, top with the mozzarella cheese, and bake until the cheese is melted, about 10 minutes.

Serves 8

Note: In parts of the Carolina lowlands eggplant is still called guinea squash.

We used to have an old-fashioned American meat loaf occasionally. It was good, but most often it had too much filler for my taste. I learned to make meat loaf in Paris. I remember—back in the sixties—this friend of mine had invited a bunch of us for dinner. She was American but she'd lived a long time in the Virgin Islands. So her kitchen was all mixed up. I had asked her, I said, "What're you fixin'?" and she said, "Oh, just a meat loaf and some vegetables." So I just watched. She fixed it in a blender. She put in the eggs and lots of garlic and onions, no green peppers, then she mixed that with the meat, and I think she used a piece of bread too and that was all, and she made it with lamb and beef . . . three-fourths lamb and one-fourth beef proportionally . . . and it was great. It was fabulous. And so that's the meat loaf I make today.

I love lamb . . . lamb breast. I make a corn bread stuffin'. But it's hard to find a lamb breast whole nowadays.

Shepherd's Pie

¼ cup bacon fat
1 medium onion, chopped
1 small green pepper, chopped
4 cloves garlic, minced
1¼ pounds ground lamb
¾ pound ground beef (see Note)
½ teaspoon thyme
¼ teaspoon rosemary
¼ teaspoon ground cumin
salt, to taste
freshly ground pepper, to taste
2 pounds potatoes
8 tablespoons (1 stick) butter
1 cup milk

Heat the bacon fat in a large cast-iron skillet. Add the onion, green pepper, and garlic. Sauté for 5 minutes. Add the meat, stirring to break up any large lumps. Add the thyme, rosemary, cumin, salt, and pepper. Cook for 20 minutes, stirring to keep the bottom from sticking. Meanwhile, peel and quarter the potatoes. Place them in a pot of salted water to cover and bring to a boil. Lower the flame and cook for 15–20 minutes. Drain the potatoes and add the butter and milk. Mix and mash the potatoes until the butter is melted and the potatoes are mashed. The potatoes should be fluffy but slightly stiff. Spread the potatoes over the meat in the skillet, peaking the potatoes in the middle. Place the skillet in an oven preheated to 350° and cook for 20 minutes until the pie is set and the potatoes are brown on top.

Serves 6

Variations: To spice up the pie, try adding mushrooms, carrots, or other vegetables to the meat.

Any leftover meat will also work well.

So what I did was take black-eyed peas, which is very Afro-American, very traditional, and I cooked them with green chili peppers—I didn't have any meat in them —and lots of onions and spices and some Mexican oregano, things like that, and thyme, and I put dumplings in. Boy, was that somethin'.

Turnip Greens with Corn Bread Dumplings

2 bunches turnip greens, about 3 pounds
water
1 large onion, chopped
3 beef bouillon cubes
salt, to taste
freshly ground pepper, to taste
½ teaspoon red pepper flakes
¼ cup vinegar
1 10-ounce box corn bread mix

Select tender young greens. Pick the greens, removing all old or yellow leaves and the heavy stems. Wash the greens in several changes of water, making sure all sand and grit are washed out. Place the greens in a large pot and cover with water. Bring the pot to a boil. Add the onion, bouillon cubes, salt, pepper, red pepper flakes, and vinegar. Lower the heat and simmer for 25 minutes. Mix the corn bread according to the directions on the box, making sure the mixture is nice and thick. Spoon heaping tablespoons of it into the greens, as many as you would like. Cover and simmer 20–30 minutes more.

Serves 6–8

Note: Some purists insist that greens can be cooked only with ham hocks, bacon, or smoked neck bones for seasoning. I am not a purist. Greens can be seasoned with herbs, spices, onion, and so forth. For my greens, I go by my vibrations.

I also like—it's a very old dish, it's not new at all, but people don't make it much anymore—turnip greens and corn bread dumplings. Very, very classic Afro-American . . . an old dish. I love it.

I think that's why my friends keep me around. They want me to cook. But when I love 'em, I love to cook for 'em.

Leah

LEAH LUCY LANGE CHASE

January 6, 1923 New Orleans, Louisiana.

I met Leah Chase and her husband Dooky in 1979 when I went to New Orleans to photograph a story on Creole cooking for *Essence* magazine. Leah was so warm, sincere, and charming, and her food sooooo delicious, that I fell madly in love. Over the years our friendship developed, and we began to talk often. We talked of cooking, family, and art. She became my teacher and mentor in the ways of Creole cooking and life, and with each trip to New Orleans I would learn more and more. In fact, the girls who work in her restaurant kid me, asking, "What you and Mrs. Chase got going? What you-all doing back in that kitchen? You know she don't let no one back there." Leah welcomed me into her life, and she has now become an important part of my family, and I have become one of her sons.

Leah

Red Beans and Rice
Rice Dressin'
Panée Meat
Veal Steaks
Creole Gumbo
Okra Gumbo
Chicken and Shrimp Soup
Tomato Gravy
Gumbo Z'Herbes
Mother's Vegetable Soup
Soup Meat and Potatoes with Vinaigrette
Jelly Cake
Creole Sauce
Creole Wieners with Spaghetti
Creole Potato Salad
Breakfast Shrimp
Stuffed Veal Shoulder Roast
Orange Pecan Pancakes
Sautéed Chicken Livers
Redfish with Pecans
Veal Roast with Tuna and Anchovies
Rice Salad

I was born in New Orleans, but I grew up in Madisonville. You have to know Madisonville, just a skip and a jump from New Orleans. Then it was about an hour's ride because we didn't have the causeway and the bridge. My mother was from New Orleans, but we were brought up out there. I'm a country girl really.

At that time Madisonville was pretty country, it's still pretty country. But small towns are good. You learn a lot growing up in small towns. You learn to survive. You learn how to really do things for yourself.

I'm the top of the line of eleven living children, eleven! My mother had a dozen girls before she had a boy.

We all had little jobs to do. Yes, in big families everybody had to do somethin'. If you consider all those children Mama had, you know that she was always pregnant and always needed help one way or the other so we all had somethin' to do.

My darling Mama was not a good cook. She could cook, I guess, but then who could take an interest in cooking for a whole army every day? You put on some greens and this, that, and the other, and you fed those kids and your husband. This was Depression time, so you were doin' no gourmet cooking. You cooked what you had.

Red Beans and Rice

1 pound red beans (dried kidney beans)
2 quarts water
1 large onion, coarsely chopped
1 small green pepper, chopped
½ pound Creole sausage, sliced, or other spicy sausage
½ pound smoked ham, cubed
1 teaspoon salt
freshly ground pepper, to taste
1 bay leaf
1 teaspoon thyme
¼ cup bacon fat or vegetable oil
3 cloves garlic, minced
2 tablespoons chopped parsley

Pick the beans, making sure all rocks are removed. Place the beans in a pot and add the water and onion. Bring the pot to a boil, lower the heat, and cook for 1–1½ hours, until the beans get tender. Add the green pepper, sausage, ham, and salt. Stir and then add the pepper, bay leaf, thyme, and bacon fat. Add a little more water if needed. Cook about 30 minutes more or until the beans are soft and the mixture is creamy. Add the garlic and parsley. Turn off the heat, cover, and allow to sit for several minutes before serving.

Serves 8–10

That's one thing about a big family in the country, you learn how to fend for yourself early. You learn how to keep yourself out of trouble, 'cause there was a razor strap settin' on the doorknob, and even at twenty, I didn't think I was too big.

Rice Dressin'

½ pound chicken gizzards
½ pound chicken necks
¼ cup bacon fat
1 medium onion, chopped
1 small green pepper, chopped
2 stalks celery, chopped
1 pound chicken livers
1 pound Creole sausage or other spicy sausage
3 cups stock or water
1 bay leaf
1 teaspoon salt
½ teaspoon freshly ground pepper
2 cups rice

Place the gizzards and necks in a pot, cover with water, and bring to a boil. Boil gently until the gizzards are done, about 45 minutes. Chop the gizzards and remove the meat from the necks and chop. Place the bacon fat in a large frying pan. Add the onion, green pepper, and celery. Finely chop the chicken livers and add them to the pan. Slice the sausage and add that to the pan. Cook until the chicken livers and sausage are done and the onion has wilted. Place 3 cups of the gizzard stock in a pot. Add the chicken liver sausage mixture, the gizzards and neck meat, bay leaf, salt, and pepper. Bring to a boil and add the rice. Turn the flame to low, cover, and cook until the rice is done. Mix well and serve.

Serves 12

Panée Meat

8 veal cutlets, ½ inch thick
2 eggs
½ cup milk
salt, to taste
freshly ground pepper, to taste
2 cups Italian bread crumbs
½ cup vegetable oil
1 lemon, cut into wedges

Place the cutlets two at a time between waxed paper and pound very thin. Beat the eggs and milk together. Sprinkle the meat with salt and pepper, then dip in the egg mixture and bread crumbs. Turn several times to make sure the meat is breaded well. Heat the oil in a frying pan. When it is good and hot, add the meat, several pieces at a time. Fry until golden brown. Serve with lemon wedges.

Serves 4

In the country where I came up, there was one little butcher shop. The owner was this deaf man, but you went to his shop and you waited in line on Sunday morning till he cut the meat. It was all baby veal. There was never, ever a piece of beef in his shop. Never! You never asked for anything by the pound. You said, "Give me one of this" or "four of that" or whatever. And he'd cut it up for you, but it was always baby veal.

Veal Steaks

- 4 1-pound veal steaks
- salt, to taste
- freshly ground pepper, to taste
- 1 tablespoon olive oil
- 1 tablespoon vegetable oil
- 3 cloves garlic, minced
- ¼ cup chopped parsley
- 1 large onion, halved and sliced lengthwise

Allow the steaks to come to room temperature. Sprinkle with salt and pepper. Heat the olive and vegetable oils in a large cast-iron skillet and fry the steaks quickly over a high flame. Mix the minced garlic and chopped parsley together and as each steak is done to taste sprinkle with the garlic-parsley mixture and set aside on a warming platter. When the steaks are done, quickly sauté the onion in the skillet and serve alongside the steaks.

Serves 4

My father raised hogs, and there was nothin' too good for the hogs, my dear. We had a certain kind of grass that we had to pull for them. Those hogs were somethin' else. You know, usually you throw a hog trash or slop or whatever, not those hogs that Charlie Lange had. He would kill ya. We had to wash sweet potatoes and boil them! Those hogs didn't eat anything raw. We had to cook for those dumb hogs, would you believe! We were fetchin' water and cleanin' pigpens. Those hogs didn't wallow in any mud! No indeed, they lived as good as I did.

Oh, we ate plenty of pork, but the primary purpose for the pigs was the lard. If you got yourself two hundred pounds of lard off those hogs, that was really good, 'cause you had lard for cooking for the whole year. See, we didn't buy any lard or anything like that. When they would slaughter the hogs and hang up the meat, they would smoke some of the hams. But most of the hams were fresh, they weren't salt cured. We would get our liver puddin' and all kinds of stuff. We would hurry to catch the blood while it was still warm so we could make what you call boudin, blood sausage.

Mama used to always run and hide when the killin' time came, she couldn't stand that. I'll never forget one time, we had this huge hog, about four hundred pounds I guess. Well, when Daddy shot him, the bullet just flattened out. Daddy had to shoot him again. I guess that's what you'd call a hard head.

We ate a lot of game 'cause people in the country hunted—deer, rabbit, squirrels, duck, you name it—and a lot of fish. My father had a drag line. Do you know what a drag line is? See, you hook the line on one end of the bayou to the other end, and then you run 'em twice a day. You run 'em in the morning; you get in your boat and lift those lines and take your fish off. See, you catch great big catfish like that on drag lines. So you would go and run your lines and nobody ever touched your line, and you didn't touch anybody else's.

Every Sunday God send, you had gumbo, filé gumbo—the crabs, the shrimp, sausage, stew meat, everything in that filé gumbo. Every Sunday. And you had stewed chicken—for sure, if you had a big family—and baked macaroni and rice. Would you believe? And potato salad. You had all the vegetables because you grew 'em. My father grew most of the things we ate.

Creole Gumbo

4 blue crabs, split and cleaned
¾ pound Creole hot sausage, cut into bite-size pieces
¾ pound smoked sausage, cut into bite-size pieces
½ pound stew beef, cut into ½-inch pieces
1 pound smoked ham, cut into ½-inch pieces
½ cup vegetable oil
4 tablespoons flour
1 large onion, chopped
3 quarts water
10 chicken wings, cut in half (tips discarded)
1 pound shrimp, peeled and cleaned
1 tablespoon paprika
1 teaspoon salt
1 tablespoon filé powder
2 dozen oysters, with their liquid
⅓ cup chopped parsley
3 cloves garlic, minced
1 teaspoon ground thyme

Place the crabs, sausages, stew beef, and ham in a large pot. Cover and cook in its own juices for 25 minutes, stirring often. Heat the oil in a skillet until hot. Add the flour, stirring constantly, until it makes a golden brown roux. Add the onion and cook over low heat until the onion wilts. Pour the onion mixture over the ingredients in the large pot. Add the water, chicken wings, shrimp, paprika, salt, and filé powder. Bring to a boil and simmer for 30 minutes. Add the oysters, parsley, garlic, and thyme. Lower the heat and cook for 10 minutes before serving.

Serves 10

Well, Thanksgiving dinner was never turkey. We didn't have turkey. We didn't raise them, and we couldn't buy them. We ate what we had. Thanksgiving when I was a child wasn't a big day for us. It was still a workin' day. Everybody worked.

Easter was a big day, and Good Friday! You did nothing on Good Friday. And my father wouldn't dare put a shovel in the soil on Good Friday, that was a terrible thing. The thing for some people was to fish on Good Friday. So everybody would go fishin'. But my father didn't let us go fishin' on Good Friday. No way! No pleasure, for sure. It was a day to be solemn. You didn't eat anything until noon on Saturday.

We had the special meal on Easter Sunday. You had okra gumbo, 'cause okra was in season then. When okra was in season, you had okra gumbo on Sundays. You didn't have the filé gumbo. Then you had turtle 'cause that was the season to catch what we call the cowands—the snappin' turtles. Easter Sunday would not be Easter Sunday if you didn't have turtle. And my dear, was it great. Put in lots of fresh thyme, a lot of peppercorns, and the whole red peppers. You see, the thing was to get that female turtle with all those eggs. With the eggs still inside. You'd take those eggs and when your gravy was making, you'd just stick 'em with a pin and drop 'em in your gravy. Best tasting gravy you ever had. That was your Easter Sunday dinner all the time . . . still is.

Okra Gumbo

2 pounds fresh okra, sliced
6 crabs, cleaned
2 tablespoons tomato paste
1 large onion, chopped
1 green pepper, chopped
3 stalks celery, chopped
3 quarts water
1 bay leaf
1 teaspoon thyme
2 teaspoons salt
1 teaspoon freshly ground pepper
½ teaspoon cayenne pepper
3 cloves garlic, minced
1 pound shrimp, shelled

Place the okra, crabs, tomato paste, onion, green pepper, and celery in a large heavy pot. Add the water and bring the pot to a boil. Add the bay leaf, thyme, salt, and pepper. Boil for 30–40 minutes over a medium flame, until the okra is done and all the ropiness is cooked out. Add the cayenne pepper, garlic, and shrimp. Simmer 15 minutes more.

Serves 10

Chicken was your Sunday! Stewed mostly, but sometimes fried. These were not your young chickens. You stew only hens . . . you didn't dare stew your spring chicken. You just used those to fry. But most of the time it was stewed chicken.

Chicken and Shrimp Soup

1 3-pound chicken, quartered
2 quarts water
1 medium onion, chopped
2 stalks celery, chopped
1 teaspoon thyme
salt
1 teaspoon white pepper
8 tablespoons (1 stick) butter
2 tablespoons flour
6 large mushrooms, minced
¼ teaspoon cayenne pepper
1 pound small shrimp, shelled

Place the chicken in a large pot, cover with the water, and add the onion, celery, thyme, 1 teaspoon of salt, and white pepper. Bring the pot to a boil, lower the heat, and boil gently for 1 hour. Remove the chicken and set aside to cool. Strain and reserve the cooking liquid. In a saucepan slowly heat the butter until melted, making sure not to burn it. Add the flour, stirring constantly; brown the flour slightly and cook the pastiness out. Slowly add the flour mixture to the cooking liquid, whisking constantly to prevent lumping. Remove the chicken meat from the bones, coarsely chop the meat, and return it to the cooking liquid. Add the chopped mushrooms and cayenne pepper. Check for salt and add more if necessary. Bring to a slow boil and add the shrimp. Serve garnished with chopped parsley.

Serves 8

You had all the matchin' things on the Sunday table. And on Sunday you had a starched and ironed tablecloth that my mother had made out of flour sacks and embroidered.

I look at those days, and I look at today and how people can't cope and are hungry, particularly in New Orleans and in the South. You have no right to be hungry. You can take what we used to call a dinner stick, just a straight cane pole, and put a hook on there and go catch you a dinner. The ponds are all around you.

We didn't throw anything away, not anything at all.

I just love thyme. I wish I could grow that fresh thyme like my mother used to. She always had a giant tub of it. You didn't do any cooking without fresh thyme and bay leaves that you could just go break off. You always had parsley and the peppers. My mother used very little cayenne pepper because my daddy raised red peppers. Some of the peppers she dried but mostly she used them fresh. You always had your basil too. That was just a thing you always had growing in your garden because it smelled good all the time and you wanted that basil for your tomato gravy and things like that.

Tomato Gravy

3 tablespoons bacon fat or oil
¼ cup chopped onion
2 tablespoons flour
3 medium tomatoes, seeded and chopped (about 2 cups)
½ teaspoon thyme
1 teaspoon sugar
salt, to taste
freshly ground pepper, to taste

Heat the bacon fat or oil in a large frying pan. Sauté the onion until it starts to brown. Add the flour and cook, stirring constantly, until the flour forms a nice brown roux. Add the tomatoes and stir well. Add the thyme, sugar, salt, and pepper. Simmer over a low flame for 20–30 minutes. Serve over chicken, meat, or rice.

Makes approximately 2 cups

Creoles use a lot of paprika. If you're making gravy and you want a good brown gravy, it's hard to get a pretty brown. I don't like what I call muddy lookin' gravy. But if you take a little paprika and put that in your brown gravy, it's going to come out nice and golden. A nice lookin' gravy.

Daddy made up his own filé. My sister still does. You have a sassafras tree and you just dry the leaves, dry 'em real good and grind 'em. Daddy had a little hand grinder, and he'd just grind those leaves on down.

And you know what we ate a lot of? Onions. The whole town used to tease us about eating onions. Because my father was just a fanatic on growing onions . . . and garlic! Never, ever did we go to a grocery to buy an onion; never in my whole life, till I got to be married. Daddy would bunch 'em and tie 'em in bunches, and we had a shed where they hung, along with the garlic.

Gumbo Z'Herbes

1 bunch each of collard greens, turnip greens, mustard greens, kale, carrot tops, radish tops, watercress
2 bunches beet tops
1 head cabbage
3 pounds spinach
½ pound slab bacon, cut into 4 pieces
2 medium onions, chopped
1 bunch parsley (flat leaf type)
1 cup chopped celery
1 cup chopped green pepper
10 cups water
1 cup sliced scallions
5 cloves garlic, chopped
1 pound smoked ham, cut into ½-inch cubes
1 pound Creole sausage, cut into bite-size pieces
4 tablespoons vegetable oil
1 teaspoon thyme
1 teaspoon sweet basil
3 bay leaves
½ teaspoon allspice
⅛ teaspoon ground cloves
½ teaspoon cayenne pepper
salt
freshly ground pepper
8 tablespoons (1 stick) butter
5 tablespoons flour

Pick the greens, removing all dead leaves and large stems. Wash them several times. Place the greens in a large pot along with the bacon, onions, parsley, celery, green pepper, and water. Bring to a boil; lower the heat and cook, covered, for 1½–2 hours, until the greens are tender. Drain the greens, reserving the cooking liquid. Discard the bacon. When the greens are cool enough to handle, divide them in half. Coarsely chop half and put the other half through a blender or food processor. In the pot, without the liquid, sauté the scallions, garlic, ham, and sausage in the vegetable oil for 10 minutes. Return the liquid and the greens to the pot. Add the thyme, basil, bay leaves, allspice, cloves, cayenne pepper, salt, and pepper. Simmer for 30 minutes. With a fork cut the butter into the flour to form a paste. Spoon the paste, a little at a time, into the pot until all the butter mixture is dissolved in the greens. Simmer over a low flame 15 minutes more. Serve in deep soup bowls over a little hot rice.

Serves 12

So we learned how to live off the land—off what you could get. You learned how to pick strawberries all through the woods. You learned how to pick muscadines. You know how to pick crab apples, you knew how to pick May haws. I love crab apples, all green and sour, and you make crab apple jelly, May haw jelly. You can make muscadine or blackberry jelly. Or if you want, you can make wine with any of 'em. There was no way to go hungry in the country. You knew how to work with all those things.

Mama would make this big pot of soup with all the vegetables. She liked to make it in the spring when we had all the carrots in the garden and all. You always had the big soup bone with the meat on it, and then she'd put the potatoes in it. She wouldn't give you the meat and potatoes in your bowl. She'd take all the meat off the bone and put that with the potatoes and make like a vinegar sauce to put over the meat and the potatoes. You ate that like a second course.

Mother's Vegetable Soup

1 large soup bone with some meat on it
3 quarts water
1 cup chopped onion
1 32-ounce can tomatoes, coarsely chopped
3 large carrots, chopped
2 pounds beef brisket
1 large turnip, cut into eighths
½ head cabbage, coarsely chopped
2 stalks celery, chopped
1 cup string beans, cut into 1-inch pieces
½ pound pumpkin, coarsely chopped
2 ears of corn, cut off the cob
1 small red pepper
3 large potatoes, peeled and cut into eighths
salt, to taste
freshly ground pepper, to taste

Place the soup bone and the water in a large pot and bring slowly to a boil. Just as the pot begins to boil, add the onion, tomatoes, and carrots. Return to a boil and lower the flame. Simmer for 30 minutes. Add the beef brisket. Cover and simmer for 20 minutes. Add the turnip, cabbage, celery, string beans, pumpkin, corn, and red pepper. Cover loosely and simmer for 30 minutes. Add the potatoes, salt, and pepper, and simmer 15–20 minutes more. Remove the soup bone, the beef brisket, and the potatoes. Serve the soup as a first course.

Serves 6–8

Soup Meat and Potatoes with Vinaigrette

2 pounds beef brisket (from the vegetable soup)
1 large soup bone (from the vegetable soup)
potatoes (from the vegetable soup)
vegetable oil
vinegar
salt, to taste
freshly ground pepper, to taste
2 tablespoons chopped parsley

Slice the beef brisket and place along with the potatoes on a platter. Crack the soup bones and remove the marrow. Add the marrow to the platter. Make a vinaigrette sauce of vegetable oil and vinegar (one to one). Pour over the meat, potatoes, and marrow. Toss. Sprinkle with salt, freshly ground pepper, and chopped parsley. Serve as a second course.

Serves 6–8

You sat at the table and you ate, and the subject matter at the table had to be just so. You didn't talk about ugly things at all. If you did, my dad would give you one look. You better not even talk about anything. The bathroom, the toilet door, you couldn't say it. "You'd better watch what you're sayin'," he'd say. So it was always a pleasant thing to sit down and eat.

You had to do somethin' for somebody every single day of your life. You would go to the grocery for somebody, and you'd better not say you were tired and you better not take a nickel for it. Daddy believed in that. You always do for others.

Mother hated to wash dirty dishes. So if you had a bowl of soup, she gave it to you in the bowl. Then you had the meat on the back side of the bowl. On the weekdays when she made mustard greens and rice—and maybe she would make blackberry pie—well, you had to eat the greens and rice on your plate and you had to clean your plate, then flip that plate over for your pie. Mother wasn't washin' but one plate, don't care what.

Jelly Cake

16 tablespoons (2 sticks) butter
2 cups sugar
3 cups self-rising flour
4 eggs
1 cup milk (approximately)
1 coconut
1 cup strawberry jelly

Using a knife and fork, cut the butter and sugar together. Add the flour, eggs, and milk. Mix well and pour into two round baking pans. Bake at 325° for 20 minutes, until done. Remove from the pans and set aside to cool. While the cake is cooling, crack the coconut and discard the water. Pry the meat away from the brown shell. When the cake is cool, place the first layer on a plate and cover the top with jelly. Grate coconut over the top. Place the second layer on top of the first and spread jelly over this also. Add the grated coconut. Spread jelly around the sides and add the grated coconut.

If you had a cold or a fever, my dear, you got a cup of hot wine with cinnamon and sugar. You went to bed and covered up, and you sweated that fever out.

Daddy made the best strawberry wine you ever tasted. The whole of Madisonville —they'd have the bingo parties at the church, and when his wine was the prize, well everybody would go play. Because this was Charles Lange's wine. He made the best. Clear as a crystal. Not sweetie, sweetie sweet but really good. A really good fruit wine.

My first love was sewing, just like my mother, but I couldn't see myself workin' in that at all; I guess I was militant from way back. So I went out to look for a job in, of all places, the French Quarter. Well, my dear, that was a no-no for sure. That just wasn't the thing a respectable young woman did. My aunt called my mother, and she threw a fit. But I liked it anyway. I went to work for a woman waitin' tables. It was the Colonial Restaurant on Chartres and Toulouse. It was really a nice little restaurant. Not too many seats; sat maybe about fifty people. But it was nice. I learned a lot there. I learned to like the business. Later she moved the restaurant and changed the name to the Coffee Pot. We really ran that place. She was getting old and tired, I guess. So there we were, just three of us. Do you believe it was just three young girls? I was eighteen, Lucia was eighteen, and Estelle was just sixteen. Eighteen, eighteen, and sixteen—and I'm tellin' you, we ran that restaurant. We did everything. We'd open that restaurant, close it, ordered, put the money in the bank, and went about our business. When you look at today and you see so many people who can't cope at that age, you wonder what happened. But those were good days, those were fun days!

I remember the first lunch we did. Lucia said, "Let's put on lunch. Let's do somethin' different every day." So we did. Our first lunch—I'll never forget—was Creole wieners and spaghetti. All it is . . . you make a real good Creole sauce, a lot of seasonings in the sauce, and you drop those wieners in there. And serve it over spaghetti.

Creole Sauce

½ cup olive oil
1 medium onion, chopped
1 medium green pepper, chopped
3 cloves garlic, minced
2 ribs celery, chopped
1 16-ounce can crushed tomatoes
1 teaspoon thyme
1 tablespoon sugar
1 bay leaf
salt, to taste
freshly ground pepper, to taste
1 4-ounce can tomato paste
1 cup chicken stock

Heat the oil in a large saucepan. Add the onion, green pepper, garlic, and celery. Sauté until the onion starts to wilt. Add the crushed tomatoes, stir, and add the thyme, sugar, bay leaf, salt, and pepper. Add 2 tablespoons of the tomato paste and ½ cup of the chicken stock. Simmer for 15 minutes. Adjust the seasonings and add tomato paste and stock as needed to get a nice thick sauce.

Makes about 4 cups

Creole Wieners with Spaghetti

8 beef wieners
4 cups Creole sauce (see page 224)
3 quarts water
1 tablespoon butter
1 teaspoon salt
8 ounces spaghetti

Place the wieners in a saucepan with the Creole sauce and bring to a simmer. Cook slowly for 15 minutes, until the wieners are done. Meanwhile, bring the water to a boil in a large pot. Add the butter and salt, then add the spaghetti and cook for 10 minutes. Serve the wieners and Creole sauce over the spaghetti.

Serves 4

I really learned to love the French Quarter. It was so different than it is today. It's really raunchy today. You always had the strippers, but in those days people did things . . . well, even the strippers had class. Strippers were strippers, and when they came out they were dressed nice. When they got to their jobs they were whatever they were, but they were behind those doors. Nobody saw them unless they wanted to go in. But today you have all these raunchy shops on Bourbon Street and the people. But you still find the finest music on Bourbon Street.

Creole Potato Salad

3 pounds Idaho potatoes
salt
6 eggs, hard-boiled and coarsely chopped
2 stalks celery, chopped
3 green onions (scallions), finely chopped
4 sprigs parsley, chopped
¼ cup chopped small sweet pickles
1 cup mayonnaise
freshly ground pepper, to taste

Peel the potatoes and cut into ¾-inch cubes. Place them in a large pot and cover with water. Add 1 teaspoon of salt. Bring to a boil, lower the heat, and cook until a fork will just pierce the potatoes. Drain and rinse in cold water. Place the potatoes in the refrigerator for about 30 minutes until cool. Add the eggs, celery, green onions, parsley, pickles, mayonnaise, salt, and pepper. Mix well. Chill before serving.

Serves 6

Dooky's folks had the restaurant then. It was just a little place on the corner where the bar is. They had a smaller bar and had some tables. They served sandwiches and fried chicken and that kind of thing. They did well too. They opened that in 1941. I met Dooky in 1946. And my dear, in three months I was married.

✤

Breakfast Shrimp

3 cups Creole sauce (see page 224)
1 pound small shrimp, shelled

Heat the Creole sauce in a saucepan and add the shrimp. Cook until the shrimp just turn pink. Serve over grits with breakfast.

Serves 4

Dooky worked like a son of a gun on that bar. He built up a bar business, but then there was no emphasis put on food. It was just that when you came to drink, you ate a little too. Then in 1946 his mother lived here, and they were too many for that little room so they moved up the street where I'm livin' now. That room became the dining room, and people still came and on Saturday nights they'd bring what they call set-ups: half pints and mixers or somethin' like that. While you were drinkin' your half pints, you'd buy yourself some fried chicken or somethin'. But I really wanted to build up a dining room business, so I said I was goin' to do a lunch. So then we did good with our lunch every day. That caught on real big. Everybody was comin' for lunch. Then I was cookin' two days a week. Hittin' and missin', because I was just doin' the things I knew. I had no training other than that home thing. So I was just gettin' it in my brains and workin' at it.

In 1957 I started comin' every day. When the kids were in school I started comin' every day. I've been at it ever since; you know, every day.

Stuffed Veal Shoulder Roast

1 quart oysters with their liquid
6 slices stale French bread
¼ cup vegetable oil
1 medium onion, chopped
2 green onions (scallions), chopped
2 stalks celery, chopped
2 cloves garlic, minced
½ teaspoon salt
½ teaspoon freshly ground pepper
¼ teaspoon cayenne pepper
½ teaspoon thyme
1 bay leaf
4 sprigs parsley, chopped
1 8-pound veal shoulder roast, shoulder blade removed (veal pocket roast)
½ cup flour
1 cup white wine
1 cup water
2 carrots, peeled and sliced
3 large potatoes, peeled and cut into eighths

Pour the oyster water over the stale bread and set aside. In a large skillet heat the oil and then add the onion, green onions, celery, and garlic. Sauté for 10 minutes, until the onions wilt. Break up the soggy bread and add to the pan. Coarsely chop the oysters and add them to the pan. Add the salt, pepper, cayenne pepper, thyme, bay leaf, and parsley. Cook until the oysters are done and the mixture thickens. Sprinkle the inside of the roast with salt and pepper and stuff with the oyster mixture. Close the opening and secure. If the back end of the roast is thick, stick extra pieces of garlic into the meat. Sprinkle with flour and place in a lightly greased baking pan. Place the pan in an oven preheated to 350°. Brown on all sides. Pour the wine and water around the roast and scatter the carrots and potatoes around it. Cook another 1–2 hours, until the meat is done to your taste. Serve hot or cold.

Serves 6

When people have supported you all these years, you owe them. I'm hung up on that. You owe your public. They've been comin' through thick and thin. This has been the hub of the black community through all those years, and now the black community has grown. You have judges. At one time you had a handful of doctors. Now you have so many you can't count 'em all, so many black attorneys you don't know them all. You just can't know them all anymore. So you've got to move with 'em or you're going to lose them. You can't expect your people to grow and still come to your little hole in the wall now. So we had to make that move. Emily [her daughter] came in—it always takes that young blood to say, "Go." And that's what we did. Things worked out well too. It really worked out well. Worked the heck outta me, but it worked out. Now we have a beautiful restaurant and a better menu, and we can do private parties and all. I would like to grow just a little more, though; maybe have several menus and do a Sunday brunch . . . that kind of thing. I hope I'm still able to change when the times say "change." I hope I'm still able to do that.

Orange Pecan Pancakes

1 cup pancake mix
½ cup orange juice
½ cup water
1 egg
1 tablespoon vegetable oil
½ cup chopped pecans

In a large bowl mix together the pancake mix, orange juice, water, egg, and vegetable oil. Mix until the large lumps are gone. Let the mixture stand for several minutes to thicken. For each pancake pour ¼ cup of the mixture onto a lightly oiled griddle. When the first side begins to set and the surface bubbles, sprinkle with pecan pieces. Turn and cook the other side.

Serves 4

Chicken livers? Well, grill 'em, bread 'em or fry 'em . . . whatever. Put them under a broiler, even do 'em up with a gravy.

Sautéed Chicken Livers

- 4 tablespoons butter
- 1 pound chicken livers
- 1 small onion, thinly sliced
- 1 small green pepper, thinly sliced
- 1 clove garlic, minced
- salt, to taste
- freshly ground pepper, to taste
- 1 tablespoon white wine
- 1 tablespoon chopped parsley

Melt the butter in a large frying pan. Add the chicken livers, onion, green pepper, garlic, salt, and pepper. Cover and simmer over a low flame for 10 minutes or until the livers are almost done. Stir occasionally. Add the wine, turn up the heat, and cook out most of the liquid. Sprinkle with the parsley and serve.

Serves 4

Redfish with Pecans

¼ cup pecan pieces
1 cup flour
¼ teaspoon cayenne pepper
½ teaspoon salt
1 egg
¼ cup heavy cream
¼ cup water
8 tablespoons (1 stick) butter
1 tablespoon vegetable oil
4 redfish fillets, about 2 pounds
1 cup orange juice
2 tablespoons chopped pecan pieces

Place the ¼ cup of pecans in the bowl of a blender or food processor. Blend or chop for just a few moments so the pieces are small but not powdery. In a bowl mix together the pecans, flour, cayenne pepper, and salt. Beat the egg with the heavy cream and water in another bowl. Meanwhile, heat 4 tablespoons of butter and the vegetable oil in a skillet. Dip the fish in the egg mixture and then in the flour-pecan mixture, making sure to coat well. Fry over a medium flame. While the fish are frying, melt the other 4 tablespoons of butter in a small saucepan and brown the butter over a low flame, stirring constantly. When the butter is brown, add the orange juice and simmer to reduce by about half. Add 2 tablespoons of chopped pecans and serve over the fish.

Serves 4

I hardly ever cook for my family. What family? They're all gone, they all have their own families. But when I do cook for 'em, it's in mass, it's cooking for a crowd. I'll do duck, barbecued ribs, baked hams—the works. And of course I do the gumbo.

I like to do different things. I will do those crawfish or rainbow trout and stuff it with salmon mousse. I like to do special things. I love to stuff veal. I like to work with fish. You make a pecan breading for the redfish with pecans, and you put a little Mornay sauce on them. Make it, not with lemon, but with orange. You make it with the orange juice and the orange. It's good with popcorn rice and a little fresh spinach. That's really great.

I don't always have the time to create because there's always so much to do. I get the time about once a month to create special dishes such as a roast veal I did that was really interesting. You cook the veal in anchovies and tuna. You put a lot of anchovies and tuna in there. After you let it brown, you take those drippings and make a sauce with 'em. You whip it up with a little mayonnaise and a bit of horseradish, to give it a little bite. You serve the veal cold with that sauce on it and rice salad. It's unbelievable. A good summer dish.

Veal Roast with Tuna and Anchovies

1 6-pound veal leg or rib roast (bone removed)
8 anchovies
3 cloves garlic, minced
freshly ground pepper, to taste
½ cup flour
¼ cup vegetable oil
1 cup chopped onions
¼ cup chopped green pepper
2 cups water
1 8-ounce can tuna in oil
1 tablespoon horseradish
3 tablespoons heavy cream

Using a small knife punch 9 to 10 holes in the roast. Fill those holes with anchovies and garlic. Sprinkle with the pepper and dust the roast with the flour. Heat the oil in a large heavy Dutch oven. Brown the roast on all sides over a medium-high flame. Lower the flame and add the onions and green pepper. Cook for several minutes and then add the water. Cover tightly and cook slowly for 1 hour. Add more water if necessary. Add the tuna and cook 15 minutes more. Remove the meat and set aside to cool. Pour the tuna including all the liquid into a blender. Add the horseradish and the heavy cream. Blend well to make a sauce to serve over the sliced meat.

Serves 6

Rice Salad

3 cups cooked rice
½ cup chopped green onion (scallion)
½ cup chopped green pepper
½ cup chopped celery
3 tablespoons chopped parsley
1 cup pecans
2 tablespoons white wine vinegar
6 tablespoons olive oil
freshly ground pepper, to taste

Mix all ingredients together and serve at room temperature or slightly chilled.

Serves 8–10

Note: Substitute fish stock or chicken stock for water when making the rice.

B. G.

BRENDA MARIE GOODWIN

October 23, 1951 Boston, Massachusetts.

I met Brenda in New Orleans, on the same trip that I met Leah Chase. I had gone to the art museum and was standing in line to buy a poster. There was a young lady in front of me who had on a pair of jeans that really caught my eye. We conversed and discovered that we both lived in the East, me in New York and she in Philadelphia. We exchanged numbers and eventually started going out. The first time we got together, she cooked. When I suggested adding something, she looked at me as if I were crazy. "Do you really think you know what you're talking about?" She discovered I did. After that we became fast friends and loved to talk about cooking, spices, and recipes. A date for us could be shopping for spices, and of course we loved to be in the kitchen together.

B. G.

Chicken Wings Like Mom Used to Make
Cucumber and Beet Salad
Dad's Egg Salad
Roasted Potato Slices
Stewed Turkey Necks
Codfish Balls
Bob's Cod Steaks
Grilled Cheese Sandwiches on an Iron
Spicy Deviled Eggs
Old Bay Fried Chicken
Chicken Livers with Onions and Mushrooms
Roast Chicken
Seasoned Rice
Country-Style Vegetables
Vincent's Quick Flan
Broiled Salmon Cakes
My Stir-Fry Cabbage
Quick Pizza Rounds
Mushrooms Stuffed with Crab Meat
French Toast Supreme
Quick Quiche

I always liked to cook, even when I was small. I guess I first started cooking when I was four or five years old. My mother was very pleased that I liked to cook because she didn't like to cook. So she was very good at encouraging me, and she didn't get upset when I burned something. The first time I cooked the bacon, I burned it, but she said, "That's all right, baby, you just keep right on trying."

Chicken Wings Like Mom Used to Make

2 pounds chicken wings
2 teaspoons celery salt
1 tablespoon garlic powder
½ teaspoon red pepper (cayenne)
salt, to taste
freshly ground pepper, to taste
¾ cup vinegar

Wash the chicken wings and pat dry. Season with celery salt, garlic powder, red pepper, salt, and pepper. Toss to mix well. Season the wings in the pan they are to be cooked in. Pour the vinegar in and place the pan in an oven preheated to 450° and bake for 10 minutes. Turn the wings and bake 10 minutes more. Brown the wings under the broiler, about 5 minutes on each side.

Serves 4

The first things I learned to cook well were oatmeal and biscuits. I had a crush on my cousin's boyfriend. Whenever she brought him over, I would cook him biscuits. It was a big joke because they were always little and very hard. Everything I cooked was little. I liked to make little things . . . very little.

My father loved to cook. He had traveled extensively, being in the navy for twenty years. He knew a lot about cooking from all over the world. His favorites were Italian and German, but Daddy really had learned a lot about all kinds of cooking. He really liked spices, and some of the things he cooked seemed quite exotic to us, like avocados and anchovies, but everything was always great.

Cucumber and Beet Salad

6 medium beets
water
1 large cucumber
1 red onion, sliced
½ cup mayonnaise
½ teaspoon Dijon mustard
1 teaspoon white wine vinegar
½ teaspoon sugar
salt, to taste
freshly ground pepper, to taste

Cut off the beet tops. Wash the beets and place them in a deep saucepan or pot. Cover with water. Bring to a boil and simmer for 20 minutes or until the beets are easily pierced with a fork. Drain and cool under running water. Place in the freezer for 10 minutes, until nice and cool. Peel the beets and slice them into ¼-inch slices. Place them in a large bowl. Peel and slice the cucumber and add along with the sliced red onion. Mix together the mayonnaise, mustard, wine vinegar, sugar, salt, and pepper. Pour over the beets. Mix and chill well before serving.

Serves 6

My parents divorced when I was four, but I spent the summers with my father and we always seemed to wind up cooking together. He'd cook something I'd had many times before, but he'd always seem to do something different with it. That's a technique I've incorporated into my cooking, always trying something different. I remember once he made egg salad—he knew how much I loved egg salad—and he used the basics like eggs, mayonnaise, and a little onion. But then he added some crumbled bacon, some diced sweet pickle, and pickle juice, and it made all the difference in the world. It was wonderful.

Dad's Egg Salad

5 hard-boiled eggs
½ cup mayonnaise
1 tablespoon minced onion
2 tablespoons minced sweet pickles
1 teaspoon sweet pickle juice
2 pieces crisply fried bacon

Peel the eggs and chop them. In a bowl mix together the chopped egg, mayonnaise, onion, pickles, and pickle juice. Chill the egg mixture. To serve: Crumble the crisp bacon over the egg mixture and mix again. Serve over toasted bread.

Serves 4

When I was little, Mom and I used to make these roasted potatoes. We'd just peel and slice the potatoes, the long way so they'd be kinda thick and big. You take a black cast-iron skillet and put salt in it and then put the potatoes in. And you just roast the whole thing. I really used to love those. We'd make 'em and just watch TV.

Roasted Potato Slices

6 large Idaho potatoes
salt

Wash and scrub the potatoes. Slice them into ¼-inch slices. Sprinkle a layer of salt on a baking pan or cookie sheet. Spread the potatoes over the salt and sprinkle more salt over them. Place the pan or cookie sheet in the broiler and broil for 3–5 minutes on each side, until the potatoes are just brown.

Serves 4

Note: May be served as a side dish or as a snack.

Variations: Brush the potatoes with oil and follow original directions. Sprinkle the potatoes with garlic salt, rosemary, and pepper instead of plain salt. Cook over an outdoor grill instead of broiling.

You know, turkey necks are real inexpensive, so we used to always have them. You can make soups and stuff out of 'em. You can always afford turkey necks. Plus they say that the meat next to the bone is always the sweetest. I happen to think they're right. I always liked the bony parts of the chicken. A real country girl, you know.

Stewed Turkey Necks

- salt
- freshly ground pepper
- ½ cup flour
- 3 pounds fresh turkey necks
- 2 tablespoons vegetable oil
- 1 cup chopped onions
- 4 cloves garlic, minced
- 1 celery stalk, chopped
- 1 bay leaf
- 2 cups water
- 4 chicken bouillon cubes
- ¼ teaspoon red pepper flakes
- 2 teaspoons celery seeds

Add salt and pepper to the flour and dredge the turkey necks in the seasoned flour. Heat the oil in a large heavy Dutch oven and brown the necks on all sides. Add the onion, garlic, celery, and bay leaf. Sauté for 3 minutes. Add the water, stir, and add the bouillon cubes, red pepper flakes, and celery seeds. Cover and cook over a medium-low flame for 2 hours or until the meat is very tender. Adjust the seasonings, adding salt and pepper if needed.

Serves 4

I got interested in working with cod because growing up in Baltimore we used to have these little things called "coddies." They were little fish cakes made with codfish . . . mainly potatoes but with a little codfish. You'd get 'em everywhere, in all the little neighborhood stores and stuff. You'd buy 'em, they'd be like five cents. You'd get two crackers, mustard, and this little coddie. I used to love those things. You know it's a taste I've never been able to duplicate, but I sure have tried.

My stepmother lived in Queens [New York] and her landlady was from Trinidad, and she used to make these codfish cakes which of course were totally different from the coddies that I knew. I really loved the taste, and I wanted to make some codfish. I didn't know what to do. I was in a market once and ran across this little box, and it was this dried codfish. And you take this codfish and you reconstitute it and add mashed potatoes to it. So I thought, well, I can do this myself. So I bought the whole codfish and boiled it to remove the salt and kinda did it my way. I would put Old Bay Seasoning in it and onions and sometimes even milk. I'd make these balls and fry 'em, and people really liked them. I made them once for a friend's party, and just everybody wanted the recipe.

Codfish Balls

1 pound dried salted codfish
1 small onion, chopped
2 stalks celery, chopped
6 cups water
2 large potatoes
¾ cup minced onion
¼ cup minced celery
¼ cup minced parsley
1 teaspoon red pepper flakes, or to taste
1 teaspoon Old Bay Seasoning, or to taste (see page 250)
½ teaspoon salt
2 teaspoons baking powder
2 eggs, beaten
oil for frying

Wash the codfish. Place the fish in a large pot and add the chopped onion, celery, and water. Bring to a boil. Lower the flame and simmer gently until the fish is tender. Drain the fish and place in a mixing bowl. Using a fork, flake and shred the fish, removing all bones. Rinse the fish again in fresh water and squeeze dry. Meanwhile, boil and mash the potatoes. Add them to the flaked fish. Mix well. Add the minced onion, celery, parsley, red pepper flakes, Old Bay Seasoning, and salt if necessary. Mix well. Sift in the baking powder and mix again. Add the beaten eggs and mix thoroughly. Form the mixture into small balls. Heat the oil and deep-fry the balls until golden brown.

Serves 6–8

Note: These codfish balls are great served with stewed tomatoes.

Bob's Cod Steaks

3 tablespoons olive oil
1 medium onion, chopped
1 cup chopped celery
2 cloves garlic, minced
1 28-ounce can crushed tomatoes
1 teaspoon poultry seasoning
1 teaspoon celery salt
1 tablespoon vinegar
½ teaspoon sugar
½ teaspoon red pepper
4 ½-pound cod steaks
1 lemon
salt, to taste
freshly ground pepper, to taste

Heat the oil in a large heavy frying pan. Add the onion, celery, and garlic. Sauté until the onion begins to wilt. Add the tomatoes, poultry seasoning, celery salt, vinegar, sugar, and red pepper. Stir and simmer for 15–20 minutes. Meanwhile, wash the cod steaks and pat dry. Squeeze the lemon over the cod steaks and sprinkle with salt and pepper. Allow them to sit for about 1 hour. Place the steaks in the sauce, making sure that they get covered and that they have liquid beneath them. Cover and simmer for 15 minutes. Turn and cook until the fish just flakes.

Serves 4

I majored in chemistry in college, and I often think that I majored in chemistry because of its similarity to cooking. In fact, when you're working in the lab, it's often called "cooking."

I went to Morgan and, well, college students never seem to have any money, and I was no exception. The food in the cafeteria was so bad that when you got back to your room you were still hungry. We would bring things from home that could stay out a long time, things like cheese. Late at night studying, you'd get hungry. Anyway, I would make these cheese sandwiches, and the only way I had of making them hot was to use my iron. So I'd just make them on my iron—put them on some foil and put the iron on high. You have to be inventive, you know.

Grilled Cheese Sandwiches on an Iron

8 slices bread
butter
8 slices American cheese
aluminum foil

Butter the bread on both sides. Place 2 slices of cheese between each two slices of bread. Cover a clothes iron tightly with aluminum foil. Set the iron on high. Turn the iron upside down and brown the sandwiches on both sides. Or place the sandwiches on a plate and place the iron on top, brown, and turn.

Serves 4

Once I got out of school I began to develop my own way with things—recipes of my own. I found this book on herbs and spices, and it was a real adventure trying to learn what spices seemed to go with what dishes. I liked to try different things, even things that weren't in the book. I really learned to use my imagination.

Spicy Deviled Eggs

8 large eggs
water
1 tablespoon vinegar
2 teaspoons finely chopped onion
2 teaspoons finely chopped green pepper
2 small sweet pickles, finely chopped
1 teaspoon finely chopped parsley
½ teaspoon celery salt
½ teaspoon onion powder
⅛ teaspoon cayenne pepper
1 teaspoon Dijon mustard
½ cup mayonnaise
salt, to taste
freshly ground pepper, to taste
paprika, to taste

Place the eggs in a saucepan and cover with water. Turn the flame to medium-high. When the water gets hot, add the vinegar. Allow to come to a gentle boil and cook for 8–10 minutes. Place the eggs in cold water and allow them to cool. Peel the eggs and cut them in half. Scoop out the yolks and place them in a mixing bowl. Add the onion, green pepper, sweet pickles, and parsley. Mash the yolks and mix well. Add the celery salt, onion powder, cayenne pepper, Dijon mustard, mayonnaise, salt, and pepper. Mix thoroughly until thick and creamy. Add more mayonnaise if the mixture is too dry. Spoon the mixture into the egg halves and sprinkle with paprika.

Serves 4 (16 egg halves)

I always use fresh herbs and spices if I can get them. Garlic is definitely the one seasoning I can't do without. I've always heard that garlic and onion are great for promoting better health. Fresh lemon is also a staple in my kitchen. I put it on almost everything. As my mother would say, "It gives everything a nice clean taste."

Lately I've incorporated two new herbs into my cooking. I get excited when I find something new like cilantro or ginger, I think they're just wonderful. Although the first time I tasted cilantro I didn't like it—it's definitely an acquired taste. Now I use it all the time.

I make a wonderful ginger tea, putting fresh ginger root in the blender with honey, then adding the resulting mixture to hot water. Add a little more honey and some lemon, and you've got it—delicious.

There's a special spice mixture that I find indispensable: Old Bay Seasoning. Coming from Baltimore as I do, I just naturally learned to cook with Old Bay. It's made in Baltimore, and everyone there uses it, mainly to steam crabs or put on seafood, but it's also wonderful on lots of things, like fried chicken.

Old Bay Fried Chicken

1 frying chicken, cut into 12 pieces
1 cup flour
1 tablespoon Old Bay Seasoning
1 teaspoon lemon pepper
1 teaspoon salt
2 teaspoons paprika
vegetable oil

Wash the chicken and pat dry. Place the flour, Old Bay Seasoning, lemon pepper, salt, and paprika in a brown bag. Heat the vegetable oil, about 1 inch deep, in a large heavy skillet. Dredge the chicken in the flour. Place the chicken pieces in the hot oil, a few at a time. Cover, lower the flame, and fry for approximately 15 minutes per side or until the chicken is crisp and golden brown.

Serves 4

✤

Spices can be used for many things . . . if only you know what you're doing. We used to have to wait on the corner for the bus, and my friend Barbara and I would always complain about our feet getting cold . . . so her mother told us to put red pepper in our shoes to keep our feet warm. Sure 'nough, it works. Red pepper increases the circulation, so put red pepper in your shoes to keep your feet warm. I know you don't believe me, but it's the truth. It'll burn your feet up.

✤

Chicken Livers with Onions and Mushrooms

2 tablespoons butter
1 medium onion, sliced
3 cloves garlic, minced
1 pound chicken livers
1 pound mushrooms, sliced
½ cup cream sherry wine
1 tablespoon cornstarch
salt, to taste
freshly ground pepper, to taste

Heat the butter in a large cast-iron frying pan. Add the onion and garlic. Sauté until the onion begins to wilt. Add the livers and mushrooms. Cook over a medium-high flame for 10 minutes, stirring gently to keep things moving. Add the cornstarch to 3 tablespoons of the wine. Mix and add the remaining wine. Add the sherry-cornstarch mixture to the pan and cook for 15–20 minutes, stirring often, until the livers are done and the mixture thickens. Season with salt and pepper.

Serves 4

Roast Chicken

1 3–4-pound chicken
1 lemon
salt, to taste
fresh ground pepper, to taste
1 teaspoon garlic powder
2 teaspoons rosemary
1 small onion, chopped
2 cloves garlic, minced
1 stalk celery, chopped
1 tablespoon chopped parsley
4 tablespoons butter

Wash the chicken inside and out, then pat dry. Squeeze lemon juice all over the chicken, inside and out. Turn the lemon halves inside out and use them to rub the juice into the chicken. Chop the used lemons and reserve them. Season the chicken inside and out with the salt, pepper, garlic powder, and rosemary. Mix together the chopped lemon, onion, garlic, celery, and parsley. Place the vegetables inside the chicken. Place the chicken on its side in a greased roasting pan. Dot with butter. Place the pan in an oven preheated to 450°. Roast for 10 minutes. Turn the chicken on its other side and roast 10 minutes more. Place the chicken on its back and baste with the butter and juices. Lower the heat to 350° and continue roasting 30–40 minutes.

Serves 4

Even for those who don't like to cook, I think a few basics will make their food taste much better. Use as little water as possible; steam or poach food instead of boiling everything. Use flavored liquids instead of water where possible, such as stocks, broths, and bouillon. Experiment with herbs and spices. Don't be afraid to try something. And take your time.

In the winter when I'm making a lot of soups and things like that, I like to save everything. When I cut up my fresh vegetables, I keep all the trimmings and ends. I just throw them in a plastic bag and freeze them, then next time I need some flavoring I just chop them up real fine, boil them, then strain and use like a stock or to cook other vegetables or rice.

Seasoned Rice

2 tablespoons butter
½ cup minced onion
1 tablespoon minced green pepper
1 tablespoon minced celery
¼ teaspoon garlic powder
2 cups water
2 chicken bouillon cubes
1 tablespoon minced parsley
1 cup rice

Heat the butter in a saucepan. Add the onion, green pepper, and celery. Sauté until the onion has wilted and just begins to brown on the edges. Add the garlic powder and water. Mix well. Add the bouillon cubes and bring to a boil. Add the rice and parsley. Stir well and cover. Turn the flame to low and cook for 30–40 minutes, until all the liquid is absorbed and the rice is done.

Serves 4

I used to cook frozen vegetables all the time. They were quick and easy, and you could always spice them up. But now that I've learned to cook with fresh vegetables, I'll never go back to frozen.

Country-Style Vegetables

½ cup water
½ 10-ounce box frozen greens (collards, turnip, or mustard)
⅔ cup chopped onion
½ cup chopped celery
1 clove garlic, minced
3 chicken bouillon cubes
1 8-ounce can tomato sauce
2 cups chopped fresh tomatoes
1 10-ounce box frozen okra
2 cups frozen corn
salt, to taste
freshly ground pepper, to taste

Place the water in a large saucepan and bring to a boil. Add the frozen greens, onion, celery, garlic, and bouillon cubes. Cover and cook over a medium flame for 10 minutes. Add the tomato sauce, chopped tomatoes, okra, and corn. Cover and cook for 20 minutes or until the okra is tender. Season with salt and pepper.

Serves 4

✤

I have a friend who's from Jamaica, and he doesn't waste anything. One time he had a pineapple, and he cut the skin off, cored it, and cut up the pieces for us to have after dinner. Well, then he took all the scraps and trimmings and put 'em in the food processor with some water and sugar, and came up with this delicious drink. That man uses everything: ground lemon rind for seasoning, everything. He never throws anything away.

✤

Vincent's Quick Flan

4 eggs
1 8-ounce can condensed milk
1 8-ounce can evaporated milk
⅛ teaspoon cinnamon
⅛ teaspoon nutmeg
1 teaspoon vanilla extract
1 cup sugar

In a large bowl beat the eggs. Add the condensed milk and evaporated milk. Stir. Add the cinnamon, nutmeg, and vanilla extract; set aside. In a heavy frying pan heat the sugar, stirring constantly, until the sugar is completely melted and has become a dark caramel color. Be careful not to burn the sugar. Pour the caramelized sugar into custard cups, then pour the egg-milk mixture into the cups. Place the cups in a deep baking pan and pour in warm water to a depth of about 2 inches. Cover the whole thing with aluminum foil and place in an oven preheated to 350°. Bake for 1 hour. Allow to cool to room temperature and then refrigerate for 4 hours, until well chilled. To serve: Turn the custard cup over onto a small plate, being careful not to break the flan. Pour any remaining flan liquid over the top.

Serves 4

When I cook I usually prepare poultry and fish. But if I go to someone's house and they prepare red meat, it's okay, I'll eat it. But I love my chicken and fish—and lots of fresh vegetables. I've switched from butter to olive oil in lots of my cooking. I stir-

fry and steam my vegetables. If I make a sauce, it's something simple such as wine and lemon juice with a little cornstarch. Or I'll just reduce it, adding a little butter. But generally I like to keep things simple and healthy . . . but interesting.

Broiled Salmon Cakes

1 10–12-ounce can salmon
⅓ cup minced onion
¼ cup minced green pepper
1 tablespoon chopped parsley
⅛ teaspoon Old Bay Seasoning
1 teaspoon baking powder
1 teaspoon Dijon mustard
1 teaspoon lemon juice or vinegar
1 egg, beaten
½ cup mayonnaise
½ cup seasoned bread crumbs

Drain the salmon and remove all the bones and hard pieces. Place the salmon in a bowl and, using a fork, break up the salmon. Add the onion, green pepper, parsley, red pepper, Old Bay Seasoning, and baking powder. Mix well. Add the mustard, lemon juice or vinegar, egg, mayonnaise, and bread crumbs. Form into 3-inch patties and place them in a greased broiling pan or nonstick broiling pan. Broil for 3–5 minutes on each side until brown.

Serves 4

I had some friends stop by one day, and when I went to look to see what I could cook for dinner, there weren't enough vegetables of any one kind to do a stir-fry. So I was there trying to think of what I could cook. I decided to combine everything that I had, all the ingredients. I came up with this vegetable gumbo, and everybody loved it.

My Stir-Fry Cabbage

- 1 tablespoon vegetable oil
- 2 cups coarsely shredded cabbage
- 1 small green pepper, cut into thin strips
- 1 small onion, cut into thin slices
- 1 teaspoon chopped parsley
- ¼ teaspoon red pepper flakes
- ½ teaspoon minced ginger
- ½ teaspoon sugar
- 1 teaspoon soy or tamari sauce

Heat the oil in a wok or a large frying pan over a medium-high flame. When the oil is hot, add the cabbage. Stir to coat the cabbage with oil. Add the green pepper and onion. Cook for 3–5 minutes, tossing constantly. Sprinkle with the parsley, red pepper flakes, ginger, and sugar. Fry 3 minutes more or until the cabbage is done but still crisp. Pour the soy or tamari sauce over the cabbage. Mix well and serve.

Serves 4

Quick Pizza Rounds

1 loaf French or Italian bread, sliced
1 small (4-ounce) can tomato paste
Spice Island pizza seasoning
1 cup coarsely grated cheese (mozzarella or any other cheese)
chopped onions
chopped green pepper
chopped mushrooms
chopped shrimp
chopped olives

Slice the bread. Spread the top of each slice with tomato paste and sprinkle to taste with pizza seasoning. Sprinkle each piece with cheese and top with a little of each topping. Brown for 15–20 minutes in an oven preheated to 450°. Serve hot.

Serves 4

One of my old boyfriends—from St. Thomas—was a good cook, and he gave me this idea for cooking crab cakes. When you cook crab or when you pick the crab meat, take maybe three-fourths of a teaspoon of cider vinegar and sprinkle it over the crab meat. It gives it a real clean taste and makes the crab meat sweet. It makes a difference. You don't really taste the flavor of the vinegar, but you do taste what it does.

Mushrooms Stuffed with Crab Meat

1 pound crab meat
1 small onion, minced
1 tablespoon minced parsley
1 teaspoon Old Bay Seasoning
1 teaspoon vinegar
1 teaspoon Dijon mustard
⅔ cup mayonnaise
2 pounds large mushrooms
½ cup cream sherry or white wine
3 cloves garlic, minced
4 tablespoons melted butter

Pick the crab meat, removing all shells and hard pieces. Add the onion, parsley, Old Bay Seasoning, vinegar, mustard, and mayonnaise. Mix well, making sure to break up the larger pieces of crab meat. Wash and stem the mushrooms. Stuff the caps with the crab mixture. Mix together the wine, garlic, and butter. Pour into a small baking pan. Add the mushrooms. Place the pan in a preheated broiler and broil for 5 minutes or until brown. Serve with a little of the sauce spooned over each mushroom.

Serves 6–8

I didn't know anything about grilling until I lived in Philadelphia. We never had a grill. When I was growing up, we lived in the projects and you just couldn't do that. When I lived in Philadelphia, I had a fire escape off the kitchen. I had gone to one of these things where if you go they give you something . . . something for $29.99. Well they gave me this little gas grill. It actually worked, so I started cooking out a lot. And, of course, I started making barbecue sauce. But when I make barbecue sauce I always make it from whatever's there. You know, like tomato paste, tomato sauce, ketchup . . . whatever . . . lemon juice and vinegar. Always mustard, dry mustard. And something sweet like prune juice or sugar or orange marmalade or even rum. I only sorta had a recipe—at least until now.

I used to love to cook for my boyfriends, and they all loved it—all except one. I just couldn't believe it. He came by one Sunday, and I cooked potatoes with sausage and peppers and onions. Then I made these beautiful fluffy scrambled eggs and French toast: My French Toast Supreme with Grand Marnier, vanilla, cinnamon, and nutmeg. I served him this, and he looked at me and said, "Don't you ever cook regular food?" You know to him it was just too many different tastes, just too exotic. He wanted, like, just bacon and eggs.

French Toast Supreme

1 cup half and half or evaporated milk
4 eggs, lightly beaten
⅛ teaspoon cinnamon
⅛ teaspoon nutmeg
1 teaspoon vanilla extract
1 teaspoon Grand Marnier or rum
8 thick slices bread
4 tablespoons butter
confectioners' sugar
¼ cup chopped hazelnuts

In a shallow bowl mix together the half and half or evaporated milk, eggs, cinnamon, nutmeg, vanilla extract, and Grand Marnier. Dip the bread slices in the mixture, thoroughly covering both sides. Heat the butter in a large frying pan and brown the bread slices 2 at a time; add more butter for the next batch as needed. Sprinkle each completed piece with confectioners' sugar and a little of the chopped nuts. Serve with maple syrup.

Serves 4

Quick Quiche

- 2 tablespoons vegetable oil
- 1 cup any combination of the following: mushrooms, ham, shrimp, bacon, crab meat, tomatoes, scallops, onions, green peppers, red peppers
- 2 eggs
- 1 cup milk
- 1 teaspoon chopped parsley
- ½ teaspoon celery salt
- freshly ground pepper, to taste
- ⅛ teaspoon cayenne pepper
- 1 prepared pie crust
- 1 cup grated cheese (Swiss, cheddar, or Monterey Jack)

Heat the oil in a frying pan. Add any combination of the desired ingredients. Sauté for 5 minutes. In a bowl beat the eggs, then add the milk, parsley, celery salt, pepper, and cayenne pepper. Mix well. Add the sautéed mixture and mix well again. Pour into a pie crust and top with the grated cheese. Place in an oven preheated to 375° and bake for 30–40 minutes or until the quiche is well set and a knife inserted in the center comes out clean. Allow the quiche to cool to room temperature before serving.

Serves 4

I've always believed that if someone enjoys food and is not afraid to cook, he can be a good cook. If you relax into it, like anything else, cooking can be fun and very rewarding. Cooking is never fun when it's something you feel you must do.

Chickie

ENRIQUE ANTONIO RIGGS

June 3, 1943 Panama City, Panama.

Chickie and I met over a bid whist game, so you know we immediately had a lot to say to each other. Over the years a true friendship developed, and when we learned that we both liked to cook, well, that was just sort of the icing on the cake. We love to cook together, and no session is complete without a friendly disagreement on what herb or spice to put on what. But when it finally gets to the table, well . . . the girls always eat it, and with no leftovers.

Chickie

Candied Carrots
Chicken Glazed with Ginger Preserves
Smothered Catfish
Mom's Fried Chicken
Pepper Sauce
Lobster Stuffed with Corn Bread
Mussels à la Enrique
Foil-Baked Shad
Grilled Curried Salmon Steaks
Codfish Balls
Pigeon Peas and Rice
Sorrel Beverage
Family Cake
Student Body Oxtails
Curried Goat
Fried Rice

I lived in Panama until I was almost seven years old. That's when my family migrated to New York. We took up residence in Harlem, and I lived there until it was time for me to go to college in 1964. Both my parents were born in Panama, but my father's parents were from Jamaica. My mother's parents were from Barbados. Both families had migrated to Panama to work on the Panama Canal. So I have a very interesting background: half Jamaican, half Badian, born in Panama, grew up in Harlem. I've had the best of both worlds.

I didn't actually do any cooking when I was growing up, but Mom used to have us in the kitchen while she was doing things. My sister and I would wash dirty dishes while she was preparing meals or whatever. But we used to watch her. It wasn't a real conscious thing, but you realize as you go along that you've picked up quite a bit of the knowledge that she had.

I didn't actually realize that I was interested in cooking until I went in the army in 1969. I was a cook, and so that was basically my first real effort at hanging around the kitchen and doin' anything.

I don't know that I really wanted to be a cook, but that wasn't my decision. When I joined the Reserve program back in 1968, the slot they put me in was the cook's spot. I sure didn't join specifically just to be a cook! I was forced into that. I was a ball player and affiliated with the NFL program, and they got us involved with the Reserve forces in the first place. So there I was, a cook. I went to cooking school in Fort Knox, Kentucky. I went to the cooking school for eight weeks, then went back to the Reserve unit and was a cook for six years. I got out of the Reserves for a while and went back in after I finished dental school. I'm in a completely different thing altogether now.

They didn't teach you a damn thing about cooking in the army cooking school, except in dealing with meats and poultry, things like that. They taught us the difference between moist heat cooking methods and dry heat cooking methods in certain types of meats. They taught us how to cut some meats. They taught us how to carve and how to cut up the chicken—where the natural folds are, where to cut. Those are some of the more important things I learned in that cooking school.

Candied Carrots

- 4 large carrots
- water
- 2 teaspoons sugar
- ¼ teaspoon salt
- 3 tablespoons butter
- ½ teaspoon cinnamon
- ½ teaspoon nutmeg
- ¼ teaspoon allspice
- 4 ounces pineapple preserves

Wash and peel the carrots. Slice them and place in a large saucepan. Cover with water. Add the salt and sugar. Bring the pan to a boil, lower the flame, and simmer gently for 10–15 minutes or until the carrots are tender. Drain the carrots, then return them to the pot. Add the butter and mix well. Sprinkle with the cinnamon, nutmeg, and allspice. Mix well. Spread the pineapple preserves over the top and simmer gently for 5 minutes or so, until the mixture has thickened.

Serves 4

Cooking in the army was a lot of opening of cans. The menus weren't very interesting because they were all very bland, and the minimum menu was made for one hundred. Oftentimes you were cooking for three hundred or four hundred people at a time, and that was no fun. All you'd do was double or triple the recipe.

You know as well as I do that good cooking is experimental. You try out a lot of things and see what will work. You see what you like, what other people like, what works together, and then you just go with it.

Chicken Glazed with Ginger Preserves

6 chicken breast halves
1 lemon
salt, to taste
freshly ground pepper, to taste
6 tablespoons butter
1 8-ounce jar ginger preserves

Wash chicken. Place in a glass baking dish. Squeeze the lemon over the chicken and allow to sit for 10 minutes. Season with salt and pepper. Dot with butter and place in an oven preheated to 350° for 15 minutes. Remove the chicken and cover both sides with ginger preserves. Return to the oven and bake until done, another 10–15 minutes. Cut the chicken to test for doneness.

Serves 6

Variations: Use apricot or pineapple preserves.

I started messin' around in the kitchen right after I got out of Fort Knox. That was back in 1970 or so. I enjoyed it. It's a therapeutic thing. I hooked that up with my jazz. I used to be a musician years back, and that was the most therapeutic thing I did, at least until I started cooking. A lot of times when I'm cooking I like to listen to my music so that I can get sort of a double whammy!

Smothered Catfish

4 ½-pound catfish fillets
salt, to taste
freshly ground pepper, to taste
1 lemon
½ cup vegetable oil
½ cup flour
dash of cayenne pepper
1 medium onion, sliced
1 small green pepper, sliced
1 small yellow pepper, sliced
2 cloves garlic, minced
1 cup tomato sauce
½ teaspoon thyme
½ cup white wine

Wash the catfish fillets and pat dry. Sprinkle with salt and pepper and squeeze the lemon over the catfish. Heat the oil in a large cast-iron skillet. Dredge the fillets in a mixture of flour, salt, pepper, and cayenne pepper. Brown quickly on both sides. Remove and set aside. Pour off some of the oil and then sauté the onion, peppers, and garlic. When the onion wilts, add the tomato sauce, thyme, and wine. Simmer for 10 minutes. Place the fish back in, cover with the sauce, and simmer for 15 minutes, until the fish is done. Adjust the seasonings.

Serves 4

I always kid about my mother, about her being able to cook anything she can see. I always say to people that she can pick grass, pick a leaf off a tree, and can cook that up and nobody would know what it was because it would be so good.

At first I started trying to emulate what my mother did, and in some cases I came pretty close. But it never tasted quite the way hers did. As I cooked more I found my own way. I wouldn't say it was equally as good as hers because I don't think I'll ever be able to cook as good as my mother. But I found some interesting things that I could do. My sister and I get together and compare notes on different things that we do around the kitchen. I find it all fascinating.

What Mom has and what she does are unique. I still say to this day that she fries the only chicken in this country—this whole world. It's the only chicken that I have tasted that just flips me out! There's no way I could ever emulate that, though I have tried.

Mom's Fried Chicken

1 whole chicken, cut into 10–12 pieces
2 lemons
water
1 tablespoon Dijon mustard
1 tablespoon soy sauce
1 tablespoon vinegar
1½ teaspoons salt
1 teaspoon freshly ground pepper
1 teaspoon crushed allspice
1 teaspoon paprika
1 cup flour
peanut oil

Place the chicken in a large bowl and cover with water. Squeeze the juice of 1 lemon over and toss the chicken. Soak for 3 hours. Drain and pat dry. Mix together the juice of the remaining lemon, the mustard, soy sauce, vinegar, 1 teaspoon of salt, ½ teaspoon of pepper, ½ teaspoon of allspice, and paprika. Pour over the chicken and toss to cover. Marinate for 2–3 hours, turning occasionally. In a plastic bag mix the flour with the remaining ½ teaspoon each of salt, pepper, and allspice. Flour the chicken and fry in hot peanut oil to cover for 20 minutes, until the chicken is golden brown and crisp but juicy and tender on the inside.

Serves 4

Oh yeah! That hot pepper sauce. She can whip that up in no time, no time at all. And if that bad boy sits for a while, a couple of months or so, oooh yeah! She makes that stuff with about six or seven different peppers. There are a lot of Caribbean people who make pepper sauce, but a lot of them don't know how to make it either. And she just calls it pepper or pepper sauce. That sauce is some kinda mean.

Pepper Sauce

24 hot peppers (various kinds)
1 small onion, minced
4 tablespoons dry mustard
4 tablespoons turmeric
1 teaspoon salt
1 cup vinegar

Wash the peppers. Using a knife and fork, remove the stems and any dark or rotten spots. Do not touch the peppers. Coarsely chop them and then place them in a blender or food processor. Process until the peppers are in fine pieces, then pour into a glass bottle. Add the minced onion, mustard, turmeric, salt, and vinegar. Mix well and adjust the seasonings and vinegar to suit individual tastes. Cover and set aside.

Makes about $1\frac{1}{2}$ cups sauce

Note: The longer it sits, the hotter and the better it becomes.

I started thumbing through cookbooks and magazines, looking at different recipes. On Sundays you'll get the Sunday Times *magazine. They will have recipes and pictures of different meals. I have always been attracted to exotic food, not because it is exotic but because, basically, it is different. I get tired of havin' the same ole thing all the time.*

I love to cook fish, seafood meals. We do bluefish, we do swordfish, we love to do lobster, we do mussels, we do oysters.

✤

Lobster Stuffed with Corn Bread

4 cups water
½ teaspoon salt
4 1¼-pound lobsters
1 12-ounce box corn bread mix
1 cup white wine

Place the water in a large deep pot, add the salt, and bring to a boil. Add the lobsters. Cover and cook for 5–8 minutes, until the lobsters are red but not done. Remove the lobsters, place them on a cutting board, and cut them in half. Remove the tomalley (green liver) and organs from the body cavity. Meanwhile, mix the corn bread according to the instructions on the box except substitute white wine for the water. (Note: The reserved tomalley may be mixed in with the corn bread.) Pour some of the corn bread mix into each body cavity. Do not fill to capacity since the corn bread will rise. Bake the lobster in an oven preheated to 350° for 15–20 minutes. Baste several times with white wine.

Serves 4

The mussels I do—I think I'll call them Mussels à la Enrique for lack of a better word. I went to an Italian restaurant, and the mussels I ordered were really seasoned well. So I asked the waiter how they were done. He didn't know and said the chef wouldn't tell. I said, "Fine, I'll figure it out myself." I picked out the taste of different things, and I looked at the components of it and decided to try it. I don't think I did the same thing the restaurant did, but it's a very nice recipe. I played around with different ways of doing it and came up with the best way so far.

♣

Mussels à la Enrique

5 pounds mussels
3 cups white wine
2 16-ounce bottles prepared Caesar salad dressing
1 4-ounce jar pimentos, minced
2 tablespoons minced garlic
1 small onion, chopped

Wash and scrub the mussels, discarding any that are open. Place the mussels in a deep stockpot (8–10 quarts). Pour the wine over the mussels. Cover and bring to a boil. Lower the heat slightly and steam until the mussels are all open. Meanwhile, place the Caesar dressing in a saucepan with the pimentos, garlic, and onion. Bring to a boil, then turn off and allow to just sit. When the mussels are done, pour the Caesar dressing over them and mix well. Serve in a deep bowl with plenty of the sauce along with hunks of French or Italian bread for dipping.

Serves 4

Foil-Baked Shad

1 4–5-pound whole shad, scaled and gutted
1 lemon
water
salt, to taste
freshly ground pepper, to taste
1 teaspoon garlic powder
1 teaspoon thyme
2 teaspoons soy sauce
1 pound mushrooms, sliced
1 bunch scallions, sliced

Wash the fish and squeeze half the lemon over it. Wash again and pat dry. Squeeze the other half of the lemon over the fish and inside. Season the fish inside and out with salt, pepper, garlic powder, and thyme. Season the inside with the soy sauce. Fill the cavity with the mushrooms and scallions. Wrap the fish in heavy foil, sealed well. Place in an oven preheated to 250° and bake for 7–8 hours, depending on size.

Serves 6

I use a lot of different spices. But the ones that I tend to favor are thyme, dill, garlic, black pepper, and some sort of salt. Now the reason I say some sort of salt is that I don't always used iodized salt. I might use sea salt or other condiments and spices in combination which give the taste of salt. Sometimes I use soy in place of salt.

Grilled Curried Salmon Steaks

4 ½-pound salmon steaks
1 lemon
½ cup teriyaki sauce
salt, to taste
½ teaspoon white pepper
2 teaspoons curry powder

Wash the salmon and pat dry. Place the salmon in a glass baking dish. Squeeze the lemon over the fish, turning once to cover all sides. Allow the fish to sit for 15–20 minutes. Pour the teriyaki sauce over the fish. Season all sides with salt, pepper, and curry powder. Allow to sit 15 minutes more. Grill over charcoal or broil in the stove for 3–4 minutes per side. Do not overcook; salmon should be moist.

Serves 4

The thing that makes a good cook is basically the thing that makes anybody good at whatever it is they do: They have to love it. Because if you have a real interest, you're going to do it to the max. Cooking is the ability to be able to put different combinations together and have it become something very pleasing to the palate. And it has to have eye appeal.

Codfish Balls

- 1 pound salted codfish
- 1 lime
- water
- 1 potato
- 5 tablespoons butter
- 2 eggs
- 1 medium onion, minced
- 1 small green pepper, minced
- 1 clove garlic, minced
- 1 stalk celery, minced
- 2 small hot peppers (Ancho peppers), minced
- 6 tablespoons flour
- 1 tablespoon baking powder
- 1 teaspoon salt
- 1 teaspoon freshly ground pepper
- 1 teaspoon thyme
- 2 cups vegetable oil

Cut the salted codfish into several pieces. Place them in a bowl and cover with water. Add the juice of half the lime. Soak for 1 hour. Drain and rinse the fish. Place the fish in a large saucepan and cover with fresh water. Add the juice of the other half lime. Bring to a boil and boil for 5 minutes. Drain and shred the codfish, removing all bones. Place the codfish on a chopping board and chop fine. Boil the potato, mash with 2 tablespoons of butter, and add the codfish. Beat the eggs and add them to the fish-potato mixture. Heat the remaining butter in a skillet and add the onion, green pepper, garlic, celery, and hot peppers. Sauté for 2–3 minutes. Add to the fish-potato mixture. Sift together the flour and baking powder. Add to the fish-potato mixture along with the salt, pepper, and thyme. Mix well. Heat the oil until very hot. Form the cod mixture into small balls and deep-fry until golden brown.

Serves 8–10 as appetizers

Cooking is experimentation. You just have to try to find it. You have to be able to experiment and try new things. You're gonna have some failures as well as some successes, but that's the only way you're going to find out.

I keep the heritage. Carol [his wife] is also of Caribbean background, and we have a strong thing about our background and Caribbean flavors. My mother still involves herself in Caribbean-style cooking, and a lot of it I've gotten from her. It comes with the culture. We do it Caribbean style.

Pigeon Peas and Rice

- 4 slices bacon, cut into ¼-inch dice
- 1 large onion, chopped
- 1 green pepper, chopped
- 1 stalk celery, chopped
- 3 cloves garlic, minced
- 6 cups water
- 1 pound pigeon peas
- 1 teaspoon thyme
- salt, to taste
- freshly ground pepper, to taste
- 1 tablespoon coconut oil
- 2 cups rice

In a large Dutch oven, sauté the bacon until it just begins to get crisp. Add the onion, green pepper, celery, and garlic. Sauté until the onion wilts. Add the water, pigeon peas, thyme, salt, and pepper. Bring to a boil, lower the flame, and simmer until the peas are just tender. Add the coconut oil and rice, cover, and cook slowly for 30 minutes or until all the liquid is absorbed.

Serves 8–10

We do meats. We do chicken. We do plantains. We do rice and peas. We do ackee. We do those types of things—Jamaican meals—not that much different from what you're going to get all over the world, just done differently. We do all the island beverages. We tried our hand at ginger beer, but that didn't come out too well. That is a real difficult one to do. You're dealing with cream of tartar. You gotta know how to really work with that before you can get it right. I do sorrel real well. That's a nice drink.

✤

Sorrel Beverage

5 gallons water
1 pound fresh ginger
1 tablespoon whole cloves
4 4-inch strips dried orange peel
1 2-ounce package dried sorrel leaves

Place the water in a large stock pot over a high flame. Peel and thinly slice the ginger. When the water boils, add the ginger, cloves, orange peel, and sorrel leaves. Boil for 3–4 minutes. Turn off the flame and cover the pot tightly. Allow the pot to sit for 2 days. Drain, sweeten to taste, and chill.

Note: Sorrel leaves are available in Spanish and Caribbean food stores.

Every year we do a family cake. We do a West Indian fruitcake. My mother is not a baker, so for a number of years she paid somebody to make cakes for us because she didn't know how to do it. Then I found this book up in Canada when I went to the Expo in 1967. It was a Trinidadian cookbook, and it had this recipe for what they call their Christmas cake. That's the recipe we started with, but we changed it some year after year. We never knew how easy it really was to make the cake. So we do it every year. You have to have that cake when people come to visit during the holidays.

Family Cake

16 ounces currants
16 ounces sultanas (white raisins)
4 ounces dried orange peel, shredded
8 ounces red cherries
8 ounces raisins
2 cups dark rum or brandy
1½ pounds butter
2 cups sugar
1 dozen eggs, lightly beaten
4 cups flour
4 teaspoons allspice
4 teaspoons cinnamon
1 teaspoon ground mace
1 whole nutmeg, grated
½ teaspoon baking soda
1 cup molasses
2 cups chopped nuts

Place the fruit in a large covered container and pour in the rum. Mix well and allow the fruit to sit for up to several months but no less than several weeks. Cream together the butter and sugar. Then add the eggs and mix well. Sift together the flour, allspice, cinnamon, mace, nutmeg, and baking soda. Sift the flour mixture into the butter-sugar mixture. Cream well. Add the fruit and rum. Add the molasses and chopped nuts. Pour into a greased cake pan and bake in an oven preheated to 275° for 3 hours. Place the cake in a cake tin and pour over additional rum. Cover and allow to sit for at least 1 week.

Serves 8–12

I make my own eggnog. I do that from scratch. I only do it once a year. If I'm asked to, I'll do it twice: Christmas and Thanksgiving. But I normally do it once a year. That sort of stuff is too rich to have often. So we hook that eggnog up with the family cake, and hey!

We haven't traveled all over the world yet, but we figure before we go, we are going to dine all over the world. So we pick up a lot of cookbooks that have different styles of preparing things. We might decide every now and then that, hey, I'd like to try this or that. And so we do. I'll try anything once.

Learning to cook from Mom, you had to just watch her. I had to be there to watch her because she is not a measurer. She doesn't measure a thing. According to her you put a little of this and a little of that, and she comes up with something great. We didn't know how much she was using because she didn't know herself. She just eyeballs it. And I got that eyeballing talent from her.

We have a lot of stereotypes here in this country, about things that people eat in different parts of the world. Because of what it sounds like, people will say, "I don't like that" or "I'll never eat anything like that." But you kinda got to open up your culinary experience, so to speak. It might just be that some things that we don't use in this country or eat in this country—on a routine basis anyway—you'll learn to like. I never thought that ever in my life would I be into oysters or squid. I never thought I would eat anything like that. But now I love both of them. I'll try anything once.

I love oxtails. When I was in school in Washington, D.C., my buddy and I would take a break from our studies by whippin' up a feast. We'd get the guys together, chip in, and have a feast. I would fix the oxtails. This one time we invited another friend, an undergrad, and he came over and asked what was on the menu. I said, "We got greens, salad, and oxtails." And he said, "Oxtails. I can't eat that. That's too close to the rear for me." So we said, "Fine, that's just more for us." Anyway, this guy decided to just taste a little, after he saw us digging in. Now you know, he went back four times. We thought we were going to have to roll him home.

Student Body Oxtails

4 pounds oxtails, cut into pieces
1 teaspoon salt
freshly ground pepper, to taste
½ cup flour
½ cup vegetable oil
2 medium onions, chopped
3 cloves garlic, minced
4 carrots, peeled and cut into ½-inch slices
1 16-ounce can whole tomatoes, coarsely chopped
1 bay leaf
1 teaspoon thyme
1 cup red wine
2 cups beef stock

Sprinkle the oxtail pieces with salt and pepper, then dredge in flour. Heat the oil in a large casserole. Brown the oxtail pieces in batches. As they brown remove to a platter. When all are done, return them to the pot and add the onions, garlic, carrots, tomatoes, bay leaf, thyme, red wine, and beef stock. Bring to a boil, lower the flame, cover, and simmer for about 3 hours or until the meat is very tender. Adjust the seasonings and let sit for 10 minutes before serving.

Serves 8 hungry students

Another thing that some people turn their nose up at is goat. But I love goat, curried goat. We do a great curried goat—which is a lot like cooking beef except for the bones and the time it takes—seasoned the Caribbean way.

Curried Goat

3 pounds goat meat, chopped by the butcher
8 cups water
2 medium onions, coarsely chopped
4 cloves garlic, smashed
4 scallions, sliced
½ teaspoon thyme
salt, to taste
freshly ground pepper, to taste
2 tablespoons curry powder, or slightly more to taste
2 tablespoons vegetable oil
red pepper flakes, to taste

Place the goat meat in a large bowl and cover with 4 cups of water. Soak for 1 hour. Drain the meat and return to the bowl. Add the onions, garlic, scallions, thyme, salt, pepper, and 1 tablespoon of curry powder. Mix well and let stand for several hours, turning occasionally. Pick the meat out of the bowl, shaking off as much of the herbs and spices as possible. Heat the oil in a large heavy pot over a moderately high flame. When the oil is hot, brown the meat, in batches if necessary. When all the meat is brown and in the pot, add the remaining 4 cups of water and all the seasonings from the bowl. Cook, covered, over a medium low flame until tender, about 40 minutes. Add the remaining tablespoon of curry powder and the red pepper flakes. Simmer, uncovered, until done (about 1 hour), when all the water is absorbed and the meat starts to fall off the bone.

Serves 4

One of the things I picked up from Mom was rice. She never cooks rice that sticks together, not steamed rice like the Orientals use because they use chopsticks. That's their style. Whether it was peas and rice, shrimp and rice, or just plain rice, Mom's was never sticky. "Every cop on his own beat," as I used to say. Rice is something I always tried to do, and I got it.

Fried Rice

8 cups water
1 teaspoon salt
1 cup rice
2 tablespoons bacon fat
2 scallions, sliced
1 small onion, chopped
2 carrots, sliced
½ small green pepper, chopped
1 stalk celery, chopped
½ 10-ounce package frozen green peas
salt, to taste
freshly ground pepper, to taste
1 8-ounce can crushed pineapple

Place the water and salt in a large saucepan. Bring to a boil and add the rice. Lower the flame to medium. Boil the rice until it is soft, about 15 minutes. Drain and rinse the rice. Heat the bacon fat in a cast-iron skillet. Add the scallions, onion, carrots, green pepper, celery, and green peas. Sauté until the onion wilts. Add the rice and stir to mix well. Season with salt and pepper. Sauté for 10 minutes, until the rice takes on some color. Add the pineapple, mix well, and cover. Let the rice sit for about 5 minutes before serving.

Serves 4

I think it's important that a man and a woman both do some of the cooking. People's careers being what they are, it is too demanding for one person to have to cook all the time. At my house we just pitch in and do whatever we're going to do.

Aunt Margie

MARGARET McCALL THOMAS WARD

October 30, 1918 Montgomery, Alabama.

The sweetbreads were what did it. Aunt Margie, my wife's mother's sister, has Christmas brunch for her whole family each year. That first year, after my marriage to Victoria, was the first time I had ever eaten sweetbreads, and after that . . . well, I went right into Aunt Margie's kitchen and demanded to know just how this marvelous concoction was made. Of course it took me several years to pry her secrets out of her. I learned along the way that she was a fabulous cook. And I hope she has grown to love me coming into her kitchen.

Aunt Margie

Bread and Butter Pickles
Poached Pears
Smothered Steak
Braised Ribs
Tomato Juice
Ambrosia
Baked Fish with Tomatoes and Olives
Apple Butter
Sweet Potato Scones
Smothered Chicken
Miss Bessie's Candied Yams, Arkansas Style
Beef Stew
Christmas Sweetbreads
Buffet Grits
Artillery Punch

I started learning to cook by just watching. My grandmother—my father's mother—lived with us. Mother and Dad both worked. They had the newspaper, so Grandmother took over the kitchen. I think that is the only way two women can live together in the same house; one does one thing and one does the other. Grandmother was an excellent cook, and she took that responsibility.

Bread and Butter Pickles

- 2 pounds small pickling cucumbers (Kirbys)
- 1 large onion, thinly sliced
- 2 tablespoons coarse salt
- 1 cup cider vinegar
- 1 cup brown sugar
- 1 tablespoon mustard seeds
- 1 teaspoon celery seeds
- ½ teaspoon whole allspice
- ¼ teaspoon turmeric
- ¼ teaspoon cayenne pepper

Wash the cucumbers and cut into ¼-inch slices. Place the cucumbers and onion in layers in a large bowl, sprinkling each layer with coarse salt. Cover the mixture with ice cubes, then cover the bowl and refrigerate overnight. Rinse the mixture well in cold water, drain, and set aside. Heat the cider vinegar in a large saucepan and add the sugar, mustard seeds, celery seeds, whole allspice, turmeric, and cayenne pepper. Bring to a boil, stirring almost constantly. Add the cucumbers and onion. Allow the pan to return to a boil.

Lower the flame and simmer for 2–3 minutes. Pour the cucumber mixture into a bowl and allow it to cool. Cover and chill for 12–24 hours.

Serves 6–10

Note: Pickles will keep about ten days unless sealed in jars, then they will keep indefinitely.

What my sister and I had was a composite of two philosophies in terms of cooking. Grandma, I would say, was more of the provincial cook, the stir-in-the-pot kind of thing. Mother had more formal training. She graduated from Tuskegee Institute, and she'd also been to Hampton Institute; she had the advantage of the training with New England teachers, who had come down to teach the children in the South. She had learned to cook from Fannie Farmer's Boston Cooking School Cook Book. *So when Victoria and I were little girls, Mother would teach us these little fine points of cooking, like making your own mayonnaise or a caramel frosting for a cake. You had to have certain utensils, and you had to be precise in what you did. So this was the combination of the two kinds of cooking we grew up with: the relaxed, warm, family kind of thing that Grandmother exemplified, and the more formal, doing it precisely. I think we got a perspective on cooking that has followed us the rest of our lives.*

Poached Pears

6 russet pears
1 lemon
6 tablespoons sugar
1 cup water
1 stick cinnamon
water

Spiral peel the lemon in one piece. Set the zest aside. Peel the pears and place in a bowl of water to cover to which the juice of 1 lemon has been added. Place the cup of water in a saucepan along with the sugar, cinnamon, and lemon zest. Bring to a boil, then lower the heat. Drain the pears and add them loosely in a single layer. Simmer gently for 15–20 minutes, until the pears are done. Repeat if all the pears did not fit the first time. Place the pears in a bowl, pour the syrup over, and cool. When cool, refrigerate until time to serve.

Serves 6

My parents both worked, and after Grandmother died they would come home tired. So if I came home from school early, I would start to do the kind of cooking my grandmother had done, making a stew or smothering steak in gravy and mashed potatoes. This was the way I got into cooking.

Smothered Steak

3 pounds steak
salt, to taste
freshly ground pepper, to taste
½ cup flour
2 tablespoons vegetable oil
2 tablespoons butter
1 large onion, chopped
2 cloves garlic, minced
1 small green pepper, chopped
2 cups beef stock or water
1 cup ketchup
½ teaspoon thyme
¼ teaspoon red pepper flakes

Pound the steak with a mallet to break down the tissue and cut into 4 pieces. Salt and pepper the steak on both sides, then dredge in flour. Heat the oil and butter in a large cast-iron skillet. Fry the steak, browning well on both sides. Remove the steak and set aside. Add the onion, garlic, and green pepper. Sauté for 10 minutes. Put the steak back in and add the beef stock or water, ketchup, thyme, and red pepper flakes. Stir to mix well, cover, and simmer gently for 1 hour, until the meat is tender. Taste for salt and pepper. Make a mixture of about 2 tablespoons of flour and 6 tablespoons of liquid from the skillet. Add to the skillet and simmer until desired thickness is reached. Let the steak rest for 5 minutes before serving.

Serves 4

If you got home first and everyone else would be coming home hungry, then you would make the meal, or start it at least. That was true for me and my sister Victoria. It wasn't like you had to, it's just that's what you do.

The kitchen is where most youngsters first learn to help, and I always got KP. Victoria and I had the dishes to wash and the trash to take out. Then we progressed to making fudge. I think that was next, then lemonade. Of course you made everything from scratch. I still make a good pitcher of lemonade.

Now Grandmother came from a long line of good cooks because Grandmother was born a slave. She raised a large family, and she knew how to cook. She was a very creative woman, a very artistic woman. She was an expert seamstress, and I think that this artistic quality in a person carries through in everything they do.

Grandmother had such an influence on us. She was at one time a heavy woman and would stand in the kitchen and have one hand on her left hip with the palm out and the right hand would be the one to stir, standing and stirring. It's so amusing because sometimes I'll find myself doing the same thing—a physical motion that is a part of nurturing. You know you carry this over, this warmth.

Braised Ribs

1 cup flour
salt
freshly ground pepper
6 beef short ribs (about 4 pounds), cut into pieces
1 large onion, coarsely chopped
1 medium green pepper, chopped
3 cloves garlic, minced
1 bay leaf
3 cups beef stock or water
6 carrots, cut into ½-inch dice
4 potatoes, cut into cubes
½ teaspoon thyme

Place the flour and ½ teaspoon each of salt and pepper in a bag. Shake to mix. Dredge the ribs in the flour, place them in a large Dutch oven, and put in an oven preheated to 350°. Brown the ribs for 15 minutes. Add the onion, green pepper, garlic, and bay leaf. Turn with a spoon to mix well. Add the stock or water. Cover and return to the oven. Cook about 1 hour. Add the carrots, potatoes, thyme, salt, and pepper. (If the gravy is too thick, add a little more water.) Cover and return to the oven for about 30 minutes.

Serves 6

Mother would come home from work, and if she had a little extra time, you would find that she had taken fresh tomatoes at high peak season and had squeezed those tomatoes, making the best tomato juice I've ever had. Then if she had the time, she'd make the best pot of vegetable soup you ever had in your life. You'd come home from school and all the kitchen windows would be steamed up from that soup. These are the things that make children want to be kind to other people and to do things with food that make people happy. That's what food is.

Tomato Juice

ripe tomatoes
pinch of celery salt
pinch of sugar
pinch of fresh chopped basil

The tomatoes should be very ripe. Cut the tomatoes into quarters and place them in a blender or food processor. Blend well. Pour the mixture through a strainer. Add a pinch of celery salt, sugar and basil, and drink.

Note: Vodka is optional. Make only in the summer when the tomatoes are truly vine ripened.

Mother was great at pan broiling. She would do that with steak. She would take a piece of round steak and pound it and pound it. Just pound in those seasonings, then pan broil until it was tender. There was nothing quite like it.

Grandmother taught us to make the ambrosia. You have to take fresh coconuts and crack them, and only have the fresh coconut milk and the fresh grated coconut and fresh oranges to make that ambrosia. I guess that's why when people serve ambrosia, I generally don't want it because for it to be right you just have to have fresh coconut. The oranges have to be right too.

Ambrosia

6 large navel oranges
1 large coconut
½ cup sugar

Peel and quarter the oranges, then slice into ¼-inch slices. Try to save approximately ¼ cup of juice. Crack the coconut, saving the milk. Pull the meat from the shell and peel off the brown skin. Grate the coconut meat. In a large bowl place a layer of orange, a layer of grated coconut, and a layer of half the sugar. Make another layer. Pour over this the ¼ cup of orange juice and ¼ cup of coconut milk. Add more liquid if you desire. Chill and serve.

Serves 6–8

Of course it isn't all fun and games once you get married because then you have to produce a whole meal every day. You don't have that family support system anymore except maybe to call up for a recipe or two.

All my experiences in life have added to my cooking. I've brought that kind of seasoning into the cooking that I do.

The children's father and I, when we married, we went to Cuba on our honeymoon. So Latin American things have a special place in my heart. I was so impressed with their seasoning, the way they did things—going down to the waterfront and buying a big fresh fish. Years later we went and stayed for several months. We had a cook, and I would watch everything she did. That's when I really got into their style of spicing. I just watched, learned, then tried it myself.

Baked Fish with Tomatoes and Olives

1 4-pound red snapper or other fish, scaled and gutted but with the head left on
1 lime
salt, to taste
freshly ground pepper, to taste
4 tablespoons olive oil
1 medium onion, chopped
1 small green pepper, chopped
4 cloves garlic, minced
3 cups crushed tomatoes
1 cup white wine
1 teaspoon thyme
1 bay leaf
½ teaspoon oregano
½ teaspoon red pepper flakes
1 cup pitted olives or Alcaparrado (see Note on page 161)
1 tablespoon chopped parsley

Rinse the fish in cold water. Squeeze the lime all over the fish, rinse again, and pat dry. Sprinkle the fish inside and out with salt and pepper. Place in a large greased baking pan and set aside. Heat the oil in a large cast-iron skillet. Add the onion, green pepper, and garlic. Sauté for 5 minutes. Add the tomatoes and wine. Stir and simmer for 5 minutes. Add the thyme, bay leaf, oregano, red pepper flakes, salt, and pepper. Simmer for 10 minutes. Add the olives or Alcaparrado and simmer 10 minutes more. Add the parsley, then pour the mixture over the fish. Place the fish in an oven preheated to 400° and bake for 30–45 minutes, depending on thickness.

Serves 6

Each experience influences your cooking practice, from the beginning to the last thing you did. Each experience adds that little bit of spice.

Apple Butter

1 quart apple cider or juice
4 pounds apples
1 cup sugar
1 teaspoon cinnamon
¼ teaspoon allspice
¼ teaspoon ground cloves
¼ teaspoon salt

Bring the cider to a boil in a large saucepan. Lower the flame and simmer for 15–20 minutes, until the liquid reduces by almost half. Meanwhile, wash, core, and quarter the apples. Add the apples to the cider and cook until the apples are tender, stirring almost constantly. When the apples are done, remove them from the pan and process in a food processor or force them through a sieve. Return the mixture to the pan and add the sugar, cinnamon, allspice, cloves, and salt. Simmer for 30 minutes, stirring almost constantly, until the mixture thickens. Pour into boiled jars and seal immediately.

Makes about 1 quart

Food is something to sit down to—a time to sit down and converse and have a happy time as opposed to having an awful discussion about what costs what, who did or didn't do what. Meals should be happy times and have happy memories.

Sweet Potato Scones

2 cups flour
1 tablespoon baking powder
½ teaspoon baking soda
½ teaspoon salt
½ cup sugar
½ teaspoon cinnamon
4 tablespoons butter, at room temperature
1 cup mashed sweet potatoes
1 egg, lightly beaten
½ cup milk

Sift together the flour, baking powder, baking soda, and salt. Mix together the sugar and cinnamon. Cream together the sugar-cinnamon mixture and the butter. Add the mashed sweet potatoes and beaten egg to the sugar-butter mixture. Fold in the flour mixture and add the milk. Mix well. Rub hands with flour and scoop out lumps of dough using a spoon. Flour lightly in hands and drop onto an ungreased baking sheet. The lumps should be unformed. Bake in an oven preheated to 400° for 20–25 minutes, until the tops are just brown.

Serves 6–8

The way the table is set and the way the food is presented—putting things on a platter to make them look attractive as well as taste attractive—is most important.

It's the ability to improvise that makes a good cook, to be able to experiment and not have to be glued to the cookbook. I think you need the basic cookbook when you get started, but then you learn how to try a little of this or a little of that, just to see what it will do. That's the artistic element of cooking, and it personalizes everything.

There was a time during the war when everyone was cooking with a pressure cooker. The food never tasted quite the same; the pressure cooker takes something out of it. It's supposed to leave it in, but I didn't think so. It's good for cooking dried beans, but I still think that the simmered pot is the best way to do it.

One of the little things I have learned is that even if people drop in on you, there should always be something in the icebox or up on the shelves that you can make a meal from. Back in the old days, when people did visit a lot, you just had to be ready. So one time we visited this lady with our parents and stayed to play cards. Well, Lottie, that was her name, I think, went into the kitchen for a while and came back with hot biscuits, homemade jelly, and hot cocoa. It was just such a wonderful surprise, and it was delicious. It made a wonderful memory, and it also taught what can be done with food if only you think about it and have a desire to spread a little hospitality.

When you speak of Southern cooking, I guess we think it's all black. But it isn't. It has to do with blacks and whites because they all ate pretty much the same way. It was really more a matter of taking what was there and using it.

Smothered Chicken

1 3-pound chicken, cut into pieces
salt, to taste
freshly ground pepper, to taste
2 tablespoons vegetable oil
2 tablespoons bacon drippings
½ cup flour
2 cups chicken stock or water
1 medium onion, chopped
2 cloves garlic
½ teaspoon thyme
¼ teaspoon sage
¼ teaspoon red pepper flakes

Rinse chicken and pat dry. Cut off and discard all fat. Place the neck, gizzard, heart, and fat in a saucepan with water to cover to make the stock. Sprinkle the chicken with salt and pepper. Heat the oil and bacon drippings in a large cast-iron skillet. Dredge the chicken in flour and fry until brown on all sides. Add the chicken stock, onion, garlic, thyme, sage, and red pepper flakes. Cover and simmer gently for 1 hour or until the chicken is tender. Taste for salt and pepper and add if necessary. Make a mixture of about 2 tablespoons of flour and 6 tablespoons of liquid from the skillet. Add to the skillet and simmer until the gravy reaches the desired thickness. Allow the chicken to rest for 5 minutes before serving.

Serves 4

I learned to make sweet potatoes from Bessie Miller, Miss Bessie, the children called her. She was a surrogate mother for my children because she lived with me and Sam throughout our marriage. She taught Edith to make the biscuits, and she taught

✤

I like to put the garlic in there and just let it marinate and allow the garlic to really get in there. That garlic really gives flavor to things. That's the way I cook.

Another thing, you have to keep those utensils looking just right or else the food doesn't taste as good. I have an old black iron skillet, and there's nothing like it. But that pan has to look just right. Never scrub it out with soap and Brillo, never. You have to use a lot of elbow with it, and keep that pan seasoned right.

I'm a librarian, and many librarians are excellent cooks. Are you aware of that? Number one, librarians read a lot and study a lot. I like to just sit down and read a good cookbook. I have an awful habit of getting in bed—I keep five or six books by the bed and another stack on the floor—and sometimes the next morning the light is still on and the last book I was reading is still in my hands. I love cookbooks. I collect them. Everywhere I go I always buy a cookbook, whether it's a big one or a little one put out by the Ladies Aid Society.

All of my children can cook. The boys started when they were young and wanted to make their own money. First they sold eggs, then they got into baking cakes and selling them. Then they had newspaper routes. They wanted to make a little money, and these were the things they did, and I helped them.

The first time I ever had sweetbreads was at a party, and I just fell in love with them. I was determined to learn how to prepare them. I just kept going through cookbooks until I found recipes for them. Of course I've developed my own way with them.

Christmas Sweetbreads

8 veal sweetbreads (about 3 pounds)
water
salt
1 lemon
2 cups milk
2 tablespoons butter
3 tablespoons flour
freshly ground white pepper, to taste

Sweetbreads must be very fresh, never frozen. They are extremely perishable. Gently wash and clean the sweetbreads, removing all excess fat and connective material. Cover with cold water and soak overnight in the refrigerator. Drain and cover with fresh water. Add 1 teaspoon of salt and the juice of a lemon. Bring to a gentle boil, lower the flame, and simmer for 15–20 minutes, until the sweetbreads turn white. Remove from the heat and place the sweetbreads under running water to cool them and stop the cooking process. Do not overcook them. Drain well and cut into ½-inch cubes. Meanwhile, heat the milk slowly in a saucepan. Do not allow the milk to boil. In another saucepan heat the butter over a low flame. When the butter is hot, add the flour and cook for 2–3 minutes, stirring constantly. Add the milk, whisking constantly. Simmer slowly to thicken. Add salt and white pepper and then add the cubed sweetbreads. Allow the sweetbreads to reheat before serving.

Serves 6–8

I love grits in the morning, but they also go great with any buffet. I don't know when I started adding bell peppers, but I love the contrast between the soft grits and the crunchy raw peppers.

Buffet Grits

5 cups water
1 cup grits
2 tablespoons butter
1 teaspoon salt
½ teaspoon freshly ground pepper
1 small green pepper, minced

Place the water in a saucepan and bring to a boil. Slowly stir in the grits. Return to a boil, then lower the heat and simmer gently for 15–20 minutes. When desired consistency is reached, remove from the heat. Stir in the butter, salt, pepper, and minced green pepper. Serve immediately.

Serves 6–8

But as I say, what I have learned about cooking is what I have learned about life. It's some of all of it put together.

✤

Artillery Punch

½ cup sugar
juice of 3 lemons
1 tablespoon bitters
16 ounces claret (red wine)
16 ounces sherry
16 ounces bourbon
16 ounces brandy
ice
16 ounces club soda

In a large punch bowl, stir together the sugar, lemon juice, bitters, red wine, sherry, bourbon, and brandy. Let stand 1 hour. Add 1 large block of ice and the club soda.

Serves 10

Food . . . it's more than just putting something on the table.

Liquid and Dry Measure Equivalencies

Customary	Metric
¼ teaspoon	1.25 milliliters
½ teaspoon	2.5 milliliters
1 teaspoon	5 milliliters
1 tablespoon	15 milliliters
1 fluid ounce	30 milliliters
¼ cup	60 milliliters
⅓ cup	80 milliliters
½ cup	120 milliliters
1 cup	240 milliliters
1 pint (2 cups)	480 milliliters
1 quart (4 cups, 32 ounces)	960 milliliters (0.96 liters)
1 gallon (4 quarts)	3.84 liters
1 ounce (by weight)	28 grams
¼ pound (4 ounces)	114 grams
1 pound (16 ounces)	454 grams
2.2 pounds	1 kilogram (1,000 grams)

Oven Temperature Equivalencies

Description	°Fahrenheit	°Celsius
Cool	200	90
Very slow	250	120
Slow	300–325	150–160
Moderately slow	325–350	160–180
Moderate	350–375	180–190
Moderately hot	375–400	190–200
Hot	400–450	200–230
Very hot	450–500	230–260

INDEX